Introduction to Abstract Algebra

Introduction to Abstract Algebra

*From Rings, Numbers, Groups, and Fields
to Polynomials and Galois Theory*

BENJAMIN FINE

ANTHONY M. GAGLIONE

GERHARD ROSENBERGER

Johns Hopkins University Press

Baltimore

© 2014 Johns Hopkins University Press
All rights reserved. Published 2014
Printed in the United States of America on acid-free paper
9   8   7   6   5   4   3   2   1

Johns Hopkins University Press
2715 North Charles Street
Baltimore, Maryland 21218-4363

ISBN 13: 978-1-4214-1176-7
ISBN 10: 1-4214-1176-8

Library of Congress Control Number: 2013937859

A catalog record for this book is available from the British Library

Special discounts are available for bulk purchases of this book. For more information, please contact Special Sales at 410-516-6936 or special sales@press.jhu.edu.

Johns Hopkins University Press uses environmentally friendly book materials, including recycled text paper that is composed of at least 30 percent post-consumer waste, whenever possible.

Contents

Preface

Abstract algebra is a crucial component of any mathematics curriculum. Besides the material itself, algebra and algebraic methods pervade all of mathematics. This book is an introduction to abstract algebra where we pursue a somewhat nontraditional approach to presenting the material. We call this approach the algebraic method and from classroom experience this method has proved itself to be extremely effective.

The three authors each have over thirty years experience teaching algebra at all levels of the undergraduate and graduate curriculum and we have attempted in this book to incorporate what we have learned to be the best techniques for presenting this material to first-time algebra students. One thing is clear. For a student, the first exposure to abstraction is never easy. Whereas all texts provide the appropriate definitions for the various algebraic structures and then provide examples, many students become lost as to how and why these structures arose originally. Often the basic picture is lost in a sea of abstract definitions. Our approach is meant to alleviate this and to show that the abstract structures arise naturally. We must remember and pass on to our students that all of abstract algebra arose from problems in classical algebra or number theory and that abstract structures, groups, rings, fields and vector spaces were defined to aid in the solution of these problems. The definitions arose as generalizations and abstractions of already known and well-studied systems: the integers, the rationals, the real numbers, vectors in two and three dimensions and so on. Then other examples and models were identified to show that the definitions were meaningful beyond the original system. These examples are then followed by a search for what makes our original system unique (up to isomorphism) among structures of this type. This is the gist of the algebraic method. For example, rings begin with the integers $\mathbb{Z}$ which are well-known to all students. After looking at $\mathbb{Z}$ and defining a ring we present other examples to show that the definition is wide enough. We then look at what makes $\mathbb{Z}$ unique up to isomorphism among all rings. This leads to commutative rings, ordered rings, integral domains, unique factorization domains and so on. We pursue this type of approach: well-known system to abstract structure to uniqueness of the original system throughout the book.

We have included much more material than can be covered in a standard one-semester undergraduate algebra course. We felt that this would allow flexibility for an instructor to build a course that would cover the basics yet still include various and different topics. It also allows for a full year course in algebra. We feel that in a year-long course most of the book can be covered. Once the difficulties in understanding abstraction and the algebraic method are surmounted, the other material can be covered more smoothly.

Chapters 1 through 7 with some sections omitted provide a standard first course in abstract algebra in the United States. Chapter 2 on algebraic preliminaries can be omitted, or done very rapidly, if the students have had a good bridge course, which includes an introduction to proofs. If an instructor wants to omit the material on the real and complex numbers in Chapter 5 (the construction of the real number system is usually done in analysis courses) then Chapter 8 on direct products and abelian groups can be substituted. This chapter includes the fundamental theorem of finite abelian groups. Although the text does present an elementary proof of this important theorem (a feature not usually found in books at this level) the instructor may omit the proof and just show how the theorem can be used to determine an algorithm for constructing all nonisomorphic abelian groups of a given order in terms of direct products.

Most beginning books on abstract algebra cover groups first, the idea being that one operation is simpler than two. However, we feel that rings are a much more natural structure for beginning students than groups. These students have a great deal more experience with the integers and the rationals, the motivating examples for rings, than with permutations, the motivation for groups. Hence, although we define and study some basic properties of groups in Chapter 2, we deal for the most part with rings before groups. However after doing Chapter 1 on the algebraic method and whatever parts of Chapter 2 are deemed necessary, an instructor can jump directly to Chapter 6 to do a more traditional course where group theory is covered first.

In presenting examples of algebraic structures we have assumed that the students have some familiarity with matrices and matrix algebra. In Chapter 13 we present a survey of linear algebra from the viewpoint of the algebraic method. This material would be included in a year-long course which would also include Chapter 15 on Galois theory.

Each chapter has a wide selection of exercises of varying degrees of difficulty. Hints are provided for the more difficult exercises.

We would like to thank the many people who have read versions of this book and made suggestions. Especially we would like to thank Cameron Bishop for helping us construct the diagrams.

Benjamin Fine
Anthony M. Gaglione
Gerhard Rosenberger

Introduction to Abstract Algebra

Chapter 1

Abstract Algebra and Algebraic Reasoning

1.1 Abstract Algebra

Abstract algebra or **modern algebra** can be best described as the theory of **algebraic structures**. Briefly, an **algebraic structure** is a set S together with one or more binary operations on it satisfying axioms governing the operations. There are many algebraic structures, but the most commonly studied structures are **groups, rings, fields** and **vector spaces**. Also widely used are **modules** and **algebras**. In this first chapter we will introduce some basic material on what we call the **algebraic method**.

Mathematics traditionally has been subdivided into three main areas — **analysis, algebra and geometry**. These areas overlap in many places so that it is often difficult to determine whether a topic is one in geometry, say, or in analysis. Algebra and algebraic methods permeate all these disciplines and most of mathematics has been algebraicized, that is, uses the methods and language of algebra. Groups, rings and fields play a major role in the modern study of analysis, topology, geometry and even applied mathematics. We will see these connections in examples throughout the book.

Abstract algebra has its origins in two main areas and questions that arose in these areas — the **theory of numbers** and the **theory of equations**. The theory of numbers deals with the properties of the basic number systems — integers, rationals and reals while the theory of equations, as the name indicates, deals with solving equations, in particular polynomial equations. Both are subjects that date back to classical times. A whole section of Euclid's *Elements* is dedicated to number theory. The foundations for the modern study of number theory were laid by Fermat in the 1600's and then by Gauss in the 1800's. In an attempt to prove Fermat's big theorem, Gauss introduced the complex integers $a + bi$ where a and b are integers and showed that this set has unique factorization. These ideas were extended by Dedekind and Kronecker, who developed a wide-ranging theory of algebraic number fields and algebraic integers. A large portion of the terminology used in abstract algebra; rings, ideals, and factorization, comes from the study of algebraic number fields. This has evolved into the modern discipline of algebraic number theory.

The second origin of modern abstract algebra was the problem of trying to determine a formula for finding the solutions, in terms of radicals, of an arbitrary fifth-degree polynomial (a formula like the quadratic formula for finding the solutions of a second-degree polynomial). It was proved first by Ruffini in 1800, and then by Abel, that it is impossible to find a formula in terms of radicals for such a solution. Galois in 1820 extended this and showed that such a formula is impossible for any degree five or greater. In proving this he laid the groundwork for much of the development of modern abstract algebra, especially field theory and finite group theory. Earlier, in 1800, Gauss proved the **fundamental theorem of algebra**, which says that any nonconstant complex polynomial equation must have a solution. In this book we will present a fairly comprehensive, but introductory treatment, of Galois theory and sketch a proof of the results mentioned above.

Finally **linear algebra**, although a part of abstract algebra, arose in a somewhat different context. Historically it grew out of the study of solution sets of systems of linear equations and the study of the geometry of real n-dimensional spaces. It began to be developed formally in the early 1800's with work of Jordan and Gauss and then later in the century by Cayley, Hamilton and Sylvester.

1.2   Algebraic Structures

As we mentioned in the first section an **algebraic structure** is a set with one or more binary operations defined on it whose properties are governed by axioms. In the next chapter we will try to explain each of these concepts. Modern abstract algebra deals, for the most part, with four algebraic structures: groups, rings, fields and vector spaces, and extensions of these basic structures. All these structures are motivated more or less by the standard number systems that are used in elementary mathematics so we will briefly introduce these in Section 1.4. Before this however, we will introduce what we call the algebraic method. This method depends on a knowledge of certain *known* systems such as the integers or the real numbers.

1.3   The Algebraic Method

Here we start by describing a very general method to study algebraic structures. At this point these ideas should not be very clear to the reader; however the reader should look back at this outline once we go through the process of studying rings in Chapter 3. The basic format of the study of abstract algebra and algebraic reasoning can be described as follows:

Start with some **known** system S: for example S could be the **integers** $\mathbb{Z}$, or the rationals $\mathbb{Q}$, or the reals $\mathbb{R}$, or the complex numbers $\mathbb{C}$, or the n-vectors in real Euclidean n-space, $\mathbb{R}^n$, or the set S_n of permutations on n symbols. We then ask a series of questions:

MAIN QUESTIONS

(1) What are the *essential properties* of the known system?
   (a) Once these essential properties are identified they are abstracted and then taken as the axioms for a general abstract algebraic structure.

(2) Are there any other concrete examples of structures of this type?

(3) What makes the known system S *unique* among all systems of this type? What does *uniqueness* mean? This is related to mappings between systems so we get the question;
   (a) What types of mappings preserve the structure of these systems?

(4) What additional properties are satisfied by such structures?

SECONDARY QUESTIONS

(5) What are the properties of the mappings that preserve the structures?

(6) What are **substructures**? When is a subset actually a substructure?

(7) What is the relationship between substructures and mappings?

(8) Can we classify all structures of this type?
      (a) If not (and in general classification is not entirely possible) what subclasses of such structures can be completely classified?

This whole procedure we will call the **algebraic method**.

1.4   The Standard Number Systems

In the algebraic method outlined in the previous section the starting point is some known system. Hence we need to know certain algebraic systems to begin with. In elementary mathematics for the most part we deal with four different number systems each being part of the next. These are the **natural numbers**, the **integers**, the **rational numbers** and the **real numbers**. These basic number systems will serve as the motivating examples for general algebraic structures. Here we briefly review some hopefully well-known facts about them.

The simplest of these number systems is the **natural numbers** which we will denote by $\mathbb{N}$. These consist of the counting numbers $1, 2, 3, \ldots$. Our standard arithmetic operations are addition, $+$, subtraction, $-$, multiplication, $\times$ and division $/$. Of these, only addition and multiplication are defined on $\mathbb{N}$. Subtraction is not defined on $\mathbb{N}$ since, for example, $3 - 5$ is no longer a natural number. We will have more to say later about an operation being *defined* on a set.

In order to allow subtraction, we must append to the natural numbers the number 0 and the negative counting numbers $-1, -2, \ldots$. When we do this, we develop a new number system called the **integers**. These are generally denoted by $\mathbb{Z}$, the $\mathbb{Z}$ coming from the German word for number *Zahlen*. Hence, as a set, $\mathbb{Z} = \{0, \pm 1, \pm 2, \ldots\}$ and the natural numbers are a subset of $\mathbb{Z}$. In terms of the arithmetic operations, the integers are closed under addition, subtraction and multiplication. Saying that the integers are closed under subtraction is equivalent to saying that any integer x has an **additive inverse** $-x$. The integers are not closed under division since for example $3/5$ is no longer an integer. If a and b are integers, we say a **divides** b, denoted by $a \mid b$, if there exists an integer c such that $b = ac$. If not, we say a does not divide b and write $a \nmid b$.

To permit division we must add fractions to the integers. To obtain fractions we start with all possible ratios of integers m/n with m, n integers and $n \neq 0$. We then say that two ratios m_1/n_1 and m_2/n_2 are the same fraction if $m_1 n_2 = m_2 n_1$. We will make this precise in Chapter 4. When we then take the collection of all fractions we get a new number system, the **rational numbers**, containing $\mathbb{Z}$. The rational numbers are usually denoted $\mathbb{Q}$ for *Quotienten-Zahlen*. The rationals are closed under all the standard arithmetic $+, -, \times, /$ just as long as the denominator is nonzero. So we can write

$$\mathbb{Q} = \left\{ \frac{m}{n} : m, n \text{ are integral with } n \neq 0 \right\}.$$

This is read as "$\mathbb{Q}$ equals the set of all m/n such that m and n are integral with n nonzero."

From the viewpoint of arithmetic, it would seem that the rational numbers are all we need. However it became clear to the Greeks over 2000 years ago that there were numbers that were very *real* but were not rational. First we have to describe what we mean by real. In this first chapter we will say that a positive number is real if it measures a length. We exhibit one real number that is not rational.

Theorem 1.4.1. *The number $\sqrt{2}$ is real but not rational.*

Proof. First we must show that $\sqrt{2}$ measures a length. Consider a right triangle with legs one unit by one unit. From the Pythagorean theorem the hypotenuse then has length $h = \sqrt{1^2 + 1^2} = \sqrt{2}$. It follows that $\sqrt{2}$ measures the length of the hypotenuse and hence is *real*. We must show that it cannot be rational.

We will do a proof by contradiction. That is we will suppose the opposite and show that this must lead to a contradiction, and therefore our original assumption must be false.

Let us assume that $\sqrt{2}$ is rational. Hence $\sqrt{2} = \frac{m}{n}$ for some integers m, n with $n \neq 0$. We can further assume that this fraction is in lowest terms, that is no integer with absolute value greater than 1 divides both m and n.

From $\sqrt{2} = \frac{m}{n}$ we have by squaring both sides $2 = \frac{m^2}{n^2}$ which implies that $2n^2 = m^2$. This implies that $2 \mid m^2$ and hence $2 \mid m$ or m is even (see exercises). Since m is even, $m = 2k$ for some integer k and hence $2n^2 = 4k^2$

so that $n^2 = 2k^2$. This implies that $2 \mid n^2$ and hence $2 \mid n$ or n must be even. This is the contradiction since both m and n are even which would contradict that $\frac{m}{n}$ is in lowest terms. Therefore our original assumption must be false and $\sqrt{2}$ is not rational.   □

The theorem shows that there are certain very real numbers that are not rational. We will defer a precise definition of the real numbers until Chapter 4 but for now say that the set of **real numbers**, $\mathbb{R}$, consists of all numbers measuring lengths and their additive inverses (their negatives). Later, we will show that this is equivalent to all possible decimal expansions and their inverses.

The rational numbers are a subset of the reals and in terms of arithmetic on the reals we can add, subtract, multiply and divide just as long as we don't divide by 0. On top of these operations we can take all nth roots of positive reals.

In terms of decimal expansions it is quite easy to see how the rational numbers sit inside the reals. In the exercises we give a hint how to prove the following theorem and we will return to it in Chapter 4.

Theorem 1.4.2. *A real number x is a rational number if and only if its decimal expansion either terminates or repeats.*

1.5   The Integers and Induction

The integers will serve as the primary motivating example for one of the most fundamental algebraic structures, a **ring**. We will introduce this in Chapter 3. Here we will summarize many of the known basic properties of $\mathbb{Z}$.

In the next chapter we will discuss binary operations and their properties in general. A binary operation is a function that takes two elements of a set and produces a third element of the set. The integers $\mathbb{Z}$ have two binary operations defined on them, addition denoted by $+$ and multiplication that we will denote by $\cdot$. These two operations satisfy the following:

(1) $a + b = b + a$ for any two integers a and b. We say that addition on $\mathbb{Z}$ is **commutative**.

(2) $(a+b)+c = a+(b+c)$ for any integers a, b, c. We say that addition on $\mathbb{Z}$ is **associative**.

(3) $a+0 = a$ for any integer a. We say that 0 is an **additive identity** on $\mathbb{Z}$.

(4) For each integer a there is an integer $-a$ such that $a + (-a) = 0$. We say that each integer has an **additive inverse**.

Later we will see that any set that has an operation $+$ satisfying (1) through (4) will be called an **abelian group**.

(5) $(ab)c = a(bc)$ for any integers a, b, c. We say that multiplication on $\mathbb{Z}$ is **associative**.

(6) $a(b + c) = ab + ac$ and $(b + c)a = ba + bc$ for any integers a, b, c. We say that multiplication on $\mathbb{Z}$ is **distributive** over addition.

We will see that in Chapter 3 that properties (1) through (6) will be taken as the basic axioms of a **ring**.

(7) $ab = ba$ for any two integers a and b. We say that multiplication on $\mathbb{Z}$ is **commutative**.

(8) $a \cdot 1 = a$ for any integer a. We say that 1 is a **multiplicative identity** on $\mathbb{Z}$.

The integers satisfy many other properties. Some relate to the multiplicative structure of the integers. For example the reader may be familiar with the **fundamental theorem of arithmetic** which says that any integer greater than 1 has a unique expression as a product of prime numbers written in order. This really is the starting off point for the theory of numbers, an important branch of mathematics. We will discuss this in some depth in Chapter 4.

Here we mention two very important properties of $\mathbb{Z}$ that subsequently will be shown to make the integers unique among the class of rings. These are the **principle of mathematical induction** and the **least well-ordering property**.

The Principle of Mathematical Induction. *Let S be a subset of the natural numbers $\mathbb{N}$. Suppose $1 \in S$ (this means 1 belongs to S) and S has the property that if $n \in S$ then $(n+1) \in S$. Then $S = \mathbb{N}$.*

We will abbreviate the Principle of Mathematical Induction by PMI.

The Least Well-Ordering Property. *Let S be a nonempty subset of the natural numbers $\mathbb{N}$. Then S has a least element.*

We will abbreviate the least well-ordering property by LWO.

In the theorem below we show that the PMI is equivalent to the LWO. By equivalent, we mean here that if we assume that the PMI is true then we can prove the LWO and if we assume the LWO is true then we can prove the PMI.

Theorem 1.5.1. *The principle of mathematical induction is equivalent to the least well-ordering property.*

Proof. To prove this we must assume first the principle of mathematical induction and show that the well-ordering property holds and then vice versa. Suppose that the PMI holds and let S be a nonempty subset of $\mathbb{N}$. We must show that S has a least element. Using the same set notation as in the definition of $\mathbb{Q}$ and letting $\forall$ abbreviate "for all," we let T be the set

$$T = \{x \in \mathbb{N} : x \leq s, \forall s \in S\}.$$

Now $1 \in T$ since S is a subset of $\mathbb{N}$. If whenever $x \in T$ it were to follow that $(x + 1) \in T$, then by the inductive property $T = \mathbb{N}$ but then S would be empty contradicting that S is nonempty. Therefore there exists an a with $a \in T$ and $(a+1) \notin T$. We claim that a is the least element of S. Now $a \leq s$ for all $s \in S$ because $a \in T$. If $a \notin S$ then every $s \in S$ would also satisfy $(a + 1) \leq s$. This would imply that $(a + 1) \in T$ which is a contradiction. Therefore $a \in S$ and $a \leq s$ for all $s \in S$ and hence a is the least element of S.

Conversely suppose the well-ordering property holds and suppose that S is a subset of $\mathbb{N}$ with the properties that $1 \in S$ and that whenever $n \in S$ it follows that $(n + 1) \in S$. We must show that $S = \mathbb{N}$. If $S \neq \mathbb{N}$, then the set difference, i.e., $\mathbb{N} - S = $ set of elements in $\mathbb{N}$ but not in S, would be a nonempty subset of $\mathbb{N}$. Thus by the LWO, it has a least element, say n. Hence $(n - 1)$ is not in $\mathbb{N} - S$ or $(n - 1) \in S$. But then by the assumed property that S has $(n - 1) + 1 = n \in S$ — also which is a contradiction. Therefore $\mathbb{N} - S$ is empty and $S = \mathbb{N}$. $\square$

The PMI is also called the **inductive property** and is the basis for **inductive proofs** or **proofs by induction**. These types of proofs play a large role in abstract algebra. In an inductive proof we want to prove statements $\mathcal{P}(n)$ which depend on positive integers, n. In the induction we show that $\mathcal{P}(1)$ is true, then show that the truth of $\mathcal{P}(k+1)$ follows from the truth of $\mathcal{P}(k)$. From the inductive property $\mathcal{P}(n)$ is then true for all positive integers n.

There is an alternative form of the PMI called **course of values induction**. Both forms are equivalent. The course of values induction asserts the following.

Course of Values Induction. *Let S be a subset of the natural numbers $\mathbb{N}$, such that $1 \in S$ and if for all natural numbers $m < n, m \in S$, then it is true that $n \in S$, then $S = \mathbf{N}$.*

We note that in either formulation of the PMI, the induction does not have to start at 1, but can start at any integer n. In this case, the set S is the set of all integers greater than or equal to n. We leave a proof of the equivalence of standard induction and course of values induction to the exercises.

We present a straightforward example which has an ancient history in number theory.

EXAMPLE 1.4.1. Show that $1 + 2 + \cdots + n = \frac{(n)(n+1)}{2}$.

Here we let $S_n = 1 + 2 + \cdots + n$ and we want to prove that $S_n = \frac{(n)(n+1)}{2}$ for all natural numbers n.

For $n = 1$ we have $1 = S_1 = \frac{(1)(2)}{2} = 1$. Hence the statement is true for $n = 1$.

Assume that the statement is true for $n = k$, that is

$$S_k = 1 + 2 + \cdots + k = \frac{k(k+1)}{2}$$

and consider $n = k + 1$. Then

$$S_{k+1} = 1 + 2 + \cdots + k + (k+1) = (1 + 2 + \cdots + k) + (k+1)$$

$$= \frac{k(k+1)}{2} + (k+1) = \frac{(k+1)(k+2)}{2}.$$

Therefore the statement is true for $n = k + 1$ and hence true by induction for all $n \in \mathbb{N}$.

The series of integers

$$1, 1 + 2 = 3, 1 + 2 + 3 = 6, 1 + 2 + 3 + 4 = 10, \ldots$$

are called the **triangular numbers** since they are the sums of dots placed in triangular form as in Figure 1.1. These numbers were studied by the Pythagoreans in Greece in 500 B.C.

Throughout this book we will see many other proofs by induction which use either of the two formulations.

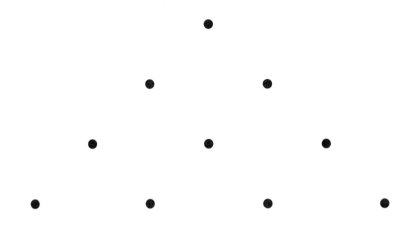

Figure 1.1 Triangular Numbers

1.6   Exercises

1.1. Recall that if a and b are integers then a divides b, denoted $a \mid b$, means there exists an integer c such that $b = ac$. Show that if m, n are positive integers such that $m \mid n$ and $n \mid m$, then $m = n$.

1.2. Show that if m is an integer and $2 \mid m^2$, then $2 \mid m$.
(HINT: Explain why any integer is of the form $2k$ or $2k + 1$ where k is an integer. Now let m be of the form $2k + 1$ and then show that $2 \nmid m^2$. Explain why this shows that m must be of the form $2k$, i.e., $2 \mid m$.)

1.3. Prove Theorem 2: A real number is rational if and only if its decimal expansion either terminates or repeats.
(HINT:($\Rightarrow$) Let $q = m/n$ be a rational number in lowest terms then show that the decimal expansion for q either terminates or repeats. To do this think about how the decimal expansion is obtained. In particular what are the possible remainders when dividing by n?
($\Leftarrow$) Suppose that $x = .a_1 \ldots a_n a_1 ... a_n ...$ is a repeating decimal. Then we need to show that x is rational. To do this consider what $10^n x$ looks like and then take $10^n x - x$.)

1.4. Use the PMI to prove the law of exponents: For a, b any real numbers and n any natural number

$$(ab)^n = a^n b^n.$$

1.5. Use the PMI to prove for any natural number n

$$1^3 + 2^3 + \cdots + n^3 = \frac{n^2(n+1)^2}{4}.$$

*1.6. Prove that standard induction (PMI) and course of values induction are equivalent. By this we mean that if we assume the standard PMI then course of values induction is true, and conversely.

1.7. Use course of values induction to prove the following part of the fundamental theorem of arithmetic: Every integer $n > 1$ can be expressed as a product of primes. Here the term product may mean just one factor.

1.8. If m and n are natural numbers, we define the greatest common divisor of m and n, denoted $\gcd(m, n)$, to be a natural number $d = \gcd(m, n)$ such that

(i) $d \mid m$ and $d \mid n$;

(ii) if k is a natural number such that $k \mid m$ and $k \mid n$, then $k \mid d$.

Then (i) says that d is a common divisor and (ii) says it is the greatest common divisor.

Use problem 1.1 to show that the gcd of two natural numbers is unique if it exists. Our next problem will show it always exists.

1.9. Use the LWO to show that the gcd of any two natural numbers, m and n, defined in problem 1.8 always exists.

(HINT: Let $A(m, n) = \{an + bm > 0 : a$ and b are integers$\}$. First say why $A(m, n)$ is nonempty. Then show the smallest element of $A(m, n)$ (you must say how you know this set has a smallest element) is the $\gcd(m, n)$. Here you will have to use the division algorithm which says that if a, b are natural numbers there always exist unique integers q and r such that $a = bq + r$ where $0 \le r < b$.)

1.10. The previous problem shows that the gcd of two natural numbers m, n is always a linear combination of m and n, e.g., $am + bn$. We say two natural numbers are relatively prime when their gcd is 1, i.e., m and n are relatively prime means $\gcd(m, n) = 1$. Use the result and proof of the previous problem to show that two natural numbers are relatively prime if and only if there exists integers a, b such that $am + bn = 1$.

1.11. Show that if m, n, k are all natural numbers and $n \mid mk$ but m and n are relatively prime, then $n \mid k$.
(HINT: Use the result of problem 1.10.)

Chapter 2

Algebraic Preliminaries

In this chapter, we will introduce some very necessary algebraic preliminaries. These include set theory, mappings between sets, equivalence relations, binary operations and sizes of sets. We begin by looking at basic set theory.

2.1 Sets and Set Theory

In the last chapter we defined an algebraic structure as a set with certain binary operations defined on it. From this description it is clear that the most basic object in algebra is a set. The concept of a set seems intuitively clear although this is in actuality far from true and an unrestricted use of the set concept has led to contradictions in mathematics. It was for this reason that an axiomatic treatment of set theory became necessary. Our discussion of set theory, however, will be strictly naive.

A **set** is a well-defined collection of objects. By well defined we mean that it is clear whether an object is, or is not, in the set. For example the collection of *older people* is *not* a set since we have not defined what we mean by older people. On the other hand the collection of people over 65 years of age is a set.

Throughout this chapter we will designate sets by capital letters such as A, B, C, etc. The individual objects within a set are called **elements** and we will denote elements by lower-case letters such as a, b, c, etc. The symbol $\in$ will denote **element of** and the notation $a \in A$ will indicate that the element a belongs to the set A. If this is not the case, we write $a \notin A$, indicating that a does not belong to the set A.

We can describe sets in different ways. First we can give a complete description verbally of the set. For example A is the set of all positive integers less than 10. If possible we can list all the elements in the set within brackets. Hence the set A above can be written as

$$A = \{1, 2, 3, 4, 5, 6, 7, 8, 9\}.$$

Finally we can list a general element of the set and then a description of the general element, for example;

$$A = \{x : x \text{ is a positive integer } < 10\}.$$

This is read "A equals the set of all x such that x is a positive integer less than 10." Sometimes before the colon we indicate the set which the elements are allowed to vary over, i.e.,

$$A = \{x \in \mathbb{Z} : x \text{ is a positive integer } < 10\}.$$

Suppose that A and B are two sets. If every element of A is also an element of B, one says that A is a **subset** of B, denoted $A \subset B$ (or $A \subseteq B$). If $A \subset B$ and $B \subset A$, then the sets A and B are said to be **equal**, denoted $A = B$. Thus to show that two sets are equal, we must show a double inclusion. Finally, if $A \subset B$ but $A \neq B$, then A is called a **proper subset** of B, denoted $A \subsetneq B$. The set consisting of no elements at all is called the **null set** or **empty set** and is designated by $\emptyset$. The empty set is considered as a subset of every other set, that is, $\emptyset \subset A$ for every set A.

If A is a set then the collection of all subsets of A is called the **power set** of A, denoted by $\mathcal{P}(A)$. Hence the power set is a set whose elements are actually sets. For example, suppose that $A = \{1, 2, 3\}$. Then the power set of A has eight elements given by

$$\mathcal{P}(A) = \{\emptyset, \{1\}, \{2\}, \{3\}, \{1, 2\}, \{1, 3\}, \{2, 3\}, \{1, 2, 3\}\}.$$

If a set A has n elements in it, then there are 2^n elements in its power set, $\mathcal{P}(A)$. This will be proved later on.

2.1.1 Set Operations

Given existing sets there are several important operations that produce new sets. Important in this regard is the concept of a **universal set**. A **universal**

set, that we will denote by $\mathcal{U}$, is the set of all elements relevant to whatever set theory problem is being considered. The universal set can change from situation to situation. As we discuss the set operations it will be implicitly understood that all sets are part of some universal set $\mathcal{U} \neq \emptyset$. This last inequality is sometimes expressed by saying the set $\mathcal{U}$ is **nontrivial**.

If A is a set its **complement**, denoted A', is the set of all elements in the universal set not in A. That is,

$$A' = \{x : x \notin A\}.$$

For example if $\mathcal{U} = \{1, 2, 3, 4, 5\}$ and $A = \{1, 2, 3\}$ then $A' = \{4, 5\}$.

If A and B are sets then the collection of elements common to both A and B is called the **intersection**, which we denote by $A \cap B$. Hence

$$A \cap B = \{x : x \in A \text{ and } x \in B\}.$$

If A and B are two sets such that $A \cap B = \emptyset$, then A and B are called **disjoint sets**.

The set of elements in either A or B (which includes those in both A and B) is called their **union**, denoted by $A \cup B$. Hence

$$A \cup B = \{x : x \in A \text{ or } x \in B\}.$$

EXAMPLE 2.1.1. Let $\mathcal{U} = \{1, 2, 3, 4, 5, 6, 7, 8, 9\}$ and $A = \{1, 3, 5, 7, 9\}, B = \{3, 6, 7, 8\}$. Find

    (1) A',
    (2) $A \cap B$,
    (3) $A \cup B$,
    (4) $A \cap B'$.

(1) The set A' are those elements not in A. Therefore

$$A' = \{2, 4, 6, 8\}.$$

(2) $A \cap B$ are those elements common to both A and B. Therefore

$$A \cap B = \{3, 7\}.$$

(3) $A \cup B$ are those elements either in A or in B. Therefore

$$A \cup B = \{1, 3, 5, 6, 7, 8, 9\}.$$

(4) $A \cap B'$ are those elements in A and B'. Therefore these are the elements that are in A but not in B. Therefore

$$A \cap B' = \{1, 5, 9\}.$$

The set $A \cap B'$ is sometimes denoted $A \setminus B$ or $A - B$ and called the **set difference**.

We need to extend the operations of union and intersection so that they can be applied to an arbitrary number of sets. If we have a collection of sets and we have one set called A_α for each element α of a set Λ with $\Lambda \neq \emptyset$, we call this an **indexed collection of sets** and Λ is called the **index set**. This will be denoted by

$$\{A_\alpha\}_{\alpha \in \Lambda};$$

for example, if $\Lambda = \mathbb{N}$, the set of natural numbers, we could write $\{A_i\}_{i \in \mathbb{N}}$ meaning that we have a countable number (we will talk more about what it means to be countable in Section 2.4) of sets or $\{A_1, A_2, \dots\}$. We should note that in general it is not even necessary that Λ be countable, e.g., Λ could be the set of all real numbers $\mathbb{R}$ which is uncountable. We will have more to say about uncountable sets in Section 2.4.

Let $\{A_\alpha\}_{\alpha \in \Lambda}$ with $\Lambda \neq \emptyset$ be a collection of sets. The union of these sets is the set of all elements which belong to *at least one* of the A_α, i.e.,

$$\bigcup_{\alpha \in \Lambda} A_\alpha = \{x : x \in A_\alpha \text{ for at least one } \alpha\}.$$

In case the index $\Lambda = \mathbb{N}$ or is finite, say $\Lambda = \{1, 2, ..., n\}$, we use the notations, $\bigcup_{i=1}^\infty A_i$, $\bigcup_{i=1}^n A_i$, respectively for

$$\bigcup_{i=1}^\infty A_i = A_1 \cup A_2 \cup \cdots \cup A_n \cup \cdots,$$

$$\bigcup_{i=1}^n A_i = A_1 \cup A_2 \cup \cdots \cup A_n.$$

Again given the collection of sets $\{A_\alpha\}_{\alpha \in \Lambda}$ with $\Lambda \neq \emptyset$, the intersection of these sets is the set of all elements which belong to all the A_α, i.e.,

$$\bigcap_{\alpha \in \Lambda} A_\alpha = \{x : x \in A_\alpha \text{ for all } \alpha \in \Lambda\}.$$

Similar notations as for unions are adopted in the case of intersections when the index set Λ is countable or finite. Sometimes we even suppress the index set if we don't know or care what it is. Thus, for example, we could write $\cup_i S_i$ for the union of sets S_i where we have suppressed the nonempty index set. Again we can similarly suppress the nonempty index set for an intersection of sets.

The symbols $\forall$ and $\exists$ are commonly used in mathematics with the meanings $\forall =$ "for all" and $\exists =$ "there exists." We will use them occasionally. For example, we could have defined $\cup$ and $\cap$

$$\bigcup_{\alpha \in \Lambda} A_\alpha = \{x : \exists \alpha \in \Lambda \text{ with } x \in A_\alpha, \Lambda \neq \emptyset\},$$

$$\bigcap_{\alpha \in \Lambda} A_\alpha = \{x : x \in A_\alpha \; \forall \, \alpha \in \Lambda, \Lambda \neq \emptyset\}.$$

We always assume in the following that the index set Λ is nonempty.

Another set operation that will play a role when we study rings is called the **symmetric difference** $A \triangle B$. These are the elements in either A or B but not both. We then have

$$A \triangle B = (A \cup B) - (A \cap B) = (A \cap B') \cup (A' \cap B).$$

All of these operations on sets — unions, intersections and complements — can be handled using a graphical procedure known as a **Venn diagram**. In a **Venn diagram** the universal set, $\mathcal{U}$, is represented by a rectangular region while portions of this region represent subsets. The Venn diagram for the case of two sets A and B is pictured in Figure 2.1.

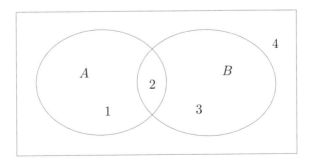

Figure 2.1 Venn Diagram for Two Sets

In the diagram, regions 1,2 are A and regions 2,3 are B. It follows that A' is regions 3,4, $A \cap B$ is region 2 and regions 1,2,3 are $A \cup B$. The symmetric difference is regions 1,3.

A collection of subsets $\{S_i\}_{i \in I}$ is said to **partition** the universal set $\mathcal{U}$ if $S_i \neq \emptyset$ for all $i \in I$; they are pairwise disjoint, that is, $S_i \cap S_j = \emptyset$ if $i \neq j$, and their union is all of $\mathcal{U}$. That is, $\mathcal{U} = \bigcup_i S_i$. Here it is understood that the union is taken over i such that the sets S_i are disjoint.

In the Venn diagram the four regions 1,2,3,4 partition the universal set. Note that any nonempty set, not just the universal set, can be so partitioned.

Finally we give one more set operation. If A and B are two sets, the **cartesian product** $A \times B$ is the set of all ordered pairs (a, b) such that $a \in A$ and $b \in B$, i.e.,

$$A \times B = \{(a, b) : a \in A, b \in B\}.$$

Here it is to be emphasized that $(a_1, b_1) = (a_2, b_2)$ if and only if $a_1 = a_2$ and $b_1 = b_2$. This means that the order really counts. Similarly we can define the cartesian product of any finite number of sets. For example, $\mathbb{R}^n = \mathbb{R} \times \cdots \times \mathbb{R}$ (n times), consists of ordered n-tuples $(x_1, \ldots, x_n)$ where each $x_i \in \mathbb{R}$, for $1 \leq i \leq n$.

2.2   Functions

In the study of algebraic structures it is very important to study functions or mappings between different structures. In this section we concentrate on functions.

Definition 2.2.1. *If A and B are two nonempty sets then a **function** or **mapping** from A to B is a correspondence that associates to each element $a \in A$ a **unique** element $b \in B$. We usually give the function a name, such as f, and use the notation $f : A \to B$ to indicate that this function maps from A into B. Sometimes we may also write $A \xrightarrow{f} B$. If $a \in A$ and $f(a) = b$ then we say that b is the **image** of a under f.*

If $f : A \longrightarrow B$ then the set A is called the **domain** of the function while B is the **codomain**. Not every element of B must be an image of an element of A. The set of images of elements of A is denoted $f(A)$ and is called the **range** or **image** of the function f, i.e.,

$$f(A) = \{f(a) : a \in A\}.$$

The range is always a subset of the codomain, $f(A) \subset B$.

Similarly we can define the image of any subset of the domain. If as above $f : A \longrightarrow B$ and E is any subset of A, $E \subset A$, then the **image set** $f(E)$ is defined by

$$f(E) = \{f(x) \in B : x \in E\}.$$

Given a subset of the codomain $F \subset B$, we can define the **preimage set**, $f^{-1}(F)$, by

$$f^{-1}(F) = \{x \in A : f(x) \in F\}.$$

We note that the image set $f(E)$ can be the empty set $\emptyset$ only if $E = \emptyset$, but a preimage set, $f^{-1}(F)$, can be the empty set even if F is nonempty — this could happen if there is nothing in A mapped to the set F in B.

If $f : A \to B$ and E is any nonempty subset of A, $E \subset A$ with $E \neq \emptyset$, then we define the **restriction** of f to E to be the mapping denoted by $f_{|E}$ and given $f_{|E}(x) = f(x)$ for all $x \in E$. We note that technically the domain of $f_{|E}$ is E rather than A so $f_{|E}$ is really a different mapping than f.

The subset of the cartesian product defined by $\{(a, f(a)) : a \in A\}$ is called the **graph** of the function. Notice that in general this is a formal definition and there is nothing necessarily visual in the graph.

EXAMPLE 2.2.1. Let $A = \{1, 2, 3, 4, 5\}$ and $B = \{a, b, c, d\}$. Consider the function

$$f : 1 \to a, 2 \to b, 3 \to a, 4 \to b, 5 \to a.$$

Then the range of f is $f(A) = \{a, b\}$. For $E = \{1, 3\}$, its image is the singleton $f(E) = \{a\}$. But the preimage of $F = \{c, d\}$ is empty, $f^{-1}(F) = \emptyset$ because there is nothing in A mapped to either c or d. The graph of f is the set of pairs

$$\{(1, a), (2, b), (3, a), (4, b), (5, a)\} \subset A \times f(A).$$

If $f : A \longrightarrow B$ and the range of f coincides with the codomain, i.e., $f(A) = B$, then we say that the function is **onto**. An **onto function** is called a **surjection**. The function in Example 2.2.1 is not a surjection because $f(A) \subsetneq B$. However if $\mathbb{Z}$ denotes the integers then the function $f : \mathbb{Z} \longrightarrow \mathbb{Z}$ given by $f(z) = z + 1$ is a surjection. Being an onto function $f : A \to B$ means that each $b \in B$ is the image of some $a \in A$ or equivalently that each $b \in B$ has a **preimage** in A or that $f(A) = B$.

If a function from A to B has the property that different elements of A have different images in B then it is called a **one-to-one function** or an **injection**. We sometimes write one-to-one as 1-1. In symbols this says that if $a_1 \neq a_2$ then $f(a_1) \neq f(a_2)$. Equivalently, and often more useful in proving that a function is one-to-one is that $f(a_1) = f(a_2)$ implies that $a_1 = a_2$. The function in Example 2.2.1 is not an injection because for example $f(1) = f(3)$. However, the function $f : \mathbb{Z} \longrightarrow \mathbb{Z}$, $f(z) = z + 1$ is an injection. See the example below.

A function that is both one-to-one and onto, that is, both an injection and a surjection, is called a **one-to-one correspondence** or a **bijection**.

EXAMPLE 2.2.2. The function from $\mathbb{Z}$ to $\mathbb{Z}$ given by $f(z) = z + 1$ for each integer z is a bijection. To show this we must show that the function is both one-to-one and onto. Suppose first that $f(z_1) = f(z_2)$. Then $z_1 + 1 = z_2 + 1$ and then clearly $z_1 = z_2$. This implies that f must be one-to-one. Now let $z \in \mathbb{Z}$ and let $y = z - 1$. Then $f(y) = y + 1 = (z - 1) + 1 = z$ and hence z is the image of some integer. This shows that f is also onto and hence a bijection.

Suppose A is a set and $I_A : A \rightarrow A$ by $I(a) = a$ for all $a \in A$, that is, I_A maps each element of A to itself. This is called the **identity function on** A. It is clearly a bijection. (See the exercises for a proof.)

A bijection is invertible. We make this precise. To do this we must define the **composition** of functions.

Definition 2.2.2. *If $f : A \rightarrow B$ and $g : B \rightarrow C$ then the **composition** $g \circ f$ or gf is the function $g \circ f = gf : A \rightarrow C$, given by $g \circ f(a) = gf(a) = g(f(a))$, symbolically $A \xrightarrow{f} B \xrightarrow{g} C$. So that $g \circ f$ has domain A and codomain C.*

If $f : A \rightarrow B$ then $g : B \rightarrow A$ is the **inverse** of f if $f \circ g = I_B$ and $g \circ f = I_A$ where I_A and I_B are the identity functions on A and B respectively. If g is the inverse of f, we will denote g by f^{-1}. If such a function g exists for f, then we say f is **invertible**. The next theorem says that invertibility is equivalent to being a bijection.

Theorem 2.2.1. *A function $f : A \rightarrow B$ is invertible if and only if f is a bijection.*

Proof. Supposing that f is a bijection we show that it has an inverse. For each $b \in B$ there is an $a \in A$ with $f(a) = b$ since f is onto. But since f is

1-1, if $f(a_1) = b = f(a)$, then $a_1 = a$. This says that $a = $ the preimage of b is unique. Let $g : B \longrightarrow A$ be given by $g(b) = a$ if $f(a) = b$. Since the preimage a is unique and since we can find such an $a \in A$ for every $b \in B$ this function g is a well-defined function $g : B \longrightarrow A$. We claim that it is the inverse of f. Consider any $b \in B$. Then we have $f \circ g(b) = f(g(b)) = f(a) = b$ by definition so $f \circ g = I_B$. Analogously $g \circ f = I_A$.

Now suppose that g is the inverse of f. We show that f is a bijection. Suppose first that $f(a_1) = f(a_2)$. Then $g(f(a_1)) = g(f(a_2))$. But $g \circ f = I_A$ and hence $g(f(a_1)) = a_1 = g(f(a_2)) = a_2$. Hence $a_1 = a_2$ and f is one-to-one. Now let $b \in B$. Then $g(b) = a$. Then $f(a) = f(g(b)) = I_B(b) = b$. It follows that b has the preimage a and hence f is onto. Therefore f is one-to-one and onto and hence a bijection.   □

From the proof of the above theorem we note that it follows that if $f : A \to B$ and $g : B \to A$ are functions then $g \circ f = I_A$ implies that f is injective and $f \circ g = I_B$ implies that g is injective. Hence we get the following lemma.

Lemma 2.2.1. *If $f : A \to B$ is a bijection and $g : B \to C$ is a bijection then $g \circ f : A \to C$ is also a bijection.*

We leave the proof of this to the exercises.

2.3   Equivalence Relations and Factor Sets

Relations and especially equivalence relations play a major role in the study of algebraic structures. Informally a relation on a set A is a statement about elements (a_1, a_2) of the cartesian product $A \times A$ that is, either true or false. If it is true for a pair (a_1, a_2) we say that a_1 is related to a_2. We will denote this by $a_1 \sim a_2$. If the statement is not true then a_1 is not related to a_2. For example equality on any set is a relation while $>$ is a relation on the set of integers. Formally a relation is just a subset of the cartesian product $A \times A$.

Definition 2.3.1. *A **relation** on the set A is a subset $R \subset A \times A$. If $(a_1, a_2) \in R$ then a_1 is **related to** a_2.*

Notice that the order is important. It may be that a_1 is related to a_2 but a_2 is **not related** to a_1. For example under the relation $>$ we have $6 > 4$ so 6 is related to 4 but 4 is not related to 6. Further an element may or may not be related to itself. We define three important properties of relations.

Definition 2.3.2. *(1) A relation* $\sim$ *on a set* A *is* ***reflexive*** *if* $a \sim a$ *for all* $a \in A$, *that is, each element of* A *is related to itself.*

(2) A relation $\sim$ *on a set* A *is* ***symmetric*** *if whenever* $a \sim b$ *then* $b \sim a$.

(3) A relation $\sim$ *on a set* A *is* ***transitive*** *if whenever* $a \sim b$ *and* $b \sim c$ *then* $a \sim c$.

EXAMPLE 2.3.1. (1) The relation $>$ defined on the integers is transitive but neither symmetric nor reflexive while $\geq$ is reflexive and transitive but not symmetric.

(2) Congruence defined on the set of Euclidean triangles in the plane is reflexive, symmetric and transitive.

(3) Equality on any set is reflexive, symmetric and transitive.

Relations, such as congruence and equality as in the last example, that are reflexive symmetric and transitive, play an especially important role.

Definition 2.3.3. *An* ***equivalence relation*** *on a set* A *is a relation that is reflexive, symmetric and transitive. We will usually use the symbol* $\equiv$ *for an equivalence relation.*

Recall that a **partition** of a set $A \neq \emptyset$ (we tacitly assume A is nonempty for the rest of this section) is a collection of mutually disjoint, nonempty subsets whose union is all of A. We now show that partitions and equivalence relations are really equivalent concepts, meaning that if we have a partition we have an equivalence relation and to every equivalence relation we have a partition.

Definition 2.3.4. *Suppose that* $\equiv$ *is an equivalence relation on a nonempty set* A *and suppose that* $a \in A$. *The set*

$$[a] = \{a_1 \in A : a_1 \equiv a\},$$

that is, the set of all elements equivalent to a, *is called the* ***equivalence class*** *of* a.

Theorem 2.3.1. *(1) Let* $\equiv$ *be an equivalence relation on* A. *Then the set of equivalence classes partitions* A.

(2) Let $\{S_i\}_{i \in I}$ *be a partition of* A. *Define* $a_1 \sim a_2$ *to mean that* a_1 *and* a_2 *are in the same partitioning subset. Then* $\sim$ *is an equivalence relation on* A *and its equivalence classes are precisely the* S_i.

Proof. We will prove (1) and leave (2) to the exercises. To show that the equivalence classes partition A we must show that each equivalence class is nonempty, that every element is in an equivalence class so that their union is all of A and that different equivalence classes are actually disjoint. This third statement is equivalent to saying that given $a, b \in A$ then either $[a] = [b]$ or $[a] \cap [b] = \emptyset$.

Each element $a \in A$ is equivalent to itself since an equivalence relation is reflexive. Hence $a \in [a]$ for every $a \in A$ and therefore each $[a] \neq \emptyset$, and also the union of all the equivalence classes is all of A.

Now suppose that $a, b \in A$ and $[a] \cap [b] \neq \emptyset$. We then will show that $[a] = [b]$. This is equivalent to the third statement above. Suppose $x \in [a] \cap [b]$. Then $x \in [a]$ so $a \equiv x$ by symmetry. But $x \in [b]$ so $x \equiv b$. By transitivity $a \equiv b$ so $a \in [b]$. But then everything equivalent to a is also equivalent to b so $[a] \subset [b]$. The argument works identically in the other direction so also $[b] \subset [a]$ and hence $[a] = [b]$. It follows that two equivalence classes are either disjoint or the same and therefore the set of equivalence classes forms a partition of A. See the exercises for (2). $\qquad\square$

Suppose that $\equiv$ is an equivalence relation on the set A. The set of equivalence classes under this relation is called the **factor set** of A modulo the relation. We denote this by $A/_{\equiv}$, i.e.,

$$A/_{\equiv} = \{[a] : a \in A\}.$$

Notice that the elements of the factor set are subsets (equivalence classes) of the original set A. We will see versions of the following theorem throughout the book. When we look at groups it is a version of what is called the **group isomorphism theorem**. When we look at rings the **ring isomorphism theorem**. For vector spaces it is a version of the **dimension theorem**. The student should look this theorem over, try to understand the proof and then look back at it when the above named theorems are introduced.

Theorem 2.3.2. *(1)Let $f : A \longrightarrow B$ be a surjection (an onto mapping). On the set A define the relation $a_1 \sim a_2$ if $f(a_1) = f(a_2)$. Then $\sim$ is an equivalence relation on A.*

(2) Let $A/_\sim$ be the factor set and define the function $\hat{f}$ from $A/_\sim$ to B by

$$\hat{f}([a]) = f(a)$$

that is, $\hat{f}$ maps an equivalence class to the image of any of its element. Then $\hat{f}$ is a bijection.

Proof. To prove (1) we must show that the defined relation is reflexive, symmetric and transitive. Clearly for any $a \in A$ we have $f(a) = f(a)$ and hence $a \in [a]$ and the relation is reflexive. If $a_1 \sim a_2$, then $f(a_1) = f(a_2)$ and then clearly $f(a_2) = f(a_1)$ and so $a_2 \sim a_1$ and the relation is symmetric. Finally if $a_1 \sim a_2$ and $a_2 \sim a_3$ then $f(a_1) = f(a_2)$ and $f(a_2) = f(a_3)$. It follows that $f(a_1) = f(a_3)$ so that $a_1 \sim a_3$ and the relation is transitive. Hence this relation is an equivalence relation.

Now for (2), consider the factor set and the map $\hat{f}$ from $A/_\sim$ to B given by

$$\widehat{f}([a]) = f(a).$$

We must show that this is a bijection. Since an equivalence class is a set we must show first that the map is a well-defined function, that is, if $[a_1] = [a_2]$ then $\hat{f}([a_1]) = \hat{f}([a_2])$. Since $[a_1] = [a_2]$ we must have $a_1 \sim a_2$ so that $f(a_1) = f(a_2)$. Hence $\hat{f}([a_1]) = \hat{f}([a_2])$ and hence the map is well defined.

If $b \in B$ then there exists an $a \in A$ with $f(a) = b$ since we assumed that f is a surjection or is onto. Then $\hat{f}([a]) = f(a) = b$ and therefore $\hat{f}$ is also a surjection.

Now suppose that $\hat{f}([a_1]) = \hat{f}([a_2])$. Then $f(a_1) = f(a_2)$. It follows that $a_1 \sim a_2$ and hence $[a_1] = [a_2]$. Therefore $\hat{f}$ is one-to-one and hence $\widehat{f}$ is a bijection. $\qquad\square$

EXAMPLE 2.3.2. Let $A = \{1, 2, 3, 4, 5\}$ and $B = \{a, b, c\}$ and suppose

$$f : 1 \to a, 2 \to b, 3 \to c, 4 \to a, 5 \to b$$

Then we have the factor set

$$[1] = \{1, 4\}, [2] = \{2, 5\}, [3] = \{3\}.$$

The bijection

$$\hat{f} : A/_\sim \ \to \ B$$

is given by

$$\hat{f} : [1] \to a, [2] \to b, [3] \to c.$$

2.4 Sizes of Sets

Intuitively we can see that $A = \{1, 2, 3\}$ and $B = \{a, b, c\}$, while not the same set, have the same size. Using the concept of a bijection between sets we can formalize the concept of **having the same size**.

Definition 2.4.1. *Two sets A and B are* **equivalent** *or* **have the same size** *if there is a bijection between them.*

The natural numbers $\mathbb{N}$ are the counting numbers $\mathbb{N} = \{\, 1, 2, 3, ...\}$. These will define the sizes of nonempty finite sets.

Definition 2.4.2. *A set A has* **size** *n if it has the same size as the subset $\{1, 2, ..., n\}$ of $\mathbb{N}$. We will denote this by $|A| = n$. A nonempty set A is* **finite** *if it has size n for some natural number n. Otherwise the set is said to be an* **infinite set**. *If $A = \emptyset$ then we define the size of A to be $|A| = 0$. The empty set $\emptyset$ is then also finite with size 0.*

If two sets A and B are equivalent or have the same size according to Definition 2.4.1, we write $|A| = |B|$.

EXAMPLE 2.4.1. The set $A = \{a, b, c\}$ is finite of size 3, i.e., $|A| = 3$, while the set of all natural numbers, $\mathbb{N}$, is itself infinite. We will have more to say about this a little later.

Lemma 2.4.1. *Having the same size is an equivalence relation on the collection of all sets.*

Proof. If A and B are sets, we define $A \sim B$ to mean $|A| = |B|$ or there exists $f : A \longrightarrow B$ such that f is 1-1 and onto. Now $A \sim A$ since the identity map on A, I_A, is a bijection on the set A and hence $|A| = |A|$. So $\sim$ is a reflexive relation. If $A \sim B$ there is a bijection f from A onto B, then there is a bijection (the inverse of f) from B to A. (We note that Theorem 2.2.2 says that if f is a bijection, then it has an inverse f^{-1}. It must also be shown that f^{-1} is a bijection. This is not hard and is left for the exercises.) So $B \sim A$. Therefore the relation $\sim$ is symmetric. Finally if $A \sim B$ and $B \sim C$ then from Lemma 2.2.1 it follows that $A \sim C$ (why?). Therefore the relation $\sim$ is transitive and hence is an equivalence relation. $\square$

From the above lemma and Theorem 2.3.1, we have that the collection of all sets is partitioned into equivalence classes by the relation of having the same size. Thus each equivalence class contains all sets of the same size. We can then call the **cardinal number** or **size** of a set the size of any set in its equivalence class.

Definition 2.4.3. *We say $|A| \le |B|$ meaning that the size of A is less than or equal to the size of B if there is a bijection from A onto a subset of B. We say $|A| < |B|$ meaning that the size of A is less than the size of B if there is a bijection from A onto a proper subset of B but there cannot be a bijection from A onto B, i.e., $|A| \ne |B|$.*

In order for our concept of size to be reasonable, the following lemma should be true.

Lemma 2.4.2. *If A and B are sets with $A \subset B$, then the size of A is less than or equal to the size of B, in symbols, $A \subset B \implies |A| \le |B|$.*

Proof. Since $A \subset B$, then $I_A : A \to A$ is a bijection from A onto a subset of B. Thus by Definition 2.4.3, $|A| \le |B|$. □

We said that a set was infinite in Definition 2.4.2 if it was not finite. But what does that mean? Guided by Definition 2.4.3, we say that a set A is infinite if there is a bijection from A onto a proper subset of itself. For example, $\mathbb{N}$ is infinite because the map $f : \mathbb{N} \longrightarrow \mathbb{N}$ given by $f(n) = 2n$ is 1-1 and from $\mathbb{N}$ onto $f(\mathbb{N}) = $ the set of even natural numbers. Since $f(\mathbb{N})$ is a proper subset of $\mathbb{N}$, $\mathbb{N}$ is infinite.

We saw in Section 2.2 that the set $A = \{1, 2, 3\}$ has precisely eight subsets, that is, $|\mathcal{P}(A)| = 8$. This is a special case of the following theorem relating the size of a set to the size of its power set.

Theorem 2.4.1. *If $|A| = n \in \mathbb{N}$ then $|\mathcal{P}(A)| = 2^n$. That is, if a finite set has n elements, it has 2^n subsets.*

Proof. The result is clear if $A = \emptyset$. If $A = \emptyset$ so that $|A| = 0$ then there is only one subset, namely $\emptyset$, and so $|\mathcal{P}(A)| = 1 = 2^0$. Now we suppose that $|A| = n > 0$. We now prove this by induction on the size. Suppose first that $A = \{a\}$ has exactly one element. Then it has two subsets, namely the whole set A and the empty set. Therefore if $|A| = 1$ we have $|\mathcal{P}(A)| = 2 = 2^1$ so the statement is true for $n = 1$. Suppose it is true for $n = k$, so that any set with k elements has 2^k subsets and let the set $A = \{a_1, ..., a_k, a_{k+1}\}$ have $k + 1$ elements. Let $A_1 = \{a_1, ..., a_k\}$. Each subset of A can either contain a_{k+1} or not contain a_{k+1}. It follows that each subset of A is either a subset $B \subset A_1$ or $B \cup \{a_{k+1}\}$ where $B \subset A_1$. Clearly there is a one-to-one correspondence between subsets of A_1 and subsets of A_1 union with the subset $\{a_{k+1}\}$. It follows that the number of subsets of A with a_{k+1} is exactly the same as the number of subsets without a_{k+1}. But as above a subset without a_{k+1} is just a subset of A_1 and by the inductive hypothesis there are 2^k of them. It follows that

The number of subsets of $A = 2^k + 2^k = 2(2^k) = 2^{k+1}$.

Therefore the theorem is true for $n = k + 1$ and hence true for all n by induction.

$\square$

At the first step we combine all nonfinite sets into the infinite category. However, surprisingly, there are many different levels of infinity.

Definition 2.4.4. *If $|A| = |\mathbb{N}|$ then we say that A is **countable** or **countably infinite**. If A is infinite but not countably infinite we say that A is **uncountably infinite**.*

In the last chapter we introduced the standard number systems $\mathbb{N}, \mathbb{Z}, \mathbb{Q}$ and $\mathbb{R}$. Since

$$\mathbb{N} \subset \mathbb{Z} \subset \mathbb{Q} \subset \mathbb{R}$$

all these number systems are infinite. However we will show that the first three are all countably infinite and hence of the same size but $\mathbb{R}$ is larger.

Lemma 2.4.3. *The sets $\mathbb{N}, \mathbb{Z}, \mathbb{Q}$ are all countably infinite.*

Proof. Clearly $\mathbb{N}$ is countably infinite. To show that the integers are countably infinite we must exhibit a bijection with the natural numbers. The table below exhibits this one-to-one correspondence

$$\begin{array}{cccccc} 1 & 2 & 3 & 4 & 5 & ... \\ 0 & 1 & -1 & 2 & -2 & ... \end{array}$$

showing that $|\mathbb{Z}| = |\mathbb{N}|$. So this table gives a 1-1, onto function from $\mathbb{N}$ to $\mathbb{Z}$. Notice that this is equivalent to giving a list of the elements in $\mathbb{Z}$, i.e., the first one is 0, the second one is 1, etc. Here the list would be

$$0, 1, -1, 2, -2, 3, -3, ...$$

To see that $\mathbb{Q}$ is countable we must exhibit a bijection with $\mathbb{N}$. As noted above, this is equivalent to presenting a list for $\mathbb{Q}$. We first do this for the positive rationals. Then we use an argument like the above for the integers to give a list for all the rationals. Actually instead of doing this for $\mathbb{Q}$, we will do this for a larger set than $\mathbb{Q}$. This will be sufficient as we will explain.

Consider the following infinite matrix where in the first row are all the fractions with denominator 1. In the second all the fractions with denominator 2 and so on. This set contains as a subset the positive rationals since two or more fractions may represent the same rational number. If we call this set $\mathbb{Q}_1^+$ and the set of positive rationals $\mathbb{Q}^+$, we have

$$\mathbb{N} \subset \mathbb{Q}^+ \subset \mathbb{Q}_1^+.$$

Thus by Lemma 2.4.2, $|\mathbb{N}| \leq |\mathbb{Q}^+| \leq |\mathbb{Q}_1^+|$. But if we can show that $\mathbb{Q}_1^+$ is countable, i.e., $|\mathbb{Q}_1^+| = |\mathbb{N}|$, then $|\mathbb{Q}^+| = |\mathbb{N}|$ and so $\mathbb{Q}^+$ is countable.

$$
\begin{array}{ccccc}
\frac{1}{1} & \frac{2}{1} & \frac{3}{1} & \frac{4}{1} & \cdots \\
\frac{1}{2} & \frac{2}{2} & \frac{3}{2} & \frac{4}{2} & \cdots \\
\vdots & & & & \\
\frac{1}{n} & \frac{2}{n} & \frac{3}{n} & \frac{4}{n} & \cdots \\
\vdots & & & &
\end{array}
$$

If we start to list across the first row we will never get to the second row so we use a diagonal procedure to list left to right and down at the same time. The first number on the list is the fraction $\frac{1}{1}$. Now we list to the right and down, what we call the second diagonal. So the second number on the list is $\frac{2}{1}$ followed by $\frac{1}{2}$. Now the third diagonal is $\frac{3}{1}, \frac{2}{2}, \frac{1}{3}$ and so on. Every element of $\mathbb{Q}_1^+$ will eventually be listed providing a list for the larger set $\mathbb{Q}_1^+$. This list being

$$\frac{1}{1}, \frac{2}{1}, \frac{1}{2}, \frac{3}{1}, \frac{2}{2}, \frac{1}{3}, \frac{4}{1}, \dots .$$

Then to list all of $\mathbb{Q}_1$, we proceed as follows — similar to what we did with the integers:

$$0, \frac{1}{1}, -\frac{1}{1}, \frac{2}{1}, -\frac{2}{1}, \frac{1}{2}, -\frac{1}{2}, \frac{3}{1}, \dots .$$

This gives us a list of $\mathbb{Q}_1$ and hence a bijection $f : \mathbb{N} \longrightarrow \mathbb{Q}_1$, so that $|\mathbb{Q}_1| = |\mathbb{N}|$. Since $\mathbb{N} \subset \mathbb{Q} \subset \mathbb{Q}_1$, this shows by Lemma 2.4.2 that $|\mathbb{Q}| = |\mathbb{N}|$ or that $\mathbb{Q}$ is countable. $\qquad\qquad\qquad\qquad\qquad\qquad\qquad\qquad\qquad\qquad\qquad\qquad\quad\square$

This result may seem surprising since between any two integers there are infinitely many rationals but in totality the integers are the same size as the rationals. This was originally proposed by G. Cantor and provided a fair amount of controversy until it was fully understood.

Now we come to an even more surprising result. The real numbers are actually larger than the rationals.

Lemma 2.4.4. *The real numbers* $\mathbb{R}$ *are uncountably infinite and therefore* $|\mathbb{Q}| < |\mathbb{R}|$

Proof. Since $\mathbb{Q} \subset \mathbb{R}$ we must have $|\mathbb{Q}| \leq |\mathbb{R}|$ by Lemma 2.4.2. Therefore if we can show that $|\mathbb{Q}| \neq |\mathbb{R}|$ we establish the lemma. We need to show that there cannot be a 1-1 and onto map from $\mathbb{Q}$ onto $\mathbb{R}$.

Let $I = [0,1] = \{x \in \mathbb{R} : 0 \leq x \leq 1\}$, the unit interval. This is a subset of $\mathbb{R}$ so if we can show that I is uncountably infinite, that is, there cannot be a bijection between I and $\mathbb{Q}$, then $\mathbb{R}$ is also uncountable again since $I \subset \mathbb{R}$ by Lemma 2.4.2, $|I| \leq |\mathbb{R}|$.

If I were countable then there would be a list of the elements in I, that is, a 1-1 correspondence between I and $\mathbb{N}$. We show that no such list can be complete or that no such map can be a bijection. Suppose there was such a list $r_1, r_2, \ldots, r_n, \ldots$. Each r_i has a decimal expansion

$$r_1 = .x_{11}x_{12} \ldots x_{1n} \ldots$$

$$r_2 = .x_{21}x_{22} \ldots x_{2n} \ldots$$

$$\ldots$$

$$r_n = .x_{n1}x_{n2} \ldots x_{nn} \ldots$$

$$\ldots$$

Here x_{ij} is the jth decimal digit in number r_i.

We will construct an element of $[0,1]$ that is, not on the list — a contradiction showing that the list cannot be complete. Look at x_{11} and choose a digit $x_1 \neq x_{11}$ or 0 or 9. We choose it not equal to 0 or 9 because the decimal ending in a string of 9's is the same as a decimal ending in a string of 0's. Let $r = .x_1 \ldots$ that is, the first decimal digit of r is x_1. Then $r \neq r_1$ since they differ in the first digit. Now choose x_2 not equal to $x_{22}, 0, 9$ and let $r = .x_1x_2 \ldots$. Then $r \neq r_2$ since they differ in the second digit. Continue in this manner. Then r cannot be on the list since it is not equal to any

r_j for all j because r and r_j differ in the jth digit. However r is a decimal expansion less than 1 so it is in $[0, 1]$. Therefore $\mathbb{R}$ is uncountably infinite.

$\square$

We say that a countable set has size $\aleph_0$, pronounced aleph-null, while the reals have size c, the size of the continuum. Therefore we have in terms of sizes

$$0 < 1 < 2 < \cdots < n < \cdots < \aleph_0 < c.$$

An obvious question is whether there is anything bigger than c. Surprisingly, the answer is yes, and in fact given any size set there is a strictly bigger set.

Theorem 2.4.2. *Given a set A then $|A| < |\mathcal{P}(A)|$. That is, the size of a set is always strictly smaller than the size of its power set.*

Proof. The result is clear if $A = \emptyset$. Now suppose that A is a nonempty set. The map $a \to \{a\}$, that is, mapping an element of A to the singleton subset with just that element, provides a bijection of A with a subset of $\mathcal{P}(A)$. Therefore $|A| \leq |\mathcal{P}(A)|$ by Lemma 2.4.2. We show that there cannot be a bijection from A onto $\mathcal{P}(A)$, completing the proof. Suppose then there is a bijection $\rho : A \longrightarrow \mathcal{P}(A)$. Hence $\rho(a)$ is a subset of A for each $a \in A$. Call an element $a \in A$ a *good element* if $a \notin \rho(a)$ and a *bad element* if $a \in \rho(a)$. Clearly each element is either good or bad. Let G be the set of all good elements of A, i.e.,

$$G = \{a \in A : a \notin \rho(a)\}.$$

Since ρ is a bijection and $G \subset A$, ρ must be onto so there is a $g \in A$ with $\rho(g) = G$. Suppose that g is a good element. Then by definition $g \notin \rho(g) = G$ and hence $g \in G$, i.e., g is a bad element giving a contradiction. If g were a bad element then $g \in \rho(g) = G$ and hence g is a good element again a contradiction. Therefore ρ cannot be a bijection and the power set must be larger than the set. $\square$

We saw that if $|A| = n$ then $|\mathcal{P}(A)| = 2^n$. For any size t we will denote the size of the power set by 2^t. We then have the following infinitely many infinite sizes:

$$c < 2^c < 2^{2^c} < 2^{2^{2^c}} < \cdots$$

2.5 Binary Operations

Fundamental to all algebraic structures are **binary operations** defined on a set. These are operations that take two elements of a set and produce a new element of the set. Formally we define them as functions from the cartesian product of the set with itself to the set.

Definition 2.5.1. *A **binary operation** on a nonempty set S is a function from $S \times S$ to S. That is, a binary operation is an operation on the set S which takes any two elements of S and assigns a third element of S. The word **binary** refers to the fact that these operations are done two at a time. For the time being we will denote the image of the ordered pair $(a, b) \in S \times S$ by either $a \cdot b$ or just by ab — juxtaposition. We also commit the abuse of notation of calling the binary operation just by this image, the binary operation is $ab \in S$. We also say that the binary operation is defined on the set S.*

In other words, a binary operation on a set is given when to every ordered pair (a, b) of elements of the set a unique element $c \in S$ is associated. The fact that $c \in S$ is sometimes expressed by saying a binary operation on a set S must be **closed**. For the remainder of the chapter we always assume that S is nonempty.

Binary operations may or may not satisfy certain important properties such as associativity. We now define the important properties. The first is associativity. This is important because a binary operation is done on two elements at a time. We could be interested in what would happen if we had three elements.

Definition 2.5.2. *A binary operation on S is **associative** if*

$$s_1(s_2 s_3) = (s_1 s_2)s_3 \text{ for any } s_1, s_2, s_3 \in S$$

For example, addition and multiplication of integers are associative, but subtraction is not. Both union and intersection of subsets are associative operations on the power set $\mathcal{P}(S)$ of a set S. The operation of composition of functions when defined (see Definition 2.2.2 and the exercises) is associative.

Associativity is important because in general $s_1 s_2 s_3$ can be computed in two different ways. First we could determine $s_1 s_2$ and then operate on the result with s_3 on the right. This is $(s_1 s_2)s_3$. Alternatively we could find

s_2s_3 and then operate on the left by s_1. This is $s_1(s_2s_3)$. If the operation is associative these two will be equal and we could write $s_1s_2s_3$ without ambiguity.

If a binary operation is associative then it is generalized associative, that is, in forming $s_1s_2...s_n$ it doesn't matter how you place the parentheses (as long as you don't change the order). We shall not prove this here, but just assume that it is clear.

The next important property is **commutativity**.

Definition 2.5.3. *A binary operation on S is* **commutative** *if*

$$s_1s_2 = s_2s_1 \; for \; any \; s_1, s_2 \in S$$

For example, addition and multiplication of integers are commutative but subtraction is not. Both union and intersection of subsets are commutative operations on the power set $\mathcal{P}(S)$ of a set S.

If a binary operation is associative and commutative then it is generalized commutative, that is, in forming $s_1s_2...s_n$ the order does not matter. Again, we will not prove this.

Next we come to the idea of an identity.

Definition 2.5.4. *An element s_0 of S is an* **identity** *for the binary operation $\cdot$ defined on S if*
$$s \cdot s_0 = s_0 \cdot s = s \; for \; all \; s \in S.$$

The element 0 is an additive identity in the integers while the element 1 is the multiplicative identity. The empty set $\emptyset$ is an identity for union on the power set $\mathcal{P}(S)$ of a set S while the whole set S is an identity for intersection (see the exercises).

In general an identity on a set, if it exists, must be unique.

Lemma 2.5.1. *An identity for a binary operation on a set S must be unique, that is, if e and e' are both identities for the binary operation on S then $e = e'$.*

Proof. Suppose that e and e' are both identities for a binary operation on a set S. Then $ee' = e$ since e' is an identity. On the other hand $ee' = e'$ since e is also an identity. Therefore since a binary operation is a function, $e = e'$.
$\square$

If a set S has an identity for a binary operation then we can define inverses.

Definition 2.5.5. *If s_0 is an identity for a binary operation $\cdot$ on S and $s \in S$ then an element $s_1 \in S$ is an **inverse** for s if*

$$s \cdot s_1 = s_1 \cdot s = s_0.$$

*If an element $s \in S$ has an inverse, then we say that s is **invertible**.*

Note that an element commutes with the identity and with its inverse — this doesn't say that in general the operation is commutative. Sometimes we can have left and right identities and or left and right inverses. That is, $se = s$ for all $s \in S$, but there is some $s_0 \in S$ such that $es_0 \neq s_0$. This would be a right identity. While $ss_1 = s_0$ but $s_1 s \neq s_0$ would be a right inverse. Analogously for left identities and inverses.

An identity which is only left or right but not two-sided may not be unique. For an associative binary operation inverses are unique (only full inverses not necessarily left or right inverses).

Lemma 2.5.2. *For an associative binary operation on a set inverses, when they exist, are unique.*

Proof. If x has an inverse there must by definition be an identity. Let e be the identity and x_1, x_2 both inverses for x. Consider $x_2 x x_1 = (x_2 x)x_1 = x_2(x x_1)$ by associativity. Now

$$(x_2 x)x_1 = ex_1 = x_1 \text{ since } x_2 \text{ is an inverse for } x.$$

On the other hand

$$x_2(x x_1) = x_2 e = x_2 \text{ since } x_1 \text{ is an inverse for } x.$$

It follows, since a binary operation is a function, that $x_1 = x_2$. $\qquad\square$

On the integers relative to addition every element has an inverse. However relative to multiplication only ± 1 have inverses.

On the power set $\mathcal{P}(S)$ of a set S there are no inverses in general for either union or intersection although both have identities.

If there are two operations on a set the following property relates them.

Definition 2.5.6. *Suppose that $\cdot$ and $*$ are two different binary operations on a set S. Then $\cdot$ is **left distributive** over $*$ if*

$$s_1 \cdot (s_2 * s_3) = (s_1 \cdot s_2) * (s_1 \cdot s_3) \text{ for any } s_1, s_2, s_3 \in S$$

*and $\cdot$ is **right distributive** over $*$ if*

$$(s_2 * s_3) \cdot s_1 = (s_2 \cdot s_1) * (s_3 \cdot s_1) \text{ for any } s_1, s_2, s_3 \in S$$

*If $\cdot$ is both left and right distributive over $*$ then it is **distributive** over $*$.*

In the basic number systems $\mathbb{Z}, \mathbb{Q}, \mathbb{R}$ multiplication is distributive over addition but not vice versa. In the power set, $\mathcal{P}(S)$, of a set S, union is distributive over intersection and intersection is distributive over union. that is,

$$x \cdot (y + z) = xy + xz \text{ for all } x, y, z \in \mathbb{Z}, \mathbb{Q}, \mathbb{R}$$

but

$$x + (yz) \neq (x + y)(x + z).$$

On the other hand for sets A, B, C

$$A \cup (B \cap C) = (A \cup B) \cap (A \cup C)$$

and

$$A \cap (B \cup C) = (A \cap B) \cup (A \cap C).$$

2.5.1   The Algebra of Sets

If S is a set then the set operations $\cap, \cup$ are binary operations on the power set $\mathcal{P}(S)$. These possess many nice properties that we will use as examples later in the book. We list these here.

Theorem 2.5.1. *Let S be a set and $\mathcal{P}(S)$ be its power set. Then*
   (1) Union is commutative; that is, $A \cup B = B \cup A$ for all $A, B \in \mathcal{P}(S)$.
   (2) Union is associative; that is, $(A \cup B) \cup C = A \cup (B \cup C)$ for all $A, B, C \in \mathcal{P}(S)$.
   (3) Intersection is commutative; that is, $A \cap B = B \cap A$ for all $A, B \in \mathcal{P}(S)$.
   (4) Intersection is associative; that is, $(A \cap B) \cap C = A \cap (B \cap C)$ for all $A, B, C \in \mathcal{P}(S)$.
   (5) Intersection is distributive over union; that is, $A \cap (B \cup C) = (A \cap B) \cup (A \cap C)$ for all $A, B, C \in \mathcal{P}(S)$.
   (6) Union is distributive over intersection; that is, $A \cup (B \cap C) = (A \cup B) \cap (A \cup C)$ for all $A, B, C \in \mathcal{P}(S)$.
   (7) The empty set is an identity for union; that is, $A \cup \emptyset = \emptyset \cup A = A$ for all $A \in \mathcal{P}(S)$.

*(8) The set S is an identity for intersection; that is, $A \cap S = S \cap A = A$
for all $A \in \mathcal{P}(S)$.*

Proof. There are several ways to prove results concerning set relations. We
will prove (2) and (5) to exhibit these techniques and then leave the remaining
proofs to the exercises.

We now consider (2) and want to show that $(A \cup B) \cup C = A \cup (B \cup C)$.
To show that two sets are equal we must show that the first is a subset of
the second and then the second is a subset of the first.

Let $x \in (A \cup B) \cup C$. Then by definition either $x \in A \cup B$ or $x \in C$.

If $x \in A \cup B$ then either $x \in A$ or $x \in B$. If $x \in A$ then $x \in A \cup (B \cup C)$.
On the other hand if $x \in B$ then $x \in B \cup C$ and hence $x \in A \cup (B \cup C)$.

If $x \in C$ then $x \in B \cup C$ and then $x \in A \cup (B \cup C)$. It follows then that if
$x \in (A \cup B) \cup C$ then $x \in A \cup (B \cup C)$. Therefore $(A \cup B) \cup C \subset A \cup (B \cup C)$.

Now suppose that $x \in A \cup (B \cup C)$. Then $x \in A$ or $x \in B \cup C$. If $x \in A$
then $x \in A \cup B$ and then $x \in (A \cup B) \cup C$. If $x \in B \cup C$ then $x \in B$ or
$x \in C$. If $x \in B$ then $x \in A \cup B$ and hence $x \in (A \cup B) \cup C$. If $x \in C$ then
$x \in (A \cup B) \cup C$ and hence in every case $x \in (A \cup B) \cup C$. It follows that
$A \cup (B \cup C) \subset (A \cup B) \cup C$.

Since $(A \cup B) \cup C \subset A \cup (B \cup C)$ and $A \cup (B \cup C) \subset (A \cup B) \cup C$ we
must have set equality and hence $(A \cup B) \cup C = A \cup (B \cup C)$ proving (2).

Now consider (5). We want to show that $A \cap (B \cup C) = (A \cap B) \cup (A \cap C)$.
We will do this by appealing to a Venn diagram. The relevant Venn diagram
is given in Figure 2.2.

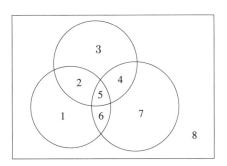

Figure 2.2. Venn diagram

Notice that there are eight regions. The set A comprises regions 1,2,5,6,
the set B comprises regions 2,3,4,5 and the set C comprises regions 4,5,6,7.
The union $B \cup C$ then is comprised of regions 2,3,4,5,6,7 and hence $A \cap (B \cup C)$

is regions 2,5,6. The intersection $A \cap B$ is regions 2,5 while $A \cap C$ is regions 5,6. Therefore $(A \cap B) \cup (A \cap C)$ is precisely regions 2,5,6 exactly the same as $A \cap (B \cup C)$. Hence these sets are equal proving (5).

Either of these two methods can then be employed to prove the other parts of Theorem 2.5.1.1. We note that doing this by Venn diagrams is not a rigorous proof because who is to say the sets must look like the ones used in the Venn diagrams. We will, however, allow such proofs.   □

2.6   Algebraic Structures and Isomorphisms

We have defined an algebraic structure as a set with certain binary operations defined on it and axioms governing the properties of these operations. In the next section we will introduce our first specific example of an algebraic structure called a **group**. Here we look at some very general properties and concepts concerning algebraic structures. By $(S, \cdot)$ we will denote any algebraic structure on a nonempty set S and with $\cdot$ an operation on S.

Lemma 2.6.1. *If $(S, \cdot)$ is an algebraic structure and e is an identity for $\cdot$ on S then e is unique, that is, e is the only identity for $\cdot$.*

Proof. This was proven in Lemma 2.5.1.   □

Notice that if something is true in the most general context then it is true in every specific example. Hence the identity in a group is unique and the additive identity in a ring is unique and there is no need to prove these for the particular structures. The same is true for the following lemma.

Lemma 2.6.2. *If x has an inverse in $(S, \cdot)$ and $\cdot$ is associative then its inverse is unique. That is, if x_1, x_2 are both inverses for x on $(S, \cdot)$ then $x_1 = x_2$.*

Proof. This was proven in Lemma 2.5.2.

□

Mappings between algebraic structures that preserve the operations are called **morphisms** or **homomorphisms**. Formally:

Definition 2.6.1. *Suppose $(S, \cdot)$ and $(T, *)$ are two algebraic structures and $f : S \to T$ is a mapping. Then f is a **homomorphism** if f **preserves the operations**; that is,*

$$f(s_1 \cdot s_2) = f(s_1) * f(s_2) \text{ for all } s_1, s_2 \in S.$$

For readers who have already seen vector spaces a vector space homomorphism is a **linear transformation**. Suppose that V and W are vector spaces over $\mathbb{R}$. Then a homomorphism is a mapping $T : V \to W$ such that

$$T(av_1 + bv_2) = aT(v_1) + bT(v_2)$$

for each $v_1, v_2 \in V$ and $a, b \in \mathbb{R}$.

Homomorphisms that are bijections are called **isomorphisms**.

Definition 2.6.2. *An **isomorphism** is a homomorphism which is also a bijection, that is, one-to-one and onto. We say that algebraic structures are **isomorphic** if there is an isomorphism between them.*

Isomorphic structures are *algebraically the same*. By this we mean that in terms of algebra their only difference is in the names of the elements.

Suppose that $(S, \cdot)$ is an algebraic structure. Then a subset T of S is a **substructure** if $(T, \cdot)$ forms exactly the same kind of algebraic structure as $(S, \cdot)$ where the binary operation in $(T, \cdot)$ is the binary operation of $(S, \cdot)$ restricted to T. We note that this means that if the algebraic structure $(S, \cdot)$ satisfies certain axioms relative to $\cdot$, then $(T, \cdot)$ must also satisfy these axioms.

Lemma 2.6.3. *Let $(S, \cdot)$ be an algebraic structure. If we assume nothing more than that $\cdot$ is a binary operation on S, then a subset $T \subset S$ is a substructure if and only if T is nonempty and closed under $\cdot$.*

2.7 Groups

We close this chapter by defining our first basic algebraic structure, a **group**. Groups turn out to be fundamental objects in many areas of mathematics, not just abstract algebra. Groups were originally defined as part of the solution by Galois of the unsolvability by radicals of polynomial equations of any degree five or greater. However once defined they found their way into geometry, analysis and topology. In this section we will give the definition, some examples and some very basic properties. In Chapters 6 and 7, we will look at group theory, the study of groups, in depth.

Definition 2.7.1. *A **group** G is a nonempty set with one binary operation (which we will denote by multiplication) such that*

(1) The operation is associative.

(2) There exists an identity for this operation.

(3) Each $g \in G$ has an inverse for this operation.

*If, in addition, the operation is commutative, the group G is called an **abelian group**. The **order** of G is the number of elements in G or the size of the set G in our previous terminology, denoted by $|G|$. If $|G| < \infty$, G is a **finite group**, otherwise G is an **infinite group**.*

Our number systems provide numerous examples of groups. The integers, the rationals and the reals all form groups under addition. Further these are all abelian groups. The elements of $\mathbb{Q}$ do not form a group under multiplication since 0 does not have a multiplicative inverse. However the set of nonzero elements of $\mathbb{Q}$ does form a group as does the set of nonzero real numbers under multiplication. These are again abelian groups. Further the set of positive rationals and the set of positive reals also form groups under multiplication (see the exercises).

Groups most often arise from invertible mappings of a set onto itself. Such mappings are called **permutations**. Recall from Theorem 2.2.1 that an invertible mapping must be a bijection so a permutation is a bijection from a set to itself and hence just a rearrangement of the elements of the set. We will prove the next theorem in Chapter 6 when we look carefully at group theory.

Theorem 2.7.1. *The set of all permutations on a set A with the operation of composition of functions forms a group called the symmetric group on A which we denote by S_A. If A has more than two elements then S_A is nonabelian.*

Groups also appear prominently in geometry. Let $\mathcal{E}$ denote the standard Euclidean plane. An **isometry** is a surjective mapping from $\mathcal{E}$ to $\mathcal{E}$ that preserves distance. that is, a map $T : \mathcal{E} \to \mathcal{E}$ such that $d(T(x), T(y)) = d(x, y)$ for any two points $x, y \in \mathcal{E}$. Let

$$G = \{T : T \text{ is an isometry on } \mathcal{E}\}.$$

In the exercises we will outline a proof of the following theorem;

Theorem 2.7.2. *G as defined above with the operation of composition of functions is a group.*

Modern geometry is heavily involved with studying the above isometry group. We will look at this more closely in Chapter 9.

The following lemma is a direct translation to groups of some of the general results from the last section. Since they were proved in general, and groups are associative, they hold for groups.

Lemma 2.7.1. *Let G be a group. Then the identity of G is unique. Further if $g \in G$ then the inverse of g is unique.*

Substructures for groups are called **subgroups**, that is, a subset H of a group G is a **subgroup** if $H \neq \emptyset$ and H forms a group under the same operation as G. We then have immediately from the last section.

Lemma 2.7.2. *Let G be a group and $H \subset G$. Then H is a subgroup of G if and only if $H \neq \emptyset$ and H is closed under the operation and inverses.*

The following is another useful criterion for being a subgroup

Lemma 2.7.3. *Let G be a group and $H \subset G$. Then H is a subgroup of G if and only if $H \neq \emptyset$ and $a, b \in H$ implies that $ab^{-1} \in H$.*

We leave the proof of this lemma to the exercises.

Mappings between groups that preserve the group operations are called **group homomorphisms**.

Definition 2.7.2. *Let G_1 and G_2 be groups. Then a mapping $f : G_1 \to G_2$ is a **(group) homomorphism** if*

$$f(g_1 g_2) = f(g_1)f(g_2) \text{ for any } g_1, g_2 \in G_1.$$

We further define:
*(1) f is an **epimorphism** if it is surjective.*
*(2) f is an **monomorphism** if it is injective.*
*(3) f is an **isomorphism** if it is bijective, that is, both surjective and injective. In this case G_1 and G_2 are said to be **isomorphic groups**, which we denote by $G_1 \cong G_2$.*

If two groups are isomorphic then they are really **algebraically the
same**. That is, one group is the same as the other but with the elements
renamed.

We mention the next results to provide a feel of how one proves results
about groups and subgroups. We will repeat them in Chapter 6.

Lemma 2.7.4. *(Cancellation Law for Groups) If G is a group and $g_1g = g_2g$
or $gg_1 = gg_2$ for elements $g_1, g_2, g \in G$ then $g_1 = g_2$*

Proof. Suppose that $g_1g = g_2g$ and e is the identity of G. Then $(g_1g)g^{-1} =
(g_2g)g^{-1}$. This implies that $g_1(gg^{-1}) = g_2(gg^{-1})$ by associativity. But then
$g_1e = g_2e$ and hence $g_1 = g_2$. An identical argument handles when $gg_1 = gg_2$.
$\square$

Lemma 2.7.5. *Let G_1 and G_2 be groups and $f : G_1 \to G_2$ a group homomorphism. Suppose that e is the identity in G_1 and e' is the identity in G_2 then
$f(e) = e'$. Hence a group homomorphism will always map the identity of the
first group to the identity of the second group. Also note $f(g^{-1}) = f(g)^{-1}$,
i.e., a group homomorphism will map the inverse of an element to the inverse
of the image of the element.*

Proof. Let $g \in G_1$. Then $f(g) = f(ge) = f(g)f(e)$ since f is a homomorphism. But $f(g) = f(g)e'$ since e' is the identity in G_2. Therefore
$f(g)f(e) = f(g)e'$ and hence $f(e) = e'$ by the cancellation law.

Now from $gg^{-1} = g^{-1}g = e$, we have from applying f and the fact that
$f(e) = e'$ that $f(g)f(g^{-1}) = f(g^{-1})f(g) = e'$. But this means that $f(g)^{-1} =
f(g^{-1})$.
$\square$

Definition 2.7.3. *Let G_1 and G_2 be groups and $f : G_1 \longrightarrow G_2$ a group
homomorphism. Let $e' \in G_2$ be the identity in G_2. Then the **kernel** of f is
the subset of G_1 consisting of all elements of G_1 mapping to the identity of
G_2. That is,*

$$ker(f) = \{g \in G_1 : f(g) = e'\}.$$

*The **image** of f is the range of f, i.e.,*

$$im(f) = f(G_1) = \{f(g) : g \in G_1\}.$$

Lemma 2.7.6. *Let G_1 and G_2 be groups and $f : G_1 \to G_2$ a group homomorphism. Then*
(1) $ker(f)$ is a subgroup of G_1.
(1) $im(f)$ is a subgroup of G_2.

Proof. (1) To show that $ker(f)$ is a subgroup we must show that $ker(f)$ is nonempty and by Lemma 2.7.5 that if $g_1, g_2 \in ker(f)$ then $g_1 g_2^{-1} \in ker(f)$.

From Lemma 2.7.5 $f(e) = e'$ so $e \in ker(f)$ and therefore $ker(f)$ is nonempty. Suppose that $g_1, g_2 \in ker(f)$. Then $f(g_1) = f(g_2) = e'$. Consider $f(g_1 g_2^{-1})$. Since f is a homomorphism we have using Lemma 2.7.5 again that

$$f(g_1 g_2^{-1}) = f(g_1) f(g_2^{-1}) = f(g_1)(f(g_2))^{-1}.$$

But the inverse of the identity in any group is the identity, thus

$$f(g_1)(f(g_2))^{-1} = e' e'^{-1} = e' e' = e'.$$

Putting these two equations together gives $f(g_1 g_2^{-1}) = e'$. Therefore $g_1 g_2^{-1} \in ker(f)$ and hence $ker(f)$ is a subgroup.

(2) Since $f(e) = e'$ it follows that $e' \in im(f)$ and hence $im(f)$ is nonempty. Now let $h_1, h_2 \in im(f)$. Then h_1, h_2 must have preimages g_1, g_2 respectively in G_1. Since f is a homomorphism, we have using again the fact from Lemma 2.7.5 that $f(g_2^{-1}) = f(g_2)^{-1}$,

$$f(g_1 g_2^{-1}) = f(g_1) f(g_2^{-1}) = f(g_1)(f(g_2))^{-1} = h_1 h_2^{-1} \in im(f)$$

Therefore $im(f)$ is a subgroup. □

2.8 Exercises

EXERCISES FOR SECTION 2.1

2.1.1. Prove: For any set A, $\emptyset \subset A$.

For each of exercises 2-13, let the universal set be $\mathcal{U} = \mathbb{Z}$, i.e., the integers, $A = \{x : x \geq 0\}$, $B = \{x : x \leq 10\}$, $C = \{x : x \leq -1\}$.

2.1.2. Find $A \cup B$.
2.1.3. Find $A \cup C$.
2.1.4. Find $B \cup C$.
2.1.5. Find $A \cap B$.
2.1.6. Find $B \cap C$.

2.1.7. Find $A \cup B \cup C$.

2.1.8. Find $A \cap B \cap C$.

2.1.9. Find A'.

2.1.10. Find $C \cap B'$ (sometimes written as $C - B$ or $C \setminus B$).

2.1.11. Do A, B, C partition the set of integers, $\mathbb{Z}$? Why or why not?

2.1.12. Are any of the sets A, B, or C pairwise disjoint? Which are and which are not?

2.1.13. Find $A \triangle B$.

2.1.14. Give an example of three (3) sets which partition $\mathbb{Z}$.

2.1.15. Let $\mathbb{R}$ be the set of all real numbers and I_n the closed interval $= [-\frac{1}{n}, \frac{1}{n}] = \{x \in \mathbb{R} : -\frac{1}{n} \leq x \leq \frac{1}{n}\}$ for each $n \in \mathbb{N}$.

(a) Find $\bigcup_{n=1}^{\infty} I_n$

(b) Show that $\bigcap_{n=1}^{\infty} I_n = \{0\}$ using a double inclusion.

(c) Suppose in the intersection in part (b) we change from closed intervals $[-\frac{1}{n}, \frac{1}{n}]$ to open intervals $(-\frac{1}{n}, \frac{1}{n})$, i.e., the endpoints are not included in each interval. Does this change the value of the intersection?

2.1.16. Using Venn diagrams verify that for sets A, B :   (a) $(A \cup B)' = A' \cap B'$ and (b) $(A \cap B)' = A' \cup B'$.

(These are called DeMorgan's laws and hold for any number of sets — not just two.)

2.1.17. Let $A = \{1, 2, 3, 4\}$ and $B = \{a, b\}$.

(a) Find $B \times A$.

(b) Find $\mathcal{P}(A)$, i.e., the power set of A.

EXERCISES FOR SECTION 2.2

For problems 1–3, suppose that $f : A \to B$.

2.2.1. Let $E \subset F \subset A$ and $G \subset H \subset B$. Then show that

(i) $f(E) \subset f(F)$

(ii) $f^{-1}(G) \subset f^{-1}(H)$

2.2.2. Let X and Y be any subsets of A. Using Problem 2.2.1 (i), prove

(a) $f(X \cup Y) = f(X) \cup f(Y)$

(b) $f(X \cap Y) \subset f(X) \cap f(Y)$

(NOTE: This not only works for two subsets of the domain, A, of f, but it also works for any number of subsets of A.)

2.2.3. Let W and Z be any subsets of B. Using Problem 2.2.1(ii), prove
(a) $f^{-1}(W \cup Z) = f^{-1}(W) \cup f^{-1}(Z)$
(b) $f^{-1}(W \cap Z) = f^{-1}(W) \cap f^{-1}(Z)$
(NOTE: This not only works for two subsets of the codomain, B, of f, but it also works for any number of subsets of B.)

2.2.4. Given a set $A \neq \emptyset$, prove that the identity map $I_A : A \to A$ is a bijection.

2.2.5. Construct examples of the following:
(a) A mapping which is not 1-1 and not onto.
(b) A mapping which is not 1-1 but is onto.
(c) A mapping which is 1-1 and not onto.
In each case, you must give the domain, codomain, and the rule of correspondence for your function.

2.2.6. Construct an example of a mapping $f : A \to B$ such that $f(E \cap F) \neq f(E) \cap f(F)$ where $E \subset A$ and $F \subset A$.

Here you must give the domain, codomain, and the rule of correspondence for your function together with the sets E and F. You must also show what $f(E \cap F)$ is and what $f(E) \cap f(F)$ is.

2.2.7. Using the notation of Problem 2.2.2, show that if f is 1-1, then

$$f(X \cap Y) = f(X) \cap f(Y).$$

(Again this works for as many subsets of the domain as you want.)

2.2.8. Suppose that $f : A \to B$. Let $E \subset A$ anf $F \subset B$. Prove that
(a) if f is 1-1, then $f^{-1}(f(E)) = E$;
(b) if f is onto, then $f(f^{-1}(F)) = F$.

2.2.9. Prove Lemma 2.2.1.

2.2.10. Let $\mathbb{R}$ be the set of real numbers and $f : \mathbb{R} \to \mathbb{R}$ be given by the rule $f(x) = x^2$. Let
$A = [1, 2] = \{x \in \mathbb{R} : 1 \leq x \leq 2\}$,
$B = (-1, 1) = \{x \in \mathbb{R} : -1 < x < 1\}$.
$C = (4, 9) = \{x \in \mathbb{R} : 4 < x < 9\}$,
$D = [0, 9] = \{x \in \mathbb{R} : 0 \leq x \leq 9\}$.
Find
(a) $f(A)$,

(b) $f(B)$,
(c) $f^{-1}(C)$,
(d) $f^{-1}(D)$ and
(e) A nonempty set $F \subset \mathbb{R}$ such that

$$f^{-1}(F) = \emptyset.$$

2.2.11. Show that if $f : A \to B$ is a bijection, then its inverse $f^{-1} : B \to A$ defined by $f^{-1}(b) = a$ if and only if $f(a) = b$ where b is any element of B and $a \in A$ is also a bijection. This can be done using Theorem 2.2.1 and also without Theorem 2.2.1. Here you must do it both ways.

EXERCISES FOR SECTION 2.3

2.3.1. Give examples of relations of a set, S, which satisfy all but one of the axioms for an equivalence relation on S. (This means you must give three examples.)

2.3.2. Consider the relation on the integers $\mathbb{Z}$ defined by for $a, b \in \mathbb{Z}$

$$a \equiv b \text{ means } (a - b) \text{ is divisible by } 3$$

(or $3 \mid (a - b)$). This is usually written $a \equiv b \bmod 3$.
(a) Show that this is an equivalence relation on $\mathbb{Z}$.
(b) Find all the equivalence classes for this relation and tell what the factor set of $\mathbb{Z}$ is modulo this relation.

2.3.3. Consider the set $A = \{1, 2, 3, 4, 5\}$.
(a) Determine which of the following are partitions of A :
$\{\{1, 2\}, \{3, 4\}, \{5\}\}$, $\{\{1, 2, 5\}, \{3\}\}$, $\{\{1, 2, 4, 5\}, \{3, 4\}\}$,
$\{\{1, 2, 5\}, \{3, 4\}\}$, $\{\{1, 2, 5\}, \{1, 2\} \cap \{3\}, \{3, 4\}\}$
(b) For those in part (a) which are partitions of A give the equivalence relation on A that Theorem 2.3.1 says exists.

2.3.4. Prove part (2) of Theorem 2.3.1.

2.3.5. Give an example of an equivalence relation $\sim$ on the set $\mathbb{R}$ of real numbers with the property that the closed interval $[0, 1]$ is one of the equivalence classes.

2.3.6. Suppose that the mapping $f : \mathbb{Z} \to \mathbb{Z}_3$ is defined by the following: For any $n \in \mathbb{Z}$ let $f(n) =$ the remainder upon division of n by 3. Thus $f(n) =$

$0, 1, 2$. Let $\mathbb{Z}_3 = \{0, 1, 2\}$. So f is a surjection. Show that the equivalence relation defined in Theorem 2.3.2 is the same as that of Problem 2.3.2 above.

2.3.7. Let $\mathbb{R}$ be the set of all real numbers. Consider the following relation defined on the usual xy-plane $= \mathbb{R} \times \mathbb{R} = \{(x, y) : x, y \in \mathbb{R}\} : (x_1, y_1) \sim (x_2, y_2)$ if and only if (x_1, y_1) and (x_2, y_2) both lie on the same graph of $y = x^2 + c$ for some real number c. Show that this is an equivalence relation on $\mathbb{R} \times \mathbb{R}$ and tell what the equivalence classes are. (This is an example of a **foliation.**)

EXERCISES FOR SECTION 2.4

2.4.1. Prove that any subset of a countable set is countable. (Here a countable set can be finite or countably infinite.)

2.4.2. In the text we have said (see Proof of Lemma 2.4.3) that we can show a set is countable (or is in 1-1 correspondence with $\mathbb{N}$) if we can list the elements of the set. So explain what is wrong with the following fallacious argument that the set of real numbers $\mathbb{R}$ is countable.

Take an arbitrary real number say $x \in \mathbb{R}$. Now start listing real numbers one by one, and after some number of steps, include x as the next element in the list. Since you wrote x within a finite number of steps, the set of real numbers is countable.

2.4.3. In the Proof of Lemma 2.4.3, it is shown that the set of all integers, $\mathbb{Z}$, is countable (in 1-1 correspondence with the set of natural numbers, $\mathbb{N}$) by means of a table which exhibits the 1-1 correspondence between $\mathbb{N}$ and $\mathbb{Z}$. Let's call this mapping $f : \mathbb{N} \to \mathbb{Z}$.

(a) Find a formula for f.

(HINT: Think in terms of even and odd natural numbers and use a piecewise defined function.)

(b) Use the formula which you found in (a) to show that f is 1-1 and onto.

2.4.4. The proof of Lemma 2.4.4 contains the following statement "the decimals ending in a string of 9's are the same as some decimals ending in a string of 0's." To show that you understand that statement prove that

$$1.0000... = .9999...$$

by letting $x = .9999...$ and then multiply x by 10, subtract and solve for x.

2.4.5. A given repeating decimal can always be converted into a fraction $(m/n$, i.e., a rational number). For example

$$\text{Let } x = .3333333333...,$$
$$10x = 3.3333...,$$
$$9x = 3$$
$$x = 1/3.$$

Also consider,

$$x = .8363636363636363...$$
$$100x = 83.636363...$$
$$99x = 82.8$$
$$x = 46/55.$$

Write the following repeating decimals as ratios of two integers:

    (a) 7.48181818...   (b) .001001001...

2.4.6. Repeating decimals can be expressed as infinite series, i.e., a repeating decimal can be shown to be a sum of a sequence of numbers. For example, show the decimal

$$0.1111... = 0.\bar{1} = \sum_{n=1}^{\infty} \frac{1}{10^n}.$$

Now sum the series on the right-hand side to express the repeating decimal as a fraction.

(HINT:This is a geometric series.)

2.4.7. To better understand the proof of Theorem 2.4.2, let us suppose the set A is countable. As a matter of fact let $A = \mathbb{N} = \{1, 2, 3...\}$. The map $n \to \{n\}$ provides a bijection between $\mathbb{N}$ and a subset of $\mathcal{P}(\mathbb{N})$ being $\{\{1\}, \{2\}, \{3\}...\} \subsetneq \mathcal{P}(\mathbb{N})$. Thus explain why Lemma 2.4.2 implies that $|\mathbb{N}| \leq |\mathcal{P}(\mathbb{N})|$. To show that $|\mathbb{N}| < |\mathcal{P}(\mathbb{N})|$ we use a proof by contradiction and assume there is a $\rho : \mathbb{N} \to \mathcal{P}(\mathbb{N})$ and ρ is a bijection. Thus for any $n \in \mathbb{N}$, $\rho(n)$ is a subset of $\mathbb{N}$. Call $n \in \mathbb{N}$ a good element if $n \notin \rho(n)$ and a bad element if $n \in \rho(n)$. Thus any $n \in \mathbb{N}$ must be either good or bad. But if we let $G = \{n \in \mathbb{N} : n \text{ is good}\}$, explain why there must be a $g \in \mathbb{N}$ such that $\rho(g) = G$. Thus g is either good or bad. But in either case, explain why we are led to a contradiction. Moreover, it is possible that $G = \emptyset$. Explain using the element g whose existence was already established why this also leads to a contradiction.

EXERCISES FOR SECTION 2.5

2.5.1. Given a set there can be more than one identity because an identity not only concerns the set but the binary operation on the set.
(a) Show that 0 is an identity on $\mathbb{Z}$ with respect to addition.
(b) Show that 1 is an identity on $\mathbb{Z}$ with respect to multiplication.
(c) Show that the empty set $\emptyset$ is an identity for union on $\mathcal{P}(\mathcal{S})$
(d) Show that the set S is an identity for intersection on $\mathcal{P}(\mathcal{S})$
(e) Do the examples (a) through (d) contradict Lemma 2.5.1?

2.5.2. Given $x, y \in \mathbb{Z}$
(a) Define $x \# y := |x - y|$. Find $6 \# 8$.
(b) Define $x \square y := 2x + y$. Find $(2 \square 3) \square 5$.

2.5.3. (a) Is division, $\div$, a binary operation on the set $\mathbb{Q}$ of rationals? Why or why not?
(b) Is subtraction, $-$, a binary operation on the set of integers, $\mathbb{Z}$? Why or why not?

2.5.4. Let S be a nonempty set. Define for $A, B \in \mathcal{P}(S)$:
$A + B := A \cap B$
$A \bullet B := A \triangle B$ (symmetric difference).
(a) Show that $+$ and $\bullet$ are commutative.
(b) What if any is the identity for $+$?
(c) Show that $+$ and $\bullet$ are associative.
(d) What if any is the identity for $\bullet$?

2.5.5. Let $A \xrightarrow{f} B \xrightarrow{g} C \xrightarrow{h} D$. Show that

$$(hg)f = h(gf).$$

This shows that composition of functions is associative — when defined. Note if, all functions under consideration have the same domain and codomain, then composition is always defined.

2.5.6. In Theorem 2.5.1, prove (1), (3), (4) and (6).

2.5.7. Consider the binary operation defined on the set $\{m, a, t, h\}$ by the following table

$\odot$	m	a	t	h
m	h	m	a	t
a	m	a	t	h
t	a	t	h	m
h	t	h	m	a

(a) Is this operation commutative?
(b) Name the identity element or explain why none exists.
(c) For each element having an inverse, name the element and its inverse.
(d) True or false:
$(m \odot a) \odot t = m \odot (a \odot t)$.

EXERCISES FOR SECTIONS 2.6-2.7

2.6.1. Prove Lemma 2.6.3.

2.6.2. Suppose that S is an algebraic structure with three (3) binary operations defined on it. These are

$$*, \odot, \square.$$

Tell what a homomorphism from S to S should be.

2.6.3. Show that $\mathbb{Q}^+$ and $\mathbb{R}^+$, i.e., the sets of positive rationals and positive reals respectively, are groups with respect to usual multiplication. Do either of these form a group with respect to addition?

2.6.4. Let $\mathcal{E}$ be the standard Euclidean plane, i.e., $\mathcal{E} = \mathbb{R} \times \mathbb{R}$ where $\mathbb{R}$ is the set of real numbers. If $\mathbf{x}_1 = (x_1, y_1)$ and $\mathbf{x}_2 = (x_2, y_2)$ where both $\mathbf{x}_1$ and $\mathbf{x}_2$ belong to $\mathcal{E}$, then the usual distance between $\mathbf{x}_1$ and $\mathbf{x}_2$ is given by

$$d(\mathbf{x}_1, \mathbf{x}_2) = \sqrt{(x_2 - x_1)^2 + (y_2 - y_1)^2}.$$

A mapping $T : \mathcal{E} \to \mathcal{E}$ is called an **isometry** if it preserves distance, i.e.,

$$d(T(\mathbf{x}_1), T(\mathbf{x}_2)) = d(\mathbf{x}_1, \mathbf{x}_2)$$

for all $\mathbf{x}_1, \mathbf{x}_2 \in \mathcal{E}$ and T is onto. Putting

$$G = \{T : T \text{ is an isometry on } \mathcal{E}\}$$

and taking the operation to be composition, $\circ$, of functions, prove that G is a group with respect to $\circ$.

(HINT: You must first show that composition, $\circ$, of functions is a valid operation on G, i.e., if T_1 and T_2 are isometries, then so is $T_1 \circ T_2$. Associativity follows from Problem 2.5.5 by taking $A = B = C = D = \mathcal{E}$. You must show there is an identity. Finally, you need to show that any isometry has an inverse and that the inverse is itself an isometry. From Theorem 2.2.1, to show that that such a map T has an inverse it is sufficient to show that it is a bijection. Since such a T is assumed to be onto, you need to show that T is injective. This 1-1 property can be shown using the fact that T preserves distance. Finally, you need to show that the inverse of an isometry is also an isometry. Examples of isometries on $\mathcal{E}$ are translation and rotation of the plane.)

2.6.5. Prove Lemma 2.7.2.

2.6.6. Prove Lemma 2.7.3.

2.6.7. In Lemma 2.7.4, show that if $gg_1 = gg_2$, then $g_1 = g_2$.

*2.6.8. Prove that G is a group if we have a binary operation on G (denoted by multiplication) such that:

  i. the operation is associative

 ii. there is a left identity, i.e., $\exists e \in G$ such that $ea = a$ for all $a \in G$

iii. every element of G has a left inverse, i.e., $\forall g \in G \; \exists g^{-1}$ such that $g^{-1}g = e$.

(Note: The above remains true if we replace all occurrences of the word *left* with the word *right*.)

2.6.9. Let $\mathcal{G}$ be the class of all groups. Define a relation on $\mathcal{G}$ as follows: for $G, H \in \mathcal{G}$, $G \cong H$ if and only if G and H are isomorphic.

Prove that $\cong$ is an equivalence relation on $\mathcal{G}$.

Chapter 3

Rings and the Integers

3.1 Rings and the Ring of Integers

At the end of Chapter 1 we briefly discussed the basic number systems: the integers $\mathbb{Z}$, the rational numbers $\mathbb{Q}$ and the real numbers $\mathbb{R}$. These serve as the primary motivating examples for algebraic structures. Each of these number systems has two basic operations, addition and multiplication, and form what is called a ring. We formally define this.

Definition 3.1.1. *A **ring** is a set R with two binary operations defined on it. These are usually called addition, denoted by $+$, and multiplication denoted by $\cdot$ or just by juxtaposition, satisfying the following six axioms:*

(1) Addition is commutative: $a + b = b + a$ for each pair a, b in R.

(2) Addition is associative: $a + (b + c) = (a + b) + c$ for $a, b, c \in R$.

(3) There exists an additive identity, denoted by 0 such that $a + 0 = a$ for each $a \in R$.

(4) For each $a \in R$ there exists an additive inverse denoted by $-a$ such that $a + (-a) = 0$.

(5) Multiplication is associative: $a(bc) = (ab)c$ for $a; b; c \in R$.

(6) Multiplication is left and right distributive over addition:

$$a(b + c) = ab + ac$$
$$(b + c)a = ba + ca$$

for $a, b, c \in R$.

If in addition:

*(7) Multiplication is commutative: $ab = ba$ for each pair a, b in R, then R is a **commutative ring**.*

Further, if:

*(8) There exists a multiplicative identity denoted by 1 such that $a \cdot 1 = a$ and $1 \cdot a = a$ for each a in R, then R is a **ring with identity** or a **ring with unity** if R satisfies (1) through (6) and (8).*

*(9) If R satisfies (1) through (8) then R is a **commutative ring with an identity** or a **unity**.*

We saw in Section 2.5 that a set G with one operation, $+$, defined on it and satisfying axioms (1) through (4) is called an **abelian group**. Therefore relative to the operation of addition, $+$, each ring is an abelian group.

We note at the beginning of this discussion of rings, that even though the operations are called "addition" and "multiplication," they may be very different from the usual addition and multiplication of numbers. We shall see this in our examples of Section 3.3. It is important to keep in mind that to be a ring all that is necessary is that these operations satisfy the axioms (1) - (6) for a ring.

A definition in abstract algebra is really only as good as its examples. We want our definitions to be quite general and cover many different types of examples. We will stress this fact throughout the book. Notice first, though, that each of the basic number systems, $\mathbb{Z}, \mathbb{Q}$, and $\mathbb{R}$, provides an example of a commutative ring with identity. In Section 3.3 we will provide many examples of rings, both commutative and noncommutative, and show how prevalent this structure is in mathematics.

A ring R with only one element is called **trivial**. A ring R is trivial if and only if $0 = 1$ in R. We will usually assume that our rings are nontrivial. To avoid trivialities, if we have a ring with identity or unity (i.e., a multiplicative identity or 1), we shall always assume that the ring is nontrivial, i.e., $1 \neq 0$.

A **finite ring** is a ring R with only finitely many elements in it. Otherwise R is an **infinite ring**. $\mathbb{Z}, \mathbb{Q}$, and $\mathbb{R}$ are all infinite rings. We will present examples of finite rings in Section 3.3.

Notice that in the definitions we do not assume that a ring R with a multiplicative identity necessarily has inverses. In the integers $\mathbb{Z}$ only the numbers 1 and -1 have multiplicative inverses while in $\mathbb{Q}$ and $\mathbb{R}$ every nonzero element has an inverse with respect to multiplication. An element $u \in R$, where R is a ring with identity, is called a **unit** if it has a multiplicative inverse in R, i.e., there exists an $s \in R$ such that $us = su = 1$.

A **field** F is a commutative ring with an identity in which each nonzero element has a multiplicative inverse or is invertible. In the terminology just introduced, a commutative ring with identity is a field if every nonzero element is a unit. The rationals $\mathbb{Q}$ and the reals $\mathbb{R}$ are fields. We will discuss fields in great depth later in the book. First we look at some general properties of rings and then at examples.

3.2 Some Basic Properties of Rings and Subrings

Since a ring is an algebraic structure and addition and multiplication are binary operations several properties become immediate. First we have that identities and inverses are unique.

Lemma 3.2.1. *Let R be a ring. Then the additive identity 0 is unique. If R has a multiplicative identity 1, then it too is unique. If $a \in R$ then its additive inverse $-a$ is unique.*

Next since R is associative relative to both addition and multiplication, it is generalized associative relative to these operations. It is further generalized commutative with respect to addition. From the distributive property, we obtain that multiplication by zero must be zero and obtain the well-known *laws of signs*.

Lemma 3.2.2. *Let R be a ring. Then if $a, b \in R$,*
    (1) $(a)(0) = (0)(a) = 0$;
    (2) $(-a)(b) = (a)(-b) = -(ab)$;
    (3) $(-a)(-b) = ab$.

Proof. We will prove (1) and (2) and leave the proof of (3) to the exercises. Suppose that $a \in R$. Notice first that if $a + x = a$ then $x = 0$ (why?). Now consider $b \in R$. We have $b + 0 = b$ so that $a(b + 0) = ab$. By the distributive property, then $ab + (a)(0) = ab$ and hence $(a)(0) = 0$. Using distributivity in the opposite direction we get that $(0)(a) = 0$ proving (1).

Now $a + (-a) = 0$ so $(a + (-a))b = 0 \cdot b = 0$ where $0 \cdot b = 0$ follows from part (1). By distributivity then

$$ab + (-a)b = 0$$

and hence $(-a)b$ is the additive inverse of ab. By the uniqueness of additive inverses then $(-a)(b) = -(ab)$. Similarly, $b + (-b) = 0$ so $0 = a(b + (-b)) = ab + a(-b)$. Thus $a(-b)$ is also the additive inverse of ab. So again by uniqueness of additive inverses, $a(-b) = -(ab)$. This completes (2).   □

If S is a subset of a ring R and S is also a ring with respect to the same operations as in R then S is a **subring** of R. For example the integers $\mathbb{Z}$ are a subring of both the rationals $\mathbb{Q}$ and the reals $\mathbb{R}$. Using our general result on algebraic structures we have the following result.

Lemma 3.2.3. *Suppose that R is a ring and $S \subset R$. Then S is a subring if and only if S is nonempty and closed under addition, additive inverses and multiplication. That is, if $a, b \in S$ then $a + b, -a$, and ab are also in S.*

Let n be a natural number and consider the subset $n\mathbb{Z}$ of the integers, $\mathbb{Z}$, consisting of all multiples of n. That is,

$$n\mathbb{Z} = \{nz : z \in \mathbb{Z}\}.$$

For example $3\mathbb{Z} = \{0, \pm 3, \pm 6, \dots\}$. We show that $n\mathbb{Z}$ for any natural number n is a subring of $\mathbb{Z}$.

Lemma 3.2.4. *For any natural number n the subset $n\mathbb{Z}$ is a subring of $\mathbb{Z}$.*

Proof. Let $S = n\mathbb{Z}$. Now $n = n \cdot 1 \in S$ so S is nonempty. Let $a, b \in S$. Then $a = nz_1$, $b = nz_2$ for some integers z_1, z_2. It follows that $a + b = nz_1 + nz_2 = n(z_1 + z_2) \in S$, $-a = -nz_1 = n(-z_1) \in S$ and $ab = (nz_1)(nz_2) = n(nz_1z_2) \in S$. Although the fact that $ab = n(nz_1z_2)$ and so belongs to $n\mathbb{Z}$ follows from associativity and commutativity of multiplication in $\mathbb{Z}$, that is, *not* what we have in mind here. We should think of nz_1 and nz_2 as sums of elements in $\mathbb{Z}$. Then to show $(nz_1)(nz_2) = n(nz_1z_2)$, we use induction on the natural number n together with associativity and distributivity in $\mathbb{Z}$. In particular we don't use commutativity of multiplication in $\mathbb{Z}$ (see the exercises). Therefore S is a subring by Lemma 3.2.3.   □

Notice that if $n \neq 1$ then the ring $n\mathbb{Z}$ does not have an identity but is commutative; so the set of rings $n\mathbb{Z}$ with $n \neq 1$ provide examples of commutative rings without identities. Later in the book we will show that every subring of $\mathbb{Z}$ is of the form $n\mathbb{Z}$ for some natural number n.

In any ring we can define multiplication by integers and exponents. We do this in the following way. Let R be a ring and r an element of R. We define $1 \cdot r = r$ and then $2r = r + r$. Inductively, having defined $n \cdot r$, then $(n+1)r = nr + r$. This defines multiplication of a ring element by any natural number. This is called a pseudoproduct since in general it does not define a product **within** R. However if R has an identity then $nr = (n \cdot 1)r$ is a product of elements of R since $n \cdot 1 \in R$. We then say that $(-n)r = -(nr) = n(-r)$ and $0r = 0$. We then have defined multiplication by any integer.

It must be understood that if $r \in R$ and $n \in \mathbb{Z}$, nr is just a shorthand notation for indicating a certain sum of elements in R. Furthermore, since the integer n is not necessarily an element of R it is *not* correct to think of nr as the product of two elements of R. This is why it is called a pseudoproduct.

In the lemma below we list the basic properties of multiplication by integers. Notice, for example, that (3) in the Lemma is not a consequence of the distributive law in R since this is not a product of elements in R.

Lemma 3.2.5. *Let R be a ring and let $r, s \in R$ and $m, n \in \mathbb{Z}$. Then:*
    (1) $mr + nr = (m + n)r$;
    (2) $m(nr) = (mn)r$;
    (3) $m(r + s) = mr + ms$;
    (4) $m(rs) = (mr)s = r(ms)$;
    (5) $(mr)(ns) = (mn)(rs)$.

Proof. We will prove (5) and leave all the others to the exercises. First let us assume that m, n are positive integers. Then we will show that (5) is true by induction on n. When $n = 1$ then $ns = 1 \cdot s = s$ so the left-hand side of (5) becomes

$$(mr)(1 \cdot s) = (mr)s.$$

But by definition $mr = (r + \cdots + r)$, That is, r added to itself m times. Thus

$$(mr)s = (r + \cdots + r)s = rs + \cdots + rs$$

by the distributive law in R. But rs added to itself m times is $m(rs)$ so we obtain

$$(mr)(1 \cdot s) = m(rs).$$

This proves (5) for any $m > 0$ and $n = 1$. Suppose now that $(mr)(ns) = (mn)(rs)$ for all $m \in \mathbb{N}$ and $n = k \in \mathbb{N}$. Consider $(m(k+1))(rs)$ and

$(mr)((k+1)s)$. By definition

$$mr((k+1)s) = mr((ks)+s).$$

By the distributive law in R this then gives us

$$(mr)(ks+s) = (mr)(ks) + (mr)s.$$

But the inductive hypothesis then implies that $(mr)(ks) = (mk)(rs)$ so that we have

$$mr((k+1)s) = (mk)(rs) + (mr)s.$$

Then just as in the case where $n = 1$ we get that $(mr)s = m(rs)$ so that we get

$$(mr)((k+1)s) = (mk)(rs) + m(rs) = (mk+m)rs = (m(k+1))(rs),$$

This establishes (5) when $m \in \mathbb{N}$ and $n \in \mathbb{N}$.

There are all together nine cases for m, n since m and n can either be positive, negative or zero. We note that if either one or both of m and n is 0, then we get zero on both sides of (5). This takes care of five cases. We still need to treat the cases where one of m or n is positive and the other is negative and the case where both are negative. We consider $m > 0$ and $n < 0$. The case $m < 0$ and $n > 0$ is treated in the same manner. We leave the case $m < 0, n < 0$ to the exercises.

Then $n = -k$ where $k > 0$ and so

$$(mr)(ns) = (mr)(-ks).$$

By definition $(-ks) = -(ks)$ and using Lemma 3.2.2 we get

$$(mr)(-(ks)) = -(mr)(ks).$$

Now m and k are both positive so that by the above inductive proof we have

$$(-(mr)(ks)) = -((mk)(rs)).$$

But by the definition of negative multiples

$$-((mk)(rs)) = (m(-k))(rs) = (mn)(rs).$$

This establishes (5) for this case.

$\square$

As in $\mathbb{Z}$ we then have the following result.

Lemma 3.2.6. *Let R be a ring with an identity and n a natural number. Define*

$$nR = \{nr : r \in R\}.$$

Then nR is a subring of R

The proof of this is identical to the proof that $n\mathbb{Z}$ is a subring of $\mathbb{Z}$ where we need to use the properties established in Lemma 3.2.5 (see the exercises). This can be extended in a very important way. Let R be a ring with an identity and r be a fixed element of R. We define a set which has in it all the multiples of r by elements of R.

$$rR = \{rr_1 : r_1 \in R\}.$$

In a similar manner as for $n\mathbb{Z}$ we can prove that rR is a subring of R (see the exercises). The subring rR is called the **subring generated** by r.

Lemma 3.2.7. *Let R be a ring with identity and $r \in R$. Then rR is a subring of R.*

We will return to this idea later on when we discuss ideals.

Finally, in much the same way that we defined multiplication by integers, we can define positive exponents. As before, let r be an element of a ring R. We let $r^1 = r$ and inductively having defined r^m, we define $r^{m+1} = r^m r$. This defines all nonnegative exponents. We don't define negative exponents since these would require multiplicative inverses. If R has an identity then we define $r^0 = 1$. Exponents in a ring satisfy the well-known laws of exponents with the restriction that there is no division.

Lemma 3.2.8. *Let R be a ring. Then*
    (1) $(r^m)(r^n) = r^{m+n}$ for $r \in R$ and $m, n \in \mathbb{N}$;
    (2) $(r^m)^n = r^{mn}$ for $r \in R$ and $m, n \in \mathbb{N}$.

The proofs are left for the exercises.

3.3   Examples of Rings

As we have stressed, a definition in abstract algebra is only good if there are many examples. In this section we provide a raft of examples, from many areas of mathematics, of rings, both commutative and noncommutative, finite and infinite.

We again mention that our basic number systems $\mathbb{Z}, \mathbb{Q}, \mathbb{R}$ are all infinite commutative rings with identities. The rationals $\mathbb{Q}$ and the reals $\mathbb{R}$ are also fields but the integers $\mathbb{Z}$ is not a field. The class of rings $n\mathbb{Z}$, with n a natural number and $n > 1$, introduced in the last section, provide examples of infinite commutative rings without identities.

To get examples in a different direction, consider from calculus the set R of continuous functions on the closed interval $[a, b]$ where $a, b \in \mathbb{R}$. That is,

$$R = \{f(x) : f : [a, b] \to \mathbb{R} \text{ and } f(x) \text{ is continuous on } [a, b]\}.$$

Since the sum, difference and product of two continuous functions are still continuous functions it follows that R becomes a ring with respect to the following operations. For $f, g \in R$, we define

$$
\begin{aligned}
(f + g)(x) &= f(x) + g(x); \\
(f \cdot g)(x) &= f(x)g(x).
\end{aligned}
$$

The zero is the zero function $f(x) = 0$ for all $x \in [a, b]$, while the multiplicative identity is the function that is identically 1. That is, $f(x) = 1$ for all $x \in [a, b]$. We call this ring $C^0[a, b]$. Let $C^1[a, b]$ be the set of continuously differentiable functions on $[a, b]$. That is, $C^1[a, b] = \{f(x) : f : [a, b] \to \mathbb{R}$ and $f'(x)$ is continuous on $[a, b]\}$. Then $C^1[a, b]$ is a subring of $C^0[a, b]$. We leave the proofs of all these facts to the exercises. That is we leave the proof of the following lemma to the exercises.

Lemma 3.3.1. *Let $a, b \in \mathbb{R}$, with $a \leq b$. Then $C^0[a, b]$ is an infinite commutative ring with unity and $C^1[a, b]$ is a subring of $C^0[a, b]$.*

The ring $C^0[a, b]$ has a very fundamental difference from the basic number systems. Recall that in $\mathbb{R}$, and hence in $\mathbb{Q}$ and $\mathbb{Z}$, if the product of two numbers is zero then at least one of the two numbers is zero. That is, if $a, b \in \mathbb{R}$ and $ab = 0$ then either $a = 0$ or $b = 0$. This is not true in $C^0[a, b]$.

Suppose that $[a, b] = [0, 2]$. Let

$$f(x) = \begin{cases} x - 1 & \text{if } 0 \leq x \leq 1 \\ 0 & \text{if } 1 \leq x \leq 2 \end{cases}$$

and

$$g(x) = \begin{cases} 0 & \text{if } 0 \leq x \leq 1 \\ 1 - x & \text{if } 1 \leq x \leq 2 \end{cases}.$$

These are both continuous functions and therefore in $C^0[0, 2]$. Neither is the zero function and hence neither is the zero of $C^0[0, 2]$. However, $f(x)g(x)$ is identically 0, that is, the zero function, on $[0, 2]$. Thus their product is zero, i.e., $f(x)g(x) = 0$. We will return to this example in Section 3.5.

To provide examples of finite rings we introduce a set of rings dependent on the integers called the modular rings.

3.3.1 The Modular Rings: The Integers Modulo n

For each natural number n we will construct a ring called the **ring of integers modulo n,** which we will denote by $\mathbb{Z}_n$, and which will be finite with n elements. Each of these rings will be a commutative ring with an identity.

In order to construct this ring we must introduce the relation of **congruence modulo n** on the integers $\mathbb{Z}$. The concept of congruence plays a major role in number theory and we will discuss it again in the next chapter.

Definition 3.3.1. *Suppose that n is a positive integer. If x, y are integers such that $x - y$ is a multiple of n or equivalently that n divides $x - y$, we say that x is* **congruent to** *y* **modulo n** *and denote this by $x \equiv y \mod n$. If n does not divide $x - y$ then x and y are* **incongruent modulo n.**

If $x \equiv y \pmod{n}$, then y is called a **residue** of x modulo n. Given $x \in \mathbb{Z}$ the set of integers

$$\{y \in Z : x \equiv y \pmod{n}\}$$

is called the **residue class** for x modulo n. We denote this by $[x]$. Notice that $x \equiv 0 \pmod{n}$ is equivalent to x being a multiple of n. We first show that the residue classes partition $\mathbb{Z}$, that is, that each integer falls in one and only one residue class.

Theorem 3.3.1. *Given $n > 0$, an integer, then congruence modulo n is an equivalence relation on the integers. Therefore the residue classes partition the integers.*

Proof. Recall that a relation $\sim$ on a set S is an **equivalence relation** if it is **reflexive**, that is, $s \sim s$ for all $s \in S$; **symmetric**, that is, if $s_1 \sim s_2$ then $s_2 \sim s_1$; and **transitive**, that is, if $s_1 \sim s_2$ and $s_2 \sim s_3$, then $s_1 \sim s_3$. If $\sim$ is an equivalence relation then the equivalence classes

$$[s] = \{s_1 \in S : s_1 \sim s\}$$

partition S.

We recall that the notation $m|n$ indicates that m divides n, so that $m = nk$ for some integer k. Consider $\equiv (\bmod\ n\ )$ on $\mathbb{Z}$. Given $x \in \mathbb{Z}$, $x - x = 0 = 0 \cdot n$ so $n|(x - x)$ and so $x \equiv x \pmod{n}$. Therefore $\equiv (\bmod\ n)$ is reflexive.

Suppose that $x \equiv y\ (\bmod\ n)$. Then $n|(x - y) \implies x - y = an$ for some $a \in \mathbb{Z}$. Then $y - x = -an$ so $n|(y - x)$ and so $y \equiv x \pmod{n}$. Therefore the relation $\equiv (\bmod\ n\ )$is symmetric.

Finally suppose $x \equiv y\ (\bmod\ n)$ and $y \equiv z\ (\bmod\ n)$. Then $x - y = a_1 n$ and $y - z = a_2 n$. But then $x - z = (x - y) + (y - z) = a_1 n + a_2 n = (a_1 + a_2)n$. Therefore $n|(x - z)$ and $x \equiv z \pmod{n}$. Therefore $\equiv (\bmod\ n)$ is transitive and the theorem is proved.   □

Hence, given $n > 0$, every integer falls into one and only one residue class. We now show that there are exactly n residue classes modulo n. In order to prove this, we need another result, called the division algorithm, which is another basic property of the integers. We will prove this in Chapter 4 (see Theorem 4.2.2). We state it here as follows:

Lemma 3.3.2. *(Division Algorithm) If $x, n \in \mathbb{Z}$ with $n > 0$, then there exist unique integers q and r such that $x = qn + r$ where $0 \le r < n$.*

Theorem 3.3.2. *Given an integer $n > 0$ there exist exactly n residue classes. In particular,$[0], [1], \ldots, [n-1]$ gives a complete set of residue classes.*

Proof. We show that given $x \in \mathbb{Z}$, x must be congruent modulo n to one of $0, 1, 2, \ldots, n - 1$. Further these are all incongruent modulo n. As a consequence

$$[0], [1], \ldots, [n-1]$$

give a complete set of residue classes modulo n and hence there are n of them.

To see these assertions suppose $x \in \mathbb{Z}$. By the division algorithm (see Lemma 3.3.2 above) we have

$$x = qn + r \quad \text{where} \ \ 0 \le r < n.$$

This implies that $x - r = qn$ or in terms of congruences that $x \equiv r \pmod{n}$. Therefore x is congruent to one element of the set $\{0, 1, 2, \ldots, n-1\}$.

Suppose $0 \le r_1 < r_2 < n$. Then $n \nmid (r_2 - r_1)$ so r_1 and r_2 are incongruent modulo n. Therefore every integer is congruent to one and only one of $0, 1, \ldots, n-1$, and hence $[0]$, $[1]$, $\ldots$, $[n-1]$ give a complete set of residue classes modulo n. $\square$

We will construct the ring $\mathbb{Z}_n$ on the set $\{[0], [1], \ldots, [n-1]\}$ of residue classes modulo n. We first need the following.

Lemma 3.3.3. *If $a \equiv b$ (mod n) and $c \equiv d$ (mod n), then*
(a) $a + c \equiv b + d$ (mod n);
(b) $ac \equiv bd$ (mod n).

Proof. Suppose $a \equiv b \pmod{n}$ and $c \equiv d \pmod{n}$, then $a - b = q_1 n$ and $c - d = q_2 n$ for some integers q_1, q_2. This implies that $(a + c) - (b + d) = (q_1 + q_2)n$ or that $n | [(a + c) - (b + d)]$. Therefore $a + c \equiv b + d \pmod{n}$.

We leave the proof of (b) to the exercises. $\square$

We now define operations on the set of residue classes.

Definition 3.3.2. *Consider the complete residue system $\{[0], \ldots, [n-1]\}$ modulo n. On this set of residue classes define*
(a) $[x_i] + [x_j] = [x_i + x_j]$;
(b) $[x_i][x_j] = [x_i x_j]$.

Theorem 3.3.3. *Given a positive integer $n > 0$, the set of residue classes,*

$$\{[0], \ldots, [n-1]\},$$

*forms a commutative ring with an identity under the operations defined above. This is called the **ring of integers modulo n** and is denoted by $\mathbb{Z}_n$. The zero element is $[0]$ and the identity element is $[1]$.*

Proof. Notice that from Lemma 3.3.3, it follows these operations are well-defined on the set of residue classes, that is, if we take two different representatives for a residue class, the operations defined by Definition 3.3.2 give the same answer. To be more precise, suppose that $a \equiv b \pmod{n}$ and $c \equiv d \pmod{n}$; then the definition of addition is that $[a] + [c] = [a + c]$. Now $[a] = [b]$ and $[c] = [d]$. So we could also write $[b] + [d] = [b + d]$. By Lemma 3.3.3 (a), $a + c \equiv b + d \pmod{n}$, so that $[a + c] = [b + d]$. This says that addition defined in this way is well defined. Similarly, we can use Lemma 3.3.3 (b) to show that multiplication of congruence classes as defined by Definition 3.3.2 (b) is also well defined. We leave the details of this to be verified in the exercises.

To show that $\mathbb{Z}_n$ is a commutative ring with an identity we must show that it satisfies, relative to the defined operations, all the ring properties. Basically, $\mathbb{Z}_n$ inherits these properties from $\mathbb{Z}$. We show commutativity of addition and leave the other properties to the exercises.

Suppose $[a], [b] \in \mathbb{Z}_n$. Then

$$[a] + [b] = [a + b] = [b + a] = [b] + [a]$$

where $[a + b] = [b + a]$ since addition is commutative in $\mathbb{Z}$.   □

This theorem is actually a special case of a general result in abstract algebra. In the ring of integers $\mathbb{Z}$ the set of multiples of an integer n forms what is called an ideal which is denoted $n\mathbb{Z}$. The ring $\mathbb{Z}_n$ is the **quotient ring** of $\mathbb{Z}$ modulo the ideal $n\mathbb{Z}$, that is, $\mathbb{Z}/n\mathbb{Z} \cong \mathbb{Z}_n$. We will examine these ideas more fully later in the book.

Hence for each n we have an example of a finite ring of size n. Each of these rings is commutative with identities.

We usually consider $\mathbb{Z}_n$ as consisting of $0, 1, \ldots, n - 1$ with addition and multiplication **modulo n**. When there is no confusion we will denote the element $[a]$ in $\mathbb{Z}_n$ as just a. We emphasize, however, that even though we are denoting the elements of $\mathbb{Z}_n$ by $0, 1, 2, \ldots, n - 1$ they are still residue classes or equivalence classes. This is just to simplify our notation. Below we give the addition and multiplication tables modulo 5, that is, in $\mathbb{Z}_5$. We note that, in line with the comment that these elements are different from usual integers consider $3 + 4$ and $3 \cdot 4$. They are both equal to 2 in $\mathbb{Z}_5$. This is certainly not what happens if we add 3 and 4 in $\mathbb{Z}$.

EXAMPLE 3.3.1. Addition and Multiplication Tables for $\mathbb{Z}_5$

+	0	1	2	3	4		.	0	1	2	3	4
0	0	1	1	3	4		0	0	0	0	0	0
1	1	2	3	4	0		1	0	1	2	3	4
2	2	3	4	0	1		2	0	2	4	1	3
3	3	4	0	1	2		3	0	3	1	4	2
4	4	0	1	2	3		4	0	4	3	2	1

Notice for example that modulo 5, $3 \cdot 4 = 12 \equiv 2 \bmod 5$ so that in $\mathbb{Z}_5$, $3 \cdot 4 = 2$. Similarly $4 + 2 = 6 \equiv 1 \bmod 5$ so in $\mathbb{Z}_5$, $4 + 2 = 1$.

Given the ring $\mathbb{Z}_n$ and an integer m with $1 < m < n - 1$ we can form the subring $m\mathbb{Z}_n$ in the manner described in the last section. However if m and n have no common divisors except ± 1 this subring will contain the identity and hence be the whole ring $\mathbb{Z}_n$. If m and n do have a common divisor $k > 1$ then $m\mathbb{Z}_n$ is a proper subring. We will prove this in the next chapter when we discuss number theory; however, we state it as a lemma now. If two integers a, b have only ± 1 as common divisors we say that a and b are **relatively prime**.

Lemma 3.3.4. *Let $n \in \mathbb{N}$ and $1 < m < n - 1$. Then if m and n are not relatively prime then $m\mathbb{Z}_n$ is a proper subring of $\mathbb{Z}_n$ without an identity. If m and n are relatively prime then $m\mathbb{Z}_n = \mathbb{Z}_n$.*

3.3.2 Noncommutative Rings

All of the examples that we have examined so far have been commutative. To provide examples of noncommutative rings we turn to matrices. For this section we assume that the reader has been introduced to matrices and matrix multiplication.

Let R be a ring. Then an $m \times n$ **matrix** over R consists of a rectangular array of elements of R with m row and n columns. If $m = n$ then it is a **square matrix**. For example

$$A = \begin{pmatrix} 2 & 4 & 3 \\ 1 & 0 & 1 \end{pmatrix}$$

is a 2×3 matrix over the integers $\mathbb{Z}$. If A is an $m \times n$ matrix we use $A_{i,j}$ to indicate the element in the ith row and jth column.

Recall that if A is an $m \times n$ matrix and B is an $n \times p$ matrix then the product AB can be formed and it has size $m \times p$ while in this case unless $m = p$ the product BA cannot be formed. However if A and B are both $n \times n$ then the two products AB and BA can be found; however, they are not necessarily equal.

For a ring R we let $M_n(R)$ denote the set of $n \times n$ matrices with entries from R. Hence $M_3(\mathbb{Z})$ would be all 3×3 integral matrices while $M_4(\mathbb{R})$ would be all 4×4 real matrices. If $A, B \in M_n(R)$ then we define addition componentwise, that is $(A + B)_{i,j} = A_{i,j} + B_{i,j}$. With this definition $M_n(R)$ becomes an abelian group. The zero element is the $n \times n$ matrix with all zeros.

On $M_n(R)$ we define multiplication by using matrix multiplication. Recall that the $n \times n$ **identity matrix** is the matrix with 1's down the main diagonal and zeros everywhere else assuming that R has an identity 1. Hence

$$I = \begin{pmatrix} 1 & 0 & 0 \\ 0 & 1 & 0 \\ 0 & 0 & 1 \end{pmatrix}$$

is the 3×3 identity matrix. If I is the $n \times n$ identity matrix and A is any $n \times n$ matrix then $AI = IA = A$ and therefore I is an identity for matrix multiplication. A square matrix A is **invertible** if there exists another square matrix B such that $AB = BA = I$. There are many criteria for invertibility but for now we only say that not all square matrices are invertible.

We then have the following result which provides us with noncommutative rings.

Theorem 3.3.4. *If R is a ring with an identity then $M_n(R)$ is a noncommutative ring with an identity. If R is infinite then $M_n(R)$ is infinite while if R is finite then $M_n(R)$ is finite.*

As specific examples then, $M_5(\mathbb{R})$ is an infinite noncommutative ring with an identity while $M_4(\mathbb{Z}_5)$ is a finite noncommutative ring with identity.

3.3.3   Rings Without Identities

We have seen that in the integers $\mathbb{Z}$, the subset $n\mathbb{Z} = \{nz : z \in \mathbb{Z}\}$, for a natural number n, forms a subring. If $n \neq 1$ then $n\mathbb{Z}$ does not have an identity (see the exercises). Hence the subrings $n\mathbb{Z}$ for natural numbers n provide examples of infinite commutative rings without identities.

To obtain examples of finite rings without identities we look at Lemma 3.3.4. In the modular ring $\mathbb{Z}_n$ if m and n do have at least one common positive factor other than 1 (recall that we then call them not relatively prime) then $m\mathbb{Z}_n$ is a proper subring of $\mathbb{Z}_n$ without an identity. These then are finite commutative rings without identities.

To get noncommutative rings without identities we again turn to matrix rings. Consider the rationals $\mathbb{Q}$ and the matrix ring $M_n(\mathbb{Q})$. Recall that this has an identity given by the identity matrix I. Let A be a noninvertible matrix in $M_n(\mathbb{Q})$. This means that there is no matrix B such that $AB = I$. Note that if $AB = I$ then it can be shown that also $BA = I$ so that $B = A^{-1}$. For those readers who have studied matrix theory this means that the rank of A is less than n or equivalently that $\det(A) = 0$. Now consider the subset of $M_n(\mathbb{Q})$ given by

$$AM_n(\mathbb{Q}) = \{AT : T \in M_n(\mathbb{Q})\}.$$

This forms a subring of $M_n(\mathbb{Q})$ and therefore provides an example of an infinite noncommutative ring without an identity.

In an identical manner if we start with $M_n(\mathbb{Z}_m)$ for some natural number m and let A be a noninvertible matrix in $M_n(\mathbb{Z}_m)$ then $AM_n(\mathbb{Z}_m)$ provides an example of a finite noncommutative ring without an identity.

3.3.4 Rings of Subsets: Boolean Rings

Let S be a nonempty set and let $\mathcal{P}(S)$ be its power set, that is the set of all subsets of S. We have seen that intersection and union provide binary operations on $\mathcal{P}(S)$ which satisfy a whole collection of nice properties. We now construct a ring structure on $\mathcal{P}(S)$.

First we define multiplication by intersection, that is, if $A, B \in \mathcal{P}(S)$ then $AB = A \cap B$. To define addition we use the operation of symmetric difference. Recall that this is the *exclusive or* so that

$$A \triangle B = (A \cup B) - (A \cap B) = (A \cap B') \cup (A' \cap B).$$

We now define on $\mathcal{P}(S)$ addition by

$$A + B = A \triangle B \text{ for } A, B \in \mathcal{P}(S)$$

With these operations we get the following result.

Theorem 3.3.5. *Let S be a set and $R = \mathcal{P}(S)$. Then R, with the definitions of addition and multiplication given as above, forms a commutative ring with an identity. The additive identity is $\emptyset$, the empty set, and the multiplicative identity is S the whole set. The additive inverse of any subset $A \in \mathcal{P}(S)$ is A itself. The ring R is finite if and only if the set S is finite.*

Proof. The proof that $\mathcal{P}(S)$ forms a ring under these definitions consists of verifying the ring axioms. This is very straightforward and we leave this to the exercises. The other assertions also follow directly from the set properties. If A is any subset of S then $A + \emptyset = (A \cup \emptyset) - (A \cap \phi) = A$ so that ϕ provides a zero element for this ring and $AS = A \cap S = A$ so that S provides a multiplicative identity. Recall that S is finite of size n then $\mathcal{P}(S)$ has size 2^n and so is still finite. If S is infinite then $\mathcal{P}(S)$ is bigger and hence infinite also. $\qquad\square$

Any subring of $\mathcal{P}(\mathcal{S})$ is called a **ring of sets**. Recall that a nonempty subset of a ring forms a subring if it is closed under addition, multiplication and additive inverse. For subsets of $\mathcal{P}(\mathcal{S})$ to be a subring it is sufficient to being closed under union, intersection and complement since symmetric difference is defined in terms of union and complement.

Lemma 3.3.5. *Let $R_1 \subset \mathcal{P}(S)$ be a nonempty collection of subsets of S. Then R_1 forms a subring of $\mathcal{P}(S)$ and hence a ring of sets if it is closed under union, intersection and complement.*

Let $A \in R = \mathcal{P}(\mathcal{S})$. Then $AA = A^2 = A \cap A = A$. Hence the square of any element in R is itself. Further A is its own additive inverse so $A + A = 2A = \emptyset = 0$ so that in R we have $2A = 0$ for any element A. These two properties would then hold also in any ring of sets.

Lemma 3.3.6. *Let R_1 be any ring of sets. Then $x^2 = x$ for any $x \in R_1$ and $2x = 0$ for any $x \in R_1$.*

We abstract this property to define a special kind of ring.

Definition 3.3.3. *A ring R is a **Boolean ring** if $x^2 = x$ for all $x \in R$.*

From Lemma 3.3.6 it follows directly that any ring of sets is a Boolean ring. We show next that the Boolean property $x^2 = x$ implies that the ring is commutative and that $2x = 0$ for any x in the ring.

Lemma 3.3.7. *Let R be a Boolean ring. Then R is commutative and $2x = 0$ for any $x \in R$.*

Proof. If R consists of just a zero element then the assertions are obvious so we assume that R has at least two elements. Let $x, y \in R$. Then $x^2 = x$ and $y^2 = y$. We first show that $2x = 0$ or equivalently that $x = -x$ for each $x \in R$. Consider

$$(x + x)^2 = x + x$$

which follows from the Boolean property. On the other hand multiplying out using the distributive law and using that $x^2 = x$ we have

$$(x + x)^2 = x^2 + x^2 + x^2 + x^2 = x + x + x + x.$$

Equating these we have

$$x + x = x + x + x + x \implies x + x = 0 \text{ or } 2x = 0.$$

Now we show that $xy = yx$. Consider $(x+y)^2$. From the Boolean property we have

$$(x + y)^2 = x + y.$$

However from the distributive property we have

$$(x + y)^2 = x^2 + xy + yx + y^2 = x + xy + yx + y.$$

Equating these we obtain

$$x + xy + yx + y = x + y \implies xy + yx = 0 \implies xy = -(yx) = yx$$

since $x = -x$, $\forall x \in R$ from above.   □

From above we have that any ring of sets is a Boolean ring. The following important result whose proof is beyond the scope of this book says that the converse is also true. That is, any Boolean ring is isomorphic (see Section 3.4) to a ring of sets. This result is known as **Stone's representation theorem**.

Theorem 3.3.6. *A ring R is a Boolean ring if and only if it is a ring of sets.*

3.3.5   Direct Sums of Rings

There is a very nice construction from which we can build new examples of rings from existing examples. This is known as the direct sum of rings. Recall that the cartesian product of two sets $A \times B$ consists of the collection of ordered pairs $\{(a, b) : a \in A, b \in B\}$.

If R and S are rings we can form a new ring out of the cartesian product in the following manner. Let $T = R \times S$ the cartesian product of R and S considered just as sets. On the set $R \times S$ define

$$(r_1, s_1) + (r_2, s_2) = (r_1 + r_2, s_1 + s_2)$$

and

$$(r_1, s_1).(r_2, s_2) = (r_1 r_2, s_1 s_2).$$

By directly verifying the ring properties we obtain the next theorem.

Theorem 3.3.7. *Let R and S be rings. Under the operations defined above $R \times S$ becomes a ring. The zero element is $(0, 0)$ where the first 0 is the zero element of R and the second 0 is the zero element of S. We call this ring the* **direct sum** *of the rings R and S denoted by $R + S$. $R + S$ is commutative if and only if both R and S are commutative.*

We note that this theorem can be extended to the direct sum of any finite number of rings. For if $R_1, \ldots, R_n$ are rings, then we can take the direct sum as having the set the cartesian product $R_1 \times \cdots \times R_n$ and define the operations componentwise as done for two rings above. The details will be left to the exercises. This concept of direct sum can even be extended to the case where there are infinitely many rings but that is beyond the scope of this book.

3.3.6   Summary of Examples

In this section we summarize the examples of rings we have discussed. This list is in no way complete, that is, there are many more examples of rings than these. However, the very fact that there are so many examples showcases the importance of the ring concept.

Infinite Commutative Rings with Identity (Unity)

(1) $\mathbb{Z}$ the integers.
(2) $\mathbb{Q}$ the rationals.
(3) $\mathbb{R}$ the reals.
(4) $C^0[a, b]$ the set of continuous functions on the interval $[a, b]$.
(5) $C'[a, b]$ the set of continuously differentiable functions on the interval $[a, b]$.
(6) $\mathcal{P}(S)$ where S is an infinite set.

Finite Commutative Rings with Identity (Unity)

(1) The modular rings $\mathbb{Z}_n$ where n is a natural number and $n > 1$.
(2) $\mathcal{P}(S)$ where S is a finite set.

Infinite Commutative Rings without Identity (Unity)

(1) $n\mathbb{Z}$ where n is a natural number and $n > 1$.
(2) rR where R is an infinite commutative ring with an identity and r is not invertible.

Finite Commutative Rings without Identity (Unity)

(1) $m\mathbb{Z}_n$ where n, m are not relatively prime natural numbers.

Noncommutative Rings

(1) $M_n(R)$ where R is a ring and $M_n(R)$ is the set of $n \times n$ matrices over R. This is infinite if R is infinite and finite if R is finite. It has an identity, the identity matrix, just as long as R has an identity.

(2) $TM_n(R)$ is a noncommutative ring without an identity if T is a non-inverible matrix in $M_n(R)$. This is finite if R is finite and infinite if R is infinite.

(3) $M_n(R)$ where R is a ring without an identity, as in (1). This would be a noncommutative ring without an identity.

3.4   Ring Homomorphisms and Isomorphisms

Recall that in Section 2.6 we defined isomorphic algebraic structures and noted that, from an algebraic point of view, these are essentially the same. In Section 2.7 we applied this to groups and defined group homomorphisms and group isomorphisms. Here we extend this to ring homomorphisms.

As with groups mappings between rings that preserve the ring operations are called **homomorphisms**

Definition 3.4.1. *Let R and S be rings. Then a mapping $f : R \to S$ is a (ring) homomorphism if*

$$f(r_1 + r_2) = f(r_1) + f(r_2) \text{ and } f(r_1 r_2) = f(r_1)f(r_2)$$

for any $r_1, r_2 \in R$

Again, exactly as in the group situation we add the further definitions:

*(1) f is an **epimorphism** if it is surjective;*

*(2) f is a **monomorphism** if it is injective;*

*(3) f is a **ring isomorphism** if it is bijective, that is, both surjective and injective. In this case R and S are said to be **isomorphic rings**, which we denote by $R \cong S$.*

If two rings are isomorphic then they are really **algebraically the same**. When we want to characterize a structure, we want to characterize it **up to isomorphism**. Hence in the previous section if we said that any finite Boolean ring is a ring of subsets then we mean that any finite Boolean ring is isomorphic to some ring of subsets. In Section 3.6, we will provide a characterization of the integers $\mathbb{Z}$. By this characterization we mean that any ring with these properties must be isomorphic to $\mathbb{Z}$.

The next results are analogous to group results.

Lemma 3.4.1. *Let R and S be rings and $f : R \to S$ be a ring homomorphism. Then $f(0) = 0$, that is, a ring homomorphism will always map the zero element of the first ring to the zero element of the second ring. Further $f(-r) = -f(r)$, that is, a ring homomorphism will map the additive inverse of an element to the additive inverse of the image of the element.*

Proof. Let $r \in R$. Then $f(r) = f(r + 0) = f(r) + f(0)$ since f is a homomorphism. But then $f(0) = 0$ the zero element in S.

Now from $r - r = 0$ we have from applying f and the fact that $f(0) = 0$ that

$$f(r + (-r)) = f(r) + f(-r) = f(0) = 0$$

and hence $f(-r) = -f(r)$. □

The next definition extends to rings the concept of the kernel and range of a homomorphism.

Definition 3.4.2. *Let R and S be rings and $f : R \to S$ a ring homomorphism. Then the **kernel** of f is the subset of R consisting of all elements of R mapping to the zero element of S. That is,*

$$ker(f) = \{r \in R : f(r) = 0\}.$$

The image of f is the range of F, i.e.,

$$im(f) = f(R) = \{f(r) : r \in R\}.$$

We next give a lemma whose proof we leave as an exercise.

Lemma 3.4.2. *Let R and S be rings and $f : R \to S$ an epimorphism, that is, an onto homomorphism. Then:*
    (1) If R is commutative then S is also commutative.
    (2) If R has an identity 1 then S also has an identity given by $f(1)$. If we denote the identity of S also by 1 this says that $f(1) = 1$.

Lemma 3.4.3. *Let R and S be rings and $f : R \to S$ be a ring homomorphism. Then*
    (1) $ker(f)$ is a subring of R.
    (2) $im(f)$ is a subring of S.

Proof. The proofs are identical to the proofs for groups.

(1) To show that $ker(f)$ is a subring we must show that $ker(f)$ is nonempty and closed under addition, multiplication and additive inverse. Now $f(0) = 0$ so $0 \in ker(f)$ and hence $ker(f)$ is nonempty. If $r_1, r_2 \in ker(f)$ then

$$f(r_1 + r_2) = f(r_1) + f(r_2) = 0 + 0 = 0$$

and hence $r_1 + r_2 \in ker(f)$ and so $ker(f)$ is closed under addition. Identically

$$f(r_1 r_2) = f(r_1)f(r_2) = 0 \cdot 0 = 0$$

and
$$f(-r_1) = -f(r_1) = -0 = 0$$
and therefore $ker(f)$ is closed under multiplication and additive inverse. Hence $ker(f)$ is a subring.

(2) The proof that $\text{im}(f)$ is a subring is straightforward and analogous to the proof for groups and so we leave it to the exercises. $\square$

We close this section with an example of a ring homomorphism and its kernel.

EXAMPLE 3.4.4. Let $R = \mathbb{Z}$ the integers and let $S = \mathbb{Z}_n$ where $n \neq 1$ is a natural number. For each $z \in \mathbb{Z}$ let

$$f(z) = [z]$$

where $[z]$ is the equivalence class of z under the equivalence relation congruence modulo n. From the definition of the operations in $\mathbb{Z}_n$ we have $[z_1 + z_2] = [z_1] + [z_2]$ and $[z_1 z_2] = [z_1][z_2]$ for any two integers z_1, z_2. Using the definition of f this says

$$f(z_1 + z_2) = f(z_1) + f(z_2) \text{ and } f(z_1 z_2) = f(z_1)f(z_2).$$

Therefore $f(z)$ defines a ring homomorphism from $\mathbb{Z}$ onto $\mathbb{Z}_n$ and hence an epimorphism.

What is the kernel of this map? The zero element of $\mathbb{Z}_n$ is

$$[0] = \{x : x \equiv 0 \bmod n\}$$

and therefore $[0]$ consists of all multiples of n. Since $f(z) = [z]$ it follows that the kernel of f is precisely those z with $[z] = [0]$ and hence those $z \in [0]$. These are precisely the multiples of n and so $n\mathbb{Z}$. Therefore $ker(f) = n\mathbb{Z}$.

3.5   Integral Domains and Ordering

In this section and the next we are going to characterize the integers $\mathbb{Z}$ up to isomorphism. To do this we must examine an important property of the integers. Notice first that in any ring, multiplication by zero yields zero. To see this notice that

$$r(s + 0) = rs + r0 = rs \text{ and hence } r0 = 0.$$

The same argument using distributivity on the right yields $0r = 0$. However in the integers $\mathbb{Z}$ more is true. If m, n are integers and $mn = 0$ then either m or n must be zero. This is not true in all rings. We have already seen an example in the ring $R = C^0[0, 2]$ of continuous functions on the interval $[0, 2]$ of two functions $f(x), g(x)$ neither of which is the zero function for which $f(x)g(x) \equiv 0$. Another example is provided by the modular ring $\mathbb{Z}_6$. The elements $2, 3$ are not zero but their product is 0 (in $\mathbb{Z}_6$). Elements like 2 and 3 in $\mathbb{Z}_6$ and the functions in $C^0[0, 2]$ are called **zero divisors** or **divisors of zero**.

Definition 3.5.1. *If R is a ring then a **zero divisor** or a **divisor of zero** in R is a nonzero element $r \in R$ such that there exists a nonzero element $s \in R$ with $rs = 0$ or there exists a nonzero element $t \in R$ with $tr = 0$. Note that the elements s and t may not be equal and only at least one must exist.*

In the integers and in the rationals and reals there are no zero divisors. We single out rings with this property.

Definition 3.5.2. *An **integral domain** D is a commutative ring with an identity that has **no zero divisors**.*

It is easy to show that this is equivalent to the following — so we leave it to the exercises:

Lemma 3.5.1. *A commutative ring D with an identity is an integral domain if and only if it satisfies the property that whenever $rs = 0$ for elements $r, s \in D$ then either $r = 0$ or $s = 0$.*

Since $\mathbb{Z}$, $\mathbb{Q}$, and $\mathbb{R}$ are all commutative rings with identities and have no divisors of zero, it follows that they are all integral domains. These are all infinite integral domains. To get finite integral domains, we proceed as follows. For each natural number n, the modular ring $\mathbb{Z}_n$ was a commutative ring with an identity. We need the following result called **Euclid's lemma**, which will be proved in the next chapter. We say that a positive integer $p > 1$ is a **prime** if its only divisors are ± 1 and $\pm p$.

Lemma 3.5.2. *(Euclid's Lemma) Let $a, b \in \mathbb{Z}$. Suppose that $p|ab$ where p is a prime. Then $p|a$ or $p|b$.*

The following lemma answers the question of when the modular rings are integral domains and so also gives us examples of finite integral domains.

Lemma 3.5.3. *The modular ring $\mathbb{Z}_n$ is an integral domain if and only if n is prime.*

Proof. Suppose first that n is not prime. Then $n = n_1 n_2$ with $1 < n_1 < n, 1 < n_2 < n$. Then in $\mathbb{Z}_n$ the corresponding elements n_1, n_2 are both nonzero since neither can be a multiple of n. However in $\mathbb{Z}_n$, $n_1 n_2 = n = 0$ and therefore both n_1 and n_2 are zero divisors. It follows that if n is not prime then $\mathbb{Z}_n$ cannot be an integral domain.

Now suppose that $n = p$, a prime. We show that $\mathbb{Z}_p$ is an integral domain. Now also suppose that $ab = 0$ in $\mathbb{Z}_p$. Considered as integers the product of a and b must then be a multiple of p and hence $p|ab$. From Euclid's lemma then either $p|a$ or $p|b$. If $p|a$ then a is a multiple of p and therefore $a = 0$ in $\mathbb{Z}_p$. Identically if $p|b$ then $b = 0$ in $\mathbb{Z}_p$. Hence if p is a prime then $\mathbb{Z}_p$ has no zero divisors by Lemma 3.5.1 and therefore is an integral domain. Note that we have proven an if and only statement. You should explain why this does it — see the exercises.   □

Recall that groups satisfy the cancellation law. That is, if $g_1 g_2 = g_1 g_3$ then $g_2 = g_3$. Since a ring is an abelian group with respect to addition, rings satisfy the cancellation law for addition. However if there are zero divisors they will not satisfy the cancellation law for multiplication. In $\mathbb{Z}_6$ for example $2 \cdot 3 = 0 = 2 \cdot 0$ but $3 \neq 0$. In integral domains, however, there is the cancellation law for multiplication.

Lemma 3.5.4. *An integral domain D satisfies the **cancellation law** for multiplication. That is,*

$$ab = ac \text{ with } a \neq 0 \implies b = c.$$

Proof. Suppose that $ab = ac$ with $a \neq 0$. Then $ab - ac = a(b - c) = 0$. Since $a \neq 0$ and D has no zero divisors it follows that $b - c = 0$ or $b = c$.

□

We will need the following fact in Chapter 5.

Lemma 3.5.5. *Let R and S be rings with identities. We will denote the identities of both of these rings by 1. Also suppose that S is an integral domain. Let $f : R \to S$ be a nontrivial homomorphism. Then $f(1) = 1$.*

Proof. Saying that f is a nontrivial homomorphism means that f does not take all elements of R into 0, the additive identity of S. Now consider $f(1) = 1f(1)$. But $f(1) = f(1 \cdot 1) = f(1)f(1)$ since f is a homomorphism. Thus we have that

$$1 \cdot f(1) = f(1)f(1).$$

This implies that

$$(1 - f(1))f(1) = 0 \text{ in } S.$$

Since S is an integral domain the last equality implies that either $f(1) = 0$ or $f(1) = 1$. However, if $f(1) = 0$ then if $r \in R$ we have

$$f(r) = f(r \cdot 1) = f(r)f(1) = 0$$

which contradicts the fact that f is nontrivial.

Hence $f(1) = 1$. □

An element of a ring may or may not have a multiplicative inverse. As we have already pointed out, those that do have multiplicative inverses are called **units**.

Definition 3.5.3. *In a ring R with an identity, an element which has a multiplicative inverse is called a **unit**. So an element $a \in R$ is a unit or is **invertible** if there exists a $b \in R$ such that $ab = ba = 1$.*

The only units in $\mathbb{Z}$ are ± 1 while in the rationals and reals every nonzero element is a unit. A unit cannot be a zero divisor.

Lemma 3.5.6. *A unit in a ring cannot be a zero divisor.*

Proof. Let R be a ring with an identity and $u \in R$ a unit. Then u cannot be a zero divisor. Suppose that $uv = 0$ for some $v \in R$. Since u is a unit there exists a u_1 with $u_1 u = 1$. Then $u_1(uv) = u_1(0) = 0$. But $u_1(uv) = (u_1 u)v = 1v = v$ and therefore equating the two expressions we get that $v = 0$. This argument can be repeated with $wu = 0$ to get in the same way that $w = 0$. We leave this for the exercises. □

The units in a ring R will form a group called the **unit group** or **group of units** of R denoted by $U(R)$.

Lemma 3.5.7. *Let R be a ring with an identity then the set of units in R forms a group under multiplication.*

Proof. Let $U(R)$ denote the set of units in R. First we show that $U(R)$ is nonempty and closed under multiplication. The multiplicative identity 1 satisfies $1 \cdot 1 = 1$ and hence is a unit and therefore $1 \in U(R)$. Then $U(R) \neq \phi$ and 1 will serve as a multiplicative identity under multiplication for $U(R)$. If $a, b \in U(R)$ then let a^{-1}, b^{-1} denote the multiplicative inverses of a, b which exist since a, b are units. Then $(ab)(b^{-1}a^{-1}) = a(bb^{-1})a^{-1} = aa^{-1} = 1$. It follows that $(ab)^{-1} = b^{-1}a^{-1}$ and therefore ab is also a unit. We note that $(b^{-1}a^{-1})(ab) = 1$ is shown in an identical fashion. $U(R)$ is thus closed under multiplication. Associativity follows from associativity of multiplication in the ring R while the existence of inverses follows from the definition of a unit. $\qquad \square$

A commutative ring with an identity (or unity) such that every nonzero element is a unit is called a **field**. Hence the rationals $\mathbb{Q}$ and the reals $\mathbb{R}$ are fields. We will do a formal introduction to fields in Chapter 5. Here we look at some very basic properties. If F is a field then every nonzero element is a unit. Hence from the lemma above there can be no zero divisors. Hence as a corollary to the lemma we obtain.

Corollary 3.5.1. *Any field must be an integral domain.*

The question immediately arises, "What about in the other direction?" That is, when is an integral domain a field? Clearly not always since the integers $\mathbb{Z}$ form an integral domain but not a field. However, the following beautiful result shows that in the case of finite rings, fields and integral domains coincide.

Theorem 3.5.1. *A finite integral domain must be a field.*

Proof. Let D be a finite integral domain and denote the elements of D by $\{d_1, \ldots, d_n\}$. To show that D is a field we must show that every nonzero element of D is a unit, that is, has a multiplicative inverse. Suppose that $d \neq 0$ and consider the set $\{dd_1, \ldots, dd_n\}$ consisting of all products of the elements of D by the particular element d. By the cancellation law for multiplication in D we have that $dd_i = dd_j$ only if $i = j$ and hence the n products dd_i are all distinct and therefore the set $\{dd_1, \ldots dd_n\}$ coincides with the set $\{d_1, \ldots, d_n\}$. The identity 1 is included among the $d_1, \ldots, d_n$ and so is included among the $dd_1, \ldots, dd_n$. It follows that there exists a d_j with $dd_j = 1$ and hence d has a multiplicative inverse and therefore D is a field. Note that since D is commutative that $dd_j = d_j d$. $\qquad \square$

The modular rings, $\mathbb{Z}_n$, are all finite rings hence they will be fields whenever they are integral domains. We saw in Lemma 3.5.3 that this was precisely when n is a prime.

Corollary 3.5.2. *The modular ring $\mathbb{Z}_n$ is a field if and only if n is prime.*

This implies that if $m \in \mathbb{Z}_n$, n is a prime, and $m \neq 0$ then m has a multiplicative inverse. In the next chapter on number theory we will show how to find inverses and do arithmetic in the modular rings.

In our quest to completely characterize $\mathbb{Z}$, notice next that the integers can be ordered. By this we mean that given any two integers x, y we can say that either $x = y$, or $x > y$, or $x < y$. We formalize this in an algebraic setting.

Definition 3.5.4. *A ring R is an **ordered ring** if there exists a distinguished set R^+, $R^+ \subset R$, called the **set of positive elements**, with the properties that:*

(1) The set R^+ is closed under addition and multiplication.

(2) If $x \in R$ then exactly one of the following is true: (trichotomy law)

(a) $x = 0$,
(b) $x \in R^+$,
(c) $-x \in R^+$.

*If further R is an integral domain we call R an **ordered integral domain**.*

In any ordered integral domain D, there is a standard way of introducing an order.

Definition 3.5.5. *If D is an ordered integral domain then for $x, y \in D$, we define $x < y$ or $y > x$ to mean that $y - x = y + (-x) \in D^+$.*

We note that with this ordering D^+ can be identified with those $x \in D$ such that $x > 0$ because $x > 0$ if and only if $x = x - 0 \in D^+$.

We then get in a direct manner.

Lemma 3.5.8. *If D is an ordered integral domain then*

(a) $x < y$ and $y < z$ implies $x < z$.

(b) If $x, y \in D$ then exactly one of the following holds:

$$x = y \ or \ x < y \ or \ y < x.$$

(c) $x^2 > 0$ for any $x \in D, x \neq 0$. In particular the multiplicative identity 1 must be positive.

Proof. (a) Suppose that $x < y$ and $y < z$. Then $y - x \in D^+$ and $z - y \in D^+$. Since D^+ is closed under addition we have $(y - x) + (z - y) = z - x \in D^+$ and hence $x < z$.

(b) If $x, y \in D$ then $y - x \in D$ and therefore exactly one of $y - x \in D^+$, $y - x = 0$ or $-(y - x) \in D^+$ is true. If $y - x \in D^+$ then $x < y$. If $y - x = 0$ then $x = y$. Finally if $-(y - x) \in D^+$ then $x - y \in D^+$ and therefore $y < x$.

(c) Suppose that $x \in D$ with $x \neq 0$. Then either x or $-x$ is in D^+. If $x \in D^+$ then since D^+ is closed under multiplication $x^2 \in D^+$ and hence $x^2 > 0$. If $-x \in D^+$ then $(-x)(-x) \in D^+$. But $(-x)(-x) = x^2$ so $x^2 \in D^+$ and as before $x^2 > 0$.

$\square$

It is clear that our standard number systems $\mathbb{Z}, \mathbb{Q}$ and $\mathbb{R}$ are all ordered integral domains. Again the question arises as to the modular rings.

Lemma 3.5.9. *If R is an ordered ring and $a \in R$ is a positive element, then the set $\{na; n \in \mathbb{N}\} \subset R^+$.*

The proof is by induction on n (see the exercises).

Theorem 3.5.2. *An ordered ring must be infinite.*

Proof. Let R be an ordered ring and $a \in R^+$. By Lemma 3.5.8 above the set $S = \{na : n \in \mathbb{N}\}$ is a subset of R^+ which is contained in R. Suppose that any two distinct elements of this set are equal, i.e., $n_1 a = n_2 a$ for two natural numbers n_1, n_2 with $n_1 < n_2$. Since $n_1 < n_2$ the difference $n_2 - n_1$ is a natural number and the equality $n_1 a = n_2 a$ gives $(n_2 - n_1)a = 0$. This is impossible because a is a positive element and $n_2 - n_1$ is a natural number, so that $(n_2 - n_1)a > 0$. Thus no two elements of the set S are equal. This implies that all the elements of S are distinct and hence S forms an infinite subset of R and so R itself must be infinite. $\square$

Since for each natural number n the ring $\mathbb{Z}_n$ is finite we have.

Corollary 3.5.3. *The modular ring $\mathbb{Z}_n$ cannot be an ordered ring.*

3.6 Mathematical Induction and the Uniqueness of $\mathbb{Z}$

We are now at the point where we can characterize the integers. In Chapter 1 we mentioned the following two properties of the integers $\mathbb{Z}$.

(1) **Principle of Mathematical Induction** (PMI) Let S be a subset of the natural numbers $\mathbb{N}$. Suppose $1 \in S$ and S has the property that if $n \in S$ then $n + 1 \in S$. Then $S = \mathbb{N}$.

(2) **The Least Well-Ordering Property** (LWO) Let S be a nonempty subset of the natural numbers $\mathbb{N}$. Then S has a least element.

One of these must be taken as an axiom for the integers. However, once this is done the other property must hold. This was the content of the following theorem proved in Section 1.5.

Theorem 3.6.1. *The inductive property is equivalent to the well-ordering property. That is, PMI is equivalent to LWO. Hence if PMI is adopted as an axiom then LWO is a theorem. But if LWO is adopted as an axiom then PMI is a theorem.*

Notice that the inductive property can be defined in any ordered integral domain D if we replace the natural numbers by the set of positive elements in D. Hence it makes sense to say that an ordered integral domain D satisfies the PMI and hence the LWO. We show that the inductive property is precisely what is needed to completely characterize $\mathbb{Z}$ among rings.

We introduce the following, extending the PMI to general ordered integral domains.

Definition 3.6.1. *Let D be an ordered integral domain. Let us denote the multiplicative identity of D by $1 \in D$. Then D satisfies the* **principle of mathematical induction** *or PMI provided the following is true. Let S be a subset of D^+. Suppose that $1 \in S$ and whenever $d \in S$ then $d + 1 \in S$. Then $S = D^+$.*

We need the following lemma.

Lemma 3.6.1. *Let D be an ordered integral domain that satisfies the PMI. Then the only subring of D that contains the identity is all of D.*

Proof. Suppose that R is a subring of D and $1 \in R$. We show that $R = D$. Consider the set $N = \{n \cdot 1 : n \in \mathbb{N}\}$. We note that $N \subset D^+$ since $1 \in D^+$ by Lemma 3.5.7 (c) and then $n \cdot 1 \in D^+$ by Lemma 3.5.9 for all natural numbers n. Since $1 \in R$ we have that $N \subset R$ since R is a subring and hence is closed under addition. Now $1 \in N$ and whenever $r \in N$ it follows that $r + 1 \in N$ because $r = n_1 \cdot 1$ but then $r + 1 = n_1 \cdot 1 + 1 = (n_1 + 1) \cdot 1$ where $(n_1 + 1) \in \mathbb{N}$. So by the PMI in D (see Definition 3.6.1) it follows that N is the set of all the positive elements of D, that is, $N = D^+$. Hence R contains all the positive elements of D. Since R is a subring it must then contain 0 and all the negative elements of D. However, we claim these are all the elements of D and therefore $R = D$. Since D is ordered every element of D is either positive, negative or zero and we have just shown all of these belong to R. Therefore the claim is true. $\square$

Theorem 3.6.2. *Let D be an ordered integral domain satisfying the PMI. Then $D \cong \mathbb{Z}$, that is, D is isomorphic to $\mathbb{Z}$.*

Proof. Let $f : \mathbb{Z} \to D$ by $f(z) = z \cdot 1$. It is straightforward from Lemma 3.2.5 that this is a homomorphism and so we leave its verification to the exercises. We show that the map is injective. Suppose that $f(z_1) = f(z_2)$ with $z_1, z_2 \in \mathbb{Z}$. Suppose that $z_1 \neq z_2$, Then without loss of generality we may assume that $z_1 > z_2$ so that $z_1 - z_2 \in \mathbb{N}$. But $f(z_1) = f(z_2)$ which implies that $f(z_1 - z_2) = 0$ since f is a ring homomorphism. Now $f(z_1 - z_2) = (z_1 - z_2) \cdot 1 = 0$ where $1 \in D$. Since D is an ordered integral domain we have $1 > 0$ by Lemma 3.5.7 and hence Lemma 3.5.8 implies that $(z_1 - z_2) \cdot 1 > 0$ which contradicts $(z_1 - z_2) \cdot 1 = 0$. Therefore $z_1 = z_2$ and the map f is one-to-one. Then $\text{im}(f)$ is a subring of D and since $f(1) = 1$ it follows that $1 \in \text{im}(f)$. Then from Lemma 3.6.1 above $\text{im}(f)$ is a subring of D containing the identity so $\text{im}(f)$ is all of D and hence f is onto and therefore an isomorphism. $\square$

The above result completely characterizes $\mathbb{Z}$ up to isomorphism.

Corollary 3.6.1. *Up to isomorphism the integers are the unique ordered integral domain satisfying the principle of mathematical induction.*

We close this section and this chapter by giving an alternative characterization of $\mathbb{Z}$.

Definition 3.6.2. *The **characteristic** of a ring R denoted char(R) is the least positive integer n such that $nr = 0$ for all $r \in R$. If no such n exists then R has **characteristic zero**.*

Lemma 3.6.2. *If R is a ring with an identity then char(R) is the least positive integer n such that $n \cdot 1 = 0$. In particular this is true for integral domains.*

Proof. Suppose that $n \cdot 1 = 0$ with n the minimal such natural number. Then if $r \in R$ we have $n \cdot r = (n \cdot 1)r = 0$. Thus since char$(R)$ is the least such positive integer, char$(R) \leq n$. But since $1 \in R$, we must have $(\text{char}(R)) \cdot 1 = 0$, but since n hence n is the least such positive integer, $n \leq \text{char}(R)$. Hence n is the characteristic, i.e., $n = \text{char}(R)$ □

Lemma 3.6.3. *The characteristic of an integral domain must be zero or a prime.*

Proof. Suppose that D is an integral domain and char$(D) = n$ so that n is the minimal natural number such that $n \cdot 1 = 0$. Suppose that n is not prime so that $n = n_1 n_2$ with $1 < n_1 < n, 1 < n_2 < n$. Then $n \cdot 1 = (n_1 \cdot 1)(n_2 \cdot 1) = 0$. Since D has no zero divisors it follows that either $n_1 \cdot 1 = 0$ or $n_2 \cdot = 0$ contradicting the minimality of n. Therefore n must be a prime.   □

The standard number systems $\mathbb{Z}$, $\mathbb{Q}$, $\mathbb{R}$ all have characteristic zero. For each natural number $n > 1$ the modular ring $\mathbb{Z}_n$ has characteristic n.

Definition 3.6.3. *A ring R with identity is a **prime ring** if the only subring of R which contains 1 is R.*

From Lemma 3.6.1, it follows that $\mathbb{Z}$ is a prime ring. In an analogous manner it follows that for each natural number n the ring $\mathbb{Z}_n$ is also a prime ring. The proof of this last fact is left to the exercises.

Theorem 3.6.3. *If R is a prime ring of characteristic zero then R is isomorphic to $\mathbb{Z}$. If R is a prime ring of characteristic n then R is isomorphic to $\mathbb{Z}_n$.*

Proof. The proof is similar to the proof of Theorem 3.6.2. Let R be a prime ring of characteristic 0. Now define $f : \mathbb{Z} \to R$ by $f(z) = z \cdot 1$. Since R is a prime ring this must be an isomorphism. We leave it to the exercises to fill in the rest of the details here.   □

It is interesting to note that R was not assumed to be a commutative ring in the statement of the above theorem. However, the theorem implies that it is.

3.7   Exercises

EXERCISES FOR SECTIONS 3.1 AND 3.2

3.1.1.   If R is a ring and $a, b \in R$, show that $(-a)(-b) = ab$. This completes the proof of Lemma 3.2.2.

(HINT: Consider the equations $(-a) + a = 0$ and $(-b) + b = 0$. Use distributivity to multiply these two equations together, then use (2) of this lemma to cancel appropriate terms. By the way, this is certainly not the only way of doing this!)

3.1.2. If R is a ring with identity 1 and R is not the trivial ring, that is, R has more than one element, then prove that $1 \neq 0$. Use this to verify the statement in the text that a ring is trivial if and only if $1 = 0$.

3.1.3. Give a proof of Lemma 3.2.3.

3.1.4.   Let $\mathbb{Z}[\sqrt{3}] = \{a + b\sqrt{3} : a, b \in \mathbb{Z}\}$. With the usual operations of addition and multiplication in $\mathbb{R}$ — the ring of real numbers — show that $\mathbb{Z}[\sqrt{3}]$ forms a commutative ring with unity.

(NOTE: Make sure to first show that $\mathbb{Z}[\sqrt{3}]$ is closed with respect to addition and multiplication.)

3.1.5. Explain why the ring of integers, $\mathbb{Z}$, is not a field.

3.1.6. Use Lemma 3.2.2 to show that if R is a ring and we define $a - b = a + (-b)$ and if $a, b, c \in R$, then $a(b - c) = ab - ac$.

3.1.7. Let R be a ring with an identity and let $r \in R$. Prove that rR is a subring of R. This is called, as noted in the text, the subring generated by r. Then show that rR contains r. If R does not have an identity, then must rR contain r?

3.1.8. Let R be a ring and let $r, s \in \mathbb{R}$.
(a) Then show that:
   (1) $(r^m)(r^n) = r^{m+n}$
   (2) $(r^m)^n = r^{mn}$
for all $m, n \in \mathbb{N}$.

(HINT: Here you will have to use induction.)

(b) Does the following common property of the ring of real numbers hold for any ring R:

$$(rs)^m = r^m s^m \text{ for } m \in \mathbb{N} ?$$

3.1.9. Suppose that R is any ring and $n \in \mathbb{N}$, $z_1, z_2 \in R$. Then show $(nz_1)(nz_2) = n(nz_1 z_2)$ as described in the text, i.e., using induction on n together with associativity and distributivity in R. In particular we do not need R to be commutative.

3.1.10. If R is a ring and $r, s \in R$ and $m, n \in \mathbb{N}$ then prove that

$$((-m)r)((-n)s) = ((-m)(-n))rs = (mn)rs.$$

This is the last case in the verification of (5) from Lemma 3.2.5.

3.1.11. Prove Lemma 3.2.5 (1).

3.1.12. Prove Lemma 3.2.5 (2).

3.1.13. Prove Lemma 3.2.5 (3).

3.1.14. Prove Lemma 3.2.5 (4).

3.1.15. Let R be any ring and set

$$Z(R) = \{z \in R : zr = rz \text{ for all } r \in R\}.$$

Then $Z(R)$ is called the **center** of R. Prove that $Z(R)$ is a subring of R.

EXERCISES FOR SECTION 3.3

3.3.1. Let $C^0[a, b]$ be the set of functions $f : [a, b] \to \mathbb{R}$ which are continuous on $[a, b]$. With the operations defined in the text before Lemma 3.3.1, show that $C^0[a, b]$ is a commutative ring with identity. (NOTE: You must explicitly say what the identities are with respect to $+$ and $\cdot$.)

3.3.2. Let $C^1[a, b]$ be the set of functions $f : [a, b] \to \mathbb{R}$ which have a continuous derivative on $[a, b]$; i.e., $f'(x)$ is continuous on $[a, b]$ and let the operations of $+$ and $\cdot$ be those in Problem 1. Prove that $C^1[a, b]$ is a subring of $C^0[a, b]$.

Does $C^1[a, b]$ have an identity with respect to $\cdot$?

3.3.3. The two functions $f(x)$ and $g(x)$ defined in the text by

$$f(x) = \begin{cases} x - 1 & \text{if } 0 \le x \le 1 \\ 0 & \text{if } 1 < x \le 2 \end{cases} \text{ and } g(x) = \begin{cases} 0 & \text{if } 0 \le x \le 1 \\ 1 - x & \text{if } 1 < x \le 2 \end{cases}$$

have the property that $f(x)g(x) = 0$ where 0 is the zero function, i.e., $0(x) = 0$ for all x and neither $f(x)$ nor $g(x)$ is the zero function. Show that $f(x), g(x) \in C^0[0,2]$. Do $f(x)$ and $g(x)$ lie in $C^1[0,2]$? Why or why not?

3.3.4. Show part (b) of Lemma 3.3.3.

3.3.5. Use Lemma 3.3.3 to show that multiplication of congruence classes is well defined by Definition 3.3.2.

3.3.6. Complete the verification that $\mathbb{Z}_n$ is a commutative ring with identity. That is, in the proof of Theorem 3.3.3, it was only shown that addition of congruence classes is commutative. Here you must verify *all* the other axioms for addition and multiplication of a commutative ring with identity.

3.3.7. Write out addition and multiplication tables for $\mathbb{Z}_6 = \{0,1,2,3,4,5\}$.

3.3.8. Using the concept of the GCD, defined in the exercises for Chapter 1, we can say that the integers m, n are relatively prime if and only if $\gcd(m,n) = 1$. Let $n \in \mathbb{N}$ and $1 < m < n - 1$. If $\gcd(n,m) = 1$, then show that in $\mathbb{Z}_n$, m has a multiplicative inverse.

(HINT: Use that fact, also contained in the exercises for Chapter 1, that the gcd of two integers can always be written as a linear combination of the two integers.)

3.3.9. Let $R = M_2(\mathbb{Z}_2)$.
(a) What is the order of R? i.e., $|R|$?
(b) Let $A = \begin{pmatrix} 0 & 1 \\ 1 & 1 \end{pmatrix}$ and $B = \begin{pmatrix} 1 & 0 \\ 0 & 0 \end{pmatrix}$. Compute AB and BA.
(c) What does (b) tell you about the ring R?
(d) Does R have an identity? If so what is it; if not why not?

3.3.10. (a) Generalize what you did in the previous example to $R = M_n(\mathbb{Z}_m)$ for $n >$. What is the order of R? The answer will be in terms of m, n.
(b) Is R commutative? why or why not?
(c) Does R have an identity. If so what is it?

3.3.11. In the ring $M_2(\mathbb{Z})$:
(a) Find the multiplicative inverse of each of the following elements

$$(i) \quad \begin{pmatrix} 2 & 5 \\ 1 & 3 \end{pmatrix} \text{ and } \begin{pmatrix} 1 & 6 \\ 3 & 17 \end{pmatrix}.$$

Is each multiplicative inverse in $M_2(\mathbb{Z})$?

(b) Show that the element of $M_2(\mathbb{Z})$: $\begin{pmatrix} 1 & -1 \\ 0 & 2 \end{pmatrix}$ does not have a multiplicative inverse in $M_2(\mathbb{Z})$.

3.3.12. Let $R = M_3(\mathbb{Q})$. Consider $A = \begin{pmatrix} 1 & -1 & 1 \\ 2 & 0 & 1 \\ 3 & -1 & 2 \end{pmatrix}$.

(a) Show that A is not invertible in R, i.e., there is no matrix B such that $AB = I$.

(NOTE: Here you may use any facts you know from linear algebra!)

(b) What are the elements of AR and why is AR a subring of $M_3(\mathbb{Q})$?

(c) Show AR is noncommutative.

(d) Explain why the ring AR does not have an identity.

3.3.13. Show that if $n \in \mathbb{N}$ and $n > 1$, then the ring $n\mathbb{Z}$ does not have an identity.

3.3.14. Let S be any nonempty set. With addition and multiplication defined on the power set $\mathcal{P}(S)$ by

$$
\begin{aligned}
A + B &= A \triangle B = A \cup B - (A \cap B) \\
A \cdot B &= A \cap B
\end{aligned}
$$

show that $R = \mathcal{P}(S)$ forms a commutative ring with unity.

(HINT: Use Venn diagrams.)

3.3.15. Let $S = \{a, b\}$. How many elements are there in $\mathcal{P}(S)$? List all the elements in $\mathcal{P}(S)$ and give them names. Using the definitions of $+$ and $\cdot$ given in the text, write the addition and multiplication table for this ring of sets.

3.3.16. Explain why we would not get a ring if we defined the operations on $\mathcal{P}(S)$ by

$$
\begin{aligned}
A + B &= A \cup B \\
A \cdot B &= A \cap B.
\end{aligned}
$$

3.3.17. Prove Lemma 3.3.4.

*3.3.18. In the text, we proved that any Boolean ring is commutative. In other words, if R is a ring such that $x^2 = x$ for all $x \in R$, then R is commutative. Prove that if R is a ring such that $x^3 = x$ for all $x \in R$, then R is commutative.

3.3.19. Prove Theorem 3.3.7.

3.3.20. If $R_1, R_2, \ldots, R_n$ are finitely many rings, use the definitions in the text for $+$ and $\cdot$ in the case of two rings to define $+$ and $\cdot$ on $R = R_1 + R_2 + \cdots + R_n$ where the set for R is $R_1 \times R_2 \times \cdots \times R_n$. Tell when the ring R is commutative.

3.3.21. If a ring R_1 has order m, i.e., $|R_1| = m$ and a ring R_2 has order k, then what is the order of $R_1 + R_2$?

3.3.22. (a) Give an example of a commutative ring with 16 elements.
(b) Give an example of a noncommutative ring with 16 elements.

3.3.23. Give an example of a ring with ten elements which does not have a unity, i.e., a multiplicative identity.

3.3.24. Show that if R is a ring with identity and $u \in R$ is a unit, then if $wu = 0$, then $w = 0$.

3.3.25. If $a, b \in \mathbb{Z}$, let us define addition, $\oplus$, and multiplication, $\odot$, as follows:

$$a \oplus b = ab$$
$$a \odot b = a + b.$$

Is the set of integers $\mathbb{Z}$ with these operations a ring? Why or why not?

3.3.26. Show that neither of the following addition tables can define the addition of a ring of order 4, $R = \{a, b, c, d\}$:

(i)

+	a	b	c	d
a	a	b	c	a
b	b	c	d	a
c	c	d	a	b
d	a	a	b	c

(ii)

+	a	b	c	d
a	a	b	c	d
b	b	c	a	d
c	c	a	d	b
d	b	c	d	a

3.3.27. Again let $R = \{a, b, c, d\}$ and define addition and multiplication on R by the tables below:

+	a	b	c	d
a	a	b	c	d
b	b	a	d	c
c	c	d	a	b
d	d	c	b	a

·	a	b	c	d
a	a	a	a	a
b	a	b	c	d
c	a	c	d	b
d	a	d	b	c

(a) Show that R with respect to the above $+$ and $\cdot$ forms a ring. Here you don't have to go through all verifications for associative and distributive laws. (How many would there be?) But just do three verifications for each law.

(b) Is R a commutative ring? Does R have a identity? If so, what element is it?

(c) Is R a field?

3.3.28. Let $R = M_2(\mathbb{Z})$. What is $Z(R)$, i.e., the center of R?

(HINT: See Problem 15 from Sections 3.1 and 3.2. Also you may restrict your attention to diagonal matrices.)

EXERCISES FOR SECTION 3.4

3.4.1. Let the mapping $f : \mathbb{Z} \to \{0, 1\}$ be defined by $f(n) = \begin{cases} 0 & \text{if } n \text{ is even} \\ 1 & \text{if } n \text{ is odd} \end{cases}$.

Assuming that f is a ring homomorphism, do the following.

(a) Write out the addition and multiplication tables for the ring which is the range of f, i.e., $f(\mathbb{Z}) = \{0, 1\}$.

(b) What ring is $f(\mathbb{Z})$?

(c) What is the kernel of f, $ker(f)$?

3.4.2. Let R_1 and R_2 be arbitrary rings. Consider the mapping

$$\Pi_1 : R_1 + R_2 \to R_1$$

defined by $\Pi_1(a, b) = a$ where $a \in R_1$ and $b \in R_2$. This map Π_1 is called the **projection** of $R_1 + R_2$ onto R_1.

Prove that Π_1 is an epimorphism.

(HINT: Of course, there is also a Π_2 which is the **projection** of $R_1 + R_2$ onto R_2.)

3.4.3. Let R_1 and R_2 be arbitrary rings. Consider the mapping

$$f_1 : R_1 \to R_1 + R_2$$

defined by $f_1(r) = (r, 0)$ where $r \in R_1$ and $0 \in R_2$.

(a) Prove that f_1 is a monomorphism. This map is called an **embedding** of R_1 into $R_1 + R_2$

(b) What subring of $R_1 + R_2$ does this show that R_1 is isomorphic to?

(HINT: Of course, there is also an f_2 which is the **embedding** of R_2 into $R_1 + R_2$.)

3.4.4. Prove Lemma 3.4.2 (2), i.e., for any ring homomorphism $f : R \to S$. $im(f) = f(R)$ is a subring of S.

3.4.5. The first part of this problem generalizes Problem 3.4.4. Let f, R, S be as in Problem 3.4.4.

(a) Prove that if E is any subring of R then $f(E)$ is a subring of S.

(b) Prove that if F is any subring of S then $f^{-1}(F)$ is a subring of R.

3.4.6. If R and S are arbitrary rings, then prove that $R + S \cong S + R$.

3.4.7. Let R and S be rings and let $f : R \to S$ be an epimorphism.

(a) Prove that if R is a ring with identity, which we shall call 1, then S is also a ring with unity and its identity is $f(1)$, i.e., a ring homomorphism takes the multiplicative identity into the multiplicative identity.

(b) Prove that if R is a commutative ring, then so is S.

(c) Would (a) and (b) still be true if the map f is not onto?

3.4.8. Let R_1, R_2, R_3 be rings and suppose that $f : R_1 \to R_2$, $g : R_2 \to R_3$. If f and g are ring homomorphisms, then prove that the composition

$$h = g \circ f : R_1 \to R_3$$

is also a ring homomorphism.

3.4.9. Suppose that E and F are fields and that $f : E \to F$ is an isomorphism. Prove that if $x \in E$, $x \neq 0$, then $f(x^{-1}) = f(x)^{-1}$.

3.4.10. Let R and S be rings and let $f : R \to S$ be a homomorphism. Prove that f is a monomorphism if and only if $ker(f) = \{0\}$.

EXERCISES FOR SECTION 3.5

3.5.1. Suppose that D is a commutative ring with identity that has no zero divisors, i.e., an integral domain. Then show this is true if and only if D is a commutative ring with identity such that for any $r, s \in D$, $rs = 0$ implies that either $r = 0$ or $s = 0$.

3.5.2. Prove that if r is a zero divisor in a commutative ring R, then ar is also a divisor of zero for any $a \in R$ just as long as $ar \neq 0$.

3.5.3. Give an example to show that the sum of two zero divisors in a ring need not be a zero divisors.

(HINT: Use $\mathbb{Z}_6$.)

3.5.4. Let $R = M_2(\mathbb{Z})$. Verify that each of the following matrices is a zero divisor in R:

(a) $\begin{pmatrix} 0 & 1 \\ 0 & 0 \end{pmatrix}$

(b) $\begin{pmatrix} 1 & 2 \\ 0 & 0 \end{pmatrix}$

(c) $\begin{pmatrix} 2 & 3 \\ 4 & 6 \end{pmatrix}$

3.5.5. In Lemma 3.5.3, we wanted to show that $\mathbb{Z}_n$ is an integral domain if and only if (iff) n is a prime. This means that we have to show a double implication. What we did show was that if n is not a prime, then $\mathbb{Z}_n$ is not an integral domain and if n is a prime then $\mathbb{Z}_n$ is an integral domain. Explain why this is sufficient for the iff.

3.5.6. Which elements in $\mathbb{Z}_{10}$ are zero divisors?

3.5.7. Verify that the the element $(1,0)$ is a zero divisor in $\mathbb{Z}_3 + \mathbb{Z}_5$.
(HINT: This shows that the direct sum of two integral domains need not be an integral domain.)

3.5.8. Consider the ring $\mathbb{Z}[\sqrt{3}] = \{a + b\sqrt{3} : a, b \in \mathbb{Z}\}$ defined in Problem 3.1.3. Is $\mathbb{Z}[\sqrt{3}]$ an integral domain? Why or why not?

3.5.9. Go through the proof of Theorem 3.5.1 with D replaced by $\mathbb{Z}$ and show where the argument breaks down.

3.5.10. Prove that if S is any nonempty set and $R = \mathcal{P}(S)$ is the ring of subsets or Boolean ring on S with the usual operation of $+$ and $\cdot$, then every nonzero element of R except for the identity (S), is a divisor of zero.

3.5.11. (a) What are the units in $\mathbb{Z}_{10}$?
(b) Write a multiplication table for the units which you found in part (a) to verify that it is a group $U(\mathbb{Z}_{10})$.

3.5.12. Here are some containment relations between the classes of rings which we have studied:

Fields $\subset$ Integral Domains $\subset$ Commutative Rings $\subset$ Rings.

Give examples to show that all of the above containments are proper.

In problems 3.5.13–3.5.18, suppose that D is an ordered integral domain with $a, b, c, d \in D$.

3.5.13. Prove that if $a > b$, then $-a < -b$.

3.5.14. Prove that if $a > b$ and $c > d$, then $a + c > b + d$.

3.5.15. Prove that if $a, b, c, d \in D^+$ and $a > b, c > d$, then $ac > bd$.

3.5.16. Prove that if $a > 0$ and $b > c$ then $ab > ac$.

3.5.17. Prove that if $a < 0$ and $b > c$ then $ab < ac$.

3.5.18. Prove that if $a > 0$ and $ab > ac$, then $b > c$.

3.5.19. Suppose that in a ring R the following cancellation law holds

$$xy = xz \implies y = z$$

for every nonzero element $x \in R$. Prove that R cannot have any zero divisors.
(HINT: R is not assumed to be commutative!)

3.5.20. (a) Prove that if D is an integral domain and $f : D \to E$ is an isomorphism then $E = f(D)$ is also an integral domain.
(That is, the isomorphic image of an integral domain is an integral domain.)

(b) If instead of an isomorphism f was just an epimorphism, then would it still be true that $E = f(D)$ is an integral domain?

*3.5.21. Can there be an integral domain of order 6? Why or why not?

(HINT: You may assume here that the addition is given just as that in $\mathbb{Z}_6$. We also note that later on when we know more about fields this will become a much easier problem!)

3.5.22. Give an induction proof of Lemma 3.5.9.

(HINT: Induct on the natural number n and use the definition of integer multiple of an element in a ring.)

EXERCISES FOR SECTION 3.6

3.6.1. Let D be an ordered integral domain and let $f : \mathbb{Z} \to D$ be given by $f(z) = z \cdot 1$. Here you must understand that z is an integer and 1 may **not** be. In particular 1 is the multiplicative identity of D.
(a) Explain why $z \cdot 1$ makes sense as an element of D for all $z \in \mathbb{Z}$.
(b) Prove that f is a ring homomophism.
(HINT: Use Lemma 3.2.5.)

3.6.2. In the text a **prime ring** was defined to be a ring with identity which has the property that the only subring which contains the identity, 1, is the whole ring.
(a) Prove directly that $\mathbb{Z}$ is a prime ring.
(b) Prove directly that for each positive integer n, $\mathbb{Z}_n$ is a prime ring.

3.6.3. Let D be an integral domain and let $a_1, a_2, \ldots, a_n \in D$ such that

$$a_1 a_2 \cdots a_n = 0,$$

then at least one of the a's must be 0.

(HINT: Use mathematical induction.)

3.6.4. If $f : R \rightarrow S$ is a ring homomorphism and $x \in R$, prove that

$$f(x^n) = f(x)^n$$

for every $n \in \mathbb{N}$.

(HINT: Use mathematical induction.)

3.6.5. Let $\mathbb{Q}$ be the set of rational with the usual definitions of $+, \cdot, <$. Thus $\mathbb{Q}$ is an ordered integral domain. Show that $\mathbb{Q}$ does not satisfy the LWO.

3.6.6. Prove that if R is a prime ring of characteristic 0, then R is isomorphic to $\mathbb{Z}$.

3.6.7. Prove that if R is a prime ring of characteristic n, then $R \cong \mathbb{Z}_n$.

(HINT: To obtain that the map is one-to-one use the division algorithm, Lemma 3.3.2.)

3.6.8. Prove that an ordered integral domain must have characteristic zero. Of course this means that ordered field has characteristic zero.

(HINT: Use the proof of Theorem 3.5.2 and Lemma 3.5.8 (c). Note that from Lemma 3.5.8 (c) we must have $1 > 0$.)

Chapter 4

Number Theory and Unique Factorization

4.1 Elementary Number Theory

Algebraically we have completely classified the integers as an ordered integral domain which satisfies the principle of mathematical induction or the least well-ordering property. From this point of view the structure of the integers is relatively simple. Its additive structure is just an infinite cyclic group (We will discuss this in Chapter 6.). However, the multiplicative structure becomes surprisingly complex. This is the basis for **elementary number theory** which is concerned for the most part with the multiplicative properties of the integers. Many results are almost magical and further, number theory serves as a breeding ground for results in general ring theory. In this chapter we give an introduction to number theory and prove several algebraic results that are closely tied to number theory.

Modern number theory has essentially three branches, which overlap in many areas. The first is **elementary number theory**, which can be quite nonelementary, and which consists of those results concerning the integers themselves which do not use analytic methods. The cornerstone of elementary number theory is the **fundamental theorem of arithmetic** which we will present in Section 4.4. This branch has many subbranches: the theory of congruences, diophantine analysis, geometric number theory, quadratic residues to mention a few. The second major branch is **analytic number theory**. This is the branch of the theory of numbers that studies the integers by using methods of real and complex analysis. The final major branch is **algebraic number theory** which extends the study of the integers to other algebraic number fields. For a comprehensive treatment and explanation of all these terms we refer the reader to [FR].

Number theory has a long and glorious history that we will briefly review. Number theory also has served as a motivating discipline for abstract algebra. Many purely algebraic concepts such as the concept of a ring originally arose in what is now known as algebraic number theory.

Number theory originated from arithmetic and computations with whole numbers. Every culture and society has some method of counting and number representation. However, it wasn't until the development of a place value system that symbolic computation became truly feasible. The numeration system that we use is called the Hindu-Arabic numeration system and was developed in India most likely during the period 600–800 A.D. This system was adopted by Arab cultures and transported to Europe via Spain. The adoption of this system in Europe and elsewhere was a long process and it wasn't until the Renaissance and after that symbolic computation widely superseded the use of abaci and other computing devices. We should remark that although mathematics is theoretical it often happens that abstract results are delayed without proper computation. Calculus and analysis could not have developed without the prior development of the concept of an irrational number.

Much of the beginnings of number theory came from straightforward observation. A great deal of number theoretic information was known to the Babylonians, Egyptians, Greeks, Hindus and other ancient cultures. Greek mathematicians, especially the Pythagoreans (around 450 B.C.), began to think of numbers as abstractions and deal with purely theoretical questions. The foundation material of number theory — divisors, primes, gcd's, lcm's, the Euclidean algorithm, the fundamental theorem of arithmetic and the infinitude of primes — although not always stated in modern terms, are all present in Euclid's *Elements*. Three of Euclid's books, Book VII, Book VIII and Book IX treat the theory of numbers. It is interesting that Euclid's treatment of number theory is still geometric in its motivation and most of its methods. It wasn't until the Alexandrian period, several hundred years later, that arithmetic was separated from geometry. The book *Introductio Arithmeticae* by Niomachus in the second century A.D. was the first major treatment of arithmetic and the properties of the whole numbers without geometric recourse. This work was continued by Diophantus of Alexandria about 250 A.D. His great work *Arithmetica* is a collection of problems and solutions in number theory and algebra. In this work he introduced a great deal of algebraic symbolism as well as the topic of equations with indeterminate quantities. The attempt to find integral solutions to algebraic equations is now called **diophantine analysis** in his honor. Fermat's big theorem of solving $x^n + y^n = z^n$ for integers is an example of a diophantine problem. Perhaps the most famous problem in diophantine analysis was Fermat's big theorem. This stated that the equation $x^n + y^n = z^n$ has no integer solutions with $xyz \neq 0$ for $n \geq 3$. Fermat stated this about 1650 but supplied no proof. He said that he had found a marvelous proof of this but the margin was too small to contain it. This marvelous proof was never found. This was an open problem until 1996 when it was proved by Andrew Wiles.

The improvements in computational techniques led mathematicians in the 1500's and 1600's to look more deeply at number theoretical questions. The giant of this period was Pierre Fermat who made enormous contributions to the theory of numbers. It was Fermat's work that could be considered the beginnings of number theory as a modern discipline. Fermat professionally was a lawyer and a judge and essentially only a mathematical amateur. He published almost nothing and his results and ideas are found in his own notes and journals as well as in correspondence with other mathematicians. Yet he had a profound effect on almost all branches of mathematics, not just number theory. He, as much as Descartes, developed analytic geometry. He did major work, prior to Newton and Leibniz, on the foundations of calculus. A series of letters between Fermat and Pascal established the beginnings of probability theory. In number theory, the work he did on factorization, congruences and representations of integers by quadratic forms determined the direction of number theory until the nineteenth century. He did not supply proofs for most of his results but almost all of his work was subsequently proved (or shown to be false). The most difficult proved to be his big theorem which remained unproved until 1996. The attempts to prove this big theorem led to many advances in number theory including the development of algebraic number theory. From the time of Fermat in the mid-seventeenth century through the eighteenth century a great deal of work was done in number theory but it was basically a series of somewhat disconnected, but often brilliant and startling, results. Important contributions were made by Euler, who proved and extended much of Fermat's results. During this period, certain problems were either stated or conjectured which became the basis for what is now known as **additive number theory**. The Goldbach conjecture and Waring's problem are two examples. We will not touch on this topic in this book but refer an interested reader to [NZ].

In 1800 Gauss published a treatise on number theory called *Disquitiones Arithmeticae*. This book not only standardized the notation used but set the tone and direction for the theory of numbers up until the present. It is often joked that any new mathematical result is somehow inherent in the work of Gauss and in the case of number theory this is not really that far-fetched. Tremendous ideas and hints of things to come are present in Gauss's *Disquisitones*. Gauss's work on number theory centered on three main concepts: the theory of congruences, the introduction of algebraic numbers and the theory of forms, especially quadratic forms, and how these forms represent integers. Gauss, through his student Dirichlet, was also important in the infancy of analytic number theory. In 1837 Dirichlet proved, using analytic methods, that there are infinitely many primes in any arithmetic progression $\{a + nb\}$ with a, b relatively prime. Dirichlet's use of analysis really marks the beginning of analytic number theory. The main work in analytic number theory though, centered on the prime number theorem, also conjectured by Gauss among others, including Euler and Legendre. This result deals with the asymptotic behavior of the function

$$\pi(x) = \text{ number of primes } \leq x.$$

The actual result says that

$$\lim_{x \to \infty} \frac{\pi(x)}{x/\ln x} = 1$$

and was proved in 1896 by Hadamard and independently by de la Valle Poussin. Both of their proofs used the behavior of the **Riemann zeta function**

$$\zeta(z) = \sum_{n=1}^{\infty} \frac{1}{n^z}$$

where $z = x + iy$ is a complex variable. Using this function, Riemann in 1859 attempted to prove the prime number theorem. In the attempted proof he hypothesized that all the zeros $z = x + iy$ of $\zeta(z)$ in the strip $0 \leq x \leq 1$ lie along the line $x = \frac{1}{2}$. This conjecture is known as the **Riemann hypothesis** and is still an open question.

Algebraic number theory also started basically with the work of Gauss. Gauss did an extensive study of the complex integers, that is, the complex numbers of the form $a + bi$ with a, b integers. Today these are known as the **Gaussian integers**. Gauss proved that they satisfy most of the same properties as the ordinary integers including unique factorization into primes. In modern parlance he showed that they form a **unique factorization domain** (see Section 4.5). Gauss's algebraic integers were extended in many ways in attempt to prove Fermat's Big Theorem, and these extensions eventually developed into algebraic number theory. Kummer, a student of Gauss and Dirichlet, introduced in the 1840's a theory of algebraic integers and a set of ideal numbers from which unique factorization could be obtained. He used this to prove many cases of the Fermat theorem. Dedekind, in the 1870's developed a further theory of algebraic numbers and unique factorization by ideals which extended both Gaussian integers and Kummer's algebraic and ideal numbers. Further work in the same area was done by Kronecker in the 1880's. The close ties between number theory, especially diophantine analysis, and algebraic geometry led to Wiles's proof of the Fermat theorem and to an earlier proof by Faltings of the Mordell conjecture, which is a related result. The vast areas of mathematics used in both of these proofs is phenomenal.

Probabilistic methods were incorporated into number theory by P. Erdos and studies in this area are known as **probabilistic number theory**. A great deal of recent work has gone into primality testing and factorization of large integers. These ideas have been incorporated extensively into cryptography (see [Ko]).

4.2   Divisibility and Primes

The starting point for the theory of numbers is **divisibility**. We have already seen the first part of the following definition.

Definition 4.2.1. *If a, b are integers we say that a **divides** b, or that a is a **factor** or **divisor** of b, if there exists an integer q such that $b = aq$. We denote this by $a|b$. b is then a **multiple** of a. If $b > 1$ is an integer whose only factors are $\pm 1, \pm b$ then b is a **prime**, otherwise $b > 1$ is **composite**.*

We will see in Section 4.4 that the set of primes will play a fundamental role in the structure of the integers.

The following properties of divisibility are straightforward consequences of the definition

Theorem 4.2.1. *Let a, b and c be integers.*
    (1) $a|b$ implies that $a|bc$ for any integer c.
    (2) $a|b$ and $b|c$ implies $a|c$.
    (3) $a|b$ and $a|c$ implies that $a|(bx + cy)$ for any integers x, y.
    (4) $a|b$ and $b|a$ implies that $a = \pm b$
    (5) $a|b$ and $a > 0, b > 0$ then $a \leq b$.
    (6) $a|b$ if and only if $ca|cb$ for any integer $c \neq 0$.
    (7) $a|0$ for all $a \in \mathbb{Z}$ and $0|a$ only for $a = 0$.
    (8) $a| \pm 1$ only for $a = \pm 1$.
    (9) $a_1|b_1$ and $a_2|b_2$ implies that $a_1 a_2|b_1 b_2$.

Proof. We prove (2) and leave the remaining parts to the exercises.

    Suppose $a|b$ and $b|c$. Then there exists $x, y \in \mathbb{Z}$ such that $b = ax$ and $c = by$. But then $c = axy = a(xy)$ and therefore $a|c$ since $x, y \in \mathbb{Z}$.   □

    If b, c, x, y are integers then an integer $bx + cy$ is called a **linear combination** of b, c. Thus part (3) of Theorem 4.2.1 says that if a is a **common divisor** of b, c then a divides any linear combination of b and c.

    Further, note that if $b > 1$ is a composite then there exists $x > 0$ and $y > 0$ such that $b = xy$ and from part (5) we must have $1 < x < b, 1 < y < b$.

    In ordinary arithmetic, given a, b we can always attempt to divide a into b. The next theorem called the **division algorithm** says that if $a > 0$ either a will divide b or the **remainder** of the division of b by a will be less than a.

Theorem 4.2.2. *(Division Algorithm) Given integers a, b with $a > 0$ then there exist unique integers q and r such that*

$$b = qa + r \text{ where either } r = 0 \text{ or } 0 < r < a.$$

    We usually denote the inequalities above by $0 \leq r < a$. One may think of q and r as the **quotient** and **remainder**, respectively, when dividing b by a.

Proof. Given a, b with $a > 0$ consider the set

$$S = \{b - qa \geq 0 \;\; : \;\; q \in \mathbb{Z}\}.$$

If $b > 0$ then $b + a > 0$ and the sum is in S with $q = -1$. If $b \leq 0$ then there exists a $q > 0$ with $-qa < b$. Then $b + qa > 0$ and is in S as $b + qa = b - (-qa)$.

Therefore in either case S is nonempty. Hence S is a nonempty subset of $\mathbb{N} \cup \{0\}$ and therefore has a least element r by the LWO. If $r \neq 0$ we must show that $0 < r < a$. Suppose $r \geq a$, then $r = a + x$ with $x \geq 0$ and $x < r$ since $a > 0$. Then $b - qa = r = a + x \implies b - (q+1)a = x$. This means that $x \in S$. Since $x < r$ this contradicts the minimality of r which is a contradiction. Therefore if $r \neq 0$ it follows that $0 < r < a$.

The only thing left is to show the uniqueness of q and r. Suppose $b = q_1 a + r_1$ also where $0 \leq r_1 < a$. Since r was minimal, $r \leq r_1$. But $0 \leq r \leq r_1 < a$. This implies that $0 \leq r_1 - r < a$. Since $r_1 = b - aq_1$ and $r = b - qa$, subtracting we get

$$r_1 - r = a(q - q_1).$$

This says that $r_1 - r$ is a multiple of the positive integer a which is nonnegative but strictly less than a. The only such multiple of a is 0. It follows that $q - q_1 = 0$ which implies that $q = q_1$ and this in turn gives $r = r_1$.

$\square$

As we remarked in the introduction in the last section, number theory both motivates abstract algebra and also provides methods of proof for abstract results. Here we give an example. In Chapter 3, we saw that for any natural number n the subset $n\mathbb{Z}$ consisting of all integer multiples of n actually formed a subring. The next result shows that any subring of $\mathbb{Z}$ must be of the form $n\mathbb{Z}$ for some natural number n. Hence the subrings $n\mathbb{Z}$ as n runs over the natural numbers provide a complete collection of the subrings of $\mathbb{Z}$.

Theorem 4.2.3. *Let S be a subring of $\mathbb{Z}$. Then $S = n\mathbb{Z}$ for some integer $n \geq 0$.*

Proof. If S is the trivial subring $S = \{0\}$, then $S = 0\mathbb{Z}$ and we are done. So suppose that $S \neq \{0\}$. Then the set $S^+ = \{x \in S : x > 0\}$ is nonempty. So since S^+ is a set of natural numbers, it has by LWO a least element. Call this least element $n_0 \in S$. We claim that $S = n_0\mathbb{Z}$. Since $n_0 \in S$ and any element of $n_0\mathbb{Z}$ is either a sum of n_0 added to itself a certain number of times, 0, or the negative of such a sum, we have

$$n_0\mathbb{Z} \subset S$$

because S being a subring is closed with respect to addition and additive inverses and it contains 0. Hence we need to show the reverse inclusion. So let $x \in S$. We then use the division algorithm (Theorem 4.2.2) to divide x by n_0 to get $x = qn_0 + r$ where $0 \leq r < n_0$. But then $r = x - qn_0$. But since x, n_0 both belong to S and since S is a subring again, we must have that $r \in S$. If $r > 0$, that would contradict the minimality of n_0. Thus $r = 0$ and so $n_0 \mid x$. This implies that $x \in n_0\mathbb{Z}$. Therefore

$$S \subset n_0\mathbb{Z}.$$

This proves our claim and so completes the proof. $\square$

4.3   Greatest Common Divisors

The next ideas that are necessary are the concepts of **greatest common divisor** and **least common multiple**.

Definition 4.3.1. *Given nonzero integers a, b their **greatest common divisor** or **GCD** $d > 0$ is a positive integer which is a common divisor, that is, $d|a$ and $d|b$, and if d_1 is any other common divisor then $d_1|d$. We denote the greatest common divisor of a, b by either, $\gcd(a, b)$, $GCD(a, b)$, or (a, b).*

The next result says that given any nonzero integers they do have a greatest common divisor and further it is unique.

Theorem 4.3.1. *Given nonzero integers a, b their GCD exists, is unique and can be characterized as the least positive linear combination of a and b.*

Proof. Given nonzero a, b consider the set

$$S = \{ax + by > 0 : x, y \in \mathbb{Z}\}.$$

Now $a^2 + b^2 > 0$ so S is a nonempty subset of $\mathbb{N}$ and hence has a least element $d > 0$ by the LWO. We show that d is the GCD.

First we must show that d is a common divisor. Now $d = ax + by$ and is the least such positive linear combination. By the division algorithm $a = qd + r$ with $0 \leq r < d$. Suppose $r \neq 0$. Then $r = a - qd = a - q(ax + by) = (1 - qx)a - qby > 0$. Hence r is a positive linear combination of a and b and therefore is in S. But then $r < d$ contradicting the minimality of d in S. It follows that $r = 0$ and so $a = qd$ and $d|a$. An identical argument shows that $d|b$ and so d is a common divisor of a and b. Let d_1 be any other common divisor of a and b. Then d_1 divides any linear combination of a and b and so $d_1|d$. Therefore d is the GCD of a and b.

Finally we must show that d is unique. Suppose d_1 is another GCD of a and b. Then $d_1 > 0$ and d_1 is a common divisor of a, b. Then $d_1|d$ since d is a GCD. Identically $d|d_1$ since d_1 is a GCD. Therefore $d = \pm d_1$ by Theorem 4.2.1 (4) and then $d = d_1$ since they're both positive. □

If $(a, b) = 1$, that is, the only positive common divisor of a and b is 1, then we say that a, b are **relatively prime**. The following is easy to show from Theorem 4.3.1 and its proof will be left to the exercises.

Lemma 4.3.1. *If a and b are integers, then a and b are relatively prime if and only if 1 is expressible as a linear combination of a and b.*

Proof. See the exercises. □

The following two results provide some properties of GCD's.

Lemma 4.3.2. *If $d = (a, b)$ then $a = a_1 d$ and $b = b_1 d$ with $(a_1, b_1) = 1$.*

Proof. If $d = (a, b)$ then $d|a$ and $d|b$. Hence $a = a_1 d$ and $b = b_1 d$. We have

$$d = ax + by = a_1 dx + b_1 dy.$$

Dividing both sides of the equation by d we obtain

$$1 = a_1 x + b_1 y.$$

Therefore $(a_1, b_1) = 1$ by Lemma 4.3.1. □

Lemma 4.3.3. *For any integer c we have that $(a, b) = (a, b + ac)$.*

Proof. Suppose $(a, b) = d$ and $(a, b + ac) = d_1$. Now d is the least positive linear combination of a and b. Suppose $d = ax + by$. d_1 is a linear combination of $a, b + ac$ so that

$$d_1 = ar + (b + ac)s = a(cs + r) + bs.$$

Hence d_1 is also a linear combination of a and b and therefore $d_1 \geq d$. On the other hand $d_1|a$ and $d_1|b + ac$ and so $d_1|b$. Therefore $d_1|d$ so $d_1 \leq d$ by Theorem 4.2.1 (5). Combining these we must have $d_1 = d$. □

The next result, called the **Euclidean algorithm**, provides a technique for both finding the GCD of two integers and expressing the GCD as a linear combination. It is not hard to show that if a, b are nonzero integers then $(a, b) = (-a, b) = (a, -b) = (-a, -b)$ (see exercises). Also it is easy to verify (see exercises) that if a, b are nonzero integers and we have written (a, b) as a linear combination of a and b then (a, b) is also a linear combination of $-a$, and b and of a and $-b$ and of $-a$ and $-b$. Thus we now assume that both a and b are positive.

Theorem 4.3.2. *(The Euclidean Algorithm) Given integers b and $a > 0$ form the repeated divisions*

$$b = q_1 a + r_1, 0 < r_1 < a,$$

$$a = q_2 r_1 + r_2, 0 < r_2 < r_1,$$

$$\vdots$$

$$r_{n-2} = q_n r_{n-1} + r_n, 0 < r_n < r_{n-1},$$

$$r_{n-1} = q_{n+1} r_n.$$

The last nonzero remainder r_n is the GCD of a, b. Further r_n can be expressed as a linear combination of a and b by successively eliminating the r_i's in the intermediate equations; in other words, by using "back" substitution.

Proof. In taking the successive divisions as outlined in the statement of the theorem each remainder r_i gets strictly smaller and still nonnegative. Hence it must finally end with a zero remainder. Therefore there is a last nonzero remainder r_n. We must show that this is the GCD.

Now from Lemma 4.3.3 the gcd $(a, b) = (a, b - q_1 a) = (a, r_1) = (r_1, a - q_2 r_1) = (r_1, r_2)$. Continuing in this manner we have then that $(a, b) = (r_{n-1}, r_n) = r_n$ since r_n divides r_{n-1}. This shows that r_n is the GCD.

To express r_n as a linear combination of a and b notice first that

$$r_n = r_{n-2} - q_n r_{n-1}.$$

From the immediately preceding division we similarly get $r_{n-1} = r_{n-3} - q_{n-1} r_{n-2}$. We then substitute this in the above to get

$$r_n = r_{n-2} - q_n (r_{n-3} - q_{n-1} r_{n-2}) = (1 + q_n q_{n-1}) r_{n-2} - q_n r_{n-3}.$$

Doing this successively we ultimately express r_n as a linear combination of a and b.

□

EXAMPLE 4.3.1. Find the GCD of 270 and 2412 and express it as a linear combination of 270 and 2412.

We apply the Euclidean algorithm

$$2412 = (8)(270) + 252,$$

$$270 = (1)(252) + 18,$$
$$252 = (14)(18).$$

Therefore the last nonzero remainder is 18 which is the GCD. We now must express 18 as a linear combination of 270 and 2412.

From the next to the last equation

$$18 = 270 + (-1)252.$$

The first equation gives

$$252 = 2412 - (8)(270).$$

Now substituting this for 252 gives

$$18 = 270 + (-1)(2412 - (8)(270)) = (-1)(2412) + (9)(270).$$

We need the following result known as **Euclid's lemma** that we stated in the last chapter. In the next section we will use a special case of this applied to primes. We note that this special case is traditionally also called Euclid's lemma.

Lemma 4.3.4. *(Euclid's lemma) Let $a, b, c \in \mathbb{Z}$. Suppose $a|bc$ and $(a, b) = 1$, then $a|c$.*

Proof. Suppose $(a, b) = 1$ then 1 is expressible as a linear combination of a and b by Lemma 4.3.1. That is,

$$ax + by = 1.$$

Multiply through by c, so that

$$acx + bcy = c.$$

Now $a|a$ and $a|bc$, so a divides the linear combination $acx + bcy$ and hence $a|c$.   □

Now suppose that $d = (a, b)$ where $a, b \in \mathbb{Z}$ and $a \neq 0, b \neq 0$. Then we note that given one integer solution of the equation

$$ax + by = d$$

we can easily obtain all integer solutions.

Suppose without loss of generality that $d = 1$, that is, a, b are relatively prime. If not we can divide through by $d > 1$. Suppose that we are given that $x_1, y_1 \in \mathbb{Z}$ is a solution to $ax + by = 1$ and $x_2, y_2 \in \mathbb{Z}$ is any other solution. That is,

$$ax_1 + by_1 = 1$$
$$ax_2 + by_2 = 1.$$

Then

$$a(x_1 - x_2) = -b(y_1 - y_2).$$

Since $(a, b) = 1$ we get from Lemma 4.3.4 that $b|(x_1 - x_2)$ and hence $x_2 = x_1 + bt$ for some $t \in \mathbb{Z}$. Substituting back into the equations we then get

$$ax_1 + by_1 = a(x_1 + bt) + by_2 \implies by_1 = abt + by_2,$$

since $b \neq 0$, $y_2 = y_1 - at$. Hence all solutions are given by

$$x_2 = x_1 + bt,$$
$$y_2 = y_1 - at$$

for some $t \in \mathbb{Z}$

We next introduce the **least common multiple**.

Definition 4.3.2. *Given nonzero integers a, b their **least common multiple** or **LCM** $m > 0$ is a positive integer which is a common multiple, that is, $a|m$ and $b|m$, and if m_1 is any other common multiple then $m|m_1$. We denote the least common multiple of a, b by either $lcm(a, b)$, $LCM(a, b)$, or $[a, b]$.*

As for GCD's given any nonzero integers they do have a least common multiple and it is unique.

Theorem 4.3.3. *Given nonzero integers a, b their LCM exists and is unique. Further we have*

$$(a, b)[a, b] = \pm ab.$$

Proof. We get a $+$ sign if both a and b have the same sign but if a and b have opposite signs then we need the $-$ sign because both the GCD and the LCM were defined to be positive. Let $d = (a, b)$ and let $m = \frac{ab}{d}$. We show that m is the LCM. Now $a = a_1 d, b = b_1 d$ with $(a_1, b_1) = 1$ by Lemma 4.3.2. Then $m = a_1 b_1 d$. Since $a = a_1 d$, $m = b_1 a$ so $a|m$, Identically $b|m$ so m is a common multiple. Now let m_1 be another common multiple so that $m_1 = ax = by$. We then get

$$a_1 dx = b_1 dy \implies a_1 x = b_1 y \implies a_1 | b_1 y$$

But $(a_1, b_1) = 1$ so from Lemma 4.3.4 $a_1|y$. Hence $y = a_1 z$. It follows then that

$$m_1 = b_1 d(a_1 z) = a_1 b_1 dz = mz$$

and hence $m|m_1$. Therefore m is an LCM.

The uniqueness follows in the same manner as the uniqueness of GCD's. Suppose m_1 is another LCM, then $m|m_1$ and $m_1|m$ so $m = \pm m_1$ and since they are both positive $m = m_1$.

$\square$

EXAMPLE 4.3.2. Find the LCM of 270 and 2412.
From Example 2.2.1 we found that $(270, 2412) = 18$. Therefore

$$[270, 2412] = \frac{(270)(2412)}{(270, 2412)} = \frac{(270)(2412)}{18} = 36180.$$

We close this section by using the idea that GCD's can be linearly expressible to present a proof of Lemma 3.3.4. Recall that in any ring R the set of multiples of a single element $r \in R$ forms a subring. We called this subring rR. The question answered in Lemma 3.3.4 was in the modular ring $\mathbb{Z}_n$ for which m is the subring $m\mathbb{Z}_n$ all of $\mathbb{Z}_n$. This is true precisely when $(m, n) = 1$, that is, m and n are relatively prime.

Lemma 4.3.5. *Let $n \in \mathbb{N}$ and $1 < m < n - 1$. Then if m and n are not relatively prime then $m\mathbb{Z}_n$ is a proper subring of $\mathbb{Z}_n$ without an identity. If m and n are relatively prime then $m\mathbb{Z}_n = \mathbb{Z}_n$.*

Proof. Suppose that $(m, n) = 1$. Then there exists integers x, y such that $mx + ny = 1$. Considering this in $\mathbb{Z}_n$ we then have $mx = 1$ since $ny = 0$. Hence m is a unit in $\mathbb{Z}_n$. Let r be any element of $\mathbb{Z}_n$. Then

$$r = 1 \cdot r = (mx) \cdot r = m(xr) \in m\mathbb{Z}_n.$$

Therefore each element of $\mathbb{Z}_n$ is in $m\mathbb{Z}_n$. Thus we have shown that $\mathbb{Z}_n \subset m\mathbb{Z}_n$. Since $m\mathbb{Z}_n$ is a subring of $\mathbb{Z}_n$, we must have $m\mathbb{Z}_n \subset \mathbb{Z}_n$. Therefore we have $m\mathbb{Z}_n = \mathbb{Z}_n$.

Conversely suppose that $m\mathbb{Z}_n = \mathbb{Z}_n$. Then there exists an $x \in \mathbb{Z}_n$ with $mx = 1$ so that m is a unit. This implies that $mx - 1 = 0$ or in terms of integers $mx - 1 = ny$ for some integer y. This implies that $mx - ny = 1$ and hence 1 is linearly expressible in terms of m and n. Therefore $(m, n) = 1$ completing the proof. Do you see why? (see the exercises)

$\square$

4.4 The Fundamental Theorem of Arithmetic

In this section we prove the fundamental theorem of arithmetic which is really the most basic number theoretic result. This results says that any integer $n > 1$ can be decomposed into prime factors in essentially a unique manner. First we show that there always exists such a decomposition into prime factors. Here we use the form of mathematical induction called course of values induction.

Lemma 4.4.1. *Any integer $n > 1$ can be expressed as a product of primes, perhaps with only one factor.*

Proof. The proof is by course of values induction (see Chapter 1). $n = 2$ is prime so it's true at the lowest level. Suppose that any integer $k < n$ can be decomposed into prime factors, we must show that n then also has a prime factorization.

If n is prime then we are done. Suppose then that n is composite. Hence $n = m_1 m_2$ with $1 < m_1 < n, 1 < m_2 < n$. By the inductive hypothesis both m_1 and m_2 can be expressed as products of primes. Therefore n can also be expressed as a product of primes using the primes from m_1 and m_2. This completes the proof. $\square$

Before we continue to the fundamental theorem we mention that this result can be used to prove that the set of primes is infinite. The proof we give goes back to Euclid and is quite straightforward. There are many proofs of this result. For a survey of these see [FR].

Theorem 4.4.1. *There are infinitely many primes.*

Proof. Suppose that there are only finitely many primes $p_1, \ldots, p_n$. Each of these is positive so we can form the positive integer

$$N = p_1 p_2 \cdots p_n + 1.$$

From Lemma 4.4.1, N has a prime decomposition. Now N itself is either prime or composite. But $N > p_i$ for all $i = 1, \ldots, n$. So if N were a prime, that would contradict the assumption that $\{p_1, \ldots, p_n\}$ contains all the primes. So N must be composite. In particular then there is a prime p which divides N. Then

$$p | (p_1 p_2 \cdots p_n + 1).$$

Since again it is assumed that $\{p_1, p_2, \ldots, p_n\}$ contains all the primes, then it follows that $p = p_i$ for some $i = 1, .., n$. But then $p | p_1 p_2 .. p_i \cdots p_n$ so p cannot divide $p_1 \cdots p_n + 1$ which is a contradiction. Therefore p is not one of the given primes showing that the list of primes must be endless. □

A variation of Euclid's argument gives the following proof of Theorem 4.4.1. Suppose there are only finitely many primes $p_1, \ldots, p_n$. Certainly $n \geq 2$. Let $P = \{p_1, \ldots, p_n\}$. Divide P into two disjoint nonempty subsets P_1, P_2. Now consider the number $m = q_1 + q_2$ where q_i is a product of primes from P_1 and q_2 is a product of primes from P_2. Let p be a prime divisor of m. Since $p \in P$ it follows that p divides either q_1 or q_2 but not both. But then p does not divide m, a contradiction. Therefore p is not one of the given primes and the number of primes must be infinite.

Although there are infinitely many primes, a glance at the list of primes shows that they appear to become scarcer as the integers get larger. If we let

$$\pi(x) = \text{ number of primes } \leq x$$

a basic question concerns the asymptotic behavior of this function. This question is the basis of the famous prime number theorem. We refer the reader to [FR] for a description and discussion of this famous result. It is easy, however, to show that there are arbitrarily large spaces or gaps within the set of primes.

Theorem 4.4.2. *Given any positive integer k there exists k consecutive composite integers.*

Proof. Consider the sequence

$$(k+1)! + 2, (k+1)! + 3, \ldots, (k+1)! + k + 1.$$

Suppose n is an integer with $2 \le n \le k + 1$. Then $n | ((k+1)! + n)$. Hence each of the integers in the above sequence is composite.   □

To show the uniqueness of the prime decomposition we need Euclid's lemma, from the previous section, applied to primes.

Lemma 4.4.2. *(Euclid's Lemma) If p is a prime and $p|ab$ then $p|a$ or $p|b$.*

Proof. Suppose $p|ab$. If p does not divide a then a and p must be relatively prime, that is, $(a, p) = 1$ because the only divisors of p are ± 1 and $\pm p$. Then from Lemma 4.3.4, $p|b$.   □

We also need the following extension of Euclid's lemma (see the exercises for a proof).

Corollary 4.4.1. *Let p be a prime and k an arbitrary positive integer. If $p|a_1 a_2 \cdots a_k$ where $a_i \in \mathbb{Z}$ then $p|a_i$ for at least one i with $1 \le i \le k$.*

We now state and prove the **fundamental theorem of arithmetic**.

Theorem 4.4.3. *(The Fundamental Theorem of Arithmetic) Given any integer $n \ne 0$ there is a factorization*

$$n = c p_1 p_2 \cdots p_k$$

where $c = \pm 1$ and $p_1, \ldots, p_n$ are primes. Further, this factorization is unique up to the ordering of the factors. Here we allow the cases where there are no primes, i.e., $k = 0$, or one prime, i.e., $k = 1$.

Proof. We assume that $n \geq 1$. If $n \leq -1$ we use $c = -1$ and the proof is the same. The statement certainly holds for $n = 1$ with $k = 0$. Now suppose $n > 1$. From Lemma 4.4.1, n has a prime decomposition

$$n = p_1 p_2 \cdots p_m.$$

We must show that this is unique up to the ordering of the factors. Suppose then that n has another such factorization $n = q_1 q_2 \cdots q_k$ with the q_i all prime. We must show that $m = k$ and that the primes are the same. Now we have

$$n = p_1 p_2 \cdots p_m = q_1 q_2 \cdots q_k.$$

Without loss of generality we may assume that $k \geq m$. From

$$n = p_1 p_2 \cdots p_m = q_1 q_2 \cdots q_k$$

it follows that $p_1 | q_1 q_2 \cdots q_k$. From Corollary 4.4.1 then we must have that $p_1 | q_i$ for some i. But q_i is prime and $p_1 > 1$ so it follows that $p_1 = q_i$. Therefore we can eliminate p_1 and q_i from both sides of the factorization to obtain

$$p_2 \cdots p_m = q_1 \cdots q_{i-1} q_{i+1} \cdots q_k.$$

Continuing in this manner we can eliminate all the p_i from the left side of the factorization to obtain

$$1 = q_{m+1} \cdots q_k.$$

If $q_{m+1}, \ldots, q_k$ were primes this would be impossible. Therefore $m = k$ and each prime p_i was included in the primes $q_1, \ldots, q_m$. Therefore the factorizations differ only in the order of the factors, proving the theorem. $\square$

For any positive integer $n > 1$ we can combine all the same primes to write

$$n = p_1^{m_1} p_2^{m_2} \cdots p_k^{m_k} \text{ with } p_1 < p_2 < \cdots < p_k \text{ and all } m_i > 0.$$

This is called the **standard prime decomposition**. Note that given any two positive integers a, b we can always write the prime decomposition with the **same** primes by allowing zero exponents. Note, however, once we do this we lose the uniqueness of the standard prime decomposition.

There are several easy consequences of the fundamental theorem

Theorem 4.4.4. *Let a, b be positive integers > 1. Suppose*

$$a = p_1^{e_1} \cdots p_k^{e_k}$$

$$b = p_1^{f_1} \cdots p_k^{f_k}$$

where we include zero exponents for noncommon primes. Then

$$(a, b) = p_1^{min(e_1, f_1)} \cdot p_2^{min(e_2, f_2)} \cdots p_k^{min(e_k, f_k)},$$

$$[a, b] = p_1^{max(e_1, f_1)} \cdot p_2^{max(e_2, f_2)} \cdots p_k^{max(e_k, f_k)}.$$

Corollary 4.4.2. *Let a, b be positive integers > 1, then $(a, b)[a, b] = ab$.*

We leave the proofs to the exercises but give an example.

EXAMPLE 4.4.1. Find the standard prime decompositions of 270 and 2412 and use them to find the GCD and LCM.

Recall that we found the GCD and LCM of these numbers in the previous section using the Euclidean algorithm. We note that in general it is very difficult as the size gets larger to determine the actual prime decomposition or even whether it is a prime or not.

To find the prime decomposition we factor and then continue refactoring until there are only prime factors.

$$270 = (27)(10) = 3^3 \cdot 2 \cdot 5 = 2 \cdot 3^3 \cdot 5$$

which is the standard prime decomposition of 270.

$$2412 = 4 \cdot 603 = 4 \cdot 3 \cdot 201 = 4 \cdot 3 \cdot 3 \cdot 67 = 2^2 \cdot 3^2 \cdot 67$$

which is the standard prime decomposition of 2412. Hence we have

$$270 = 2 \cdot 3^3 \cdot 5 \cdot 67^0$$

$$2412 = 2^2 \cdot 3^2 \cdot 5^0 \cdot 67$$

$$\implies (a, b) = 2 \cdot 3^2 \cdot 5^0 \cdot 67^0 = 2 \cdot 3^2 = 18$$

and

$$[a, b] = 2^2 \cdot 3^3 \cdot 5 \cdot 67 = 36180.$$

Note that the fundamental theorem of arithmetic can be extended to the rational numbers. Suppose $r = \frac{a}{b}$ is a positive rational where a and b are positive integers. Then

$$r = \frac{p_1^{e_1} \cdots p_k^{e_k}}{p_1^{f_1} \cdots p_k^{f_k}} = p_1^{e_1 - f_1} \cdots p_k^{e_k - f_k}$$

Therefore we have the following theorem.

Theorem 4.4.5. *Any positive rational has a prime decomposition*

$$p_1^{t_1} \cdots p_k^{t_k} \text{ where } t_1, \ldots, t_k \text{ are integers.}$$

For example

$$\frac{15}{49} = 3 \cdot 5 \cdot 7^{-2}.$$

This then has the following interesting consequence.

Lemma 4.4.3. *If a is a positive integer which is not a perfect nth power then the nth-root of a is irrational.*

Proof. This result says for example that if an integer is not a perfect square then its square root is irrational. The fact that the square root of 2 is irrational was known to the Greeks.

Suppose b is an integer with standard prime decomposition

$$b = p_1^{e_1} \cdots p_k^{e_k}.$$

Then

$$b^n = p_1^{n e_1} \cdots p_k^{n e_k}$$

and this must be the standard prime decomposition for b^n. It follows that an integer a is an nth power if and only if it has a standard prime decomposition

$$a = q_1^{f_1} \cdots q_t^{f_t} \text{ with } n | f_i \text{ for all } i.$$

Suppose a is not an nth power then

$$a = q_1^{f_1} \cdots q_t^{f_t}$$

where n does not divide f_i for some i. Suppose that a has an nth root which is a positive rational number. Call it $a^{1/n}$. Then by Theorem 4.4.5

$$a^{1/n} = q_1^{e_1} q_2^{e_2} \cdots q_t^{e_t}$$

where the q_i are distinct primes and $e_i \in \mathbb{Z}$. But then

$$a = (a^{1/n})^n = q_1^{ne_1} q_2^{ne_2} \cdots q_t^{ne_t}$$

is an integer. None of the ne_i can be negative since the q_i are distinct and a has a prime decomposition in which all of the exponents are divisible by n. This forces a to be a perfect nth power, contradicting our assumption. Hence $a^{1/n}$ cannot be rational.   □

4.5   Congruences and Modular Arithmetic

In Section 3.3 we defined the relation of congruence modulo n for a natural number n on the integers $\mathbb{Z}$ and showed how this could be used to construct the modular ring $\mathbb{Z}_n$.

If $x \equiv y \pmod{n}$ then y is called a **residue** of x modulo n. Given $x \in \mathbb{Z}$ the set of integers $\{y \in \mathbb{Z} : x \equiv y(\bmod \text{ n})\}$, that is, the equivalence class of x, is called the **residue class** for x modulo n. We denote this class by $[x]$. In Section 3.3 we saw that each integer x falls into exactly one of the residue classes $[0], \ldots, [n-1]$.

In this section we use the number theory that we have developed to prove some algebraic results about $\mathbb{Z}_n$ and then show how to do some basic arithmetic in these rings.

First we restate the following result and prove it using the number theory that we have developed in this chapter.

Theorem 4.5.1. *(1) $\mathbb{Z}_n$ is an integral domain if and only if n is a prime.*
   (2) $\mathbb{Z}_n$ is a field if and only if n is a prime.

Proof. Since $\mathbb{Z}_n$ is a commutative ring with an identity for any n it will be an integral domain if and only if it has no zero divisors.

Suppose first that n is a prime and suppose that $ab = 0$ in $\mathbb{Z}_n$. Then in $\mathbb{Z}$ we have

$$ab \equiv 0 \pmod{n} \implies n|ab.$$

Since n is prime, by Euclid's lemma 4.4.2 $n|a$ or $n|b$. In terms of congruences then

$$a \equiv 0 \pmod{n} \implies a = 0 \text{ in } \mathbb{Z}_n \text{ or } b \equiv 0 \pmod{n} \implies b = 0 \text{ in } \mathbb{Z}_n.$$

Thus there are no zero divisors in $\mathbb{Z}_n$. Therefore $\mathbb{Z}_n$ is an integral domain if n is prime.

Suppose n is not prime. Then $n = m_1 m_2$ with $1 < m_1 < n, 1 < m_2 < n$. Then $n \nmid m_1, n \nmid m_2$ but $n | m_1 m_2$. Translating this into $\mathbb{Z}_n$ we have

$$m_1 m_2 = 0 \text{ but } m_1 \neq 0 \text{ and } m_2 \neq 0.$$

Therefore $\mathbb{Z}_n$ is not an integral domain if n is not prime. These prove part (1).

Since a field is an integral domain, $\mathbb{Z}_n$ cannot be a field unless n is prime. This shows that if $\mathbb{Z}_n$ is a field, then n must be a prime. To complete part (2) we must show that if n is prime then $\mathbb{Z}_n$ is a field. Suppose n is prime, since $\mathbb{Z}_n$ is a commutative ring with identity to show that its a field we must show that each nonzero element has a multiplicative inverse.

Suppose $a \in \mathbb{Z}_n, a \neq 0$. Then in $\mathbb{Z}$ we have $n \nmid a$ and hence since n is prime $(a, n) = 1$. Therefore in $\mathbb{Z}$ there exists x, y such that $ax + ny = 1$. In terms of congruences this says that

$$ax \equiv 1 \pmod{n}$$

or in $\mathbb{Z}_n$,

$$ax = 1.$$

Therefore a has an inverse in $\mathbb{Z}_n$, namely, the residue class $[x]$ of x, and hence $\mathbb{Z}_n$ is a field.

$\square$

Notice that in the last chapter we proved that if p is a prime then $\mathbb{Z}_p$ is a field by appealing to the fact that any finite integral domain is a field. Here we proved that $\mathbb{Z}_p$ is a field directly by showing that each nonzero element is a unit, that is, has a multiplicative inverse. We actually proved more.

Lemma 4.5.1. *Let $n > 1$ be any natural number. Then a (or more precisely $[a]$, the residue class of a) is a unit in $\mathbb{Z}_n$ if and only if $(a, n) = 1$.*

Proof. Suppose that $(a, n) = 1$. We identify a with its residue class $[a]$ in $\mathbb{Z}_n$. Then we can write 1 as a linear combination of a and n and hence there exists $x, y \in \mathbb{Z}$ such that

$$ax + ny = 1.$$

Now $[ny] = 0$ since it is a multiple of n and hence considering this equation in $\mathbb{Z}_n$ we then have

$$[a][x] = 1.$$

Therefore $[a]$ is a unit and $[x]$ is its inverse.

Conversely suppose that a is a unit in $\mathbb{Z}_n$. Then there is an integer x with $[a][x] = 1$ in $\mathbb{Z}_n$. This implies that

$$ax \equiv 1 \bmod n \implies ax = 1 + ny \text{ for some integer } y.$$

Then

$$ax - ny = 1$$

and hence 1 can be written as a linear combination of a and n and therefore by Lemma 4.3.1, $(a, n) = 1$. $\square$

The proof of the last lemma actually indicates a method to find the multiplicative inverse of an element modulo n if it exists. Suppose n is a natural number and $(a, n) = 1$. Use the Euclidean algorithm in $\mathbb{Z}$ to express 1 as a linear combination of a and n, that is,

$$ax + ny = 1.$$

The residue class for x will be the multiplicative inverse of a.

EXAMPLE 4.5.1. Find 6^{-1} in $\mathbb{Z}_{11}$.

Using the Euclidean algorithm

$$11 = 1 \cdot 6 + 5,$$

$$6 = 1 \cdot 5 + 1$$

$$\implies 1 = 6 - (1 \cdot 5) = 6 - (1 \cdot (11 - 1 \cdot 6)) \implies 1 = 2 \cdot 6 - 1 \cdot 11.$$

Therefore the inverse of 6 modulo 11 is 2, that is, in $\mathbb{Z}_{11}$, $6^{-1} = 2$.

EXAMPLE 4.5.2. Solve the linear equation

$$6x + 3 = 1$$

in $\mathbb{Z}_{11}$.

Using purely formal field algebra the solution is

$$x = 6^{-1} \cdot (1 - 3).$$

In $\mathbb{Z}_{11}$ we have

$$1 - 3 = -2 = 9 \text{ and } 6^{-1} = 2 \implies x = 2 \cdot 9 = 18 = 7$$

Therefore the solution in $\mathbb{Z}_{11}$ is $x = 7$. A quick check shows that

$$6 \cdot 7 + 3 = 42 + 3 = 45 = 1 \text{ in } \mathbb{Z}_{11}.$$

A linear equation in $\mathbb{Z}_{11}$ is called a **linear congruence** modulo 11.

The fact that $\mathbb{Z}_p$ is a field for p a prime leads to the following nice result known as **Wilson's theorem**.

Theorem 4.5.2. *(Wilson's Theorem) If p is a prime then*

$$(p - 1)! \equiv -1 \ (mod \ p).$$

Proof. Now $(p - 1)! = (p - 1)(p - 2) \cdots 1$. Since $\mathbb{Z}_p$ is a field each $x \in \{1, 2, \ldots, p - 1\}$ has a multiplicative inverse modulo p. Further suppose $x = x^{-1}$ in $\mathbb{Z}_p$. Then $x^2 = 1$ which implies $(x - 1)(x + 1) = 0$ in $\mathbb{Z}_p$ and hence either $x = 1$ or $x = -1$ since $\mathbb{Z}_p$ is an integral domain. Therefore in $\mathbb{Z}_p$ only $1, -1$ are their own multiplicative inverses. Further $-1 = p - 1$ since $p - 1 \equiv -1 \pmod{p}$.

Hence in the product $(p - 1)(p - 2) \cdots 1$ considered in the field $\mathbb{Z}_p$ each element is paired up with its distinct multiplicative inverse except 1 and $p - 1$. Further the product of each with its inverse is 1. Therefore in $\mathbb{Z}_p$ we have $(p - 1)(p - 2) \cdots 1 = p - 1$. Written as a congruence then

$$(p - 1)! \equiv p - 1 \equiv -1 (\text{mod } p).$$

$\square$

The converse of Wilson's theorem is also true, that is, if $(n - 1)! \equiv -1 \pmod{n}$, then n must be a prime.

Theorem 4.5.3. *If $n > 1$ is a natural number and*

$$(n - 1)! \equiv -1 \ (mod \ n)$$

then n is a prime.

Proof. Suppose $(n-1)! \equiv -1 \bmod n$. If n were composite then $n = mk$ with $1 < m < n-1$ and $1 < k < n-1$. If $m \neq k$ both m and k are included in $(n-1)!$. It follows that $(n-1)!$ is divisible by n so that $(n-1)! \equiv 0 \bmod n$ contradicting the assertion that $(n-1)! \equiv -1 \bmod n$. If $m = k$ so that $n = k^2$ and $2k \leq n-1$. Hence both k and $2k$ are included in $(n-1)!$ and so $(n-1)! \equiv 0 \bmod n$, again a contradiction. Therefore n must be prime.   □

If a is a unit in $\mathbb{Z}_n$ then a linear equation

$$ax + b = c$$

can always be solved with a unique solution given by $x = a^{-1}(c - b)$. Determining this solution uses the same technique as in $\mathbb{Z}_p$ with p a prime. If a is not a unit the situation is more complicated.

EXAMPLE 4.5.3. Solve $5x + 4 = 2$ in $\mathbb{Z}_6$.

Since $(5,6) = 1$, 5 is a unit in $\mathbb{Z}_6$. Therefore $x = 5^{-1}(2 - 4)$. Now $2 - 4 = -2 = 4$ in$\mathbb{Z}_6$. Further $5 = -1$ so $5^{-1} = -1^{-1} = -1$. Then we have

$$x = 5^{-1}(2 - 4) = -1(4) = -4 = 2$$

Thus the unique solution in $\mathbb{Z}_6$ is $x = 2$.

Since an element a is a unit in $\mathbb{Z}_n$ if and only if $(a, n) = 1$ by Lemma 4.5.1, it follows that the number of units in $\mathbb{Z}_n$ is equal to the number of positive integers less than or equal to n and relatively prime to n. This number is given by what is called the **Euler Phi Function**.

Definition 4.5.1. *For any $n > 0$,*

$$\phi(n) = \text{ number of positive integers less than or equal to } n$$

and relatively prime to n.

EXAMPLE 4.5.4. $\phi(6) = 2$ since among $1, 2, 3, 4, 5, 6$ only $1, 5$ are relatively prime to 6.

The next result follows immediately from our characterization of units.

Lemma 4.5.2. *The number of units in $\mathbb{Z}_n$, which is the order of the unit group $U(\mathbb{Z}_n)$, is $\phi(n)$.*

Definition 4.5.2. *Given $n > 0$ a **reduced residue system** modulo n is a set of integers $x_1, \ldots, x_k$ such that each x_i is relatively prime to n, $x_i \neq x_j$ mod n unless $i = j$ and if $(x, n) = 1$ for some integer x then $x \equiv x_i$ mod n for some i.*

Hence a reduced residue system is a complete collection of representatives of those residue classes of integers relatively prime to n. Hence it is a complete collection of units (up to congruence modulo n) in $\mathbb{Z}_n$. It follows that any reduced residue system modulo n has $\phi(n)$ elements.

EXAMPLE 4.5.5. A reduced residue system modulo 6 would be $\{1, 5\}$.

We now develop a formula for $\phi(n)$. We first determine a formula for prime powers and then paste them back together via the fundamental theorem of arithmetic.

Lemma 4.5.3. *For any prime p and $m > 0$,*

$$\phi(p^m) = p^m - p^{m-1} = p^m\left(1 - \frac{1}{p}\right).$$

Proof. Recall that if $1 \le a \le p - 1$ then $\gcd(a, p) = 1$. Now let's list the numbers from 1 to p^m

$$1, 2, \ldots, p-1, p, p+1, p+2, \ldots, p+(p-1), 2p, 2p+1, \ldots, 2p+(p-1), 3p, \ldots, p^{m-1} \ldots$$

$$p^{m-2}p, \ldots, p^{m-1} \cdot p.$$

Thus we note from Lemma 4.3.3 that for $1 \le a \le p - 1$, $1 = \gcd(p, a) = \gcd(p, a + p)$ where we have taken $c = 1$ in that lemma. But if we take $c = 2$ in Lemma 4.3.3 again with $1 \le a \le p - 1$, we get that $1 = \gcd(p, a) = \gcd(p, a + 2p)$. Clearly, we can continue this to find that the positive integers less than or equal to p^m which are not relatively prime to p^m are precisely the multiples of p, that is, $p, 2p, 3p, \ldots, p^{m-1}p$. All other positive $a < p^m$ are relatively prime to p^m. Hence the number relatively prime to p^m is

$$p^m - p^{m-1}.$$

$\square$

Lemma 4.5.4. *If $(a, b) = 1$ then $\phi(ab) = \phi(a)\phi(b)$.*

Proof. Let $R_a = \{x_1, \ldots, x_{\phi(a)}\}$ be a reduced residue system modulo a, $R_b = \{y_1, \ldots, y_{\phi(b)}\}$ be a reduced residue system modulo b, and let

$$S = \{ay_i + bx_j : i = 1, \ldots, \phi(b), j = 1, \ldots, \phi(a)\}.$$

We claim that S is a reduced residue system modulo ab. Since S has $\phi(a)\phi(b)$ elements it will follow that $\phi(ab) = \phi(a)\phi(b)$.

To show that S is a reduced residue system modulo ab we must show three things: first that each $x \in S$ is relatively prime to ab; second that the elements of S are distinct; and finally that given any integer n with $(n, ab) = 1$ then $n \equiv s \bmod ab$ for some $s \in S$.

Let $x = ay_i + bx_j$. Then since $(x_j, a) = 1$ and $(a, b) = 1$ it follows that $(x, a) = 1$. To see that note that saying two numbers are relatively prime really means that no prime divides both of them. Suppose that there was a prime p such that $p \mid x$ and $p \mid a$. Since $bx_j = x - ay_i$, this implies that $p \mid bx_j$. But then according to Euclid's lemma $p \mid b$ or $p \mid x_j$. Now note that either of these is impossible because if $p \mid b$, we have already assumed that $p \mid a$. But $(a, b) = 1$ so that is impossible. Similarly if $p \mid x_j$, we have also already assumed that $p \mid a$. But $(x_j, a) = 1$, so this is impossible too. This means that no prime can divide both x and a. Thus $(x, a) = 1$. Analogously $(x, b) = 1$. We leave this to the exercises.

Since x is relatively prime to both a and b we have $(x, ab) = 1$. Again to see this we need Euclid's lemma. Suppose there was a prime p such that $p \mid x$ and $p \mid ab$. Then from Euclid's lemma 4.4.3, $p \mid a$ or $p \mid b$. But both of these are impossible because we have already seen that $1 = (x, a) = (x, b)$. This shows that each element of S is relatively prime to ab.

Next suppose that

$$ay_i + bx_j \equiv ay_k + bx_l (\bmod\ ab).$$

Then

$$ab \mid ((ay_i + bx_j) - (ay_k + bx_l)) \implies ab \mid ((ay_i - ay_k) + (bx_j - bx_l)) \quad (*)$$

But if $ab \mid z$ where z is any integer, then $b \mid z$. (Why?) This implies from the right hand side of the display that $b \mid ((ay_i - ay_k) + (bx_j - bx_l))$. But then $b \mid (ay_i - ay_k)$ (Why?) But this means

$$ay_i \equiv ay_k (\bmod\ b).$$

Since $(a, b) = 1$ it follows that $y_i \equiv y_k \bmod b$. (Why?) But then $y_i = y_k$. But this is impossible since R_b is a reduced residue system module b. We could have just as easily used (*) to show that $bx_j \equiv bx_l \pmod{a}$ and then from the fact that $(a, b) = 1$ have found that $x_j \equiv x_l \pmod{a}$. (This is also left as an exercise.) So that $x_j = x_l$. But this is also impossible since R_a is a reduced residue system module a. This shows that the elements of S are distinct modulo ab.

Finally suppose $(n, ab) = 1$. Since $(a, b) = 1$ there exist $x, y \in \mathbb{Z}$ with $ax + by = 1$. Then:

$$anx + bny = n.$$

Since $(x, b) = 1$ (this is true because 1 can be written as a linear combination of x and b and $(n, b) = 1$ (this is true because $(n, ab) = 1$ - why?), it follows that $(nx, b) = 1$. This is true because if there was a prime p such that $p \mid nx$ and $p \mid b$, then from Euclid's Lemma 4.4.3 $p \mid n$ or $p \mid x$. But both of these are impossible because, respectively, on the one hand $(n, b) = 1$ and on the other hand $(x, b) = 1$.

Thus $\gcd(nx, b) = 1$. Therefore there is a y_i with $nx = y_i + tb$ since $\{y_1, \dots, y_{\phi(b)}\}$ is a reduced residue system modulo b.

We can show in the same manner $(ny, a) = 1$ (we leave this an exercise) and so there is an x_j with $ny = x_j + ua$ because $\{x_1, \dots, x_{\phi(a)}\}$ is a reduced residue system modulo a. Then:

$$a(y_i + tb) + b(x_j + ua) = n \implies n = ay_i + bx_j + (t + u)ab$$

$$\implies n \equiv ay_i + bx_j \pmod{ab}$$

and we are done. $\qquad\qquad\qquad\qquad\qquad\qquad\qquad\qquad\qquad\qquad\qquad\square$

We now give the general formula for $\phi(n)$.

Theorem 4.5.4. *Suppose $n = p_1^{e_1} \cdots p_k^{e_k}$ then*

$$\phi(n) = (p_1^{e_1} - p_1^{e_1-1})(p_2^{e_2} - p_2^{e_2-1}) \cdots (p_k^{e_k} - p_k^{e_k-1}) = n \prod_i (1 - 1/p_i).$$

Proof. From the previous Lemmas 4.5.4 and 4.5.3, we have since $\gcd(p_i^{e_i}, p_j^{e_j}) = 1$ for $i \neq j$ (Why?)

$$\phi(n) = \phi(p_1^{e_1}).\phi(p_2^{e_2}) \cdots \phi(p_k^{e_k})$$

$$= (p_1^{e_1} - p_1^{e_1-1})(p_2^{e_2} - p_2^{e_2-1}) \cdots (p_k^{e_k} - p_k^{e_k-1})$$

$$= p_1^{e_1}(1 - 1/p_1) \cdots p_k^{e_k}(1 - 1/p_k) = p_1^{e_1} ... p_k^{e_k} . (1 - 1/p_1) \cdots (1 - 1/p_k)$$

$$= n \prod_i (1 - 1/p_i).$$

$\square$

EXAMPLE 4.5.6. Determine $\phi(126)$. Now

$$126 = 2 \cdot 3^2 \cdot 7 \implies \phi(126) = \phi(2)\phi(3^2)\phi(7) = (1)(3^2 - 3)(6) = 36.$$

Hence there are 36 units in $\mathbb{Z}_{126}$.

4.6   Unique Factorization Domains

Divisibility of integers was defined purely formally, that is, $a|b$ if there exists a c such that $b = ac$. This definition can be applied to any integral domain and hence divisibility can be considered in any such ring. In a similar manner primes and prime decompositions can be considered and studied within any integral domain. The integers have unique factorization into primes and so this raises the question as to whether the integers are the only integral domain with this property. The answer is no. There are many integral domains with unique factorization and these are called **unique factorization domains**.

Definition 4.6.1. *A **unique factorization domain** abbreviated **UFD** is an integral domain D such that each nonzero element of D is either a unit or has a decomposition into primes that is unique up to ordering and unit factors.*

In this language the fundamental theorem of arithmetic can be phrased in the following manner.

Theorem 4.6.1. *(Fundamental Theorem of Arithmetic) The integers $\mathbb{Z}$ are a unique factorization domain.*

In Chapter 12, when we study polynomial rings we will present and prove many examples of UFD's. The next theorem mentions without proof several examples of such domains. The proofs will be given in Chapter 12.

Theorem 4.6.2. *The following are all unique factorization domains:*
   (1) The ring of polynomials $F[x]$ over a field.
   (2) The ring of complex integers $\mathbb{Z} = \{a + bi; a, b \in \mathbb{Z}\}$
   (3) The polynomial ring $R[x]$ where R is a UFD. In particular $\mathbb{Z}$.

4.7 Exercises

EXERCISES FOR SECTION 4.2

4.2.1. There are 25 primes less than 100. List them.

4.2.2. For each of the following integers, find the smallest nonnegative integer to which it is congruent to modulo 9.
   (a) 702.
   (b) 12.
   (c) -12.
(HINT: This is the remainder in the division algorithm.)

4.2.3. Prove Theorem 4.2.1 (1).

4.2.4. Prove Theorem 4.2.1 (3).

4.2.5. Prove Theorem 4.2.1 (4).

4.2.6. Prove Theorem 4.2.1 (5).

4.2.7 Prove Theorem 4.2.1 (6).

4.2.8 Prove Theorem 4.2.1(7).

4.2.9. Prove Theorem 4.2.1(8).

4.2.10. Prove Theorem 4.2.1(9).

4.2.11. Prove that if $a, b, c, n \in \mathbb{Z}$ with $n > 0$, and $a \equiv b \bmod n$ and $n \mid a$, then $n \mid b$.

4.2.12. Suppose that $x, y, z, d \in \mathbb{Z}$ and that $x = y + z$. Show that if d divides any two of x, y and z, then d also divides the third.

4.2.13. Prove that if p and q are primes and $p \mid q$, then $p = q$.

EXERCISES FOR SECTION 4.3

4.3.1. For each of the following sets of numbers find the GCD and write the GCD as a linear combination of the two numbers.

(a) $\{26, 382\}$.

(b) $\{-36, 90\}$.

(c) $\{-1492, -1776\}$.

4.3.2. Prove Lemma 4.3.1; i.e., a, $b \in \mathbb{Z}$ are relatively prime if and only if 1 can be expressed as a linear combination of a and b.

4.3.3. Show for two nonzero integers a and b that $(a, b) = (-a, b) = (a, -b) = (-a, -b)$.

4.3.4. Show that the two nonzero integers a and b such that if $d = (a, b)$ and d is a linear combination of a and b then d is also a linear combination of $-a$ and b, of a and $-b$ and of $-a$ and $-b$.

4.3.5. Find the LCM for each set of numbers in Problem 4.3.1.

4.3.6. Prove that if $m \in \mathbb{N}$ and $a, b \in \mathbb{Z}$, then $\gcd(ma, mb) = m \gcd(a, b)$.

4.3.7. Use Problem 4.3.6 to show that if $m \in \mathbb{N}$, and $m \mid a$ and $m \mid b$, then

$$\gcd(\frac{a}{m}, \frac{b}{m}) = \frac{\gcd(a, b)}{m}.$$

4.3.8. Use Problem 4.3.7 to give an alternate proof of Lemma 4.3.2. That is, if $d = (a, b)$ and $a = a_1 d, b = b_1 d$ then $(a_1, b_1) = 1$.

4.3.9. In the text we showed how, given one integer solution to $ax + by = 1$ where $a, b \in \mathbb{Z}$ and $a \neq 0, b \neq 0$, we were able to find all other integer solutions in terms of these. Prove that the same formulas show that if we have one solution to $ax + by = d$ where $d > 1$ then we can get any other solution in terms of it.

4.3.10. (a) Define the GCD of three nonzero integers.

(b) Define the LCM of three nonzero integers.

4.3.11. Establish the existence of the GCD of three nonzero integers by proving a result similar to Theorem 4.3.1.

4.3.12. In the text Theorem 4.3.13 has $(a, b)[a, b] = \pm ab$. Explain where $\pm$ signs come up in the proof.

4.3.13. Consider the proof of Lemma 4.3.5. It shows that $(m, n) = 1$ if and only if $m\mathbb{Z}_n = \mathbb{Z}_n$. This is not exactly what is proven. Explain why what is proven is sufficient.

4.3.14. Establish the existence of the LCM of three nonzero integers by proving a result similar to Theorem 4.3.12. In particular if a, b, and c are nonzero integers then

$$\gcd(a, b, c) \cdot \operatorname{lcm}(a, b, c) = \pm abc.$$

(HINT: Let $d = \gcd(a, b, c)$ and let $m = \frac{abc}{d}$. We know that $m \in \mathbb{Z}$ (why?). But $a = a_0 d, b = b_0 d$, and $c = c_0 d$ where a_0, b_0, and c_0 are all integers. (Why?) Thus

$$m = a_1 b_1 c_1 d^2.$$

Thus $m = abc_1$. This implies that $a \mid m$ and $b \mid m$. Show in a similar way that $c \mid m$. But then $\mathrm{lcm}(a, b, c) \mid m$. (Why?) This implies that

$$\mathrm{lcm}(a, b, c) \gcd(a, b, c) \mid abc \qquad (*)$$

(Why?) Next consider $\frac{abc}{lcm(a,b,c)}$. This is an integer. (Why?) Also $a \mid \mathrm{lcm}(a, b, c)$. Thus we have $lcm(a, b, c) = ax$ for some $x \in \mathbb{Z}$. Hence

$$\frac{abc}{\mathrm{lcm}(a, b, c)} = \frac{abc}{ax} = \frac{bc}{x}$$

and so bc/x is integral. (Why?) But $\frac{bc}{x} \mid b$ and $\frac{bc}{x} \mid c$ (Why?) Thus $\frac{abc}{lcm(a,b,c)} \mid b$ and $\frac{abc}{lcm(a,b,c)} \mid c$. Similarly show that $\frac{abc}{lcm(a,b,c)} \mid a$. This means that $\frac{abc}{lcm(a,b,c)}$ integer is a common divisor of a, b, and c. This implies that $\gcd(a, b, c) \mid \frac{abc}{lcm(a,b,c)}$. But this in turn imples that

$$\gcd(a, b, c)\mathrm{lcm}(a, b, c) \mid abc \qquad (**)$$

(Why?) But (*) and (**) imply that $\gcd(a, b, c)\mathrm{lcm}(a, b, c) = \pm ab$. (Why?) This concludes the proof.)

EXERCISES FOR SECTION 4.4

For each of 4.4.1 – 4.4.3 find the standard prime decomposition for the given numbers. Also write prime factorizations using the same primes for each number. Then using that find their GCD and LCM.

4.4.1. 120 and 4851.

4.4.2. 970 and 3201.

4.4.3. 684 and 1375.

4.4.4. Prove Corollary 4.4.1.
(HINT: Use induction on k.)

4.4.5. Prove Theorem 4.4.4.

(HINT: First prove for a and b as in the theorem, $a|b$ if and only if $e_i \leq f_i$ for all i.)

4.4.6. Prove Corollary 4.4.2.

4.4.7. If you are given a single natural number n to be tested to see if it is a prime, then you can try dividing by the natural numbers $2, 3, 4, 5, \ldots$. But a much more economical way is to try dividing just by the primes less than or equal to $\sqrt{n}$. If none of these divides n, then n itself must be a prime. Explain why this primality test works.

4.4.8. Show that if a and b are natural numbers such that ab is a perfect square and $\gcd(a, b) = 1$, then both a and b are perfect squares.

(HINT: Use the standard prime decomposition of ab.)

4.4.9. Prove that there do not exist nonzero integers a and b such that $a^2 = 3b^2$.

(HINT: Use standard prime decompositions!)

EXERCISES FOR SECTION 4.5

4.5.1. Solve the congruence

$$2x \equiv 1(mod7).$$

4.5.2. Solve the congruence

$$3x + 5 \equiv 6x + 6(mod8).$$

4.5.3. Use the converse of Wilson's theorem to prove that 7 is a prime.

(NOTE: This is not a good primality test because even if the natural number, n, that you want to test is not too large, $(n-1)!$ can still be too large to easily handle, e.g., use the same method to prove that 23 is prime. A better primality test is contained in Problem 4.4.7)

4.5.4. Use Theorem 4.5.4 to determine the number of units in $\mathbb{Z}_{160}$.

4.5.5. Find a reduced residue system module n for each n:
   (a) $n = 4$.
   (b) $n = 15$.
   (c) $n = 24$.

4.5.6. Find the remainder in division of 2^{20} by 7 without a calculator.
(HINT: $2^3 \equiv 1(mod7) \implies (2^3)^6 \equiv 1(mod7)$.)

4.5.7. Compute $\varphi(97)$.

4.5.8. Compute $\varphi(492)$.

4.5.9. Compute $\varphi(1000)$

4.5.10. Prove using Theorem 4.30, that if $n > 2$, then $\varphi(n)$ is even.

4.5.11. Verify that

$$\{2, 2^2, 2^3, 2^4, \ldots, 2^{18}\}$$

is a complete residue system modulo 19.

Chapter 5

Fields: The Rationals, Reals and Complexes

5.1 Fields and Division Rings

In Chapter 3 we defined a field as a commutative ring with an identity in which each nonzero element has a multiplicative inverse. The rationals $\mathbb{Q}$ and the reals $\mathbb{R}$ are both fields and provide the initial motivation for these algebraic objects. In this chapter we look more closely at fields and then provide characterizations among fields of the rational numbers and the real numbers.

Definition 5.1.1. *A **field** F is a commutative ring with an identity such that every nonzero element has a multiplicative inverse. If F is a noncommutative ring with an identity in which every nonzero element has a multiplicative inverse then F is called a division ring or a skew field.*

As the primary examples of fields we have the basic number systems $\mathbb{Q}$ and $\mathbb{R}$.

Theorem 5.1.1. *The rationals $\mathbb{Q}$ and the reals $\mathbb{R}$ are fields.*

As with rings in general, two fields are algebraically the same if there is an isomorphism between them, that is, a ring homomorphism that is also a bijection. Clearly if two fields F_1 and F_2 are isomorphic then they must have the same cardinality or size (see Chapter 2). Since $\mathbb{Q}$ is countable while $\mathbb{R}$ is uncountable it follows that $\mathbb{Q}$ and $\mathbb{R}$ are nonisomorphic fields.

To provide an example of a skew field or division ring we introduce the **quaternions H**. (The reason for the H is that they are sometimes called Hamilton's quaternions.) We start with the real numbers $\mathbb{R}$ and we add three new elements $\mathbf{i}, \mathbf{j}$ and $\mathbf{k}$ satisfying the properties

$$\mathbf{i}^2 = \mathbf{j}^2 = \mathbf{k}^2 = -1 \text{ and } \mathbf{ijk} = -1. \qquad (*)$$

We also assume that $\mathbf{i}, \mathbf{j}$ and $\mathbf{k}$ commute with any real number relative to both $+$ and $\cdot$, satisfy the associative laws with respect to both addition and multiplication and satisfy both left and right distributive laws of multiplication over addition. Multiplying the second entry in $(*)$ by $\mathbf{k}$ on the right and using the associative law together with the fact that $\mathbf{i}, \mathbf{j}$ and $\mathbf{k}$ commute with -1, we get

$$\mathbf{ij}(\mathbf{k}^2) = -\mathbf{k} \implies -\mathbf{ij} = -\mathbf{k} \text{ or } \mathbf{ij} = \mathbf{k}.$$

Again multiplying both sides of the second entry in $(*)$ by $\mathbf{i}$ on the left hand side and using the associative law together with the fact that $\mathbf{i}, \mathbf{j}$, and $\mathbf{k}$ commute with -1, we get

$$\mathbf{i}^2(\mathbf{jk}) = -\mathbf{i} \implies -\mathbf{jk} = -\mathbf{i}.$$

Then multiplying on the left by $\mathbf{j}$, using associativity, and using the fact that -1 commutes with $\mathbf{j}$ gives

$$-\mathbf{j}^2\mathbf{k} = -\mathbf{ji} \implies \mathbf{k} = -\mathbf{ji}.$$

Thus we have shown that
$$\mathbf{ij} = \mathbf{k} = -\mathbf{ji}.$$

This shows that **H** is not commutative. We note further that since $\mathbb{R} \subset \mathbf{H}$ it follows that the quaternions are infinite.

In Section 5.6 we construct the complex numbers $\mathbb{C}$ by just considering the first element $\mathbf{i}$ but here we consider all three.

It can be shown in a similar manner, just using $(*)$ and the above assumed properties that

$$\mathbf{jk} = \mathbf{i} = -\mathbf{kj},$$
$$\mathbf{ki} = \mathbf{j} = -\mathbf{ik}.$$

We leave these verifications to the exercises.

Now we let **H** be the set defined by

$$\mathbf{H} = \{a + b\mathbf{i} + c\mathbf{j} + d\mathbf{k} : a, b, c, d \in \mathbb{R}\}.$$

For two elements $u, v \in \mathbf{H}$ where $u = a + b\mathbf{i} + c\mathbf{j} + d\mathbf{k}$ and $v = e + f\mathbf{i} + g\mathbf{j} + h\mathbf{k}$, we define $u = v$ if and only if $a = e, b = f, c = g$ and $d = h$. Thus a $u \in \mathbf{H}$ is nonzero if and only if at least one of its components is nonzero. On this set **H** we define addition and multiplication by algebraic manipulation assuming the properties as above, that is, we add like terms and multiply elements of **H** as if they were polynomials and combine terms. This is harder to describe than to actually do as we see in the next example.

EXAMPLE 5.1.1. Let u = 2 + 3**i** + 4**j** - **k**; v = 1 + **i** + **j** + 2**k**. Then

$$u + v = 3 + 4\mathbf{i} + 5\mathbf{j} + \mathbf{k}$$

just adding the corresponding terms. While

$$u \cdot v = 2 + 2\mathbf{i} + 2\mathbf{j} + 4\mathbf{k} + 3\mathbf{i} + 3\mathbf{ii} + 3\mathbf{ij} + 6\mathbf{ik} + 4\mathbf{j} + 4\mathbf{ji} + 4\mathbf{jj} + 8\mathbf{jk} - \mathbf{k} - \mathbf{ki} - \mathbf{kj} - 2\mathbf{kk}.$$

Using the properties of **i**, **j** and **k** in (*) and those mentioned above, and combining terms (see exercises) we get

$$u \cdot v = -3 + 14\mathbf{i} - \mathbf{j} + 2\mathbf{k}.$$

It is straightforward under these definitions and assumptions that **H** will form a (noncommutative) ring. We leave the details to the exercises. The real number 1 is the multiplicative identity. To show that **H** is a skew field we must show that every nonzero element has a multiplicative inverse. To do this suppose $u = a + b\mathbf{i} + c\mathbf{j} + d\mathbf{k} \neq 0$. Define $N(u) = a^2 + b^2 + c^2 + d^2$ and define $|u| = \sqrt{N(u)}$. Then if $u \neq 0$, then $|u|$ is a positive real number. Define $\overline{u} = a - b\mathbf{i} - c\mathbf{j} - d\mathbf{k}$. By a direct computation (see the exercises) we have that

$$u \cdot \overline{u} = \overline{u} \cdot u = N(u) = |u|^2.$$

Hence

$$u \cdot \frac{\overline{u}}{N(u)} = \frac{\overline{u}}{N(u)} \cdot u = 1.$$

Therefore u has a multiplicative inverse. Notice we needed u to be nonzero so that we could divide by $N(u)$. It follows that **H** is a skew field. The construction of the complex numbers that we will do in Section 5.6 is very similar to this construction. We have gone through this example quickly. The reader should come back and relook at the example after reading about the complex numbers.

Lemma 5.1.1. *The quaternions* **H** *are a skew field.*

We now look at some basic properties of fields, some of which we have already examined.

Theorem 5.1.2. *A field is an integral domain.*

Proof. A field F is by definition a commutative ring with an identity. We must then show that there are no zero divisors. Suppose that $ab = 0$ with $a, b \in F$ and $a \neq 0$. Then a has an inverse a^{-1} and hence $a^{-1}(ab) = a^{-1} \cdot 0 = 0$. But then $a^{-1}(ab) = (a^{-1}a)b = b = 0$ and therefore there are no divisors of zero. $\qquad\square$

Not every integral domain is a field. A very simple example is provided by the integers $\mathbb{Z}$. However, in the finite situation, fields and integral domains coincide.

Theorem 5.1.3. *Any finite integral domain D is a field. Hence a finite ring R is a field if and only if it is an integral domain.*

Proof. This was Theorem 3.5.1. We again present the proof. Let D be a finite integral domain and denote the elements of D by

$$D = \{d_1, \ldots, d_n\}$$

To show that D is a field we must show that every nonzero element of D is a unit, that is, has a multiplicative inverse. Suppose that $d \neq 0$ and consider the set $\{dd_1, \ldots, dd_n\}$ consisting of all products of the elements of D by the particular element d. By the cancellation law for multiplication in D we have that $dd_i = dd_j$ only if $i = j$ and hence the n products dd_i are all distinct and therefore the set $\{dd_1, \ldots, dd_n\}$ coincides with the set $\{d_1, \ldots, d_n\}$. The identity 1 is included among the $d_1, \ldots, d_n$ and so is included among the $dd, \ldots, dd_n$. It follows that there exists a d_j with $dd_j = 1$ and hence d has a multiplicative inverse and therefore D is a field. Note that we have implicitly used commutativity because for d_j to be the inverse of d we need $dd_j = d_j d = 1$. $\qquad\square$

We have seen that the modular rings $\mathbb{Z}_n$ are integral domains precisely when n is prime. Combining this with the theorem above we recover the result.

Corollary 5.1.1. *$\mathbb{Z}_n$ is a field if and only if n is a prime.*

The reader may wonder if there are any finite division rings which are not fields because the example we gave, that is, the quaternions, was infinite. The answer is no. This is the content of a theorem of Wedderburn which we state without proof since its proof is beyond the scope of this book.

Theorem 5.1.4. *(Wedderburn) A finite division ring is a field.*

Next we show a special property of field homomorphisms. Any nontrivial field homomorphism must be one-to-one. Here nontrivial means the homomorphism does not take all elements to zero.

Lemma 5.1.2. *If $f : F_1 \rightarrow F_2$ is a nontrivial homomorphism between fields then f is a monomorphism (f is one-to-one).*

Proof. Recall (Lemma 3.5.5) that if a ring has an identity $1 \neq 0$ and f is a nontrivial ring homomorphism then $f(1) = 1$ assuming that the codomain is an integral domain. Hence if $f : F_1 \rightarrow F_2$ is a nontrivial homomorphism between the fields F_1 and F_2 then $f(1) = 1$ where the left-hand 1 is the unity in F_1 and the right-hand 1 is the unity in F_2. Similarly $f(0) = 0$. We now show that f is one-to-one. Suppose that $f(x) = f(y)$ then $f(x - y) = 0$ since f is a ring homomorphism. Suppose that $x \neq y$ so that $x - y \neq 0$. Then $x - y$ has an inverse $(x - y)^{-1}$ so that $(x - y)(x - y)^{-1} = 1$. Then since f is a homomorphism

$$f(x - y)f((x - y)^{-1}) = f(1) = 1$$

and it follows that $f(x - y) \neq 0$ since if it were then it times anything in F_2 would be 0 and could not be $1 \neq 0$. Therefore if $f(x - y) = 0$ we must have $x - y = 0$ or $x = y$ and hence f must be one-to-one. $\qquad \square$

Finally we saw that an invertible element (i.e., an element $u \in R$ such that there exists a $u^{-1} \in R$ with $uu^{-1} = u^{-1}u = 1$) in a ring R is called a unit and further the set of all units in R denoted $U(R)$ forms a multiplicative group. In a field F every nonzero element is a unit and hence the set of nonzero elements of a field form a multiplicative group. This is called the multiplicative group of the field. If F is a field we denote this group by F^*.

5.2   Construction and Uniqueness of the Rationals

We have assumed that $\mathbb{Q}$ is a field. In this section we show how to formally construct the rational numbers from the integers in such a way that they become the smallest field containing $\mathbb{Z}$.

We start out with the set of ordered pairs of integers

$$\{(a, b) \in \mathbb{Z} \times Z : b \neq 0\}.$$

We should think of the ordered pair (a, b) as the fraction $\frac{a}{b}$. We call this set

$$\overline{\mathbb{Q}} = \{(a, b) \in \mathbb{Z} \times \mathbb{Z} : b \neq 0\}.$$

On $\overline{\mathbb{Q}}$, we define the relation $(a, b) \equiv (c, d)$ if and only if $ad = bc$. In terms of fractions this says that $\frac{a}{b} = \frac{c}{d}$ if and only if $ad = bc$. This is an equivalence relation on $\overline{\mathbb{Q}}$ (see the exercises).

Let $\frac{a}{b}$ stand for the equivalence class $[(a, b)]$ and let $\mathbb{Q}$ denote the set of all equivalence classes. Hence

$$\mathbb{Q} = \{\frac{a}{b} : (a, b) \in \overline{\mathbb{Q}}\}.$$

We define addition and multiplication on $\mathbb{Q}$. We define addition as follows.

$$\frac{a}{b} + \frac{c}{d} = \frac{ad + bc}{bd}.$$

Since $b \neq 0$ and $d \neq 0$ and $\mathbb{Z}$ has no divisors of zero, $bd \neq 0$ and so this is an element of $\overline{\mathbb{Q}}$ (in other words, the operation is closed on $\overline{\mathbb{Q}}$).

We must show first that this is well-defined, that is, if $\frac{a}{b} = \frac{a_1}{b_1}$ and $\frac{c}{d} = \frac{c_1}{d_1}$, then $\frac{a}{b} + \frac{c}{d} = \frac{a_1}{b_1} + \frac{c_1}{d_1}$. Now if $\frac{a}{b} = \frac{a_1}{b_1}$, then $ab_1 = a_1 b$. Similarly if $\frac{c}{d} = \frac{c_1}{d_1}$, then $cd_1 = c_1 d$. Then

$$\frac{a}{b} + \frac{c}{d} = \frac{ad + bc}{bd} \quad \text{and} \quad \frac{a_1}{b_1} + \frac{c_1}{d_1} = \frac{a_1 d_1 + b_1 c_1}{b_1 d_1}.$$

It follows that

$$
\begin{aligned}
(ad + bc)(b_1 d_1) &= adb_1 d_1 + bcb_1 d_1 = ab_1 dd_1 + cd_1 bb_1 \\
&= a_1 bdd_1 + c_1 dbb_1 = (a_1 d_1 + c_1 b_1)(bd)
\end{aligned}
$$

and hence

$$\frac{a}{b} + \frac{c}{d} = \frac{a_1}{b_1} + \frac{c_1}{d_1}.$$

Therefore addition is well-defined.

Next we define multiplication by

$$\frac{a}{b} \cdot \frac{c}{d} = \frac{ac}{bd}.$$

Just as before, since $b \neq 0$ and $d \neq 0$ and $\mathbb{Z}$ has no divisors of zero, $bd \neq 0$ and so this is an element of $\overline{\mathbb{Q}}$ (in other words, the operation is closed on $\overline{\mathbb{Q}}$). Further this is also well-defined. If $\frac{a}{b} = \frac{a_1}{b_1}$ and $\frac{c}{d} = \frac{c_1}{d_1}$, then

$$\frac{a}{b} \cdot \frac{c}{d} = \frac{ac}{bd} \quad \text{and} \quad \frac{a_1}{b_1} \cdot \frac{c_1}{d_1} = \frac{a_1 c_1}{b_1 d_1}.$$

If $\frac{a}{b} = \frac{a_1}{b_1}$, then $ab_1 = a_1 b$. Similarly if $\frac{c}{d} = \frac{c_1}{d_1}$, then $cd_1 = c_1 d$. Then

$$(ac)(b_1 d_1) = (ab_1)(cd_1) = (a_1 b)(c_1 d) = (a_1 c_1)(bd).$$

But this implies that

$$\frac{ac}{bd} = \frac{a_1 c_1}{b_1 d_1}.$$

This in turn means that

$$\frac{a}{b} \cdot \frac{c}{d} = \frac{a_1}{b_1} \cdot \frac{c_1}{d_1}.$$

Thus multiplication is well-defined.

Therefore we have an addition and multiplication on the set $\mathbb{Q}$. Further it is clear that this multiplication is both commutative and associative because multiplication on $\mathbb{Z}$ is commutative and associative. The element $\frac{0}{1}$ provides a zero element while $\frac{1}{1}$ is the multiplicative identity. For $\frac{a}{b} \in \mathbb{Q}$ its additive inverse is $\frac{-a}{b}$. If $\frac{a}{b} \neq 0$, then $a \neq 0$ and hence $\frac{b}{a}$ is defined: A quick check shows that $\frac{a}{b} \cdot \frac{b}{a} = \frac{ab}{ab} = \frac{1}{1} = 1$ and therefore each nonzero element has a multiplicative inverse. Straightforward computations show that the addition is associative and multiplication is distributive over addition. (See the exercises for verification of details for all of this.) Therefore the set $\mathbb{Q}$ under these operations forms a field. Hence we have the first part of the following theorem.

Theorem 5.2.1. *The rational numbers $\mathbb{Q}$, as constructed above, form a field. The set $\overline{\mathbb{Z}}$ of all elements of $\mathbb{Q}$ of the form $\frac{z}{1}$ for $z \in \mathbb{Z}$ is a subring of $\mathbb{Q}$*

which is isomorphic to $\mathbb{Z}$. In this sense the integers $\mathbb{Z}$ are a subring of $\mathbb{Q}$. Further if F is any subfield of $\mathbb{Q}$ satisfying $\overline{\mathbb{Z}} \subset F \subset \mathbb{Q}$ then $F = \mathbb{Q}$. In this sense, $\mathbb{Q}$ is the smallest subfield containing $\mathbb{Z}$.

Proof. The discussion prior to the theorem shows that $\mathbb{Q}$ is a field (see exercises for verifications of associativity of addition and distributivity). We call an element $\frac{a}{b}$ a fraction. To each integer $z \in \mathbb{Z}$ we can associate the fraction $\frac{z}{1}$ and let $\overline{\mathbb{Z}} = \{\frac{z}{1} : z \in \mathbb{Z}\} \subset \mathbb{Q}$. It is straightforward that the map $f : \mathbb{Z} \to \mathbb{Q}$ given by $z \longmapsto \frac{z}{1}$ is a monomorphism from the integers $\mathbb{Z}$ into $\mathbb{Q}$ (see the exercises to verify the details) and therefore $im(f) = \overline{\mathbb{Z}}$ is isomorphic to $\mathbb{Z}$. Identifying $\mathbb{Z}$ with its image, $\overline{\mathbb{Z}}$, we can say that $\mathbb{Z}$ is a subring of $\mathbb{Q}$.

Finally we must show that $\mathbb{Q}$ is the smallest field containing $\mathbb{Z}$, in the sense stated in the theorem. That is, if $\overline{\mathbb{Z}} \subset F$ with $F \subset \mathbb{Q}$ and F a field then $F = \mathbb{Q}$.

Let $\frac{a}{b} \in \mathbb{Q}$. Since $b \in \mathbb{Z}$ then $\frac{b}{1} \in \overline{\mathbb{Z}} \subset F$. But $\frac{a}{b} \in \mathbb{Q}$ implies $b \neq 0$ so since F is a field and $b \neq 0$ we have $\frac{b}{1} \neq 0$. This in turn implies that $\left(\frac{b}{1}\right)^{-1}$ exists and is an element of F. From the definition of multiplication in $\mathbb{Q}$ we have $\left(\frac{b}{1}\right)^{-1} = [(1, b)]$ because $\frac{b}{1}$ is the equivalence class $[(b, 1)]$ and $[(b, 1)][(1, b)] = [(b, b)] = [(1, 1)] = $ "1," i.e. the multiplicative identity of $\mathbb{Q}$. Therefore we can write $\left(\frac{b}{1}\right)^{-1} = \frac{1}{b} \in F$. If $a > 0$ then $\frac{a}{b} = \frac{1}{b} + \cdots + \frac{1}{b}$, a times and hence $\frac{a}{b} \in F$. If $a = 0$ then $\frac{a}{b} = 0$ in $\mathbb{Q}$ and since F is a subfield $\frac{a}{b} \in F$. Finally if $a < 0$ then $a = -c$ where c is a positive integer so by the above $\frac{c}{b} \in F$. Hence since F is a subfield $-\frac{c}{b} = \frac{-c}{b} = \frac{a}{b} \in F$. Therefore in every case $\frac{a}{b} \in F$. But $\frac{a}{b}$ was an arbitrary element of $\mathbb{Q}$ and hence $\mathbb{Q} \subset F$. Combined with $F \subset \mathbb{Q}$ we have $F = \mathbb{Q}$. $\qquad\square$

Actually even more is true. We have the following theorem.

Theorem 5.2.2. *Suppose that the field F contains $\mathbb{Z}$ and is the smallest field containing $\mathbb{Z}$ in the sense that if E is any other field satisfying $\mathbb{Z} \subset E \subset F$ then $E = F$. Then $F \cong \mathbb{Q}$.*

We note that the previous theorem was the special case where F was a subfield of $\mathbb{Q}$ containing the subring $\overline{\mathbb{Z}} \cong \mathbb{Z}$. Here F is any field containing $\mathbb{Z}$.

Proof. Suppose that F is a field and $\mathbb{Z} \subset F$. Let the equivalence class $[(a, b)] \in \mathbb{Q}$ be written as $\frac{a}{b} \in \mathbb{Q}$ Since $b \neq 0$ and $b \in Z$ we must have $b^{-1} \in F$. Map $\mathbb{Q}$ to F by $\varphi : \mathbb{Q} \to F$ where $\varphi(\frac{a}{b}) = b^{-1}a$. We claim this is a monomorphism and therefore $\mathbb{Q}$ embeds into F. Since $\varphi(\mathbb{Q})$ is a field and we claim that $\mathbb{Z} \subset \varphi(\mathbb{Q})$ then by the hypothesis that F is the smallest field containing $\mathbb{Z}$, $\mathbb{Z} \subset \varphi(\mathbb{Q}) \subset F$ implies that $\varphi(\mathbb{Q}) = F$. This would show that φ is actually an isomorphism.

To see this we proceed as follows. Since $\mathbb{Z} \subset F$ and $a \in \mathbb{Z}$, $b^{-1}a \in F$. So the map φ really does go from $\mathbb{Q}$ to F. Further since $\frac{a}{b}$ is an equivalence class, we must show that φ is well-defined, i.e., we need to show that if $\frac{a}{b} = \frac{a_1}{b_1}$, then $\varphi(\frac{a}{b}) = \varphi(\frac{a_1}{b_1})$. Now $\frac{a}{b} = \frac{a_1}{b_1}$ means that $(a,b) \equiv (a_1, b_1)$ or $ab_1 = ba_1$ where both b and b_1 are nonzero. Considering $ab_1 = ba_1$ as an equation in F, we have upon multiplying both sides by b_1^{-1} and then by b^{-1}

$$ab_1 b_1^{-1} = ba_1 b_1^{-1} \implies a = bb_1^{-1}a_1$$
$$b^{-1}a = b^{-1}bb_1^{-1}a_1 \implies b^{-1}a = b_1^{-1}a_1.$$

Note here we have used associativity and commutativity of multiplication in F. But this says that

$$\varphi\left(\frac{a}{b}\right) = b^{-1}a = b_1^{-1}a_1 = \varphi\left(\frac{a_1}{b_1}\right).$$

So we have shown that φ is well-defined.

Now we need to verify that φ is a ring homomorphism. Consider

$$\varphi\left(\frac{a}{b} + \frac{a_1}{b_1}\right) = \varphi\left(\frac{ab_1 + a_1 b}{bb_1}\right) = (bb_1)^{-1} \cdot (ab_1 + a_1 b).$$

Note that $(bb_1)^{-1}$ makes sense as an element of F since $b, b_1 \neq 0$ and F as a field has no divisors of zero so $bb_1 \neq 0$ in F. But $(bb_1)^{-1} = b_1^{-1}b^{-1}$ and using distributivity, we get

$$(bb_1)^{-1} \cdot (ab_1 + a_1 b) = ab^{-1}b_1^{-1}b_1 + a_1 b_1^{-1}b^{-1}b.$$

Here we have used associativity and commutativity of multiplication in F. So this gives that

$$\varphi\left(\frac{a}{b} + \frac{a_1}{b_1}\right) = b^{-1}a + b_1^{-1}a_1 = \varphi\left(\frac{a}{b}\right) + \varphi\left(\frac{a_1}{b_1}\right).$$

Thus φ preserves addition. We must also show that φ preserves multiplication, i.e.,

$$\varphi\left(\frac{a}{b} \cdot \frac{a_1}{b_1}\right) = \varphi\left(\frac{a}{b}\right) \cdot \varphi\left(\frac{a_1}{b_1}\right).$$

We leave this as an exercise.

Next we need to show that φ is injective. Suppose that $\varphi(\frac{a}{b}) = \varphi(\frac{a_1}{b_1})$ then $b^{-1}a = b_1^{-1}a_1$. Again considering this as an equation in F and using the appropriate properties of multiplication in F, we have after multiplying both sides by b and then by b_1 that

$$
\begin{aligned}
b \cdot b^{-1}a &= b \cdot b_1^{-1}a_1 \implies a = bb_1^{-1}a_1 \implies \\
b_1a &= bb_1b_1^{-1}a_1 \implies b_1a = ba_1.
\end{aligned}
$$

But this last equality means that $\frac{a}{b} = \frac{a_1}{b_1}$. So this proves that φ is 1-1.

As mentioned earlier in this proof we can show that φ is surjective, $\varphi(\mathbb{Q}) = F$, by showing that $\mathbb{Z} \subset \varphi(\mathbb{Q})$. But $\varphi(\mathbb{Q}) = \{b^{-1}a : \frac{a}{b} \in \mathbb{Q}\}$. Suppose that $a \in \mathbb{Z}$, then $\frac{a}{1} \in \mathbb{Q}$ and $a = 1^{-1}a \in \varphi(\mathbb{Q})$. Thus $\mathbb{Z} \subset \varphi(\mathbb{Q})$ and hence φ is onto. This then shows that $F \cong \varphi(\mathbb{Q})$, completing the proof. $\square$

The next result, stated as a corollary to Theorem 5.2.2, will be needed later on in this section. This is more of a corollary to the proof of Theorem 5.2.2 than to the statement of the theorem.

Corollary 5.2.1. *Suppose that F is a field that contains an isomorphic copy of $\mathbb{Z}$, that is, there exists a $\mathbb{Z}' \subset F$ with $\mathbb{Z}' \cong \mathbb{Z}$. Then there exists a $\mathbb{Q}' \subset F$ such that $\mathbb{Z}' \subset \mathbb{Q}' \subset F$ and $\mathbb{Q}' \cong \mathbb{Q}$.*

Proof. Let $\mathbb{Z}'$ be the isomorphic copy of $\mathbb{Z}$ in F. Let e be the multiplicative identity of F. Then

$$
\mathbb{Z}' = \{me : m \in \mathbb{Z}\}.
$$

Note here that $me = 0$ in F if and only if $m = 0$. Hence if $m \neq 0$ then $(me)^{-1}$ exists in F. Consider the set $\mathbb{Q}' \subset F$ defined by

$$
\mathbb{Q}' = \{(me)^{-1}(ne); m, n \in \mathbb{Z}, m \neq 0\}.
$$

Clearly, if we take $m = 1$, then $(1e)^{-1} = e^{-1} = e$. Hence $ne \in \mathbb{Q}'$ for all $n \in \mathbb{Z}$ and so $\mathbb{Z}' \subset \mathbb{Q}'$. Moreover we claim that the map $\varphi : \mathbb{Q} \to \mathbb{Q}'$ given by $\varphi(\frac{a}{b}) = (be)^{-1}(ae)$ is an isomorphism. This is proved in the same way that the map φ in the proof of Theorem 5.2.2 is shown to be an monomorphism (see the exercises). Here of course, φ by definition is clearly surjective. Thus we have $\mathbb{Z}' \subset \mathbb{Q}' \subset F$ and $\mathbb{Q}' \cong \mathbb{Q}$. $\square$

In Chapter 3, we saw that the integers can be characterized up to isomorphism as the only prime ring of characteristic zero. We can now characterize the rationals $\mathbb{Q}$ in an analogous way. First we recall what the characteristic is in terms of fields. Recall that a field always has an identity or unity with respect to multiplication.

The characteristic of a field F denoted $\operatorname{char}(F)$ is the smallest $n \in \mathbb{N}$ such that $n \cdot 1 = 0$. If no such n exists then the characteristic is zero.

It is clear that the characteristic of $\mathbb{Q}$ is 0. Further if p is a prime the characteristic of the finite field Z_p is p. In Lemma 3.6.3. we saw that the characteristic of an integral domain is either zero or a prime. Since a field is an integral domain we have the following.

Theorem 5.2.3. *The characteristic of a field is zero or a prime.*

If a field F has characteristic zero then it must be infinite (why?). Hence if F is a finite field its characteristic must be a prime p.

Lemma 5.2.1. *If F is a finite field then $char(F) = p$ for some prime p.*

There are infinite fields which have characteristic p. Later, after we discuss vector spaces, we will show that if F is a finite field then the size of F is p^n for some power of a prime p. Hence for example then can be no finite field of order 10.

Definition 5.2.1. *A field F is a **prime field** if its only nontrivial subfield is the field itself.*

Lemma 5.2.2. *The rationals $\mathbb{Q}$ and the finite fields Z_p are prime fields. The real number field $\mathbb{R}$ is not a prime field.*

Proof. Let $\overline{\mathbb{Z}} = \left\{ \frac{z}{1} : z \in \mathbb{Z} \right\} \subset \mathbb{Q}$. In the proof of Theorem 5.2.1, we showed that $\overline{\mathbb{Z}} \cong \mathbb{Z}$. Suppose that $F \subset \mathbb{Q}$ is any nontrivial subfield. Then $1 = \frac{1}{1} \in F$. Since F is closed, $\overline{\mathbb{Z}} \subset F$. But again by Theorem 5.2.1, $\mathbb{Q}$ is the smallest field containing $\overline{\mathbb{Z}}$. Thus $\mathbb{Q} \subset F$. We have just shown a double inclusion, i.e., $F = \mathbb{Q}$ This shows that $\mathbb{Q}$ is a prime field.

In a similar manner, we can show that $\mathbb{Z}_p$ is a prime field. For if $F \subset \mathbb{Z}_p$ is any nontrivial subfield, then $1 \in F$, and so $\{z \cdot 1 : z \in \mathbb{Z}\} \subset F$ since F is closed with respect to addition and inverses. But in $\mathbb{Z}_p$, $\{z \cdot 1 : z \in \mathbb{Z}\} = \mathbb{Z}_p$. Hence $\mathbb{Z}_p \subset F$ which shows that $F = \mathbb{Z}_p$ and thus $\mathbb{Z}_p$ is a prime field.

The reals, $\mathbb{R}$, are not a prime field because they contain $\mathbb{Q}$ as a nontrivial subfield and as noted earlier in this chapter $\mathbb{Q}$ is not isomorphic to $\mathbb{R}$. $\qquad \square$

The property of being a prime field together with characteristic zero completely characterizes $\mathbb{Q}$ up to isomorphism.

Theorem 5.2.4. *(Characterization of $\mathbb{Q}$) (i) If F is a prime field of characteristic zero then $F \cong \mathbb{Q}$.*

(ii) If F is a prime field of characteristic p then $F \cong \mathbb{Z}_p$.

Proof. Let F be a prime field of characteristic zero and let 1 be the multiplicative identity of F. Since the characteristic of F is zero the set $\mathbb{Z}' = \{z \cdot 1 : z \in \mathbb{Z}\}$ can be identified with the integers $\mathbb{Z}$ and hence F contains an isomorphic copy of $\mathbb{Z}$. Then by Corollary 5.2.1 there must exist a subfield $\mathbb{Q}' \subset F$ with $\mathbb{Q}' \cong \mathbb{Q}$. However, since F is a prime field it is equal to any nontrivial proper subfield. Thus $F = \mathbb{Q}'$ and therefore $F \cong \mathbb{Q}$.

If the characteristic is p then the set $\mathbb{Z}'_p = \{z \cdot 1 : z \in \mathbb{Z}\}$ can be identified with $\mathbb{Z}_p$. The map $\varphi : \mathbb{Z}'_p \to \mathbb{Z}_p$ given by $\varphi(z \cdot 1) = [z]$ where $[z]$ is the residue class of z in $\mathbb{Z}_p$ can be shown to be an isomorphism (see the exercises). Thus $\mathbb{Z}'_p$ is such $\mathbb{Z}'_p \subset F$ and $\mathbb{Z}'_p \cong \mathbb{Z}_p$. Since F is a prime field and $\mathbb{Z}'_p$ is a nontrivial subfield it follows that $\mathbb{Z}'_p = F$ and hence $F \cong \mathbb{Z}_p$. $\square$

5.2.1   Fields of Fractions

The construction of $\mathbb{Q}$ from $\mathbb{Z}$ was purely formal. That is, it didn't depend on the integers themselves but only on the fact that the integers are an integral domain. If we start with any integral domain D we can mimic the procedure and arrive at a field $\overline{D}$ that contains D and is the smallest field containing D. This field is called the **field of fractions** or the **field of quotients** for D. Formally let D be an integral domain. The field of fractions is constructed in an identical manner to the construction of $\mathbb{Q}$ from $\mathbb{Z}$. First we consider the subset of $D \times D$ given as

$$D^* = \{(d_1, d_2) \in D \times D : d_2 \neq 0\}.$$

Then on D^* we define the relation $(d_1, d_2) \equiv (c_1, c_2)$ if and only if $d_1 c_2 = d_2 c_1$. This can be shown to be an equivalence relation on D^* (see the exercises). We let $\frac{d_1}{d_2}$ stand for the equivalence class $[(d_1, d_2)]$ and let $\overline{D}$ be the set of equivalence classes

$$\overline{D} = \{\frac{d_1}{d_2} \; : \; (dl_1, d_2) \in D^*\}.$$

The operations on $\overline{D}$ are given, as in $\mathbb{Q}$, by

$$\frac{d_1}{d_2} + \frac{c_1}{c_2} = \frac{d_1 c_2 + c_1 d_2}{d_2 c_2}$$

$$\frac{d_1}{d_2} \cdot \frac{c_1}{c_2} = \frac{d_1 c_1}{d_2 c_2}$$

taken on the equivalence classes. That this set forms a field is identical to the proof for $\mathbb{Q}$. (The reader should go through the details just to make sure the construction is understood. This is left to the exercises.) Further, as for $\mathbb{Z}$ relative to $\mathbb{Q}$ we must have $\overline{D}$ is the smallest field containing D.

Hence we have proved the first two parts of the following.

Theorem 5.2.5. *Let D be an integral domain. Then there exists a field $\overline{D}$ containing D called the **field of fractions** or equivalently **field of quotients** for D. Further, $\overline{D}$ is the smallest field containing D. If F is a field then its field of fractions is F. Hence the field of fractions for any finite integral domain D is also D.*

Proof. Suppose that F is a field then the smallest field containing F must be F itself and hence F is its own field of fractions. A finite integral domain is already a field and hence its field of fractions is itself. □

5.3 The Real Number System

From the viewpoint of arithmetic and computation it appears that the rational numbers are sufficient since we can do all the basic arithmetic operations within the rational number system. However, as people began to do various computations it became clear that there were certain numbers that were very *real* in the sense that they could measure a distance but were not rational. We give an example of one of these. We saw in the last chapter that as a consequence of the fundamental theorem of arithmetic it followed that the square root of any natural number that is not a perfect square cannot be rational. Therefore $\sqrt{2}$ is not a rational number. However, if we consider a right triangle with sides one unit each in length then from the Pythagorean theorem the hypotenuse has length $\sqrt{2}$. Therefore $\sqrt{2}$ is *real* but not rational. Therefore we need a larger number system to handle measuring distances.

As a starting point we will take the real number system $\mathbb{R}$ to be all measurable distances and their negatives. This is only a rough description and in Section 5.3.2 we will give a formal construction of the reals starting with the rationals and show that they form a field. In this description in terms of distances it is clear, as is done in high school geometry, that the real number system can be identified with the points on a line. If we designate a single point as of zero length and lay off a unit distance to the right of it we can construct all the integers and hence $\mathbb{Z} \subset \mathbb{R}$. Since we can divide each unit interval into equal parts we can lay off all the rational numbers on the number line so that $\mathbb{Q} \subset \mathbb{R}$. By continual subdividing we can express each positive distance and hence each real number as a decimal expansion (see the exercises). That is, if $r \in \mathbb{R}$ then

$$r = \sum_{i=-\infty}^{n} a_i 10^i$$

where each a_i is an integer from $\{0, 1, 2, 3, 4, 5, 6, 7, 8, 9\}$. At this first step before we give a formal construction we define the **real number system** $\mathbb{R}$ as all decimal expansions.

There is one ambiguity that we must handle. Decimal expansions ending in an infinite string of 9's must be equated with an equivalent string ending in 0's. Hence the decimal expansion $.29999999\ldots$ must be equated with $.3000000\ldots$. With these ambiguities handled we can do all the standard arithmetic and the set of decimal expansions becomes a field. Historically the real numbers were constructed in this manner by Weierstrass in the middle of the nineteenth century.

Theorem 5.3.1. *The real numbers $\mathbb{R}$ form a field which contains the rationals as a subfield. Further, $\mathbb{R}$ is not isomorphic to $\mathbb{Q}$.*

Proof. The fact that $\mathbb{R}$ is not isomorphic to $\mathbb{Q}$ follows from a look at their cardinalities or sizes. As we saw, $\mathbb{Q}$ is countably infinite, while $\mathbb{R}$ is not. If they were isomorphic there would be a bijection between them and therefore they would be the same size or the same cardinality.   □

Since $\mathbb{Q} \subset \mathbb{R}$ the question arises as to how to recognize the elements of $\mathbb{Q}$ within $\mathbb{R}$, that is, what decimal expansions actually represent rational numbers.

Theorem 5.3.2. *The decimal form of any rational either terminates or repeats. Conversely every terminating or repeating decimal is rational.*

Proof. Note first that a terminating decimal such as .5 is equal to .50000...
and hence we can consider only repeating decimals. We have to show that
each rational number has a repeating decimal and conversely that each repeating decimal represents a rational number.

To motivate how the decimal form for a rational number becomes repeating consider how to find the decimal for the fraction $\frac{1}{7}$. The standard
method is to put zeros after the 1 to form 1.0000... and then divide this
by 7. If we do this we divide first 7 into 10 to get 1 with a remainder of 3.
Carrying down the zeros we next divide 7 into 30 to get 4 with a remainder
of 2. Continuing in this manner the first 6 digits are .142857 and the last
remainder is 1. Carrying down the zeros we next divide 7 into 10 and the
whole pattern repeats. Hence

$$1/7 = .142857142857\overline{142857}$$

Notice that we put an overline above the repeating part to indicate that it
continues indefinitely. The size of the smallest repeating part for 1/7 is 6
and this is called the **period**.

We now fashion this example into a proof. Suppose that $\frac{m}{n} \in \mathbb{Q}$ and we
may suppose that $\gcd(m, n) = 1$. Further, we may suppose that $m < n$
for otherwise we could pull out the integer part of this fraction. Now as in
the case of $\frac{1}{7}$ we form $m.0000...$ and divide this by n. In each division by
n from the division algorithm, the remainder is one of $0, 1, \ldots, n-1$. If we
get a remainder of 0 then the decimal terminates (and hence repeats with all
zeros). If not, then after at most $n-1$ steps we will hit a remainder that has
already occurred. The pattern from this point on then repeats. This shows
that we must have a repeating decimal for m/n. Further, we have shown that
the maximum size of the period is $n-1$, i.e., the size of the denominator
minus 1.

Conversely suppose that we have repeating decimal x. As above to motivate why this must be a rational number we do an example, Suppose that

$$x = .672672\overline{672}.$$

Then the period of this is 3. It follows that $10^3 x = 672 + x = 672.672672\ldots$..
Therefore

$$10^3 x - x = 672 \implies (10^3 - 1)x = 672 \implies x = \frac{672}{10^3 - 1} \in \mathbb{Q}.$$

The methods of this example can now be fashioned into a proof. Suppose that x is a repeating decimal. If $x > 1$ then $x = n + y$ where n is an integer and $0 < y < 1$ and y is also repeating. Therefore we may assume that $0 < x < 1$. Further, suppose that the repeating part starts k decimal places after the decimal point, for example $x = .15672672\overline{672}$ starts two decimal places after the decimal point. Then

$$10^k x = x_1 + y_1,$$

where x_1 is an integer and y_1 is repeating with the repeating part starting at the decimal point. In the example $10^2 x = 15 + .672672\dots$. Further, since x_1 is an integer it follows that $10^k x$ is rational if and only if y_1 is rational. Therefore we may assume that $0 < x < 1$ and the repeating part starts at the decimal point. Suppose that the period is n and suppose that the repeating part is the integer m. In the example this would be 672. Then as in the example we would have

$$10^n x = m + x \implies 10^n x - x = m \implies x = \frac{m}{10^n - 1} \in \mathbb{Q}.$$

Hence any repeating decimal represents a rational number. □

The real numbers $\mathbb{R}$ are an **ordered field** (see Section 3.4). That is, on $\mathbb{R}$ there is an ordering such that if $a, b \in \mathbb{R}$ then either $a < b$ or $a = b$ or $a > b$.

Lemma 5.3.1. *In any ordered field squares must be positive, that is, $x^2 > 0$ for all $x \neq 0$. In particular in $\mathbb{R}$ the equation $x^2 + 1 = 0$ has no solution.*

Proof. Let $x \in \mathbb{R}$ with $x \neq 0$. Then either $x > 0$ or $-x > 0$. If $x > 0$ then $x^2 = xx > 0$ since the positive elements are closed under multiplication. If $x < 0$ then $-x > 0$ and $x^2 = xx = (-x)(-x)$ by the laws of signs. But then $(-x)(-x) > 0$ and hence $x^2 > 0$. □

The properties of the real numbers $\mathbb{R}$ are crucial to doing analysis, that is, the theory of calculus. For this we need the concepts of **absolute value** and **distance**.

Definition 5.3.1. *If $x \in \mathbb{R}$ then its **absolute value** is defined by*

$$|x| = \begin{cases} x & \text{if } x \geq 0 \\ -x & \text{if } x < 0 \end{cases}$$

Alternatively we have

$$|x| = \sqrt{x^2}$$

where $\sqrt{}$ means the nonnegative square root.

Lemma 5.3.2. *Absolute value on $\mathbb{R}$ satisfies the following properties:*
 (1) $|x| \geq 0$ and $|x| = 0$ if and only if $x = 0$.
 (2) $|xy| = |x|\,|y|$.
 (3) $|x + y| \leq |x| + |y|$ (triangle inequality).

Proof. (1) and (2) are direct. We prove (3). Let $x; y \in \mathbb{R}$ then $x \leq |x|,\ y \leq |y|$. Then

$$(x + y)^2 = x^2 + 2xy + y^2 \leq |x|^2 + 2\,|x|\,|y| + |y|^2 = (|x| + |y|)^2.$$

Taking squareroots and using the fact that the square root function is increasing on the nonnegative reals, we obtain

$$|x + y| \leq ||x| + |y||\,.$$

But $|x| + |y| \geq 0$ and so $||x| + |y|| = |x| + |y|$ and therefore

$$|x + y| \leq |x| + |y|\,.$$

$\square$

   Absolute value allows us to define distance on $\mathbb{R}$.

Definition 5.3.2. *If $x, y \in \mathbb{R}$ then $d(x, y) = |x - y|$.*

Lemma 5.3.3. *The distance function $d(x, y)$ on $\mathbb{R}$ satisfies*
 (1) $d(x, y) \geq 0$ and $d(x, y) = 0$ iff $x = y$;
 (2) $d(x, y) = d(y, x)$;
 (3) $d(x, y) \leq d(x, z) + d(z, y)$ (triangle inequality).

Proof. These follow directly from the absolute value properties and so we leave the proofs to the exercises. $\square$

5.3.1   The Completeness of $\mathbb{R}$ (Optional)

The material in the next four sections is usually covered in a course in real analysis and can be omitted if one wants to focus directly on abstract algebra. However, in the spirit of how we have approached both the integers and the rationals, we would like to at least present a characterization of $\mathbb{R}$. This characterization involves the completeness property and hence involves convergence. For the proofs of many of the analytical results in these sections we refer to books on real analysis.

A **sequence** in $\mathbb{R}$ is a collection $\{x_n\}$ of real numbers indexed by the natural numbers. For example the sequence $\{\frac{1}{n}\}$ would consist of the numbers $1, \frac{1}{2}, \frac{1}{3}, \ldots$.

Definition 5.3.3. *A sequence $\{x_n\}$ in $\mathbb{R}$ **converges** or has a **limit** $x \in \mathbb{R}$ denoted $x_n \to x$ if for all $\epsilon > 0$ there exists an $N = N(\epsilon)$ such that $|x_n - x| < \epsilon$ for all $n \geq N$. Further, $\{x_n\}$ is a **Cauchy sequence** (CS) if for all $\epsilon > 0$ there exists an $N = N(\epsilon)$ such that $|x_n - x_m| < \epsilon$ for all $n, m \geq N$.*

The above concepts are often abbreviated as follows. Here "st" means "such that" and "iff" means "if and only if."

$$\lim_{n \to \infty} x_n = x \text{ iff } \forall \epsilon > 0 \exists N = N(\epsilon) \text{ st } n \geq N \implies |x_n - x| < \epsilon.$$
$$\{x_n\} \text{ is a CS iff } \forall \epsilon > 0 \exists N = N(\epsilon) \text{ st } n, m \geq N \implies |x_n - x_m| < \epsilon.$$

Roughly a sequence $\{x_n\}$ converges to x if all the terms in the sequence past a certain point get close to x. It is a Cauchy sequence if all the terms past a certain point get close together. Clearly if the terms get close to a single number x they must get close together and hence we have

Lemma 5.3.4. *Any convergent sequence in $\mathbb{R}$ is a Cauchy sequence.*

Proof. Let $\epsilon > 0$ be given. Since our sequence $\{x_n\}$ converges to x there exists an N st $|x_n - x| < \frac{\epsilon}{2}$ and $|x_m - x| < \frac{\epsilon}{2}$ if $n, m \geq N$. But then $|x_n - x_m| = |(x_n - x) + (x - x_m)| \leq |x_n - x| + |x_m - x|$ (by the triangle inequality) $< \frac{\epsilon}{2} + \frac{\epsilon}{2} = \epsilon$. This shows that $\{x_n\}$ is a Cauchy sequence. $\square$

Convergence allows us to define and study all the analytic properties of functions: continuity, differentiability and integrability. From the viewpoint of analysis what is important is that in $\mathbb{R}$ the converse is also true.

Theorem 5.3.3. *A sequence in $\mathbb{R}$ converges if and only if it is a Cauchy sequence.*

A proof of this theorem depends on the least upper bound property, which we introduce below. We note that this theorem is not true in the rational numbers $\mathbb{Q}$. To see this let $\{x_n\}$ be a sequence of rational numbers that converge to $\sqrt{2}$. For example take the decimal expansion of $\sqrt{2}$ and take x_i to be be finite decimals (terminating) in the expansion. Since $\{x_n\}$ converges in $\mathbb{R}$ it is a Cauchy sequence in $\mathbb{R}$ and hence a Cauchy sequence in $\mathbb{Q}$ (why?). However, the limit of real sequences is unique (see the exercises for a proof of this) and since the limit is irrational there is no limit to this Cauchy sequence in $\mathbb{Q}$.

Because of the above theorem we say that $\mathbb{R}$ is complete. This is essentially equivalent to the fact that $\mathbb{R}$ is in 1-1 correspondence with the points on a line.

Definition 5.3.4. *Let F be any ordered field and $S \subset F$.*

*(1) Then S is **bounded** if there exists $a, b \in F$ with $a \leq s \leq b$ for all $s \in S$. The element b is called an **upper bound** for S and the element a is called a **lower bound** for S. An element $b \in F$ is a **least upper bound** or LUB for S if b is an upper bound for S and if b_1 is another upper bound for S then $b \leq b_1$.*

(2) Suppose $a, b \in F$. Then the closed interval with endpoints a, b is the set

$$[a, b] = \{x \in F : a \leq x \leq b\}.$$

Note that by reversing all the inequalities in the above definition, we could also define the greatest lower bound for S or GLB. The GLB is not necessary for our discussions.

Definition 5.3.5. *(1) An ordered field F satisfies the **least upper bound property** (the LUB property) if every nonempty subset $S \subset F$ which has an upper bound in F also has a least upper bound in F.*

*(2) An ordered field F satisfies the **nested intervals property** if whenever ($I_n = [a_n, b_n] \subset F$ where $a_n \leq b_n$ for all n) is a sequence of nested closed intervals ($I_{n+1} \subset I_n$) whose lengths go to zero then there exists a unique point in F common to all the intervals, i.e., $\bigcap_n I_n \in F$.*

The key result on the completeness of $\mathbb{R}$ is that these properties are equivalent and further equivalent to the fact that Cauchy sequences converge.

Theorem 5.3.4. *Let F be an ordered field. Then the following are equivalent*
    (1) F satisfies the LUB property.
    (2) F satisfies the nested intervals property.
    (3) Every Cauchy sequence in F actually converges.

Definition 5.3.6. *An ordered field is **complete** if it satisfies any (and hence all) of the properties in the last theorem.*

We then have:

Theorem 5.3.5. *The real number field $\mathbb{R}$ is a complete ordered field.*

5.3.2   Characterization of $\mathbb{R}$ (Optional)

We need one additional property besides completeness to characterize the reals. First note that an ordered field must have characteristic zero (see Problem 3.6.8) and hence contain a subring isomorphic to the integers.

Definition 5.3.7. *An ordered field F is **archimedean** if for each $f_1, f_2 \in F$ there exists an integer n such that $nf_1 > f_2$.*

The complete characterization of $\mathbb{R}$ is then given by completeness together with the archidemean property.

Theorem 5.3.6. *$\mathbb{R}$ is a complete archimedean ordered field. Further, any other complete archimedean ordered field is isomorphic to $\mathbb{R}$.*

5.3.3   The Construction of $\mathbb{R}$ (Optional)

In this chapter we have shown how to construct the rationals starting from the integers. We now show how to construct the real numbers starting with the rationals.

Starting with $\mathbb{Q}$ there are several constructions that arrive at the reals. One due to Weierstrass considers the reals formally as decimal expansions. A second construction, due to Dedekind called the **Dedekind cut** construction considers the reals as special subsets of rational numbers. In this section we will describe a construction known as **Cauchy completion.**

Cauchy completion is a procedure to embed an incomplete metric space M as a dense subset of a complete metric space $\overline{M}$. The complete metric space $\overline{M}$ is called the **Cauchy completion** of M. We explain these terms below.

Definition 5.3.8. *A **metric space** is a set M with a **distance function** on it, that is, a function $d : M \times M \to \mathbf{R}$ satisfying*
    (1) $d(x, y) \geq 0$ and $d(x, y) = 0$ iff $x = y$;
    (2) $d(x, y) = d(y, x)$;
    (3) $d(x, y) \leq d(x, z) + d(z, y)$ (triangle inequality).

The rational numbers $\mathbb{Q}$ and the real numbers $\mathbb{R}$ are metric spaces where $d(x, y) = |x - y|$.

In any metric space we can define sequences, convergence and Cauchy sequences exactly as in the real numbers. In general we say that metric space M is complete if every Cauchy sequence in M converges to an element of M.

A subset S in a metric space M is dense in M if given any $x \in M$ and real number $\epsilon > 0$ there is an $s \in S$ with $d(x; s) < \epsilon$. This means that any point in M is arbitrarily close to a point in S. For example the rationals are dense in the reals (see the exercises).

Notice that the equivalence of Cauchy sequence completeness to the least upper bound property that holds in an ordered field does not necessarily hold in a general metric space since we may not have any order in the metric space.

Starting with the rationals $\mathbb{Q}$ we want to construct an ordered field F which is the completion of the rationals with respect to absolute value distance. That is, we want to construct a field $\mathbb{R}$ such that $\mathbb{Q} \subset \mathbb{R}$ and $\mathbb{R}$ is complete as a metric space and $\mathbb{Q}$ is a dense subset of $\mathbb{R}$.

The **Cauchy completion** of $\mathbb{Q}$ proceeds in the following manner:

Step (1): Consider the set $\overline{\mathbb{Q}}$ of all Cauchy sequences of rationals. That is, an element of $\overline{\mathbb{Q}}$ is a Cauchy sequence $(q_1, q_2, \dots)$ of rational numbers. Define on $\overline{\mathbb{Q}}$ the relation

$$(q_1, q_2, \dots) \sim (s_1.s_2, \dots) \text{ iff } \lim(q_i\text{-}s_i) = 0.$$

That is, after a point the two sequences get arbitrarily close.

Lemma 5.3.5. *This defines an equivalence relation on $\overline{\mathbb{Q}}$.*

We leave the proof of this lemma to the exercises.

Step (2): Let $\mathbb{R}$ be the set of all equivalence classes of Cauchy sequences of rationals under the equivalence relation above. We now want to show five things:

(1) $\mathbb{R}$ is an ordered field.
(2) $\mathbb{Q} \subset \mathbb{R}$.
(3) $\mathbb{R}$ is a metric space.
(4) $\mathbb{Q}$ is dense in $\mathbb{R}$.
(5) $\mathbb{R}$ is complete.

Step (3): We have the following theorem.

Theorem 5.3.7. $\mathbb{R}$ *is an ordered field.*

Proof. To show that $\mathbb{R}$ is a field we have to show that we can define $+, -, \times$, and multiplicative inverses to satisfy the field axioms. To prove this we need to know that Cauchy sequences are bounded so we prove this first.

Lemma 5.3.6. *If* $r = r_1, r_2, \ldots\ldots$ *is a Cauchy sequence then* (r_n) *is bounded, that is, there is a* $B > 0$ *with* $|r_n| < B$ *for all* n.

Proof. Let $\epsilon = 1$. Then since (r_n) is a Cauchy sequence it follows that there exists an N such that $|x_n - x_m| < 1$ for all $n, m \geq N$. In particular if $n > N$ we have $|x_n - X_N| < 1$. Then if $n \geq N$ we have

$$|x_n| = |x_n - X_N + X_N| \leq |x_n - x_N| + |x_N| < |x_N| + 1.$$

Now let $B = \max\{|x_1|, \ldots, |x_N|, |x_N|+1\}$. Then from the above $|x_n| \leq B$ for all n. $\qquad\square$

Now let $r, s \in \mathbb{R}$, i.e., r and s are equivalence classes of Cauchy sequences. So let $r = [(q_1, \ldots q_n, \ldots)]$ and $s = [(t_1, \ldots, t_n, \ldots)]$ be the equivalence classes of Cauchy sequences of rationals $\{q_n\}$ and $\{t_n\}$, respectively. We define

$$
\begin{aligned}
r \pm s &= [(q_1 \pm t_1, \ldots, q_n \pm t_n, \ldots)], \\
r \cdot s &= [(q_1 t_1, \ldots, q_n t_n, \ldots)].
\end{aligned}
$$

For this to make sense, we have to show that $\{q_n \pm t_n\}$ and $\{q_n t_n\}$ are again Cauchy sequences and that addition and multiplication of these equivalence classes are independent of the equivalence class representative chosen. That is, $+$ and $\times$ defined this way are well-defined. Here we will show that multiplication of equivalence classes is well-defined and leave all the other verifications to the exercises. For this purpose, suppose that $\{q_n\} \sim \{q_n'\}$ and $\{t_n\} \sim \{t_n'\}$. Then we have $\lim(q_n - q_n') = \lim(t_n - t_n') = 0$. We must show that

$$\lim(q_n t_n - q_n' t_n') = 0.$$

But all sequences here are Cauchy, hence they are all bounded by Lemma 5.3.6. In particular, $\exists M_1$ and $\exists M_2$ such that $|t_n| \leq M_1$ and $|q_n'| \leq M_2$ for all $n \in \mathbb{N}$.

Now let $\epsilon > 0$. Since $q_n - q_n' \to 0$ and $t_n - t_n' \to 0$, $\exists N_1$ and N_2 such that $|q_n - q_n'| < \frac{\epsilon}{2M_1}$ for all $n \geq N_1$ and $|t_n - t_n'| < \frac{\epsilon}{2M_2}$ for all $n \geq N_2$. Taking $N = \max\{N_1, N_2\}$. We have, using properties of absolute values, that for all $n \geq N$

$$
\begin{aligned}
\left| q_n t_n - q_n' t_n' \right| &= \left| (q_n t_n - q_n' t_n) + (q_n' t_b - q_n' t_n') \right| \\
&\leq \left| q_n - q_n' \right| |t_n| + \left| t_n - t_n' \right| \left| q_n' \right| \\
&< \frac{\epsilon}{2M_1} M_1 + \frac{\epsilon}{2M_2} M_2 = \frac{\epsilon}{2} + \frac{\epsilon}{2} = \epsilon.
\end{aligned}
$$

This shows that $(q_n t_n - q_n' t_n') \to 0$. (The other verifications are done in a similar manner.)

Clearly $[(0, 0, \dots)]$ and $[(1, 1, \dots)]$ are additive and multiplicative identities, respectively. For this to make sense, $(0, 0, \dots)$ and $(1, 1, \dots)$ must be Cauchy sequences of rationals. (They are — why?) The properties of commutativity, associativity, and distributivity follow from the fact that these are true in $\mathbb{Q}$. The additive inverse of an equivalence class $r \in \mathbb{R}$ is if $r = [\{q_n\}]$, defined as $-r = [\{-q_n\}]$. It is clear that if $\{q_n\}$ is a Cauchy sequence of rationals, then so is $\{-q_n\}$. Thus $[\{-q_n\}]$ makes sense as an element of $\mathbb{R}$. Thus far we have shown that $\mathbb{R}$ is a commutative ring with unity.

It remains to show that every nonzero element of $\mathbb{R}$ has a multiplicative inverse. If $r = [\{q_n\}]$ is an equivalence class of a Cauchy sequence and $r \neq 0$ then $\lim q_n \neq 0$. We leave it as an exercise to show then that there exists an N such that for all $n \geq N$ we have that $q_n \neq 0$. Thus it makes sense to define $\frac{1}{r} = [(0, 0, \dots, 0, \frac{1}{q_n}, \frac{1}{q_{n+1}}, \dots)]$. We need to show that the sequence $(0, 0, \dots, 0, \frac{1}{q_n}, \frac{1}{q_{n+1}}, \dots)$ is a Cauchy sequence. This is also left as an exercise.

Also note that $r \cdot \frac{1}{r} = [(0, \dots, 0, 1, 1, \dots)] = [(1, 1, \dots, 1, 1, \dots)]$. Thus we have shown that the set $\mathbb{R}$ with these operations is indeed a field.

Given $r = [(q_1, \dots, q_n, \dots)] \in \mathbb{R}$, then we define $r > 0$ or r is positive to mean $r \neq 0$ (i.e., it is not equal to the equivalence class $[(0, 0, \dots, 0, \dots)]$) and $r = [\{q_n\}]$ for some Cauchy sequence of rationals such that $\lim q_n \neq 0$ and there exists an N such that $q_n > 0$ for all $n \geq N$. Again since $r > 0$ was defined on equivalence classes (r is an equivalence class), we must show this is well-defined. We leave this verification to the exercises. If $r, s \in \mathbb{R}$, define $r > s$ to mean r-$s > 0$. This defines an order on $\mathbb{R}$ and hence $\mathbb{R}$ is an ordered field. Again we leave the details to the exercises.

Step (4): We now show that $\mathbb{Q} \subset \mathbb{R}$.

Theorem 5.3.8. $\mathbb{Q} \subset \mathbb{R}$. *More accurately, we can embed $\mathbb{Q}$ as a subring of $\mathbb{R}$.*

Proof. To each $q \in \mathbb{Q}$ associate the sequence $(q, q, q, q, \dots)$. This is clearly a Cauchy sequence of rationals (why?) Hence $[(q, q, q, q, \dots)] \in \mathbb{R}$. Note that $(q, q, q, q, \dots) \sim (\dots, q, q, q, \dots)$. So that $[(q, q, q, q, \dots)] = [(\dots, q, q, q, \dots)]$. Consider the map $q \mapsto (q, q, q, q, \dots)$. This mapping embeds $\mathbb{Q}$ in $\mathbb{R}$ (see the exercises for details) and hence we can consider $\mathbb{Q}$ as a subset of $\mathbb{R}$. □

Remark 5.3.9. *This theorem implies that when we talk about a "rational" number $\bar{q}$ in $\mathbb{R}$ what is meant is $\bar{q} = [(\dots, q, q, q, \dots)]$ where q is a rational number.*

Step (5): We must show that $\mathbb{R}$ is a metric space.

Theorem 5.3.10. $\mathbb{R}$ *is a metric space.*

Proof. To make $\mathbb{R}$ a metric space we define an absolute value on $\mathbb{R}$ and then use this to define distance by $d(r, s) = |r - s|$. If $r = [(q_1, q_2, \dots)] \in \mathbb{R}$ then we define $|r| = [(|q_1|, |q_2|, \dots)]$ We must first show that $(|q_1|, |q_2|, \dots)$ is again a Cauchy sequence of rationals. To see this, consider any $\epsilon > 0$ then there exists an $N = N(\epsilon)$ such that for all $n, m \geq N$ we have $|q_n - q_m| < \epsilon$. This must be true since $(q_1, q_2, \dots)$ is a Cauchy sequence. But now

$$
\begin{aligned}
|q_n| &= |(q_n - q_m) + q_m| \leq |q_n - q_m| + |q_m| \implies \\
|q_n| - |q_m| &\leq |q_n - q_m|.
\end{aligned}
$$

Similarly, we can get $|q_m| - |q_n| \leq |q_m - q_n| = |q_n - q_m|$. But this implies $|q_n| - |q_m| \geq -|q_n - q_m|$. Combining this with the display above, gives

$$-|q_n - q_m| \leq |q_n| - |q_m| \leq |q_n - q_m| \implies ||q_n| - |q_m|| \leq |q_n - q_m|.$$

But this implies from above that $||q_n| - |q_m|| < \epsilon$ for all $n, m \geq \dot{N}$. Thus $(|q_1|, |q_2|, \ldots)$ is a Cauchy sequence of rationals and hence $|r| = [(|q_1|, |q_2|, \ldots)]$ is in $\mathbb{R}$. We also need to show that this definition of absolute value is well-defined because it was defined on equivalence classes. This is not hard using the inequality proved above and so is left for the exercises. It is also not hard to show that $|r|$ satisfies the usual absolute value properties in Lemma 3.5.1 (see the exercises). Therefore $d(x, y) = |x - y|$ defines a metric on $\mathbb{R}$. □

Lemma 5.3.7. *If $x, y \in \mathbb{R}$ and $\overline{\omega}$ is any rational number in $\mathbb{R}$, then $|x - y| < \overline{\omega}$ means that if $x = [\{x_n\}), y = [\{y_n\}]$, and $\overline{\omega} = [(\omega, \omega, \ldots.)]$ then $\exists N = N(\omega)$ such that $|x_n - y_n| < \omega$ for all $n \geq N$.*

Proof. By the above definitions, $|x - y| = [\{|x_n - y_n|\}] < \overline{\omega}$ means that $\overline{\omega} - |x - y| > 0$. Thus we must have that $\overline{\omega} - |x - y| = [\{a_n\}]$ where $\{a_n\}$ is a Cauchy sequence of rationals such that $\exists N$ with $a_n > 0$ for all $n \geq N$. But $\overline{\omega} - |x - y| = [\{w - |x_n - y_n|\}]$. Since $\{x_n\}$ and $\{y_n\}$ are Cauchy so is $\overline{\omega} - |x - y|$. Given any $\epsilon > 0$ $\exists N_1$ and $\exists N_2$ such that $|x_n - x_m| < \epsilon/2$ and $|y_n - y_m| < \epsilon/2$ for all $n, m \geq N_1$ and $n, m \geq N_2$. So that for if $n, m \geq \max\{N_1 N_2\}$, then $|w - |x_n - y_n| - (w - |x_m - y_m|)| = ||x_m - y_m| - |x_n - y_n||$ by the inequality established in the proof of the theorem above gives that this is $\leq |(x_m - x_n) + (y_n - y_m)|$ which by the triangle inequality is $\leq \epsilon/2 + \epsilon/2 = \epsilon$. Thus $\{w - |x_n - y_n|\}$ is a Cauchy sequence of rationals and so it does make sense to consider the real number given by its equivalance class. But our given inequality would imply that $[\{w - |x_n - y_n|\}] > 0$. Which in turn by definition means $\exists N = N(\omega)$ st $|x_n - y_n| < \omega$ for all $n \geq N$. □

Step (6): We must show that $\mathbb{Q}$ is dense in $\mathbb{R}$.

Theorem 5.3.11. $\mathbb{Q}$ *is a dense subset of* $\mathbb{R}$.

Proof. To prove this we have to show that any real number is arbitrarily close to a rational number. Here we must be careful about what we mean by a "rational" number and what we mean by "arbitrarily" close. A "rational" number is an equivalence class of a sequence where the elements of the sequence are all just one and the same rational number (or at least are eventually that). When we say this rational is "arbitrarily" close to a real number, we mean that we can make the distance between these two real numbers as defined above less than any preassigned positive rational. (Also as defined above.) So let $r = [(q_1, q_2, q_3, \dots)] \in \mathbb{R}$ and suppose that $\omega > 0$ is any (small) rational number — not an equivalence class yet! We need to show that there exists a rational number (here an equivalence class) call it $\overline{q}$ such that $|r - \overline{q}| < \overline{\omega}$ where $\overline{\omega} = [(\dots, \omega, \omega, \omega, \dots)]$. (Note that $\overline{\omega} > 0$ by our definition of the ordering on $\mathbb{R}$.) Since $\{q_n\}$ is a Cauchy sequence there exists an $N = N(\omega)$ such that $|q_n - q_m| < \omega$ for all $n, m \geq N$. Choose a particular k with $k \geq N$ and set $\overline{q} = [(\dots, q_k, q_k, q_k, \dots)]$. This is an equivalence class of a Cauchy sequence of rationals and so $\overline{q} \in \mathbb{R}$. By the embedding above, we associate $\overline{q}$ with the rational number, q_k. Now $|r - \overline{q}| = [(\dots, |q_n - q_k|, |q_{n+1} - q_k|, \dots)]$. Now using the above Lemma 5.3.7, since for $n \geq N$ $|q_n - q_k| < \omega$, we have that $|r - \overline{q}| < \overline{\omega}$.   □

Step (7): Finally we must show that $\mathbb{R}$ is complete. Here we show completeness by Cauchy sequences but since we have shown that $\mathbb{R}$ is an ordered field this is equivalent to the LUB property and the nested intervals property.

Theorem 5.3.12. $\mathbb{R}$ *is complete.*

Proof. To prove this we have to show that any Cauchy sequence of real numbers converges to a real number. Let $r_1, r_2, \dots, r_n, \dots$ be a Cauchy sequence of reals. We show that it has a limit which is a real number. Realize that each r_i is itself an equivalence class of Cauchy sequences of rationals. For each n choose a rational $\overline{q_n} = [(\dots, q_n, q_n, \dots)]$, i.e., rational in the sense of our embedding, such that $|r_n - \overline{q_n}| < \frac{1}{n}$ where $\frac{1}{n} = [(\dots, \frac{1}{n}, \frac{1}{n}, \dots)]$. This can be done since $\mathbb{Q}$ is dense in $\mathbb{R}$ by Theorem 5.3.11. Consider the sequence of rationals $(q_1, q_2, q_3, \dots, q_n, \dots)$. We claim this sequence is a Cauchy sequence. Fix $\epsilon > 0$ a (small) positive rational not an equivalnce class. Choose $N \in \mathbb{N}$ such that $1/N < \epsilon/3$. Since $\{r_n\}$ is Cauchy there exists $M > 0$ such that $n, m \geq M \implies |r_n - r_m| < \overline{\epsilon/3}$ (here $\overline{\epsilon/3} = [(\dots, \epsilon/3, \epsilon/3, \dots)]$) and $n \geq N$

$$|\overline{q_n} - \overline{q_m}| \leq |\overline{q_n} - r_n| + |r_n - r_m| + |r_m - \overline{q_m}|$$

by the triangle inequality. Thus we have for $n, m \geq \max\{M, N\}$, $|\overline{q_n} - \overline{q_m}| < \overline{\epsilon}$.(Here $\overline{\epsilon} = [(\dots, \epsilon, \epsilon, \dots)]$.) Using Lemma 5.3.4, this means that $|q_n - q_m| < \epsilon$ for all $n, m \geq \max\{M, N\}$. Thus $\{q_n\}$ is a Cauchy sequence of rationals. So it makes sense to consider the real number $r = [\{q_n\}] \in \mathbb{R}$. Further, we claim that $r_n \to r$. By construction, $\overline{q_n} \to r_n$. But the fact that $\overline{q_n} \to r = [\{q_n\}]$ just follows from $\{q_n\}$ being a Cauchy sequence. For we need $|r - \overline{q_m}| = |[\{q_n\}] - \overline{q_m}|$ to be made small for n, m sufficiently large. But Lemma 5.3.7 says this is true if $|q_n - q_m|$ can be made small for n, m sufficiently large. This is precisely what it means for $\{q_n\}$ to be a Cauchy sequence. Thus we have proven $\lim(\overline{q_n} - r_n) = 0$ and $\lim(r - \overline{q_n}) = 0$. So that $\lim(r - r_n) = \lim((r - \overline{q_n}) + (\overline{q_n} - r_n)) = 0$. Thus the Cauchy sequence $r_1, r_2, \ldots, r_n, \ldots$ has the limit r showing that $\mathbb{R}$ is complete. $\square$

Having completed the **Cauchy completion** of $\mathbb{Q}$, for the rest of the book we no longer consider $\mathbb{R}$ to be a set of equivalence classes of Cauchy sequences of rationals.

5.3.4 The *p*-adic Numbers (Optional)

The real numbers $\mathbb{R}$ are a completion of the rationals $\mathbb{Q}$ and are characterized as the unique (up to isomorphism) complete archimedean ordered field. The question arises as to whether there are other completions of the rationals. The answer is yes but they must be, by necessity, nonarchimedean, and further are of a very special type. Notice that the construction of $\mathbb{R}$ from $\mathbb{Q}$ used the absolute value prominently and Cauchy sequences and denseness were in terms of this distance. First we must define a different distance function on $\mathbb{Q}$. We do this in general.

Definition 5.3.9. *A* **norm** *on a field F is a function $\|\ \| : F \to \mathbb{R}$ satisfying*
    (1) $\|x\| \geq 0$ for all $x \in F$,
    (2) $\|x\| = 0$ if and only if $x = 0$,
    (3) $\|xy\| = \|x\|\|y\|$ for all $x, y \in F$,
    (4) $\|x + y\| \leq \|x\| + \|y\|$ for all $x, y \in F$ (triangle inequality).
A **normed field** *is a field F with a norm.*

For example $\mathbb{Q}$ and $\mathbb{R}$ are normed fields with the usual absolute value. Any normed field F is a metric space under $d(x, y) = \|x - y\|$. Since a normed field F is a metric space the concepts of convergence, Cauchy sequence and completeness are all defined on F.

A norm $\|\ \|$ on F is **nonarchimedean** if it satisfies

$$\|x + y\| \leq \max(\|x\|, \|y\|).$$

Otherwise it is archimedean. The inequality above is called the strong triangle inequality because it is clearly stronger than the usual triangle inequality. That is, if this is satisfied then the usual triangle inequality is but not necessarily conversely. The induced metric is called an **ultrametric** and satisfies

$$d(x, z) \leq \max(d(x, y), d(y, z)).$$

Theorem 5.3.13. *The norm $\|\|$ is nonarchimedean if and only if*

$$\|n\| \leq 1$$

for all integers n. Further, the archimedean property on F says that given $x, y \in F$ with $x \neq 0$ there exists an integer n with $\|nx\| > \|y\|$.

We now define a different norm on the rationals $\mathbb{Q}$ called the p-adic norm.

Definition 5.3.10. *Let $x \in \mathbb{Q}$ and let p be a prime. Then $x = p^\alpha(\frac{a}{b})$ with $\gcd(a, b) = 1$. Then the p-**adic norm** is defined by*

$$\|x\|_p = \begin{cases} p^{-\alpha} & \text{if } x \neq 0 \\ 0 & \text{if } x = 0 \end{cases}$$

*The map $\mathbb{Q} \to \mathbb{Z}$ given by $ord(x) = \alpha$ is called the p-**adic valuation**.*

Lemma 5.3.8. *For any prime p the p-adic norm is a nonarchimdean norm on $\mathbb{Q}$.*

If we start with $\mathbb{Q}$ and a prime p then $\mathbb{Q}$ is a metric space with respect to the p-adic norm. We can then follow exactly the Cauchy completion procedure to get a complete normed field containing $\mathbb{Q}$ as a dense (with respect to the p-adic distance) subfield. The **field $\mathbb{Q}_p$ of p-adic numbers** is the completion of $\mathbb{Q}$ with respect to the p-adic norm. Hence we have an infinite collection of complete fields all containing the rationals $\mathbb{Q}$. Further, none of these are isomorphic to $\mathbb{R}$ and for distinct primes $p_1 \neq p_2$ the fields $\mathbb{Q}_{p_1}$ and $\mathbb{Q}_{p_2}$ are nonisomorphic.

The following theorem known as Ostrowski's theorem shows that these are the only completions of $\mathbb{Q}$.

Theorem 5.3.14. *Let F be a complete normed field that contains the rationals as a dense subset. Then F is isomorphic to either the reals or to some p-adic field $\mathbb{Q}_p$.*

A proof of this is outside the scope of this book. For a proof we refer the reader to [K].

5.4 The Field of Complex Numbers

For our basic number systems we have $\mathbb{N} \subset \mathbb{Z} \subset \mathbb{Q} \subset \mathbb{R}$. Each of the basic number systems completed the previous one relative to a certain property. On the natural numbers $\mathbb{N}$ there was no zero element or additive inverses. By adjoining zero and inverses we created the integers $\mathbb{Z}$. On $\mathbb{Z}$ there were no multiplicative inverses except for ± 1. By adjoining to $\mathbb{Z}$ the multiplicative inverse of each nonzero integer we obtained the rationals $\mathbb{Q}$. On $\mathbb{Q}$ there was not always a limit for a Cauchy sequence. By adjoining all the limits of Cauchy sequences of rationals we obatined the reals $\mathbb{R}$.

In the real numbers $\mathbb{R}$, $\sqrt{-1}$ does not exist. We now create a field in which -1 does have a square root. We formally define $\mathbf{i} = \sqrt{-1}$; that is, $\mathbf{i}$ is a new element such that $\mathbf{i}^2 = -1$. Historically, $\mathbf{i}$ was called the **imaginary unit**, but as we will see in the next section, $\mathbf{i}$ has a very real geometric significance.

A **complex number** is then an expression of the form $z = x + \mathbf{i}y$ with $x, y \in \mathbb{R}$. If $x = 0, y \neq 0$, so that z has the form $\mathbf{i}y$, then z is called a (purely) **imaginary number**. The set of complex numbers, denoted by $\mathbb{C}$, is then

$$\mathbb{C} = \{x + \mathbf{i}y \; : \; x, y \in \mathbb{R}\}.$$

If we identify a real number x with the complex number $x + 0\mathbf{i}$, we see that $\mathbb{R} \subset \mathbb{C}$ as sets. For the complex number $z = x + \mathbf{i}y$, we call x the real part and y the imaginary part of z. Note that both the real and imaginary parts of z are real numbers. We say two complex numbers are equal provided that both their real parts are equal and their imaginary parts are equal. On $\mathbb{C}$ we define arithmetic by algebraic manipulation using the fact that $i^2 = -1$. To be more specific, we assume that $\mathbf{i}$ commutes with any real number, also the associative laws and distributive laws hold for any expressions containing $\mathbf{i}$. That is, if $z = x + \mathbf{i}y; w = a + \mathbf{i}b$ then:
  (i) $z = w$ iff $x = a$ and $y = b$.
  (ii) $z \pm w = (x \pm a) + \mathbf{i}(y \pm b)$.
  (iii) $z \cdot w = (xa - yb) + \mathbf{i}(xb + ya)$
  (We note that (iii) follows from the assumed properties as follows:

$$(x + \mathbf{i}y)(a + \mathbf{i}b) = xa + \mathbf{i}(ya) + \mathbf{i}(xb) + \mathbf{i}^2(yb) = (xa - yb) + \mathbf{i}(xb + ya)).$$

EXAMPLE 5.4.1. Let $z = 3 + 4\mathbf{i}$ and $w = 7 - 2\mathbf{i}$ then:

(a) $z + w = 10 + 2\mathbf{i}$.
(b) $z - w = -4 + 6\mathbf{i}$.
(c) $zw = 29 + 22\mathbf{i}$.

It is easy to verify, using the assumed laws (see the exercises), that under these definitions $\mathbb{C}$ forms a commutative ring with an identity and that $\mathbb{R}$ is a subring of $\mathbb{C}$: The zero is $0 + \mathbf{i}0 = 0 \in \mathbb{R}$, i.e., identified as $0 + 0\mathbf{i}$, and the multiplicative identity is $1 + 0\mathbf{i} = 1 \in \mathbb{R}$ identified with $1 + 0\mathbf{i}$. In order for $\mathbb{C}$ to be a field we must have multiplicative inverses. We now show how to construct these.

Definition 5.4.1. *If $z \in \mathbb{C}$ with $z = x + iy$, then the **complex conjugate** of z, denoted by $\bar{z}$, is $\bar{z} = x - \mathbf{i}y$, and the **absolute value**, or **modulus**, of z, denoted by $|z|$, is $|z| = \sqrt{x^2 + y^2}$.*

EXAMPLE 5.4.2. Let $z = 3 + 4\mathbf{i}$. Then $\bar{z} = 3 - 4\mathbf{i}$ and $|z| = \sqrt{9 + 16} = 5$.

The following lemmas give the properties of the complex conjugate and the absolute value, and from these we will be able to construct inverses.

Lemma 5.4.1. *Let $z, w \in \mathbb{C}$. Then:*
(a) $|z| \geq 0$ and $|z| = 0$ if and only if $z = 0$.
(b) $|zw| = |z|\,|w|$.
(c) $|z + w| \leq |z| + |w|$.

The proofs of these are straightforward computations and are left to the exercises. As a hint for (c), note that we can mimic the proof given in Lemma 5.3.2. Notice that these are precisely the properties of the absolute value of real numbers and so they will allow us to do calculus on $\mathbb{C}$.

Lemma 5.4.2. *Let $z, w \in \mathbb{C}$: Then*
(1) $\overline{z + w} = \bar{z} + \bar{w}$.
(2) $\overline{zw} = (\bar{z})(\bar{w})$.
(3) $\bar{\bar{z}} = z$.
(4) $|\bar{z}| = |z|$.
(5) $\bar{z} = z$ if and only if z is real.

Proof. Properties (1) through (4) are computations so we leave their proofs to the exercises. For (5) suppose $z = x + iy$.

Then $\bar{z} = x - \mathbf{i}y$, and $z = \bar{z}$ if and only if $y = -y$. But this is possible if and only if $y = 0$, and then $z = x + 0\mathbf{i}$ and z is real. □

Of course (1) and (2) of Lemma 5.4.2 hold for any finite number of complex numbers: e.g.

$$\overline{z_1 + z_2 + \cdots + z_n} = \overline{z_1} + \overline{z_2} + \cdots + \overline{z_n}.$$

As a simple application of the above lemma we note.

Theorem 5.4.1. *If* $f(z) = a_n z^n + a_{n-1} z^{n-1} + \cdots + a_0$ *is a real polynomial, i.e., all* $a_i \in \mathbb{R}$, *then complex roots must occur in complex conjugate pairs, i.e., if* $z_0 \in \mathbb{C}$ *is such that* $f(z_0) = 0$ *then* $f(\overline{z_0}) = 0$.

Proof. Since $f(z_0) = 0$ we have

$$0 = f(z_0) = a_n z_0^n + a_{n-1} z_0^{n-1} + \cdots + a_1 z_0 + a_0.$$

Taking complex conjugates of both sides of this equation and using the properties given in Lemma 5.4.2 (extended to any finite number) gives

$$\overline{0} = 0 = \overline{f(z_0)} = a_n (\overline{z_0})^n + a_{n-1} (\overline{z_0})^{n-1} + \cdots + a_1 (\overline{z_0}) + a_0.$$

Here we have also used the fact that all $a_i \in \mathbb{R}$ to get that $\overline{a_i} = a_i$ for all $i = 0, 1, \ldots, n$ (Lemma 5.4.2 (5)).

But the right-hand side of the above equation is just $f(\overline{z_0})$. Therefore $f(\overline{z_0}) = 0$. □

In the above theorem it is very important that the coefficients all be real. One may wonder if a complex polynomial (one with complex but not necessarily real coefficients) must have a root in $\mathbb{C}$. The answer is "yes," if the degree of the polynomial (the degree of f is n) is at least 1. We also note that a polynomial of degree n has n roots counting each root a number of times equal to its multiplicity (see Chapter 12). This theorem implies that an odd degree real polynomial must have a real root. We will give a proof of this in Chapter 12. We state without proof (for a proof see Chapter 15) the important and famous **fundamental theorem of algebra**.

Theorem 5.4.2. *(fundamental theorem of algebra) If* $f(z)$ *is a nonconstant complex polynomial, i.e., the degree of* $f(z)$ *is* ≥ 1 *and the coefficients are in* $\mathbb{C}$, *then there exists a* $z_o \in \mathbb{C}$ *with* $f(z_0) = 0$.

Now consider

$$z \cdot \overline{z} = (x + \mathbf{i}y)(x - \mathbf{i}y) = x^2 - \mathbf{i}^2 y^2 = x^2 + y^2 = |z|^2.$$

Therefore we have the following result.

Lemma 5.4.3. *If $z \in \mathbb{C}$, then $z\overline{z} = |z|^2$.*

From this lemma we can see that if $z \neq 0$, then

$$z \cdot \frac{\overline{z}}{|z|^2} = \frac{|z|^2}{|z|^2} = 1$$

and so we make the following definition.

Definition 5.4.2. $\frac{1}{z} = \frac{\overline{z}}{|z|^2}$ *for any $z \in \mathbb{C}$ with $z \neq 0$.*

EXAMPLE 5.4.3. Let $z = 3 + 4\mathbf{i}$ and $w = 7 - 2\mathbf{i}$. Then
(a) $\frac{1}{z} = \frac{3-4\mathbf{i}}{3^2+4^2} = \frac{3}{25} - \frac{4}{25}\mathbf{i}$.
(b) $\frac{z}{w} = z \cdot \frac{1}{w} = (3+4\mathbf{i})\frac{7+2\mathbf{i}}{53} = \frac{13+34\mathbf{i}}{53} = \frac{13}{53} + \frac{34}{53}\mathbf{i}$.

From the definition we now have division of complex numbers and thus $\mathbb{C}$ is a field.

Theorem 5.4.3. *$\mathbb{C}$ is a field and $\mathbb{R}$ is a subfield of $\mathbb{C}$. Further, $\mathbb{C}$ is not an ordered field.*

Proof. That $\mathbb{C}$ forms a field follows from the discussion prior to the theorem. That $\mathbb{C}$ cannot be an ordered field follows from the following. In any ordered field F, since $1 \neq 0$, we have $1^2 = 1$ so 1 is positive by Lemma 3.5.8. But then, from the trichotomy law -1 must be negative and therefore cannot have a square root since the square of any nonzero element of an ordered field must be positive again by Lemma 3.5.8.   □

5.4.1   Geometric Interpretation

The complex numbers have a very nice geometric interpretation as points in the two-dimensional real space $\mathbb{R}^2 = \mathbb{R} \times \mathbb{R}$. To each complex number $z = x + \mathbf{i}y$ we can identify the point (x, y) in the xy-plane $\mathbb{R}^2 = \mathbb{C}$. Conversely, to each point $(x; y) \in \mathbb{R}^2$ we can identify the complex number $z = x + \mathbf{i}y$. When thought of in this way, as consisting of complex numbers, $\mathbb{R}^2$ is called the **complex plane**. Alternatively, we can think of the complex number $z = x + \mathbf{i}y$ as the two-dimensional vector $v = (x, y)$, that is, the vector with representative starting at $(0, 0)$ and ending at (x, y). In this interpretation $||$ is just the magnitude of the vector (x, y) which is just the distance from the point (x, y) to the origin. The conjugate $\bar{z} = x - \mathbf{i}y$ is just the point $(x, -y)$, which is the point (x, y) reflected through the x-axis (Figure 5.1).

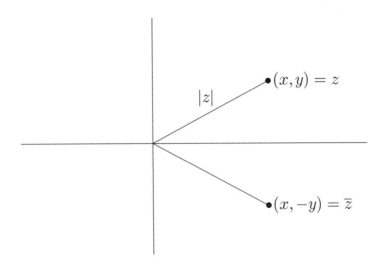

Figure 5.1. Geometrical Representation

We can describe the arithmetic operations in terms of this geometrical interpretation. Since addition and subtraction are done componentwise, addition and subtraction of complex numbers corresponds to the same vector operations as pictured in Figure 5.2.

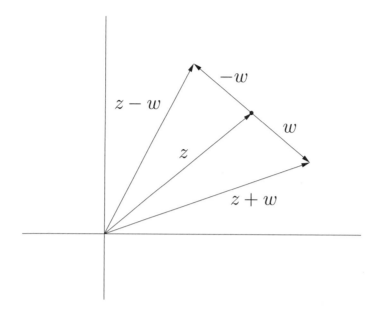

Figure 5.2. Vector Operations

Multiplication by real numbers is scalar multiplication of two-dimensional vectors. Geometrically this is a stretching or a shrinking. That is, if $z \in \mathbb{C}$ and $r \in \mathbb{R}$ then:

(1) if $r \geq 0$, $rz = w$, where w is the vector in the same direction as z with magnitude $|r|\,|z|$.

(2) if $r < 0$, $rz = w$ where w is the vector in the opposite direction as z with magnitude $|r|\,|z|$.

If $|r| > 1$, then it is a stretching; while if $|r| < 1$, then it is a shrinking. We will, however, refer to both of these as a **stretching**.

If $z = x + iy$, then $\mathbf{i}z = -y + \mathbf{i}x$. That is, multiplication by $\mathbf{i}$ takes the point (x, y) to the point $(-y, x)$. Since the inner product or dot product

$$\langle (x, y),\ (-y, x) \rangle = -xy + xy = 0,$$

these vectors are orthogonal. Since $1\mathbf{i} = \mathbf{i}$, multiplication by $\mathbf{i}$ corresponds to a counterclockwise rotation by $90°$. Therefore, i is not really "imaginary" in any sense, it corresponds to a rotation.

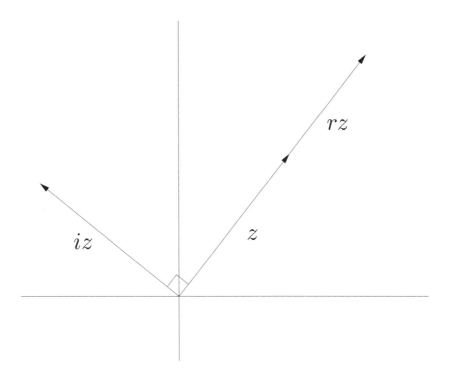

Figure 5.3. Complex Multiplication

Putting all this together we can give a complete geometric interpretation to complex multiplication. Suppose $z, w \in \mathbb{C}$ with $z = x + \mathbf{i}y$. Then consider $zw = (x + \mathbf{i}y)w = xw + \mathbf{i}yw$. Geometrically then, we first stretch the vector w by x, then stretch the vector w by y and rotate the second stretched vector by 90° counterclockwise. Finally we add the resulting vectors (see Figure 5.3).

5.4.2 Polar Form and Euler's Identity

If $P \in \mathbb{R}^2$ with rectangular coordinates (x, y), then P also has polar coordinates (r, θ) where r is the distance from the origin O to P and θ is the angle the vector $\overrightarrow{OP}$ makes with the positive x-axis (see Figure 5.4).

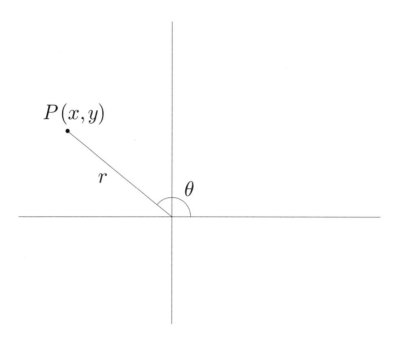

Figure 5.4. Polar Form

Here we restrict θ to be in the range $[0, 2\pi)$ so that each point in $\mathbf{R}^2$ has only one set of polar coordinates. The rectangular coordinates (x, y) of a point P are related to its polar coordinates (r, θ) by the formulas:

(i) $x = r\cos\theta$.
(ii) $y = r\sin\theta$.
(iii) $r = \sqrt{x^2 + y^2}$.
(iv) $\theta = \arctan(y/x)$ chosen in the appropriate quadrant.

If $z = x + \mathbf{i}y$ corresponds to the point (x, y) with polar coordinates (r, θ), then from the relations above, z can be written as

$$z = r(\cos\theta + \mathbf{i}\sin\theta).$$

This is called the **polar form** for z. The angle θ is the **argument of** z, denoted by Arg z, and in this context $|z|$ is called the **modulus of** z.

We note that if we were to allow θ to be an aribitrary real number (not just between 0 and 2π), two complex numbers in polar form, $z_n =$

$r_n(\cos(\theta_n) + \mathbf{i}\sin(\theta_n))$ for $n = 1, 2$, would be equal, $z_1 = z_2$, if and only if $r_1 = r_2$ and $\theta_1 = \theta_2 + 2k\pi$ where $k = 0, \pm 1, \pm 2, \dots$.

EXAMPLE 5.4.4. Suppose $z = 1 - \mathbf{i}$; then $|z| = \sqrt{2}$ and Arg $(z) = \arctan(-1) = 7\pi/4$ since $(1, -1)$ is in the fourth quadrant. Therefore

$$z = \sqrt{2}(\cos(7\pi/4) + \mathbf{i}\sin(7\pi/4)).$$

There is a very nice exponential way to express the polar form that is due to Euler. Before we describe this, we must look more closely at the powers of $\mathbf{i}$. Now, $\mathbf{i}^2 = -1$ so $\mathbf{i}^3 = \mathbf{i}^2\mathbf{i} = -\mathbf{i}$ Then $\mathbf{i}^4 = \mathbf{i}^3\mathbf{i} = -\mathbf{i}^2 = 1$ and therefore $\mathbf{i}^5 = \mathbf{i}$. From this it follows that the powers of i repeat cyclically as

$$\{1, \mathbf{i}, -1, -\mathbf{i}\}$$

and $\mathbf{i}^m = \mathbf{i}^n$ if and only if $n \equiv m \pmod 4$. For example, $\mathbf{i}^{51} = \mathbf{i}^3 = -\mathbf{i}$. Further, the multiplicative inverse of any power of i is another power of $\mathbf{i}$, and so these powers form a group under multiplication.

Lemma 5.4.4. *The powers of i form a group of order 4 under multiplication.*

This group is actually a special type of group called a cyclic group. We will discuss this in the next chapter. This result is actually a special case of a result about primitive nth roots of unity that we will discuss in the next section.

If t is a real variable, recall that the functions $e^t, \sin(t), \cos(t)$ have the following power series expansions.

$$e^t = 1 + t + t^2/2! + \dots + t^n/n! + \dots$$

$$\sin(t) = t - t^3/3! + t^5/5! - \dots + (-1)^n t^{2n+1}/(2n+1)! + \dots$$

$$\cos(t) = 1 - t^2/2! + t^4/4! - \dots + (-1)^n t^{2n}/(2n)! + \dots$$

Now consider $t = i\theta$ with θ real, and substitute into the power series expansion for e^t to find $e^{i\theta}$. (Although t is not a real variable we do this formally.)

$$e^{i\theta} = 1 + \mathbf{i}\theta + (\mathbf{i}\theta)^2/2! + \dots = 1 + (\mathbf{i}\theta) - \theta^2/2! - \mathbf{i}\theta^3/3! + \dots$$

using the rules for the powers of i. Then combining terms with and without i we get

$$
\begin{aligned}
e^{\mathbf{i}\theta} &= (1 - \theta^2/2! + \theta^4/4! - \ldots) + \mathbf{i}(\theta - \theta^3/3! + \theta^5/5! - \ldots) \\
&= \cos(\theta) + \mathbf{i}\sin(\theta).
\end{aligned}
$$

This is known as **Euler's identity**.

Lemma 5.4.5. *(Euler's identity)* $e^{\mathbf{i}\theta} = \cos(\theta) + \mathbf{i}\sin(\theta)$ *for* $\theta \in \mathbb{R}$.

Now, if $r = |z|$ and $\theta = \mathrm{Arg}(z)$, we then have from the polar form that

$$
z = r(\cos(\theta) + \mathbf{i}\sin(\theta)) = re^{\mathbf{i}\theta}.
$$

This last identity makes multiplication of complex numbers very simple. Suppose $z = r_1 e^{\mathbf{i}\theta_1}$, $w = r_2 e^{\mathbf{i}\theta_2}$. Then $zw = r_1 r_2 e^{\mathbf{i}(\theta_1 + \theta_2)}$. Breaking this into components, we then have $|zw| = |z||w|$ and $\mathrm{Arg}(zw) = \mathrm{Arg}(z) + \mathrm{Arg}(w)$.

Lemma 5.4.6. *If* $z, w \in \mathbb{C}$*, then* $|zw| = |z||w|$ *and* $Arg(zw) = Arg(z) + Arg(w)$.

Notice that $\mathrm{Arg}\,(i) = \pi/2$, and multiplication $\mathbf{i}z$ rotates z by 90°. That is, $\mathrm{Arg}(\mathbf{i}z) = \pi/2 + \mathrm{Arg}(z) = \mathrm{Arg}(i) + \mathrm{Arg}(z)$, which follows directly from the lemma.

Euler's identity leads directly to what is called **Euler's magic formula**. Suppose $\theta = \pi$. Then

$$
e^{\mathbf{i}\pi} = \cos(\pi) + \mathbf{i}\sin(\pi) = -1 + 0\mathbf{i} = -1.
$$

Put succinctly, $e^{\mathbf{i}\pi} + 1 = 0$.

Lemma 5.4.7. *(Euler's magic formula)* $e^{\mathbf{i}\pi} + 1 = 0$.

This has been called a "magic" formula because the five most important numbers in mathematics — $0, 1, e, \mathbf{i}, \pi$ — are tied together in a very simple equation. If one thinks about how diversely these five numbers appear; 0 as the additive identity, 1 as the multiplicative identity, e as the natural exponential base, π as the ratio of the circumference to the diameter of any circle, and $\mathbf{i}$ as the imaginary unit, this result is truly amazing.

5.4.3 DeMoivre's Theorem for Powers and Roots

If $z = re^{i\theta} \in \mathbb{C}$ and $n \in \mathbb{N}$ then $z^n = r^n e^{in\theta}$. Notice then that $|z^n| = |z|^n$ and $\text{Arg}(z^n) = n \text{ Arg}(z)$. This is known as **DeMoivre's theorem for powers**.

Theorem 5.4.4. *(DeMoivre's Theorem for Powers) If* $z = re^{i\theta} \in \mathbb{C}$ *and* $n \in \mathbb{N}$, *then* $z^n = r^n e^{in\theta}$. *In particular,* $|z^n| = |z|^n$ *and* $\text{Arg}(z^n) = n \text{ Arg}(z)$.

It also follows from $z = re^{i\theta}$ that if n is any integer, that $z^n = r^n e^{in\theta}$. Assuming that the usual laws for powers work for complex numbers.

EXAMPLE 5.4.5. Let $z = 1 - i$ and let us find z^{10}. Now, $|z| = \sqrt{2}$ and $\text{Arg}(z) = 7\pi/4$, so $z = \sqrt{2}e^{7i\pi/4}$. Therefore,

$$z^{10} = (\sqrt{2})^{10} e^{i70\pi/4}$$

Now, $(\sqrt{2})^{10} = 32$ while $70\pi/4 = 16\pi + 3\pi/2 \equiv 3\pi/2$ (as angles). It follows than that

$$z^{10} = 32e^{i3\pi/2} = 32(\cos(3\pi/2) + \mathbf{i}\sin(3\pi/2)) = -32\mathbf{i}.$$

If $n \in \mathbb{N}$, $z \in \mathbb{C}$ and $w^n = z$ then w is an nth-**root** of z which we denote by $z^{1/n}$. If $r \in \mathbb{R}$, $r > 0$ then there is exactly one real nth root of r. If $z = e^{i\theta}$, then $w = r^{1/n} e^{i\theta/n}$ satisfies $w^n = z$ according to DeMoivre's theorem and is one nth root of the complex number z. However, if

$$\frac{\theta + 2\pi k}{n} < 2\pi \text{ for } k \in \mathbb{N},$$

then

$$w_k = r^{1/n} e^{\mathbf{i} \cdot \frac{\theta + 2\pi k}{n}}$$

is also an nth root of z and is different from w just as long as $0 < k < n$ (see the exercises to fill in the details for this). Further, there are precisely n values of k that produce different nth roots of z for $k = 0, 1, \ldots, n-1$. If we were to allow k to be an integer either $k \geq n$ or $k < 0$ then these powers would start to repeat just like the powers of $\mathbf{i}$. (See the exercises for the details.) Thus n different nth roots of z are then

$$w_k = r^{1/n} e^{\mathbf{i} \cdot \frac{\theta + 2\pi k}{n}}, \quad k = 0, 1, \ldots, n-1.$$

We have therefore proved the following theorem, which is known as **DeMoivre's theorem for roots**.

Theorem 5.4.5. *(DeMoivre's Theorem for Roots) If $z \in \mathbb{C}$, $z \neq 0$, then there are exactly n distinct nth roots of z. If $z = re^{i\theta}$, these are given by*

$$w_k = r^{1/n} e^{i \cdot \frac{\theta + 2\pi k}{n}}, \ k = 0, 1, \ldots, n - 1.$$

EXAMPLE 5.4.6. Let $z = 1 + i$. Find the sixth roots of z. Now $|z| = \sqrt{2}$, $\mathrm{Arg}(z) = \arctan(1) = \pi/4$, so $z = \sqrt{2}e^{i\pi/4}$. The sixth roots of z are then:

(1) $w_1 = 2^{1/12} e^{i\frac{\pi/4}{6}} = 2^{1/12}(\cos(\pi/24) + i\sin(\pi/24))$

(2) $w_2 = 2^{1/12} e^{i\frac{\pi/4 + 2\pi}{6}} = 2^{1/12}(cos(9\pi/24) + i sin(9\pi/24))$.

(3) $w_3 = 2^{1/12} e^{i\frac{\pi/4 + 4\pi}{6}} = 2^{1/12}(cos(17\pi/24) + i sin(17\pi/24))$.

(4) $w_4 = 2^{1/12} e^{i\frac{\pi/4 + 6\pi}{6}} = 2^{1/12}(cos(25\pi/24) + i sin(25\pi/24))$.

(5) $w_5 = 2^{1/12} e^{i\frac{\pi/4 + 8\pi}{6}} = 2^{1/12}(cos(33\pi/24) + i sin(33\pi/24))$.

(6) $w_6 = 2^{1/12} e^{i\frac{\pi/4 + 10\pi}{6}} = 2^{1/12}(cos(41\pi/24) + i sin(41\pi/24))$.

If $z = 1$, then an nth root, w, is called an **nth root of unity**. Since $|z| = 1$, then $|w| = 1$ for any nth root of unity, and therefore the roots of unity will differ only in their angles.

Corollary 5.4.1. *There are exactly n distinct nth roots of unity given by*

$$w_k = e^{i \cdot \frac{2\pi k}{n}} = \cos(\frac{2\pi k}{n}) + i\sin(\frac{2\pi k}{n}), \ k = 0, 1, \ldots, n - 1.$$

Geometrically the n nth roots of unity all fall on a circle of radius 1 with center at the origin and are located at the vertices of an inscribed regular n-gon with one vertex on the positive real axis. In Figure 5.5 we picture the six sixth roots of unity forming the vertices of a regular hexagon. For $z \neq 0$ with $|z| \neq 1$ the nth roots are at the vertices of a regular n-gon on a circle of radius $|z|^{1/n}$.

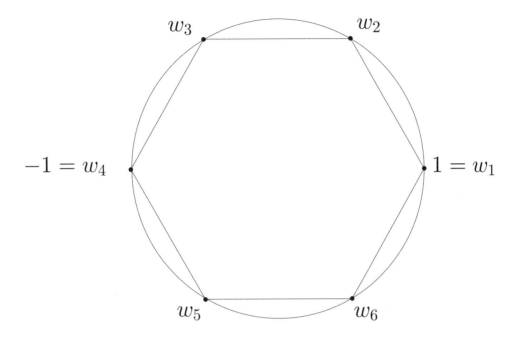

Figure 5.5. Sixth Roots of Unity

Let $w = e^{i2\pi/n}$, the nth root of unity with the smallest positive angle. This is called the **principal primitive nth root of unity**. The other nth roots are the distinct powers of w, that is, $\{w, w^2, w^3, \ldots, w^n = 1\}$. Further, these nth roots of unity form a group under multiplication. (Notice that w is not the only primitive nth root of unity, i.e., a primitive nth root is one whose powers give all the nth roots (see the exercises). Further, these roots form a special type of group, called a **cyclic group**, which we will define and discuss in Chapter 6.

Corollary 5.4.2. *The nth roots of unity form a cyclic group of order n under multiplication. They are generated by w, the nth root of unity with the smallest positive angle. (We will discuss cyclic groups in the next chapter.)*

The result we looked at before concerning the powers of **i** is just a special case of this corollary since **i** is the principal primitive fourth root of unity.

5.5   Exercises

EXERCISES FOR SECTION 5.1

5.1.1. Let $\mathbf{H}$ be the set of quaternions defined in the text. Show that for $\mathbf{i}, \mathbf{j}, \mathbf{k} \in \mathbf{H}$, $\mathbf{jk} = \mathbf{i} = -\mathbf{kj}$ and $\mathbf{ki} = \mathbf{j} = -\mathbf{ik}$ just using the facts given in the text, i.e., $\mathbf{i}^2 = \mathbf{j}^2 = \mathbf{k}^2 = -1$, $\mathbf{ijk} = -1$, and the assumptions made in the text.

5.1.2. Complete the details of Example 5.1.1 to show that $u \cdot v = -3 + 14\mathbf{i} - \mathbf{j} + 2\mathbf{k}$.

5.1.3. If $\mathbf{H}$ is the set of quaternions and $u, v \in \mathbf{H}$ with $u = a_1 + a_2\mathbf{i} + a_3\mathbf{j} + a_4\mathbf{k}$ and $v = b_1 + b_2\mathbf{i} + b_3\mathbf{j} + b_4\mathbf{k}$, write formulas for $u + v$ and $u \cdot v$.

5.1.4. Explain why if the associative law is assumed to hold (together with the other laws assumed in the text) for all choices of $\mathbf{i}, \mathbf{j}, \mathbf{k}$ in a product, this is sufficient for multiplication to be associative in $\mathbf{H}$. Further, if one were to consider all possible choices for uvw where each of u, v, w can be any of $\mathbf{i}, \mathbf{j}, \mathbf{k}$ with repetitions allowed, how many possible products are there?

5.1.5. Verify that $\mathbf{H}$ forms a noncommutative ring with unity. Here you must indicate what the identities are with repect to $+$ and $\cdot$. You may, however, assume the associative and distributive laws. You must also show why $\mathbf{H}$ is noncommutative. Also does $\mathbf{H}$ have any zero divisors? Is $\mathbf{H}$ an integral domain?

5.1.6. If $u \in \mathbf{H}$, do the computations necessary to show that $u \cdot \bar{u} = N(u)$ where if $u = a_1 + a_2\mathbf{i} + a_3\mathbf{j} + a_4\mathbf{k}$, then $N(u) = a_1^2 + a_2^2 + a_3^2 + a_4^2$.

5.1.7. There is a similarity between the **cross product** of three-dimensional vectors $\mathbf{i} = \langle 1, 0, 0 \rangle$, $\mathbf{j} = \langle 0, 1, 0 \rangle$, $\mathbf{k} = \langle 0, 0, 1 \rangle$ and the way $\mathbf{i}, \mathbf{j}, \mathbf{k} \in \mathbf{H}$ multiply. Explain it.

5.1.8. Consider the subset of $M_2(\mathbb{C})$ defined by

$$Q = \left\{ \begin{pmatrix} z & w \\ -\overline{w} & \overline{z} \end{pmatrix} : z, w \in \mathbb{C} \right\}$$

(a) Show that Q is a subring, with identity, of $M_2(\mathbb{C})$.
(b) Show this is isomorphic to the quaternions H.
(HINT: Let $z = a + ib, w = c + id$, then consider the map $\varphi : \mathbf{H} \to Q$ given by $a + b\mathbf{i} + c\mathbf{j} + d\mathbf{k} \mapsto \begin{pmatrix} a + b\mathbf{i} & c + d\mathbf{i} \\ -(c - d\mathbf{i}) & a - b\mathbf{i} \end{pmatrix}.)$

(c) Are there any advantages to presenting the quaternions as in this problem?

5.1.9. Here is a mnemonic device for remembering how $\mathbf{i}, \mathbf{j}, \mathbf{k} \in H$ multiply. Put $\mathbf{i}, \mathbf{j}, \mathbf{k}$ clockwise around a circle (see Figure 5.6). Note that a clockwise motion gives a positive product, i.e., $\mathbf{i} \cdot \mathbf{j} = \mathbf{k}$.

(a) Show that all other clockwise products of two distinct elements from $\mathbf{i}, \mathbf{j}, \mathbf{k}$ also give positive products.

(b) Verify that all possible counterclockwise products of distinct elements of $\mathbf{i}, \mathbf{j}, \mathbf{k}$ give negative products.

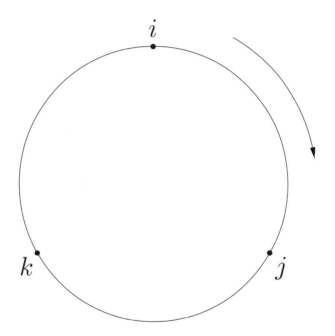

Figure 5.6.

5.1.10. Consider the set $R = \{a, b, c, d\}$ of order 4 where $+$ and $\times$ are defined by the tables below

$+$	a	b	c	d
a	a	b	c	d
b	b	a	d	c
c	c	d	a	b
d	d	c	b	a

$\times$	a	b	c	d
a	a	a	a	a
b	a	b	c	d
c	a	c	d	b
d	a	d	b	c

(a) Show that this set R together with the binary relations defined by the above tables forms a field of order 4. You do not have to prove the associative nor the distributive laws, but you do have to prove the commutative laws for both addition and multiplication. Note that if a binary relation is given by a table there is an easy way to check to see if it is commutative. What is it? (HINT: Consider the main diagonal.)

(b) Without actually doing it tell how many cases would be needed to check to verify the associative law ($(xy)z = x(yz)$) or distributive law ($x(y+z) = xy + xz$)).

(c) Can this field be isomorphic to $\mathbb{Z}_p$ for any prime p?

5.1.11. Prove: If R is a commutative ring with more than one element and which is such that $\forall\, a, b \in R\ \exists x \in R$ such that $ax = b$, then R is a field.

5.1.12. In the proof that a finite integral domain D is a field, tell exactly where the fact that $|D| < \infty$ was used.

(HINT: Try to apply this proof to the integral domain $\mathbb{Z}$ and see where it fails.)

5.1.13. (a) In the proof that a nontrivial homomorphism between fields must be injective, determine exactly at what point the fact that the domain and range were fields is used. Note that this is after we used Lemma 3.5.5.

(b) Would the same proof go through for integral domains?

5.1.14. Show that a skew field or division ring cannot have zero divisors. Does this imply that a division ring must be an integral domain?

EXERCISES FOR SECTION 5.2

5.2.1. If $\overline{\mathbb{Q}} = \{(a, b) \in \mathbb{Z} \times \mathbb{Z} : b \neq 0\}$, then prove that for $(a, b), (c, d) \in \overline{\mathbb{Q}}$, $(a, b) \equiv (c, d)$ iff $ad = bc$ is an equivalence relation on $\overline{\mathbb{Q}}$. (HINT: Only transitivity takes a little work. Write out what the equivalences mean and multiply them together.)

5.2.2. Given a fraction $\frac{a}{b} \in \mathbb{Q}$, i.e., an equivalence class, show that if you multiply numerator and denominator by the same nonzero integer, say $d \neq 0$, then the resulting fraction is equal to the original, i.e., they are equivalent under the equivalence relation defined in #1.

5.2.3. (a) Show that addition is associative as defined in $\mathbb{Q}$.

(b) Show that multiplication is distributive over addition in $\mathbb{Q}$. Explain why it is not necessary to verify both left and right distributivity in $\mathbb{Q}$.

5.2.4. In the proof of the theorem about the construction of $\mathbb{Q}$ from $\mathbb{Z}$ (Theorem 5.2.1), we did not actually prove directly that $\overline{\mathbb{Z}} = \left\{ \frac{z}{1} : z \in \mathbb{Z} \right\}$ is a subring of $\mathbb{Q}$. Explain why what we did proves it even though we did not give an explicit proof. (We say it was implicit in the proof.)

5.2.5. Given the map $f : \mathbb{Z} \to \mathbb{Q}$ by $z \mapsto \frac{z}{1}$ for any $z \in \mathbb{Z}$. Prove that this f is a monomorphism, i.e., you must show that f is a ring homomorphism which is also 1-1.

5.2.6. Complete the proof of Corollary 5.2.1 by showing that the map $\phi : \mathbb{Q} \to \mathbb{Q}'$ is an isomorphism.

(HINT: Follow the way the map ϕ in Theorem 5.2.2 is shown to be a monomorphism.)

5.2.7. Prove that the map $\phi : \{z \cdot 1 : z \in \mathbb{Z}\} \to \mathbb{Z}_p$ defined in the proof of Theorem 5.2.4 is an isomorphism.

(Note: Since $\text{cha}(F) = p$ it is possible that $z_1 \cdot 1 = z_2 \cdot 1$ where the integers z_1, z_2 are unequal. Thus you must show that this map is well-defined.)

5.2.8. Consider the map $\varphi : \mathbb{Q} \to F$ given in Theorem 5.2.2 by $\frac{a}{b} \mapsto b^{-1}a$. Show that φ preserves multiplication.

5.2.9. Prove both of the following:

Any field of characteristic 0 contains a subfield isomorphic to $\mathbb{Q}$. A field of characteristic p contains a subfield isomorphic to $\mathbb{Z}_p$.

(HINT: First prove that the intersection of subfields of a field is also a field, i.e., if $\{F_\alpha : \alpha \in \Lambda\}$ is a collection of subfields of a field F, then $\bigcap_{\alpha \in \Lambda} F_\alpha$ is a field.

Let F be any field. Let $\{F_\alpha : \alpha \in \Lambda\}$ be the collection of all nontrivial subfields of F and put $\overline{F} = \bigcap_{\alpha \in \Lambda} F_\alpha$. This is the smallest subfield of F. Show why. Show then it must be a prime field. Then to finish off consider two cases: Case (1) $\text{char}(F) = 0$ and Case (2) $\text{char}(F) = p$ for some prime p. Then use the theorem on the characterization of $\mathbb{Q}$ to finish off the proof.)

(Note: Later on in Chapter 11, we will present a much easier proof of this result.)

In the following exercises D is an arbitrary integral domain. These exercises go through the construction of the field of fractions for D. Note this is sometimes also called the **field of quotients**.

5.2.10. On D^* show that $(d_1, d_2) \equiv (c_1, c_2)$ iff $d_1 c_2 = d_2 c_1$ is an equivalence relation. Let $\frac{d_1}{d_2}$ be the equivalence class which (d_1, d_2) belongs to. Write the set whch $\frac{d_1}{d_2}$ is equal to. Finally let $\overline{D}$ be the set of all equivalences classes of elements of D^*.

5.2.11. Show that $+$ and $\times$ defined on $\overline{D}$ by

$$\frac{d_1}{d_2} + \frac{c_1}{c_2} = \frac{d_1 c_2 + c_1 d_2}{d_2 c_2}$$
$$\frac{d_1}{d_2} \cdot \frac{c_1}{c_2} = \frac{d_1 c_1}{d_2 c_2}$$

for $\frac{d_1}{d_2}, \frac{c_1}{c_2} \in \overline{D}$ are both well-defined operations on $\overline{D}$.

5.2.12. Show that with respect to the binary operations defined in Problem 5.2.11, $\overline{D}$ forms a field.

5.2.13. If we call $e \in D$ the unity of D (i.e., the multiplicative identity), then $e \neq 0$ (i.e., the additive identity of D). Consider the mapping $f : D \to \overline{D}$ given by $d \mapsto \frac{d}{e}$ for any $d \in D$. Prove that this f is a monomorphism. Thus if we identify D with its image, $im(f)$, in $\overline{D}$, then we can consider D to be a subring of $\overline{D}$.

5.2.14. Here we show that $\overline{D}$ is the smallest field containing D in the following sense. Let $D \subset F$ where F is a field. Let $\frac{d}{f} \in \overline{D}$. Then $d, f \in D$. Explain why f^{-1} exists as an element of F. Let φ be the map from $\overline{D}$ to F given by $\varphi(\frac{d}{f}) = f^{-1}d$. Prove that $\varphi : \overline{D} \to F$ is a well-defined monomorphism. So that if we identify $\overline{D}$ with its image in F, $im(\varphi)$, then we can consider $\overline{D}$ as a subring of F.

EXERCISES FOR SECTION 5.3

5.3.1. Explain why $2.\overline{9} = 2.9999999 \cdots = 3.000000 \ldots$.
(HINT: $2.\overline{9} = 2 + .999999 = 2 + (.9 + .09 + .009 + 0009 + \ldots)$ now sum the geometric series.)

5.3.2. Determine the decimal representaion, either as a terminating decimal or a repeating decimal, of each of the following:

(a) $\frac{3}{11}$. (b) $\frac{2010}{7}$.

(c) $\frac{10}{9}$.   (d) $\frac{9}{10}$.

5.3.3. Express $2.0\overline{15}$ as a fraction $\frac{m}{n}$ where $m, n \in \mathbb{Z}$ and $\gcd(m, n) = 1$.

5.3.4. Show that between any two real numbers there is a third real number, i.e., consider $\mathbb{R}$ as an ordered field if $x, y \in \mathbb{R}$ and $x < y$, then $\exists z \in \mathbb{R}$ such that $x < z < y$.

(HINT: Take $z = \frac{x+y}{2}$ then using the properties of order in an ordered integral domain — see the exercises for Section 3.4 — show z satisfies the required inequality.)

5.3.5. Show that each positive distance and hence each real number can be expressed as a decimal expansion. (Hint: place the distance x on the number line and put out integer points. It will fall $n \leq x < n+1$. This will give the first decimal digits. Now subdivide $[n, n+1)$ into ten equal parts. The distance x will be in one of these pieces. This will give the next decimal digit. Continue in this way.)

5.3.6. Prove the following properties of absolute value, $|\ |$, on $\mathbb{R}$. For $x, y \in \mathbb{R}$
  (1) $|x| \geq 0$ and $|x| = 0$ iff $x = 0$.
  (2) $|xy| = |x|\,|y|$.

5.3.7. Define $d(x, y) = |x - y|$ for $x, y \in \mathbb{R}$. Now show this distance function has the following properties using the properties of absolute values on $\mathbb{R}$:
  (1) $d(x, y) \geq 0$ and $d(x, y) = 0$ iff $x = y$.
  (2) $d(x, y) = d(y, x)$.
  (3) $d(x, y) \leq d(x, z) + d(z, y)$.

5.3.8. Show that the limit of a sequence in $\mathbb{R}$, if it exists, is unique.

(HINT: Suppose $x_n \to x$ and $x_n \to y$ where $x \neq y$. Let $\epsilon = |x - y| > 0$. Show that $|x_n - x| + |x_n - y| \geq \epsilon$ but if $x_n \to x$ and $x_n \to y$ using $\epsilon/2$ in both limit definitions, then $|x_n - x| + |x_n - y| < \epsilon$ for all sufficiently large n. This contradiction shows the limit of a sequence is unique.)

5.3.9. Define the greatest lower bound, GLB, for a bounded set $S \subset \mathbb{R}$.

5.3.10. Let X_1 and X_2 be nonempty sets of real numbers having LUB's b_1, b_2 respectively. Let $X = \{x_1 + x_2 : x_1 \in X_1, x_2 \in X_2\}$. Prove that $LUB(X) = b_1 + b_2$.

5.3.11. Suppose that $x, y \in \mathbb{R}$. Prove that if x is rational and y is irrational, then $x + y$ must be irrational.

5.3.12. Suppose that $x, y \in \mathbb{R}$. Prove that if x is rational, $x \neq 0$, and y is irrational, then xy must be irrational.

5.3.13. True or false: If $x \in \mathbb{R}$ is irrational, then x^{-1} is irrational too.

5.3.14. Give examples to show that if $x, y \in \mathbb{R}$ and x and y are both irrational, then xy may be either rational or irrational depending upon x and y.

5.3.15. Prove that a nonempty subset of an ordered field can have at most one LUB.

5.3.16. Prove: Every complete ordered field is archimedean.

(HINT: Let F be a complete ordered field and suppose that F is not archimedean. Let $\mathbb{Z}^+$ be the set of positive integers in F, i.e., if $1 \in F$ is the unity of F, then $\mathbb{Z}^+ = \{n \cdot 1 : n \text{ is a positive integer}\}$. Then since F is not archimedean $\exists a > 0$ and $\exists b > 0$ such that $\forall n \in \mathbb{Z}^+$, $na \leq b$. This implies that the nonempty subset of $F, \{na : n \in \mathbb{Z}^+\}$ has an upper bound. Explain why. Thus this set must have a LUB. Call it $z_0 \in F$. Thus $(n+1)a \leq z_0$ for all $n \in \mathbb{Z}^+$. But this implies that $na \leq z_0 - a < z_0$. Explain why this contradicts the definition of z_0. This contradiction proves F must be archimedean.)

5.3.17. Show that the rational, $\mathbb{Q}$, are dense in the reals, $\mathbb{R}$.

5.3.18. If $\overline{\mathbb{Q}}$ is the set of all Cauchy sequences of rationals and

$$(q_1, q_2, ...), (s_1, s_2, ...) \in \overline{\mathbb{Q}}$$

where $(q_1, q_2, ...) \sim (s_1, s_2, ...)$ iff $\lim(q_i - s_1) = 0$, show that $\sim$ is an equivalence relation on $\overline{\mathbb{Q}}$.

5.3.19. Let $\mathbb{R}$ be the set of all equivalence classes of Cauchy sequences of rationals under the equivalence relation in Problem 5.3.12. We define a map from $\mathbb{Q}$ to $\mathbb{R}$ as follows: for each $q \in \mathbb{Q}$ associate the sequence $(q, q, q, \ldots)$. This is a Cauchy sequence — explain why — then our map is $q \mapsto [(q, q, \ldots, q, \ldots)]$, i.e., this is the equivalence class of this Cauchy sequence. Show this map is a monomorphism, i.e., a ring homomorphism which is injective. This shows that $\mathbb{Q}$ can be embedded into $\mathbb{R}$.

5.3.20. Let $\mathbb{R}$ be the set of all equivalence classes of Cauchy sequences of rationals. If $r \in \mathbb{R}$, then r is an equivalence class of Cauchy sequences of rationals. Let's use the same notation r for such a Cauchy sequence, $r = (q_1, q_2, \ldots)$. We define the absolute value of r, as $|r| = (|q_1|, |q_2|, \ldots)$. In the text, it was shown that this was itself a Cauchy sequence. Show here that this satisfies all the usual properties of absolute values.

5.3.21. Show that the p-adic norm on the rationals is nonarchimedean.

5.3.22. Show that in the p-adic numbers the norm of any natural number is ≤ 1.

5.3.23. Let p be a prime and let Z_p denote all the p-adic numbers whose p-adic norms are less than or equal to 1. Show that this set forms a subring of $\mathbb{Q}_p$. (This is called the ring of p-adic integers).

EXERCISES FOR SECTION 5.4

5.4.1. Show that $\mathbb{C} = \{x + \mathbf{i}y \ : \ x, y \in \mathbb{R}\}$ with operations defined as follows: if $z = x + \mathbf{i}y$, $w = a + \mathbf{i}b \in \mathbb{C}$,

$$z \pm w \ = \ (x \pm a) + \mathbf{i}(y \pm b)$$
$$z \cdot w \ = \ (xa - yb) + \mathbf{i}(xb + ya)$$

forms a field.

5.4.2. Express each of the following in the form $x + \mathbf{i}y$ where $x, y \in \mathbb{R}$:
   (a) $(3 - \mathbf{i})(1 + \mathbf{i})$.   (b) $\mathbf{i}^3$.   (c) $(-\mathbf{i})^3$.
   (d) $\mathbf{i}^{25}$.   (e) $\frac{1-\mathbf{i}}{1+\mathbf{i}}$.

5.4.3. If $z, w \in \mathbb{C}$: Then show that
(a) $|z| \geq 0$, and $|z| = 0$ if and only if $z = 0$.
(b) $|zw| = |z| \, |w|$.
(c) $|z + w| \leq |z| + |w|$.

5.4.4. Prove that for $z, w \in \mathbb{C}$ the following hold:
(1) $\overline{z + w} \ = \ \overline{z} + \overline{w}$.
(2) $\overline{zw} \ = \ (\overline{z})(\overline{w})$.
(3) $\overline{\overline{z}} = z$.
(4) $|\overline{z}| \ = \ |z|$.

5.4.5. Let the map $\phi : \mathbb{C} \to \mathbb{C}$ be given by $\phi(z) = \overline{z}$ for all $z \in \mathbb{C}$. Using the properties of complex conjugates proved in the previous exercise, show that ϕ is an isomorphism. Such an isomorphism is usually called an **automorphism** because its domain and range are the same.

5.4.6.(a) For $\phi : \mathbb{C} \to \mathbb{C}$ from the previous exercise show that $\phi(z) = z$ if and only if z is real.

(b) Suppose that $\varphi : \mathbb{C} \to \mathbb{C}$ with the property that $\varphi(x) = x \ \forall x \in \mathbb{R}$ is an isomorphism. Prove that with φ as given in Problem 5.4.5 is either the identity map on $\mathbb{C}$ or ϕ, i.e., $\varphi(z) = z \ \forall z \in \mathbb{C}$, or $\varphi = \phi$.

(HINT: φ is the identity on $\mathbb{R}$, $\mathbf{i}^2 = -1$, so what are the possibilities for $\varphi(\mathbf{i})$?)

5.4.7. Express each of the following complex numbers, z, in polar form where here $\theta = \text{Arg}(z)$ is such that $\theta \in [0, 2\pi)$:

(a) $-1 - \mathbf{i}$.   (b) $-10\mathbf{i}$.

(c) -10.   (d) $-\sqrt{3} + \mathbf{i}$.

5.4.8. (Gamow's Problem — Taken from *An Imaginary Tale: The Story of $\sqrt{-1}$* by Paul J. Nahin) Gamow's problem is presented as the story of a "young and adventurous man" who discovers an ancient parchment among his late great-grandfather's papers. On it he reads the following:

"Sail to _____ North latitude and _____ West longitude where thou wilt find a deserted island. There lieth a large meadow, not pent, on the north shore of the island where standeth a lonely oak and a lonely pine. There thou wilt also see an old gallows on which we once wont to hang traitors. Start thou from the gallows and walk to the oak counting thy steps. At the oak thou must turn *right* by a right angle and take the same number of steps. Put a spike in the ground. Now must thou return to the gallows and walk to the pine counting thy steps. At the pine thou must turn *left* by a right angle and see that thou takest the same number of steps, and put another spike into the ground. Dig halfway between the spikes; the treasure is there."

We show all of this in Figure 5.7. Gamow added two footnotes: one to tell us that he has, of course, omitted values of longitude and latitude to prevent any of us from rushing off to start digging up the treasure; and a second to tell us that he knows that oak and pine trees don't grow on deserted islands but he has altered the real type of trees again to keep the actual island secret.

The young man follows the instructions, at least to the point of locating the island, where he sees the oak and pine tree. But, alas, there is no gallows! Unlike the living trees, the gallows has long since disintegrated in the weather, and not a trace of it or its location remains. Unable to carry out the rest of the instructions (or so he believes), the young man sails away with nothing for his troubles. Gamow observes that this is too bad because he could have located the treasure with no difficulty at all, if he understood just a little about the arithmetic of complex numbers.

Use complex numbers to find the treasure.

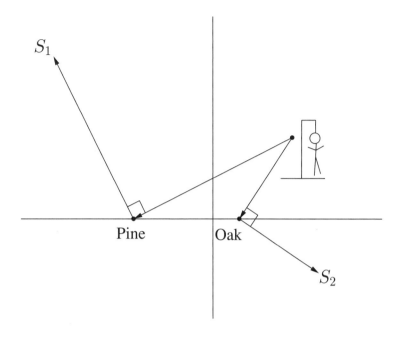

Figure 5.7.

5.4.9. Use DeMoivre's Theorem for powers to compute each of the following and express your answers in rectangular form, $x + iy$, by evaluating the necessary trigonometric functions:

(a) $(-1 - i)^6$. (b) $(1 - i\sqrt{3})^9$.

(c) $(\frac{-1}{\sqrt{2}} + \frac{i}{\sqrt{2}})^{1000}$. (d) $(\cos(21°) + i\sin(21°))^{11}$.

5.4.10. Show using trigonometric identities and the polar form of a complex number written as $z = r(\cos(\theta) + i\sin(\theta))$ the following: if $z_1, z_2 \in \mathbb{C}$ and $z_j = r_j(\cos(\theta_j) + i\sin(\theta_j))$ for $j = 1, 2$ then

(a) $z_1 z_2 = r_1 r_2(\cos(\theta_1 + \theta_2) + i\sin(\theta_1 + \theta_2))$

(b) $\frac{z_1}{z_2} = \frac{r_1}{r_2}(\cos(\theta_1 - \theta_2) + i\sin(\theta_1 - \theta_2))$ if $z_2 \neq 0$.

(Note: We showed this in the text, but we used the polar form in exponential notation, i.e., $z = e^{i\theta}$ (from Euler's formula). Here you are to do this without this exponential notation!)

5.4.11. (a) In the text we saw that from DeMoivre's theorem for powers, if $z = r(\cos(\theta) + i\sin(\theta))$ then $z^n = r^n(\cos(n\theta) + i\sin(n\theta))$ for n any integer holds just by using the form $z = re^{i\theta}$ (i.e., Euler's formula) and assuming that the usual laws of exponents hold in $\mathbb{C}$. Here we do the same without Euler's formula. If you go back to Problem 5.4.9 and allow $z = z_1 = z_2$, then note what this gives you for z^2. Then using your expression for z^2, take it one step further just using trigonometric identities, to get an expression for z^3. What does this indicate that you should get for z^n where $n > 1$ is any such integer. Use mathematical induction to prove that for any $n \in \mathbb{N}$, $z^n = r^n(\cos(n\theta) + \mathbf{i}\sin(n\theta))$.

(b) Show without Euler's formula, that $z^n = r^n(\cos(n\theta) + \mathbf{i}\sin(n\theta))$ holds for any integer. To do so, proceed as follows:

(i) Show this holds for $n = 0$.

(ii) Show this holds for any negative integer n as follows. Show it for $n = -1$ and then use the case for n positive and $n = -1$ to prove it for any integer n. Here note that if n is any negative integer, then $n = -k$ where $k \in \mathbb{N}$, then $z^n = z^{-k} = (z^{-1})^k$ assuming that the usual laws of exponents hold for $\mathbb{C}$.

5.4.12. Verify that the points with coordinates $(\cos(60°) + \mathbf{i}\sin(60°))^n$ for $n = 1, 2, 3, 4, 5, 6$ are the vertices of a regular hexgon inscribed in a circle of radius 1 with center at the orgin.

5.4.13. Find the cube roots of 1 (unity) and express them in rectangular form, i.e., $x + \mathbf{i}y$. Draw a figure showing that these numbers are the coordinates of the vertices of a regular polygon of three sides (i.e., an equilateral triangle).

5.4.14. Prove that the inverse of an nth root of unity is also an nth root of unity.

(HINT: If $w_k = e^{\mathbf{i}\cdot\frac{2\pi k}{n}} = \cos(\frac{2\pi k}{n}) + \mathbf{i}\sin(\frac{2\pi k}{n})$, $k = 0, 1, \ldots, n-1$ is any nth root of unity, consider w_k^{-1}. Use the fact that DeMoivre's theorem for powers if $z = r(\cos(\theta) + \mathbf{i}\sin(\theta))$ then $z^n = r^n(\cos(n\theta) + \mathbf{i}\sin(n\theta))$ holds for n any integer (see Problem 5.4.10), and then consider the complex number with argument $-2\pi k/n$ and the complex number with argument $\frac{2\pi(n-k)}{n}$ and note that the second complex number is included among $w_k = e^{\mathbf{i}\cdot\frac{2\pi k}{n}} = \cos(\frac{2\pi k}{n}) + \mathbf{i}\sin(\frac{2\pi k}{n})$, $k = 0, 1, \ldots, n-1$.)

5.4.15. Let $z = e^{\mathbf{i}\theta}$, then $w = r^{1/n}e^{\mathbf{i}\theta/n}$ is an nth root of z.

(a) Show that if $0 < k < n$ and $k \in \mathbb{N}$, then $w_k = r^{1/n}e^{\mathbf{i}\frac{\theta+2k}{n}}$ is also an nth root of z, but that $w_k \neq w$.

(HINT: To do this consider the arguments of both of these complex numbers and what it would mean if the two complex numbers had the same argument.)

(b) Show that no two of the complex numbers $w_k = r^{1/n}e^{\mathbf{i}\cdot\frac{\theta+2\pi k}{n}}$, $k = 0, 1, \ldots, n-1$ can be equal.

(HINT: To do this as above, consider the arguments of any two of these complex numbers and what it would mean if the two complex numbers were to be equal.)

(c) Consider $w_k = r^{1/n} e^{i\frac{\theta + 2k}{n}}$ for $k = 0, 1, \ldots, n - 1$. Take w_k^n and w_k^{-1} show that both of these complex numbers already appear in the list $w_k = r^{1/n} e^{i \cdot \frac{\theta + 2\pi k}{n}}$, $k = 0, 1, \ldots, n - 1$. This implies that, as was noted in the text, these w_k would start repeating if you allow either $k < 0$ or $k \geq n$.

5.4.16. Find the required roots and express them in rectangular form, i.e., $x + iy$:

(a) The cube roots of $-8i$.
(b) The fourth roots of -4.
(c) The sixth roots of $-i$.
(d) The fourth roots of $-1 + i\sqrt{3}$.

Chapter 6

Basic Group Theory

6.1 Groups, Subgroups and Isomorphisms

Recall from Chapter 1 that the three most commonly studied algebraic structures are groups, rings and fields. We have now looked rather extensively at rings and fields and in this chapter we consider the basic concepts of group theory. Groups arise in many different areas of mathematics. For example they arise in mathematics as groups of permutations; in geometry as groups of congruence motions; and in topology as groups of various types of continuous functions. Later in this book they will appear in Galois theory as groups of automorphisms of fields. Groups also occur in many applications outside of mathematics. Mathematicians, however, like to study groups for their own sake. This is evidenced by the large number of articles devoted solely to pure group theory which appear every year in technical journals. In this first section we look more carefully at the material introduced in Section 2.7.

Definition 6.1.1. *A* **group** *is a set G with one binary operation defined on it, which we will denote by multiplication, or juxtaposition, such that*

(1) The operation is associative, that is, $(g_1g_2)g_3 = g_1(g_2g_3)$ for all $g_1, g_2, g_3 \in G$.

(2) There exists an identity for this operation, that is, an element $1 \in G$ such that $1g = g$ and $g1 = g$ for each $g \in G$.

(3) Each $g \in G$ has an inverse for this operation, that is, for each g there exists a $g^{-1} \in G$ with the property that $gg^{-1} = 1$ and $g^{-1}g = 1$.

If in addition the operation is commutative, that is, $g_1 g_2 = g_2 g_1$ for all $g_1, g_2 \in G$, the group G is called an **abelian group**.

The **order** *of G, denoted $|G|$, is the number of elements in the group G or the* **size** *(see Chapter 2) of the set G. If $|G| < \infty$, G is said to be a* **finite group** *otherwise G is an* **infinite group**.

Let us remark immediately that since a group G has a binary operation defined on it, the operation by its very definition is closed, i.e., for any $a, b \in G$ it must be true that $ab \in G$. We also note that it is customary to talk of a group G, even though G is just a set, in a given discussion. This is actually not precise because a group, as just defined, is a set G *together* with a binary operation and it is possible that on a given set G a number of binary operations can be introduced such that G together with each of these operations is a group. In any discussion, however, the binary operation will be fixed and there will be no confusion in speaking just about the group G. It follows easily from the definiton that the identity is unique and that each element has a unique inverse.

Lemma 6.1.1. *If G is a group then there is a unique identity in G. Further if $g \in G$ its inverse is unique. Finally if $g_1, g_2 \in G$ then $(g_1 g_2)^{-1} = g_2^{-1} g_1^{-1}$.*

Proof. Suppose that 1 and e are both identities for G. Then $1e = e$ since 1 is an identity and $1e = 1$ since e is an identity. Therefore $1 = e$ since the binary operation on G is a function. Thus there is only one identity.

Next suppose that $g \in G$ and g_1 and g_2 are inverses for g. Then by associativity

$$g_1 g g_2 = (g_1 g) g_2 = 1 g_2 = g_2$$

since $g_1 g = 1$. On the other hand, again by associativity,

$$g_1 g g_2 = g_1 (g g_2) = g_1 1 = g_1$$

since $g g_2 = 1$. It follows that $g_1 = g_2$ and g has a unique inverse.

Finally consider

$$(g_1 g_2)(g_2^{-1} g_1^{-1}) = g_1 (g_2 g_2^{-1}) g_1^{-1} = g_1 1 g_1^{-1} = g_1 g_1^{-1} = 1,$$

where we have used associativity. Similarly

$$(g_2^{-1} g_1^{-1})(g_1 g_2) = 1.$$

Therefore$(g_2^{-1} g_1^{-1})$ is an inverse for $g_1 g_2$ and since inverses are unique it is the inverse of the product.   □

Groups most often arise as permutations on a set. We will see this, as well as other specific examples of groups, in the next two sections.

Finite groups can be completely described by their **group tables** or multiplication tables. These are sometimes called **Cayley tables**. In general, let $G = \{g_1, \ldots, g_n\}$ be a group with the **multiplication table** of G being:

	g_1	g_2	$\cdots$	g_j	$\cdots$	g_n
g_1	$\cdots$					
g_2	$\cdots$					
$\vdots$						
g_i	$\cdots$	$\cdots$	$\cdots$	$g_i g_j$	$\cdots$	
$\vdots$						
g_n	$\cdots$					

The entry in the row of $g_i \in G$ and column of $g_j \in G$ is the product (in that order) $g_i g_j$ in G.

Groups satisfy the **cancellation law for multiplication**.

Lemma 6.1.2. *If G is a group and $a, b, c \in G$ with $ab = ac$ or $ba = ca$ then $b = c$.*

Proof. Suppose that $ab = ac$. Then a has an inverse a^{-1} so we have

$$a^{-1}(ab) = a^{-1}(ac).$$

From the associativity of the group operation we then have

$$(a^{-1}a)b = (a^{-1}a)c \Longrightarrow 1 \cdot b = 1 \cdot c \Longrightarrow b = c.$$

The other cancellation law follows similarly by multiplication by a^{-1} on the right. $\qquad\square$

A consequence of Lemma 6.1.2 is that each row and each column in a group table of a finite group is just a permutation of the group elements. That is, each group element appears exactly once in each row and each column.

A subset $H \subset G$ is a **subgroup** of G if H is also a group under the same operation as G. Thus a subset of a group is a subgroup if it is nonempty and closed under both the group operation and inverses.

Lemma 6.1.3. *A subset $H \subset G$ of a group G is a subgroup if and only if $H \neq \emptyset$ and H is closed under the operation and inverses. That is, if $a, b \in H$ then $ab \in H$ and $a^{-1}, b^{-1} \in H$.*

We leave the proof of this to the exercises.

Let G be a group and $g \in G$; we denote by g^n, $n \in \mathbb{N}$, as with numbers, the product of g taken n times. This makes sense because of the generalized associative law. A negative exponent will indicate the inverse of the positive exponent. As usual, let $g^0 = 1$. It can be shown that group exponentiation will satisfy the standard laws of exponents. Now consider the set

$$\{1 = g^0, g, g^{-1}, g^2, g^{-2}, \ldots\}$$

of all powers of g. We will denote this set by $\langle g \rangle$.

Lemma 6.1.4. *If G is a group and $g \in G$ then $\langle g \rangle$ forms a subgroup of G called the **cyclic subgroup generated by** g. $\langle g \rangle$ is abelian even if G is not.*

Proof. If $g \in G$ then $g \in \langle g \rangle$ and hence $\langle g \rangle$ is nonempty. Suppose then that $a = g^m, b = g^n$ are elements of $\langle g \rangle$. Then $ab = g^m g^n = g^{m+n} \in \langle g \rangle$ so $\langle g \rangle$ is closed under the group operation. Further $a^{-1} = (g^n)^{-1} = g^{-n} \in \langle g \rangle$ so $\langle g \rangle$ is closed under inverses. Therefore $\langle g \rangle$ is a subgroup.

Finally $ab = g^m g^n = g^{m+n} = g^{n+m} = g^n g^m = ba$ and hence $\langle g \rangle$ is abelian independent of whether or not G is. $\square$

Suppose that $g \in G$ and $g^m = 1$ for some positive integer m. Then let n be the smallest positive integer such that $g^n = 1$. It follows that the set of elements $\{1, g, g^2, \ldots, g^{n-1}\}$ are all distinct but for any other power g^k we have $g^k = g^t$ for some $t = 0, 1, \ldots, n-1$ (see the exercises). The cyclic subgroup generated by g then has order n and we say that g has **order** n which we denote by $o(g) = n$. If no such n exists we say that g has **infinite order**. We will look more deeply at cyclic groups and cyclic subgroups in Section 6.4.

We next introduce a very useful result about orders of elements.

Lemma 6.1.5. *If G is a group and $g \in G$ with $o(g) = n \in \mathbb{N}$, then $g^k = 1$ if and only if $n \mid k$.*

Proof. If $o(g) = n$ and $n \mid k$ then $k = qn$ where $q \in \mathbb{Z}$. So $g^k = (g^n)^q = 1^q = 1$. Conversely if $g^k = 1$, then applying the division algorithm to divide

k by n gives $k = qn + r$ where $0 \leq r < n$ Thus if $r \neq 0$, we get using the assumed laws of exponents

$$g^r = g^{k-qn} = g^k \cdot (g^n)^{-q} = 1.$$

But $0 < r < n$. This contradicts the definition of n. So $r = 0$ and $n \mid k$ □

We introduce one more concept before looking at examples.

Definition 6.1.2. *If G and H are groups then a mapping $f : G \rightarrow H$ is a (group)* **homomorphism** *if $f(g_1 g_2) = f(g_1)f(g_2)$ for any $g_1, g_2 \in G$. We say if this is true that f* **preserves the operation***. If f is also a bijection then it is an* **isomorphism***. An isormorphism of the group G onto itself is an* **automorphism** *of G.*

As with rings and fields, we say that two groups G and H are **isomorphic**, denoted by $G \cong H$, if there exists an isomorphism $f : G \rightarrow H$. This means that abstractly G and H have exactly the same algebraic structure.

6.2   Examples of Groups

As already mentioned groups arise in many diverse areas of mathematics. In this section, we present specific examples of groups.

First of all any ring or field under addition forms an abelian group. Hence, for example $(\mathbb{Z}, +), (\mathbb{Q}, +), (\mathbb{R}, +), (\mathbb{C}, +)$ where $\mathbb{Z}, \mathbb{Q}, \mathbb{R}, \mathbb{C}$ are respectively the integers, the rationals, the reals, and the complex numbers, all are infinite abelian groups. If $\mathbb{Z}_n$ is the modular ring, then for any natural number n, $(\mathbb{Z}_n, +)$ forms a finite abelian group. In abelian groups the group operation is often denoted by $+$ and the identity element by 0 (zero).

In a field F, the nonzero elements are all invertible and form a group under multiplication. This is called the **multiplicative group** of the field F and is usually denoted by F^* (see Chapter 5). Since multiplication in a field is commutative the multiplicative group of a field is an abelian group. Hence $\mathbb{Q}^*, \mathbb{R}^*, \mathbb{C}^*$ are all infinite abelian groups while if p is a prime $\mathbb{Z}_p^*$ forms a finite abelian group. Recall that if p is a prime then the modular ring $\mathbb{Z}_p$ is a field.

Within $\mathbb{Q}^*, \mathbb{R}^*, \mathbb{C}^*$ there are certain multiplicative subgroups. Since the positive rationals $\mathbb{Q}^+$ and the positive reals $\mathbb{R}^+$ are closed under multiplication and inverses, they form subgroups of $\mathbb{Q}^*$ and $\mathbb{R}^*$ respectively. In $\mathbb{C}$

if we consider the set of all complex numbers z with $|z| = 1$ these form a multiplicative subgroup. Further within this subgroup if we consider the set of nth roots of unity (that is, all z such that $z^n = 1$) for a fixed n this forms a subgroup, this time of finite order.

The multiplicative group of a field is a special case of the unit group of a ring. If R is a ring with identity, recall that a **unit** is an element of R with a multiplicative inverse. Hence in $\mathbb{Z}$ the only units are ± 1 while in any field every nonzero element is a unit. In Chapter 3 we saw the following lemma giving us many examples of groups.

Lemma 6.2.1. *If R is a ring with identity then the set of units in R forms a group under multiplication called the **unit group** or **group of units** of R and is denoted by $U(R)$. If R is a field then $U(R) = R^*$.*

To present examples of nonabelian groups we turn to matrices, as we did to find noncommutative rings. If F is a field, A is an $n \times n$ matrix over F means that A has entries from F. We let

$$GL(n, F) = \{A : A \text{ is an } n \times n \text{ matrix over } F \text{ and } \det(A) \neq 0\}$$

and

$$SL(n, F) = \{A : A \text{ is an } n \times n \text{ matrices over } F \text{ and } \det(A) = 1\}$$

here of course det is the determinant.

Lemma 6.2.2. *If F is a field then for $n \geq 2$, $GL(n, F)$ forms a nonabelian group under matrix multiplication and $SL(n, F)$ forms a subgroup. $GL(n, F)$ is called the n-dimensional **general linear group** over F, while $SL(n, F)$ is called the n-dimensional **special linear group** over F.*

Proof. Recall from matrix theory that for two $n \times n$ matrices A, B with $n \geq 2$ over a field F we have

$$\det(AB) = \det(A)\det(B).$$

We also need the fact from matrix theory that the inverse of an $n \times n$ matrix A over F exists if and only if $\det(A) \neq 0$.

Now for any field, the $n \times n$ identity matrix I has determinant 1 and hence $I \in GL(n, F)$. Since determinants multiply and a field F has no zero divisors, it follows that the product of two matrices with nonzero determinant also has nonzero determinant and hence $GL(n, F)$ is closed under the usual matrix product. Further, over a field F, if A is an invertible matrix then

$$\det(A^{-1}) = \frac{1}{\det A} = (\det(A))^{-1}$$

and so if A has nonzero determinant so does its inverse. It follows that $GL(n, F)$ has the inverse of any of its elements. Since matrix multiplication is associative it follows that $GL(n, F)$ forms a group. It is nonabelian since in general matrix multiplication is noncommutative.

We leave the fact that $SL(n, F)$ forms a subgroup to the exercises. $\qquad \square$

6.2.1 Permutations and the Symmetric Group

As remarked above groups appear everywhere in mathematics. Groups arise however, most often, as groups of transformations or permutations on a set. In this section we will take a short look at permutation groups. In what follows, it is assumed that A is a nonempty set.

Definition 6.2.1. *If A is a set, a **permutation** on A is a one-to-one mapping of A onto itself, i.e., a bijection. We denote by S_A the set of all permutations on A.*

Theorem 6.2.1. *For any set A, S_A forms a group with respect to the operation of composition of functions. The group S_A is called the **symmetric group** on A. If $|A| > 2$ then S_A is nonabelian. Further if A, B have the same cardinality, then $S_A \cong S_B$.*

Proof. If S_A is the set of all permutations on the set A, we must show that composition is an operation on S_A that is associative and has an identity and inverses.

Let $f, g \in S_A$. Then f, g are one-to-one mappings of A onto itself. Consider $f \circ g : A \to A$. If $f \circ g(a_1) = f \circ g(a_2)$, then $f(g(a_1)) = f(g(a_2))$ and $g(a_1) = g(a_2)$, since f is one-to-one. But then $a_1 = a_2$ since g is one-to-one. Thus $f \circ g$ is 1-1.

If $a \in A$, there exists $a_1 \in A$ with $f(a_1) = a$ since f is onto. Then there exists $a_2 \in A$ with $g(a_2) = a_1$ since g is onto. Putting these together, $f(g(a_2)) = a$, and therefore $f \circ g$ is onto. Therefore, $f \circ g$ is also a permutation and composition gives a valid binary operation on S_A.

The identity function $1(a) = a$ for all $a \in A$ will serve as the identity for S_A, while the inverse function for each permutation will be its inverse as a function. Note the inverse of a bijection is again a bijection. (See Chapter 2.) Such unique inverse functions exist since each permutation is a bijection.

Finally, composition of functions is always associative (see Chapter 2) and therefore S_A forms a group.

Suppose that $|A| > 2$. Then A has at least three elements. Call them a_1, a_2, a_3. Consider the two permutations f and g which fix, that is leave unchanged, all of A except a_1, a_2, a_3 (i.e., if there is any $x \in A - \{a_1, a_2, a_3\}$, then $f(x) = g(x) = x$) and on these three elements we define

$$f(a_1) = a_2, f(a_2) = a_3, f(a_3) = a_1$$

$$g(a_1) = a_2, g(a_2) = a_1, g(a_3) = a_3$$

Then under composition

$$f(g(a_1)) = a_3, f(g(a_2)) = a_2, f(g(a_3)) = a_1$$

while

$$g(f(a_1)) = a_1, g(f(a_2)) = a_3, g(f(a_3)) = a_2.$$

Since $f \circ g(a_1) \neq g \circ f(a_1)$, this is sufficient for $f \circ g \neq g \circ f$ and hence S_A is not abelian.

If A, B have the same size or cardinality, then there exists a bijection $\sigma : A \to B$ (see Chapter 2). Define a map $F : S_A \to S_B$ in the following manner: if $f \in S_A$, let $F(f)$ be the permutation on B given by

$$F(f)(b) = \sigma(f(\sigma^{-1}(b))).$$

We now need to verify that F so defined is actually an isomorphism.

Since σ is a bijection, σ^{-1} exists and is itself a bijection — as noted above. But the composition of bijections is again a bijection (see Chapter 2), so $\sigma f \sigma^{-1}$ is a bijection. Now note that

$$B \xrightarrow{\sigma^{-1}} A \xrightarrow{f} A \xrightarrow{\sigma} B.$$

Thus $\sigma f \sigma^{-1} : B \to B$. This shows that $\sigma f \sigma^{-1} \in S_B$ which means that our map F really has domain S_A and codomain S_B, i.e., $F : S_A \to S_B$.

Next we need to show that F preserves the operation. Here the operations in both the domain and codomain are composition but it is possible that the two operations can be very different. This means that if $f, g \in S_A$, we need to show that

$$F(f \circ g) = F(f) \circ F(g)$$

It must be understood that not only are the f, g functions but so are $F(f)$, $F(g)$ — but they have different domains. We will show that the two functions $F(f \circ g)$ and $F(f) \circ F(g)$ are equal. Let $b \in B$ be arbitrary. We need to show that they have the same value on b. Consider

$$F(f \circ g)(b) = \sigma(f(g(\sigma^{-1}(b)))).$$

Now $F(f) \circ F(g)(b) = F(f)(\sigma(g(\sigma^{-1}(b)))) = \sigma(f(\sigma^{-1}(\sigma(g(\sigma^{-1}(b))))))$. Let's make sure that this makes sense since it's such a mess. Since $\sigma : A \rightarrow B$, $\sigma^{-1} : B \rightarrow A$ and $\sigma^{-1}\sigma : A \rightarrow A$ so $\sigma^{-1}\sigma = 1_A$. But that is all right since $g : A \rightarrow A$. Thus $\sigma^{-1}\sigma \circ g = g$. This gives

$$\sigma(f(\sigma^{-1}(\sigma(g(\sigma^{-1}(b)))))) = \sigma(f(g(\sigma^{-1}(b)))).$$

This shows that $F(f \circ g)(b) = F(f) \circ F(g)(b)$ for all $b \in B$. This means that as claimed $F(f \circ g) = F(f) \circ F(g)$ or that F preserves the operation.

Finally, we must show that F is a bijection. Here we will show that F is 1-1. So suppose that $f, g \in S_A$ and that $F(f) = F(g)$. Thus for all $b \in B$ we have $F(f)(b) = F(g)(b)$. So

$$\sigma(f(\sigma^{-1}(b))) = \sigma(g(\sigma^{-1}(b))).$$

But σ is 1-1. Thus this implies that $f(\sigma^{-1}(b)) = g(\sigma^{-1}(b))$ $\forall b \in B$. But as b varies through all of B, $\sigma^{-1}(b)$ must vary through all of A since σ^{-1} is onto. Thus we have $f = g$ as functions since they are the same on every element of their domain (which is A). It remains to show that F is onto. We will leave the verification that F is onto to the exercises. $\square$

If $A_1 \subset A$ then those permutations on A that map A_1 to A_1 form a subgroup of S_A called the **stabilizer** of A_1 denoted $stab(A_1)$. We leave the proof of the following lemma to the exercises.

Lemma 6.2.3. *If $A_1 \subset A$ then $stab(A_1) = \{f \in S_A : f : A_1 \rightarrow A_1\}$ forms a subgroup of S_A.*

A **permutation group** is any subgroup of S_A for some set A.

We now look at finite permutation groups. Let A be a finite set, say $A = \{a_1, a_2, \ldots, a_n\}$. Then each $f \in S_A$ can be pictured as

$$f = \begin{pmatrix} a_1 & \cdots & a_n \\ f(a_1) & \cdots & f(a_n) \end{pmatrix}.$$

For a_1 there are n choices for $f(a_1)$. For a_2 there are only $n-1$ choices since f is one-to-one. This continues down to only one choice for a_n. Using the multiplication principle, the number of choices for f and therefore the size of S_A is

$$n(n-1) \cdots 1 = n!.$$

We have thus proved the following theorem.

Theorem 6.2.2. *If* $|A| = n$ *then* $|S_A| = n!$.

We also observe the following trick to find f^{-1} if the permutation f is given by the two-row form above. Since f is a bijection $f(A) = \{f(a_1), \ldots, f(a_n)\} = A$. Thus we can write a two-row form for f^{-1};

$$f^{-1} = \begin{pmatrix} f(a_1) & \cdots & f(a_n) \\ a_1 & \cdots & a_n \end{pmatrix}.$$

(Do you see why this works?)

For a set with n elements we denote S_A by S_n and we call it the **symmetric group on n symbols** or the **symmetric group of degree** n. We also note that from Theorem 6.2.1 if B is any set such that $|A| = |B| = n$ then $S_A \cong S_B$. Thus there is no loss of generality just to take $A = \{1, 2, \ldots, n\}$. Thus from now on when we write S_n we will mean the group of all permutations on the set $\{1, 2, \ldots, n\}$.

EXAMPLE 6.2.1. Write down the six elements of S_3 and give the multiplication table for the group.

From the above the symmetric group S_3 is the symmetric group on the set $A = \{1, 2, 3\}$. The six elements of S_3 are then:

$$1 = \begin{pmatrix} 1 & 2 & 3 \\ 1 & 2 & 3 \end{pmatrix}, a = \begin{pmatrix} 1 & 2 & 3 \\ 2 & 3 & 1 \end{pmatrix}, b = \begin{pmatrix} 1 & 2 & 3 \\ 3 & 1 & 2 \end{pmatrix}$$

$$c = \begin{pmatrix} 1 & 2 & 3 \\ 2 & 1 & 3 \end{pmatrix}, d = \begin{pmatrix} 1 & 2 & 3 \\ 3 & 2 & 1 \end{pmatrix}, e = \begin{pmatrix} 1 & 2 & 3 \\ 1 & 3 & 2 \end{pmatrix}.$$

The multiplication table for S_3 can be written down directly by doing the required composition. For example,

$$ac = \begin{pmatrix} 1 & 2 & 3 \\ 2 & 3 & 1 \end{pmatrix} \begin{pmatrix} 1 & 2 & 3 \\ 2 & 1 & 3 \end{pmatrix} = \begin{pmatrix} 1 & 2 & 3 \\ 3 & 2 & 1 \end{pmatrix} = d.$$

To see this, we start with the right permutation and note that

$$c : 1 \to 2, 2 \to 1, 3 \to 3;$$

$$a : 1 \to 2, 2 \to 3, 3 \to 1;$$

and so

$$ac : 1 \to 3, 2 \to 2, 3 \to 1,$$

by composition of mappings

For example, $1 \to 2$ by c and then $2 \to 3$ by a. Thus $1 \to 3$ by ac. The reason for this is that we are just doing composition of functions. So $ac(1) = a(c(1)) = a(2) = 3$. These products can be carried out easily by looking at the above two-row form for each term in the product and starting from the right-hand term.

It is somewhat easier to construct the multiplication table if we make some observations. First, $a^2 = b$ and $a^3 = 1$. Next, $c^2 = 1, d = ac, e = a^2c$ and finally $ac = ca^2$.

From these relations the following multiplication table can be constructed

	1	a	a^2	c	ac	a^2c
1	1	a	a^2	c	ac	a^2c
a	a	a^2	1	ac	a^2c	c
a^2	a^2	1	a	a^2c	c	ac
c	c	a^2c	ac	1	a^2	a
ac	ac	c	a^2c	a	1	a^2
a^2c	a^2c	ac	c	a^2	a	1

To see this, consider, for example, $(ac)a^2 = a(ca^2) = a(ac) = a^2c$. More generally, we can say that S_3 has a **presentation** given by

$$S_3 = \langle a, c; a^3 = c^2 = 1, ac = ca^2 \rangle.$$

By this we mean that S_3 is **generated by** a, c, or that S_3 has **generators** a, c and the whole group and its multiplication table can be generated by using the **relations** $a^3 = c^2 = 1, ac = ca^2$.

This loosely means that every nontrivial element of S_3 can be written in terms of a and c. We will make this precise in Section 6.4.

We leave the rest of the details of verifying that this actually does give the multiplication table for S_3 to the exercises.

A theorem of Cayley actually shows that every group is a permutation group. A group G is a permutation group on the group G itself considered as a set. This result, however, does not give much information about the group.

Theorem 6.2.3. *(Cayley's Theorem) Let G be a group. Consider the set of elements of G and S_G the symmetric group on G. Then the group G is isomorphic to a subgroup of S_G. If we identify G with this subgroup of S_G then we can think of G as a permutation group on the set G itself.*

Proof. We show that to each $g \in G$ we can associate a permutation of the set G. If $g \in G$ let π_g be the map given by

$$\pi_g : g_1 \mapsto gg_1 \text{ for each } g_1 \in G.$$

It is straightforward to show that each π_g is a permutation on the set G. We also must show that the mapping $\varphi : G \to S_G$ given by $g \mapsto \pi_g$ is a 1-1 group homomorphism. We leave the details to the exercises. $\qquad\square$

6.2.2   Examples of Groups: Geometric Transformation Groups

Groups play an important role in the modern study of geometry. Classically geometry was the study of Euclidean geometry as laid out in the books of Euclid. With the discovery of non-Euclidean geometry in the nineteenth century (we refer to [WW] for any geometric concepts not explicitly defined here) the study of geometry has evolved into the study of various **geometries** or **geometric spaces**. A **geometry** is a set, whose elements are called points, and which has one or more of the following geometric notions defined on it: line, incidence, angle, distance. In this approach Euclidean geometry is just one among many.

A **metric geometry** is any geometry where the underlying space is a metric space (see Chapter 5), that is, a set where we can measure distance. In a metric geometry, an **isometry** is a mapping that preserves distance.

As we will see below the set of all isometries on such a geometry forms a group under composition. This is called the group of isometries of the geometry. Felix Klein in his famous Erlanger program (see [WW]) pointed out that to understand a metric geometry, such as Euclidean geometry, one must understand the group of isometries. We look briefly at the Euclidean geometry of the plane.

We let $\mathcal{E}^2$ stand for the Euclidean plane, that is, the set of points in a plane satisfying all the standard properties of Euclidean geometry (see [WW]). An **isometry** or **congruence motion** on $\mathcal{E}^2$ is a transformation or bijection T of $\mathcal{E}^2$ to itself that preserves distance, that is $d(a,b) = d(T(a),T(b))$ for all points $a,b \in \mathcal{E}^2$ and hence the distance between any two points a,b is the same as the distance between their images $T(a)$ and $T(b)$.

Theorem 6.2.4. *The set of congruence motions of $\mathcal{E}^2$ forms a group called the **Euclidean group**. We denote the Euclidean group by $\mathcal{E}$.*

Proof. The identity map I is clearly an isometry and since composition of mappings is associative we need only show that the product of isometries is an isometry and that the inverse of an isometry is an isometry.

Let T, U be isometries and let a, b be points in $\mathcal{E}^2$. Then $d(a,b) = d(T(a),T(b))$ and $d(a,b) = d(U(a),U(b))$ for any points a, b. Now consider

$$d(TU(a), TU(b)) = d(T(U(a)), T(U(b))) = d(U(a), U(b))$$

since T is an isometry. However,

$$d(U(a), U(b)) = d(a, b)$$

since U is an isometry. Combining these we have that TU is also an isometry.

Consider T^{-1} and points $a, b \in \mathcal{E}^2$. Then

$$d(T^{-1}(a), T^{-1}(b)) = d(TT^{-1}(a), TT^{-1}(b))$$

since T is an isometry. But $TT^{-1} = I$ and hence

$$d(T^{-1}(a), T^{-1}(b)) = d(TT^{-1}(a), TT^{-1}(b)) = d(a, b).$$

Therefore T^{-1} is also an isometry and hence $\mathcal{E}$ is a group. $\square$

A mapping in $\mathcal{E}$ will map a figure onto a congruent figure. Euclidean geometry studies those properties of $\mathcal{E}^2$ (angle, distance, shape) which are invariant under $\mathcal{E}$. An isometry **preserves orientation** if the image of any triangle ABC has the same orientation either clockwise or counterclockwise.

There are four standard congruence motions on $\mathcal{E}^2$.

Definition 6.2.2. *(a) Let v be a vector in $\mathcal{E}^2$. A* **translation** *T_v takes a point P to $P + v$. By this we mean start at the point P and place the tail of the vector at P. The image $T_v(P)$ is then the tip of the vector v. Note that a nontrivial translation (that is, a translation $T_v \neq I$) has no fixed point and that two translations commute with each other. Further a translation preserves orientation.*

(b) Let θ be an angle and O a point. A **rotation** *R of angle θ about O fixes O and takes a point $P \neq O$ to P_1 so that the angle POP_1 is θ. Note that a nontrivial rotation, where $0 < \theta < 2\pi$, has only O as a fixed point and that two rotations about the same point commute with each other. Further a rotation preserves orientation.*

(c) A **reflection** *in line L fixes each point on L and maps each point $P \notin L$ to its perpendicular mirror through L. Note that if R_L is the reflection through L then $R_L^2 = I$. Further a reflection reverses orientation.*

(d) A **glide reflection** *consists of a reflection in a line L followed by a translation parallel to L. A glide reflection also reverses orientation. Note that a reflection in a line L followed by a translation parallel to L commute so a glide reflection can also be defined as a translation parallel to L followed by a reflection in L.*

A reflection in a line can be considered as a glide reflection with the trivial translation, that is, the identity.

One of the major results concerning $\mathcal{E}$ is the following. We refer to [WW] for a more thorough treatment. Note that a glide reflection is the product of a translation and a reflection. Here product is in the sense of composition or for these motions, one motion followed by another.

Theorem 6.2.5. *If $T \in \mathcal{E}$ then T is either a translation, rotation, reflection or glide reflection. The set of translations and rotations forms a subgroup.*

For a complete proof of this we refer to [WW] (see also [L]). In the exercises we will (with hints) outline a proof. It depends on the concept of a fixed point. If T is an isometry then a point a is a **fixed point** of T if $T(a) = a$. The proof of Theorem 6.2.5 follows from an analysis of the fixed points of an isometry. This depends on the following results outside of the realm of abstract algebra which we just state.

Lemma 6.2.4. *A planar Euclidean motion is completely determined by its action on three noncollinear points — that is on a single triangle. Hence if*

T, U agree on three points then $T = U$. In particular if T fixes three points then $T = I$.

Lemma 6.2.5. *If a planar Euclidean motion has two fixed points then it is either a reflection or the identity. If it has one fixed point it's a rotation. If it has no fixed points and preserves orientation it is a translation and if it reverses orientation it is a glide reflection.*

If D is a geometric figure in $\mathcal{E}^2$, such as a triangle or square, then a **symmetry** of D is a congruence motion $T : \mathcal{E}^2 \rightarrow \mathcal{E}^2$ that leaves D in place. It may move the individual elements of D however. So, for example, a rotation about the center of a circle is a symmetry of the circle.

Lemma 6.2.6. *If D is a geometric figure in $\mathcal{E}^2$ then the set of symmetries of D forms a subgroup of $\mathcal{E}$ called the **symmetry group** of D denoted by $Sym(D)$.*

Proof. We show that $Sym(D)$ is a subgroup of $\mathcal{E}$. The identity map I fixes D so $I \in Sym(D)$ and so $Sym(D)$ is nonempty. Let $T, U \in Sym(D)$. Then U maps D to D and so does T. It follows directly that so does the composition TU and hence $TU \in Sym(D)$. If T maps D to D then certainly the inverse does. Therefore $Sym(D)$ is a subgroup of $\mathcal{E}$. $\qquad\square$

EXAMPLE 6.2.2. Let T be an equilateral triangle. Then there are exactly six symmetries of T (see the exercises). These are

$$I = \text{the identity}$$

$$r = \text{a rotation of } 120° \text{ around the center of } T$$

$$r^2 = \text{a rotation of } 240° \text{ around the center of } T$$

$$f = \text{a reflection in the perpendicular bisector of one of the sides}$$

$$fr = \text{the composition of } f \text{ and } r$$

$$fr^2 = \text{the composition of } f \text{ and } r^2$$

$Sym(T)$ is called the **dihedral group** D_3.

In Example 6.2.1 we constructed the multiplication table for the symmetric group S_3. This also had six elements. Each symmetry of an equilateral triangle is actually given by a permutation on its vertices and hence D_3 is a permutation group on three symbols. Here label the vertices $1, 2, 3$. It follows that $D_3 \subset S_3$. However, these two groups both have six elements so $D_3 = S_3$, that is, they are equal.

In general the symmetries of a regular n-gon form a group called the dihedral group which we denote D_n. As for D_3 we always have $D_n \subset S_n$ again labeling the vertices $1, 2, \ldots, n$.

6.3   Subgroups and Lagrange's Theorem

In this section we consider the relationship between a group G and a subgroup H of G. To begin this study, we first introduce certain subsets of a group which in general are not subgroups but will be important in these discussions. They are called left or right cosets.

Definition 6.3.1. *Let G be a group and let H be a subgroup of G. Then for any $a \in G$, the **left coset** of a in G with respect to H, denoted by aH, is the set*

$$aH = \{ah : h \in H\}.$$

*The **right coset** of a in G with respect to H, written Ha, is defined similarly as the set*

$$Ha = \{ha : h \in H\}.$$

If it is clear which subgroup we are discussing we will just call aH or Ha the left or right coset, respectively, of a in G.

The notation aH or Ha is meant to be suggestive of what these sets are. That is, to get aH, the left coset of H determined by a, we take all possible products of a times an element in H with a on the left-hand side. If we put a on the right, then we get Ha, that is the right coset of H determined by a. We observe that since H is a subgroup $1 \in H$, so $a = a \cdot 1 = 1 \cdot a$ and $a \in aH$ and $a \in Ha$. For this reason we could also call the left (right) coset of a in G with respect to H the left (right) coset of H in G containing a. Also if G is nonabelian, then left and right cosets in general do not have to be equal.

Our next step in this discussion is to define, using the subgroup $H \subset G$, an equivalence relation on G. The equivalence classes will all have the same size (or cardinality as sets) and will be the left cosets of H in G.

Definition 6.3.2. *Let G be a group and $H \subset G$ a subgroup. For $a, b \in G$ define*

$$a \sim b \text{ if and only if } a^{-1}b \in H.$$

Lemma 6.3.1. *Let G be a group and $H \subset G$ a subgroup. Then the relation defined above is an equivalence relation on G. The equivalence classes all have the form aH for $a \in G$ and are the left cosets of H in G. Since the left cosets are equivalence classes, G is a disjoint union of its left cosets.*

Proof. Let us show, first of all, that this is an equivalence relation. Now $a \sim a$ since $a^{-1}a = 1 \in H$. Therefore the relation is reflexive. Further $a \sim b$ implies $a^{-1}b \in H$, but since H is a subgroup of G we have $b^{-1}a = (a^{-1}b)^{-1} \in H$ and so $b \sim a$. Therefore the relation is symmetric. Finally suppose that $a \sim b$ and $b \sim c$. Then $a^{-1}b \in H$ and $b^{-1}c \in H$. Since H is a subgroup $a^{-1}b \cdot b^{-1}c = a^{-1}c \in H$, and hence $a \sim c$. Therefore the relation is transitive and hence is an equivalence relation.

For $a \in G$ the equivalence class is

$$[a] = \{g \in G : a \sim g\} = \{g \in G : a^{-1}g \in H\}.$$

But if $a^{-1}g \in H$, this means that $a^{-1}g = h$ where $h \in H$. Therefore $a \sim g$ if and only if $g = ah$ for some $h \in H$. It follows that the equivalence class for $a \in G, [a]$, being the set of all g such that $a \sim g$ is precisely the set of all ah as h varies over the elements of H. Hence we have shown that

$$[a] = \{g \in G : g = ah \text{ for some } h \in H\} = \{ah : h \in H\} = aH.$$

Put in other words the equivalence class containing a is the left coset of a in G. Since they are equivalence classes, the left cosets partition G. This means that every element of G is in one and only one left coset. In particular, $bH = H = 1H$ if and only if $b \in H$. We note that this follows in two ways. In particular, $b \in H = 1 \cdot H$ iff $b \sim 1$ or $b = 1^{-1} \cdot b \in H$. Alternatively, since H is a subgroup it is closed so if we multiply every element of H on the left by b an element of H we only get elements of H, which is the same thing that happens if we mutiply all elements of H by 1. $\square$

We call the element a a **coset representative** of the coset aH. The indexed family $\{aH\}_{a \in G}$ will give a complete collection of left cosets of H in G. There may be repetitions. We choose a set $T \subset G$ of coset representatives such that $\{aH\}_{a \in T}$ is the set of all distinct left cosets of H in G. This set

T is called a (left) **transversal** of H in G. Here it is assumed that $a \in T$ is chosen so that we get all the different left cosets. One must understand that in such a left transversal there is a unique a for each coset aH. This means that if aH and a_1H are two left cosets with $aH = a_1H$ then $a = a_1$. It follows that $G = \cup_{a \in T} aH$ and the sets in this union are disjoint.

One could define another equivalence relation by defining $a \sim b$ if and only if $ba^{-1} \in H$. Again this can be shown to be an equivalence relation on G, and the equivalence classes here are sets of the form

$$Ha = \{g \in G : g = ha \text{ for some } h \in H\} = \{ha : h \in H\}$$

called the **right cosets** of H. Also, of course, G is the (disjoint) union of distinct right cosets.

Since the left cosets partition the group, two left cosets are either equal or disjoint. This means that if they have anything in common then they must be identical. Of course, the same is true of right cosets. These remarks give rise to a very easy and useful criterion for two left cosets to be equal. There is a similar criterion for right cosets whose proof we leave to the exercises.

Lemma 6.3.2. *Let G be a group and H be a subgroup of G. Also suppose that $a, b \in G$. Then $aH = bH$ if and only if $a^{-1}b \in H$.*

Proof. Suppose first that $aH = bH$. Since $b \in bH$, then $b \in aH$. This implies that $b = ah$ where $h \in H$. Thus $a^{-1}b = h \in H$. Conversely suppose that $a^{-1}b \in H$. So that again we have $a^{-1}b = h \in H$ which in turn implies that $b = ah$. But then $b \in bH \cap aH$. This implies by the remarks preceding the statement of this lemma that $aH = bH$.   □

We note that the above lemma could have been proven just from the equivalence relation. The cosets aH and bH are the equivalence classes $[a], [b]$ respectively and therefore $aH = bH$ if and only if $a \sim b$ if and only if $a^{-1}b \in H$. We did it as above just to show how the cosets are related to the equivalence relation. We also note that the same thing can be done for right cosets. Below we just state the result but leave its verification to the exercises because it is more or less the same argument we just gave.

Lemma 6.3.3. *Let G be a group, H be a subgroup of G and $a, b \in G$. Then $Ha = Hb$ if and only if $ab^{-1} \in H$.*

It is easy to see that any two left (right) cosets have the same order (number of elements or size). To demonstrate this consider the mapping $aH \to bH$ via $ah \longmapsto bh$ where $h \in H$. It is not hard to show that this mapping is 1-1 and onto (see the exercises). Thus we have $|aH| = |bH|$. (This is also true for right cosets and can be established in a similar manner.) Letting $b \in H$ in the above discussion, we see $|aH| = |H|$, for any $a \in G$, that is, the size of each left or right coset is exactly the same as the subgroup H.

One can also see that the collection $\{aH\}$ of all distinct left cosets has the same number of elements as the collection $\{Ha\}$ of all distinct right cosets. In other words, the number of left cosets equals the number of right cosets (this number may be infinite). For consider the map

$$f : aH \to Ha^{-1}.$$

This mapping is well-defined: for if $aH = bH$, then $b = ah$ where $h \in H$. Thus $f(bH) = Hb^{-1} = Hh^{-1}a^{-1} = Ha^{-1} = f(aH)$. It is not hard to show that this mapping is 1-1 and onto (see the exercises). Hence the number of left cosets equals the number of right cosets.

Definition 6.3.3. *Let G be a group and $H \subset G$ a subgroup. The number of distinct left cosets, which is the same as the number of distinct right cosets, is called the **index** of H in G, denoted by $[G : H]$.*

Now let us consider the case where the group G is finite. Each left coset has the same size as the subgroup H and here both are finite. Hence $|aH| = |H|$ for each coset. Further the group G is a disjoint union of the left cosets, that is,

$$G = H \cup g_1 H \cup \cdots \cup g_n H.$$

Since this is a disjoint union we then have

$$|G| = |H| + |g_1 H| + \cdots + |g_n H| = |H| + |H| + \cdots + |H| = |H|[G : H].$$

This establishes the following extremely important theorem.

Theorem 6.3.1. *(Lagrange's Theorem) Let G be a finite group and $H \subset G$ a subgroup. Then*

$$|G| = |H|[G : H].$$

For a finite group this implies that both the order of a subgroup and the index of a subgroup are divisors of the order of the group. Also if G is a finite group, the number of left (or right) cosets of H in $G = [G : H]$ is

$$[G : H] = \frac{|G|}{|H|}.$$

This theorem plays a crucial role in the structure theory of finite groups since it greatly restricts the size of subgroups. For example, in a group of order 10 there can be proper nontrivial subgroups only of orders 2 and 5.

As an immediate corollary, we have the following result.

Corollary 6.3.1. *The order of any element $g \in G$, where G is a finite group, divides the order of the group. In particular if $|G| = n$ and $g \in G$ then $o(g)|n$ and $g^n = 1$.*

Proof. Let $g \in G$ and $o(g) = m$. Then by defintion m is the size of the cyclic subgroup generated by g and hence divides n from Lagrange's theorem. Then $n = mk$ for some integer k and so

$$g^n = g^{mk} = (g^m)^k = 1^k = 1.$$

$\square$

Before leaving this section we consider some results concerning general subsets of a group.

Suppose that G is a group and S is an arbitrary nonempty subset of G, $S \subset G$ and $S \neq \emptyset$; such a set S is usually called a **complex** of G. So e.g., a left or right coset is a complex.

If U and V are two complexes of G, the product UV is defined as follows:

$$UV = \{g_1 g_2 \in G : g_1 \in U, g_2 \in V\}.$$

We note that if either U or V is a singleton and the other subset is a subgroup then the complex UV becomes a left or right coset.

Now suppose that U, V are subgroups of G. When is the complex UV again a subgroup of G?

Theorem 6.3.2. *The product UV of two subgroups U, V of a group G is itself a subgroup if and only if U and V commute, that is, if and only if $UV = VU$.*

Proof. We note first that when we say U and V commute, we do *not* demand that this is so elementwise. In other words, it is not required that $uv = vu$ for all $u \in U$ and all $v \in V$; all that is required is that for any $u \in U$ and $v \in V$ $uv = v_1 u_1$, for some elements $u_1 \in U$ and $v_1 \in V$ and similarly for vu. Assume that UV is a subgroup of G. Let $u \in U$ and $v \in V$. Then $u \in U \cdot 1 \subset UV$ and $v \in 1 \cdot V \subset UV$. But since UV is assumed itself to be a subgroup, it follows that $vu \in UV$. Hence each product $vu \in UV$ and so $VU \subset UV$ since vu is the general element of VU. We next need to show that uv, which is the general element of UV, is also in VU. But since UV is a subgroup $(uv)^{-1} \in UV$ and $(uv)^{-1} = v^{-1}u^{-1}$. Since this is an element of UV, $v^{-1}u^{-1} = u_1 v_1$ where $u_1 \in U$ and $v_1 \in V$. But then $uv = ((uv)^{-1})^{-1} = v_1^{-1}u_1^{-1} \in VU$. Hence we have shown that $UV \subset VU$. The double inclusion which we have established shows the equality.

Conversely, suppose that $UV = VU$. Clearly UV is nonempty since the identity 1 of G is in both U and V and hence $1 = 1 \cdot 1 \in UV$. Let $g_1 = u_1 v_1 \in UV$, $g_2 = u_2 v_2 \in UV$. Then

$$g_1 g_2 = (u_1 v_1)(u_2 v_2) = u_1(v_1 u_2)v_2 = u_1 u_3 v_3 v_2 = (u_1 u_3)(v_3 v_2) \in UV$$

since $v_1 u_2 = u_3 v_3$ for some $u_3 \in U$ and $v_3 \in V$. Further,

$$g_1^{-1} = (u_1 v_1)^{-1} = v_1^{-1} u_1^{-1} = u_4 v_4$$

since $VU = UV$ there are some $u_4 \in U$ and $v_4 \in V$ such that $v_1^{-1} u_1^{-1} = u_4 v_4$. It follows that UV is a subgroup.   □

It is not hard to see that the intersection of two subgroups is again a subgroup.

Lemma 6.3.4. *If U and V are subgroups of a group G then their intersection $U \cap V$ is also a subgroup of G.*

Proof. Since the identity 1 of G is in both U and V we have that $1 \in U \cap V$ and hence $U \cap V$ is nonempty. Suppose that $g_1, g_2 \in U \cap V$. Then $g_1, g_2 \in U$ and hence $g_1 g_2^{-1} \in U$ since U is a subgroup. Analogously $g_1 g_2^{-1} \in V$. Hence $gg_2^{-1} \in U \cap V$ and therefore $U \cap V$ is a subgroup. (See Problem 6.1.2.)   □

Theorem 6.3.3. *(Product Formula) Let U, V be subgroups of G and let R be a left transversal of the intersection $U \cap V$ in U. Then*

$$UV = \bigcup_{r \in R} rV$$

where this is a disjoint union.

In particular if U, V are finite then

$$|UV| = \frac{|U||V|}{|U \cap V|}.$$

Proof. Since $R \subset U \subset UV$ and $V \subset UV$, we have that

$$\bigcup_{r \in R} rV \subset UV.$$

In the other direction let $uv \in UV$ where $u \in U$ and $v \in V$. Since R is a left transversal of $U \cap V$ in U we have

$$U = \bigcup_{r \in R} r(U \cap V),$$

where this is a disjoint union.

It follows that $u = rv'$ with $r \in R$ and $v' \in U \cap V \subset V$. Hence

$$uv = rv'v \in rV.$$

So

$$uv \in \bigcup_{r \in R} rV$$

and therefore $UV \subset \bigcup_{r \in R} rV$ proving the equality.

We claim, however, that more is true. In particular the union $UV = \cup_{r \in R} rV$ is disjoint. To see this suppose that for $r_1, r_2 \in R$ we have $r_1 V = r_2 V$. This would then imply that $r_2^{-1} r_1 \in V$. But recall that $R \subset U$. Hence $r_2^{-1} r_1 \in U$ since U is a subgroup. But these two in turn give that $r_2^{-1} r_1 \in U \cap V$. This would imply that $r_1(U \cap V) = r_2(U \cap V)$. But R is a left transversal so this means that $r_1 = r_2$. Since cosets are equivalance classes they are either identical or disjoint. Thus we have shown that $UV = \cup_{v \in R} rV$ is a disjoint union.

Now suppose that $|U|, |V|$ are finite. Then we have

$$|UV| = |R||V| = [U : U \cap V]|V| = \frac{|U|}{|U \cap V|}|V| = \frac{|U||V|}{|U \cap V|}$$

where we have used Lagrange's theorem 6.3.1 applied to the group U and its subgroup $U \cap V$, i.e., $|U| = [U : U \cap V]|U \cap V|$. So that $[U : U \cap V] = \frac{|U|}{|U \cap V|}$. $\square$

We now show that index is multiplicative. Later we will see how this fact is related to the multiplicativity of the degree of field extensions.

Theorem 6.3.4. *Suppose G is a group and U and V are subgroups with $U \subset V \subset G$. If G is the disjoint union*

$$G = \bigcup_{r \in R} rV$$

where R is a left transversal for V in G, and V is the disjoint union

$$V = \bigcup_{s \in S} sU$$

where S is a left transversal for U in V then we get a disjoint union for G as

$$G = \bigcup_{r \in R, s \in S} rsU.$$

In particular if $[G : V]$ and $[V : U]$ are finite then

$$[G : U] = [G : V][V : U].$$

Proof. Now

$$G = \bigcup_{r \in R} rV = \bigcup_{r \in R} r(\bigcup_{s \in S} sU) = \bigcup_{r \in R, s \in S} rsU.$$

See the exercises for verification of the last equality. Suppose that $r_1 s_1 U = r_2 s_2 U$. Then since $s_1, s_2 \in V$ we have that both $r_1 s_1 V = r_1 V$ and $r_2 s_2 V = r_2 V$. But $U \subset V$ so $r_1 s_1 U \subset r_1 s_1 V = r_1 V$ and $r_2 s_2 U \subset r_2 s_2 V = r_2 V$. But we are assuming that $r_1 s_1 U = r_2 s_2 U$ and R is a left transversal of V in G. This implies that $r_1 = r_2$. Hence from our assumption

$$r_1 s_1 U = r_1 s_2 U \implies s_1 U = s_2 U.$$

But again using the fact that S is a left transversal for U in V, we get that $s_1 = s_2$. Therefore the union is disjoint. The index formula now follows directly. $\square$

The next result says that a finite intersection of subgroups of finite index must again be of finite index.

Theorem 6.3.5. *(Poincare) Suppose that U, V are subgroups of finite index in G and that UV is a group. Then $U \cap V$ is also of finite index in G. Further*

$$[G : U \cap V] \leq [G : U][G : V].$$

If $[G : U], [G : V]$ are relatively prime then equality holds.

Proof. Let r be the number of left cosets of U in G that are contained in UV. r is finite since the index $[G : U]$ is finite and by Theorem 6.3.4 $[G : U] = [G : UV][UV : U] = [G : UV] \cdot r$. But $r = |UV| / |U|$ and from Theorem 6.3.3 $|UV| / |U| = |V| / |U \cap V| = [V : U \cap V]$. So combining all of this, we then have

$$[V : U \cap V] = r \leq [G : U].$$

Then from Theorem 6.3.4

$$[G : U \cap V] = [G : V][V : U \cap V] \leq [G : V][G : U].$$

Since both $[G : U]$ and $[G : V]$ are finite so is $[G : U \cap V]$.

Now $[G : U] | [G : U \cap V]$ and $[G : V] | [G : U \cap V]$. If $[G : U]$ and $[G : V]$ are relatively prime then

$$[G : U][G : V] | [G : U \cap V] \implies [G : U][G : V] \leq [G : U \cap V]$$

(See the exercises for a verification of the divides sign in the above display.) Therefore we must have equality.   □

The above theorem can be proven without the hypothesis that UV is a group. But our proof required that. Can you see where? (See the exercises.)

Corollary 6.3.2. *Suppose G is a finite group such that $[G : U]$ and $[G : V]$ are finite and relatively prime and UV is a group. Then $G = UV$.*

Proof. From Theorem 6.3.5, we have

$$[G : U \cap V] = [G : U][G : V].$$

From Theorem 6.3.4

$$[G : U \cap V] = [G : V][V : U \cap V].$$

Combining these we have

$$[V : U \cap V] = [G : U].$$

Using Theorem 6.3.3 the number of left cosets of U in G that are contained in UV is $|UV|/|U| = |V|/|U \cap V| = [V : U \cap V]$. But $[G : U] = |G|/|U|$. Combining this with the above equality gvies

$$\frac{|UV|}{|U|} = \frac{|G|}{|U|}.$$

Thus $|G| = |UV|$. It follows then that we must have $G = UV$. $\qquad\square$

We end this section with some applications to number theory. We have seen that the elements of $\mathbb{Z}_n$ which are less than or equal to n and relatively prime to n form a group under multiplication modulo n (see Chapter 4). This group was called the group of units in $\mathbb{Z}_n$ or $U(\mathbb{Z}_n)$. The order of this group then is the number of integers $\leq n$ and relatively prime to $n = \phi(n)$, i.e., the Euler phi function of n. But in symbols what we just observed is $|U(\mathbb{Z}_n)| = \phi(n)$. But now applying the first corollary of Lagrange's theorem, Corollary 6.3.1, we get new proofs for the theorems of Euler and Fermat given in Chapter 4.

Theorem 6.3.6. *(Euler). If $gcd(a, n) = 1$, then $a^{\phi(n)} \equiv 1$ (mod n).*

If we now take $n = p$, a prime, in the above theorem, then we note that $\phi(p) = p - 1$ and the condition of being relatively prime can be replaced by just saying $p \nmid a$, we get what is usually called Fermat's little theorem.

Corollary 6.3.3. *(Fermat) If p is a prime and $p \nmid a$, then*

$$a^{p-1} \equiv 1 \bmod p.$$

6.4 Generators and Cyclic Groups

We saw that if G is any group and $g \in G$ then the powers of g generate a subgroup of G called the cyclic subgroup generated by g. Here we explore more fully the idea of generating a group or subgroup. We first need the following.

In Lemma 6.3.4, we showed the intersection of two subroups is again a subgroup. But the argument given there can be applied in the exact same manner to show that the intersection of any number of subgroups of a group is a subgroup, e.g., it can be used to show that if $\{U_\alpha\}_{\alpha \in \Lambda}, \Lambda \neq \emptyset$ is an indexed family of subgroups of a group then their intersection, $\cap_\alpha U_\alpha$, is a subgroup (see Problem 6.1.9).

Now let S be a subset of a group G. The subset S is certainly contained in at least one subgroup of G, namely G itself. Let $\{U_\alpha\}$ be the collection of all subgroups of G containing S. Then $\cap_\alpha U_\alpha$ is again a subgroup of G from Lemma 6.3.4 and the remarks above. Further it is the smallest subgroup of G containing S (see the exercises). We call $\cap_\alpha U_\alpha$ the subgroup of G generated by S and denote it by $\langle S \rangle$ or $gp_G(S)$. (We may omit the sub G if it is clear what group the subgroup generated by the set is contained in.) We call the set S a set of **generators** for $\langle S \rangle$.

Definition 6.4.1. *A subset M of a group G is a **set of generators** for G if $G = \langle M \rangle$, that is, the smallest subgroup of G containing M is all of G. We say that G is **generated** by M and that M is a set of **generators** for G.*

Notice that any group G has at least one set of generators, namely, G itself. If $G = \langle M \rangle$ and M is a finite set then we say that G is **finitely generated**. Clearly any finite group is finitely generated. Shortly we will give an example of a finitely generated infinite group.

EXAMPLE 6.4.1. The set of all reflections forms a set of generators for the Euclidean group $\mathcal{E}$. Recall that any $T \in \mathcal{E}$ is either a translation, a rotation, a reflection or a glide reflection. It can be shown (see the exercises) that any one of these can be expressed as a product of three or fewer reflections.

We now consider the case where a group G has a single generator.

Definition 6.4.2. *A group G is **cyclic** if there exists a $g \in G$ such that $G = \langle g \rangle$.*

In this case we claim that $G = \langle g \rangle = \{g^n : n \in \mathbb{Z}\}$, that is G consists of all the powers of the element g. To see this from Definition 6.4.1, we first note that in Lemma 6.1.4 it was shown that the set $\{g^n : n \in \mathbb{Z}\}$ forms a group. Clearly $g \in \{g^n : n \in \mathbb{Z}\}$. Thus since $\langle g \rangle$ is the smallest subgroup of G containing g it must be true that $\langle g \rangle \subset \{g^n : n \in \mathbb{Z}\}$. But if H is any subgroup of G containing g then $\{g^n : n \in \mathbb{Z}\} \subset H$ since H is closed with respect to the operation. This is turn implies that $\{g^n : n \in \mathbb{Z}\} \subset \langle g \rangle$ since $\langle g \rangle$ is a subgroup containing g. Thus we have shown, as claimed, that $\langle g \rangle = \{g^n : n \in \mathbb{Z}\}$.

If there exists an integer m such that $g^m = 1$, then there exists a smallest such positive integer, say n, we have previously called this the order of g, i.e., $o(g) = n$. It follows that

Lemma 6.4.1. *Let G be a group and $g \in G$ have $o(g) = n$. Then $g^k = g^l$ if and only if $k \equiv l \bmod n$.*

Proof. See the exercises. □

Thus if $o(g) = n$ the distinct powers of g are precisely

$$\{1 = g^0, g, g^2, \ldots, g^{n-1}\}.$$

It follows that if $G = \{g^k : k \in \mathbb{Z}\}$, then $|G| = n$. We then call G a **finite cyclic group**. If no such power m exists such that $g^m = 1$ then all the powers of G are distinct and G is an **infinite cyclic group**.

We show next that any two cyclic groups of the same order are isomorphic.

Theorem 6.4.1. *(a) If $G = \langle g \rangle$ is an infinite cyclic group then $G \cong (\mathbb{Z}, +)$ that is, the integers under addition.*

(b) If $G = \langle g \rangle$ is a finite cyclic group of order n, then $G \cong (\mathbb{Z}_n, +)$ that is, the integers modulo n under addition.

It follows that for a given order there is only one cyclic group up to isomorphism.

Proof. Let G be an infinite cyclic group with generator g. We map g onto $1 \in (\mathbb{Z}, +)$. Since g generates G and 1 generates $\mathbb{Z}$ under addition this can be extended to a homomorphism. More generally, this gives rise to the map $\varphi : G \to \mathbb{Z}$ defined by $\varphi(g^n) = n \cdot 1 = n$ for any integer n. This is clearly a map from G to $\mathbb{Z}$. It is also clearly onto. We must show it is 1-1 and preserves the operations. To show it is 1-1, assume that $\varphi(g^n) = \varphi(g^k)$. This implies that $n = k$ which in turn implies that $g^n = g^k$. Next to show that φ preserves the operations, consider $\varphi(g^n \cdot g^m) = \varphi(g^{n+m}) = n + m = \varphi(g^n) + \varphi(g^m)$. This does it because in G the operation is represented by multiplication but in $\mathbb{Z}$ the operation is $+$.

Now let G be a finite cyclic group of order n with generator g. As above map g to $1 \in \mathbb{Z}_n$ and extend to a homomorphism. Again this induces the map $\varphi : G \to \mathbb{Z}_n$ defined by $\varphi(g^k) = k \cdot 1 = k$ for any integer k. But now there is an added problem which did not occur in the first part. We need to show that φ is well-defined. The reason for this is that unlike the powers g^n

in an infinite cyclic group which all must be distinct since $o(g) = \infty$, now the powers g^n repeat and so are not all distinct. Any element of G can be represented in infinitely many ways just by adding a multiple of n, e.g., qn, to the exponent because $g^{qn} = (g^n)^q = 1^q = 1$. Thus we must show that if $g^k = g^l$ in G then $\varphi(g^k) = \varphi(g^l)$. But as observed immediately before this theorem, if $o(g) = n$, as it is here, then $g^k = g^l$ if and only if $k \equiv l \bmod n$. But if $k \equiv l \bmod n$ then $k = l$ in $\mathbb{Z}_n$. Thus $\varphi(g^k) = k = l = \varphi(g^l)$. This shows that the map φ is well defined. The rest of the verification proceeds as in the case where $o(g) = \infty$.

Now let G and H be two cyclic groups of the same order. Here we use the fact that isomorphism is an equivalence relation on the class of all groups. If both are infinite then both are isomorphic to $(\mathbb{Z}, +)$ (see the exercises for Section 2.7). Since isomorphism is an equivalence relation, $G \cong \mathbb{Z}$ and $\mathbb{Z} \cong H$, hence each is isomorphic to the other, i.e., $G \cong H$. If both are finite of order n then both are isomorphic to $(\mathbb{Z}_n, +)$ and hence again using the fact that isomorphism is an equivalence relation, these groups are isomorphic to each other. □

We next consider subgroups of cyclic groups. The first and most general fact is that they must again be cyclic.

Theorem 6.4.2. *Let $G = \langle g \rangle$ be a cyclic group. Then every subgroup of G is also cyclic, i.e., every subgroup H of G is of the form $H = \langle g^t \rangle$ where $t \geq 0$ is an integer.*

Proof. Let $G = \langle g \rangle$ be a cyclic group and suppose that H is a subgroup of G. We may assume that H is nontrivial, i.e., $H \neq \{1\}$ because if it were it would be cyclic generated by g^0. Notice that if $g^m \in H$ then g^{-m} is also in H since H is a subgroup. Hence H must contain positive powers of the generator g. Thus we know that the set of all positive powers g which belong to H is a nonempty subset of $\mathbb{N}$. Thus by LWO (the least well ordering principle) this set has a least element. So let t be the smallest positive power of g such that $g^t \in H$. We claim that $H = \langle g^t \rangle$, the cyclic subgroup of G generated by g^t, i.e., every element of H is a power of g^t. Let $h \in H$ then $h = g^m$ for some integer m. Divide m by t using the division algorithm to get

$$m = qt + r \text{ where } r = 0 \text{ or } 0 < r < t.$$

If $r \neq 0$ then $r = m - qt > 0$. Now $g^m \in H$, $g^t \in H$ so $g^{-qt} \in H$ for any q since H is a subgroup. It follows that $g^m g^{-qt} = g^{m-qt} \in H$. This implies that $g^r \in H$. However, this is a contradiction since $r < t$ and t is the least positive power in H. It follows that $r = 0$ so $m = qt$. This implies that $g^m = g^{qt} = (g^t)^q$, that is, g^m is a power of g^t. Therefore every element of H is a power of g^t and therefore g^t generates H and hence H is cyclic. $\qquad \square$

We now consider what happens to the subgroups of a cyclic group if the group is either finite or infinite.

Theorem 6.4.3. *Let $G = \langle g \rangle$ be a cyclic group.*

(a) G is a finite cyclic group of order n and further if $d|n$ where $d > 0$ then there exists a unique subgroup of G of order d.

(b) $G = \langle g \rangle$ is an infinite cyclic group and further if t_1, t_2 are positive integers with $t_1 \neq t_2$ then $\langle g^{t_1} \rangle$ and $\langle g^{t_2} \rangle$ are distinct subgroups. Moreover, any nontrivial subgroup of G is itself infinite cyclic.

Proof. Suppose that $d|n$ so that $n = kd$ for $k > 0$ an integer. Let $H = \langle g^k \rangle$, that is, the subgroup of G generated by g^k. We claim that H has order d and that any other subgroup H_1 of G with order d coincides with H. Now $(g^k)^d = g^{kd} = g^n = 1$ so the order of g^k divides d by Lemma 6.1.5 and hence is $\leq d$. Suppose that $(g^k)^{d_1} = g^{kd_1} = 1$ with $d_1 < d$. Then since the order of g is n we have $n = kd|kd_1$ again by Lemma 6.1.5 with $d_1 < d$ which is impossible. Therefore the order of g^k is d and $H = \langle g^k \rangle$ is a subgroup of G of order d.

Now let H_1 be a subgroup of G of order d. We must show that $H_1 = H$. Let $H_1 = \langle g^t \rangle$ where $t > 0$. By the definition of the order of an element, we must have $o(g^t) = d$, and hence $g^{td} = 1$. But since $o(g) = n$ it follows from Lemma 6.1.5 that $n|td$ and so $kd|td$ and hence $k|t$, that is, $t = qk$ for some positive integer q since t and k are positive. Therefore $g^t = (g^k)^q \in H$. Therefore $H_1 \subset H$ and since they are of the same size $H = H_1$. This completes the proof of (a).

Now let $G = \langle g \rangle$ be an infinite cyclic group and H a subgroup of G. As in the proof of Theorem 6.4.2, $H = \langle g^t \rangle$ for $t > 0$.

From the proof of Theorem 6.4.2, in the subgroup $\langle g^t \rangle$ the integer t is the smallest positive power of g in $\langle g^t \rangle$. Therefore if t_1, t_2 are positive integers with $t_1 \neq t_2$ then $g^{t_1} \neq g^{t_2}$ because $o(g) = \infty$. But then the integer t_1 is the smallest positive power of g in $\langle g^{t_1} \rangle$ and the integer t_2 is the smallest positive power of g in $\langle g^{t_2} \rangle$. Since $g^{t_1} \neq g^{t_2}$, these two subgroups $\langle g^{t_1} \rangle$ and $\langle g^{t_2} \rangle$ are distinct.

Now if H is any nontrivial subgroup of G it follows that $H = \langle g^t \rangle$ with t the smallest positive integer such that $g^t \in H$. But then H itself must be infinite cyclic generated by g^t. For suppose that $(g^t)^k = (g^t)^m$ for some integers k, m. This would imply that $g^{tk-tn} = g^{t(k-m)} = 1$. Since g has infinite order this implies that $t(k - m) = 0$ which implies that $k = m$ since t is a positive integer. Thus the only way two powers of g^t can be the same element of H is if they have identical exponents. This shows that there are infinitely many elements in H so that H is an infinite cyclic group.   □

Theorem 6.4.4. *Let* $G = \langle g \rangle$ *be a cyclic group. Then*
   (a) If $G = \langle g \rangle$ *is finite of order* n *then* g^k *is also a generator if and only if* $gcd(k, n) = 1$. *That is, the generators of* G *are precisely those powers* g^k *where* k *is relatively prime to* n.
   (b) If $G = \langle g \rangle$ *is infinite then the only generators are* g, g^{-1}.

Proof. (a) Let $G = \langle g \rangle$ be a finite cyclic group of order n and suppose that $(k, n) = 1$. Then there exist integers x, y with $kx + ny = 1$. It follows that

$$g = g^{kx+ny} = (g^k)^x (g^n)^y = (g^k)^x$$

since $g^n = 1$. Hence g is a power of g^k that implies every element of G is also a power of g^k. Therefore g^k is also a generator.

   Conversely suppose that g^k is also a generator. Then g is a power of g^k so there exists an x such that $g = g^{kx}$. It follows from Lemma 6.4.1 that $kx \equiv 1$ modulo n and so there exists a y such that

$$kx + ny = 1.$$

This then implies that $(k, n) = 1$.
   (b) If $G = \langle g \rangle$ is infinite then any power of g other than g^{-1} generates a proper subgroup. For if g is a power of g^n for some $n \geq 2$ so that $g = g^{nx}$ it follows that $g^{nx-1} = 1$ so that g has finite order, contradicting that G is infinite cyclic.   □

   So far we have considered cyclic groups whose generating set S consists of a single element, i.e., S is a singleton, e.g., $S = \{a\}$. In this case, the group generated by S or $gp(S)$ just consists of all the distinct powers of this element. We now consider what happens if the generating set S is larger. In particular, S can have any cardinality ≥ 1. If $S = \{g_1, g_2, \ldots, g_n\} \subset G$ is finite we sometimes write for the subgroup generated in G by S, $gp_G(S) = \langle g_1, \ldots, g_n \rangle$.

Theorem 6.4.5. *Let G be a group and let $S \subset G$ be a nonempty subset. Let E be the set of all finite products of elements of S and their inverses (this includes single elements of S and their inverses). Then*

$$E = gp_G(S) = \langle S \rangle.$$

Proof. It is not hard to show that E is a subgroup of G (see the exercises). Now $S \subset E$ and E is a subgroup, then we must have

$$gp_G(S) \subset E$$

because by definition $gp_G(S)$ is the smallest subgroup of G containing S. However, since $gp_G(S)$ is itself a subgroup and $S \subset gp_G(S)$, it must be the case that $gp_G(S)$ contains all finite products of elements of S and their inverses, i.e.,

$$E \subset gp_G(S).$$

Thus $E = gp_G(S)$. $\qquad\square$

Recall that for positive integers n the Euler phi function (see Chapter 4) is defined as follows.

Definition 6.4.3. *For any $n > 0$,*

$$\phi(n) = \text{ number of positive integers less than or equal to } n$$

$$\text{relatively prime to } n.$$

EXAMPLE 6.4.2. $\phi(6) = 2$ since among $1, 2, 3, 4, 5, 6$ only $1, 5$ are relatively prime to 6.

Corollary 6.4.1. *If $G = \langle g \rangle$ is finite of order n then there are $\phi(n)$ generators for G where ϕ is the Euler phi-function.*

Proof. From Theorem 6.4.4, the generators of G are precisely the powers g^k where $(k, n) = 1$. The numbers relatively prime to n are counted by the Euler phi function. Thus we have $\phi(n)$ generators of G. $\qquad\square$

Recall that in an arbitrary group G, if $g \in G$, then the order of g, denoted $o(g)$, is the order of the cyclic subgroup generated by g. Given two elements $g, h \in G$ in general there is no relationship between $o(g), o(h)$ and the order of the product gh. However, if they commute there is a very direct relationship.

Lemma 6.4.2. *Suppose that G is an arbitrary group and $g, h \in G$ with both of finite order $o(g), o(h)$. If g and h commute, that is, $gh = hg$, then $o(gh)$ divides $lcm(o(g), o(h))$. In particular if G is an abelian group then $o(gh)|lcm(o(g), o(h))$ for all $g, h \in G$. Further if $\langle g \rangle \cap \langle h \rangle = \{1\}$ then $o(gh) = lcm(o(g), o(h))$*

Proof. Suppose $o(g) = n$ and $o(h) = m$ are finite. If g, h commute then for any k we have $(gh)^k = g^k h^k$. Let $t = lcm(n, m)$ then $t = k_1 m, t = k_2 n$. Hence

$$(gh)^t = g^t h^t = (g^m)^{k_1}(h^n)^{k_2} = 1.$$

Therefore the order of gh is finite and divides t by Lemma 6.1.5. Suppose that $\langle g \rangle \cap \langle h \rangle = \{1\}$, that is, the cyclic subgroup generated by g intersects trivially with the cyclic subgroup generated by h. Let $k = o(gh)$ which we know is finite from the first part of the lemma. Let $t = lcm(n, m)$. We then have $(gh)^k = g^k h^k = 1$ which implies that $g^k = h^{-k}$. Since the cyclic subgroups have only trivial intersection this implies that $g^k = 1$ and $h^k = 1$. But then $n|k$ and $m|k$ and hence $t|k$. Since $k|t$ and all numbers involved are positive it follows that $k = t$.   □

Recall that if m and n are relatively prime then $lcm(m, n) = mn$. Further if the orders of g and h are relatively prime it follows from Lagrange's theorem that $\langle g \rangle \cap \langle h \rangle = \{1\}$. We then get the following.

Corollary 6.4.2. *If g, h commute and $o(g)$ and $o(h)$ are finite and relatively prime then $o(gh) = o(g)o(h)$.*

Besides the order of the product of two elements, it is also nice to have a description of the orders of powers of an element.

Theorem 6.4.6. *Let G be a group and let $a \in G$ such that $o(a) = m$. Then*

$$o(a^k) = \frac{m}{gcd(m, k)}.$$

Proof. Let $t = o(a^k)$. Then $(a^k)^t = a^{kt} = 1$. Thus by Lemma 6.1.5, $m \mid kt$. Let $m = gcd(m, k)m'$ and $k = gcd(m, k)k'$ where $gcd(m', k') = 1$. Hence, $gcd(m, k)m' \mid gcd(m, k)k' \, t$. So that $m' \mid k't$. But since $gcd(m', k') = 1$, $m' \mid t$. But $m' = \frac{m}{gcd(m,k)}$. Thus we have

$$\frac{m}{gcd(m, k)} \mid t.$$

Now

$$(a^k)^{\frac{m}{\gcd(m,k)}} = (a^m)^{\frac{k}{\gcd(m,k)}} = 1.$$

This implies

$$t \mid \frac{m}{\gcd(m,k)}.$$

Combining these gives $o(a^k) = \frac{m}{\gcd(m,k)}$ since all numbers involved are positive.

$\square$

Definition 6.4.4. *If G is a finite abelian group then the **exponent** of G is the LCM of the orders of all elements of G. That is,*

$$exp(G) = LCM\{o(g) : g \in G\}.$$

Previously we defined the least common multiple of two integers. This can be easily generalized to define the least common multiple of any finite set of nonzero integers (see the exercises for the details).

Another way of defining the exponent is to note that by Lagrange's theorem every element in the group raised to a power equal to the order of the group is 1. Thus by LWO there is a least positive integer such that any element in the group raised to that power is 1. This integer is also the exponent of the group.

As a consequence of Theorem 6.4.6 we obtain

Lemma 6.4.3. *Let G be a finite abelian group. Then G contains an element of order $exp(G)$.*

Proof. Suppose that $exp(G) = p_1^{e_1} \cdots p_k^{e_k}$ with p_i distinct primes. Recall from Chapter 4 that one way of computing the lcm is to write its standard prime power expression by taking the maximum exponent that each prime occurs to in the numbers whose lcm is being computed. Again this was done previously for only two positive integers but it can be easily generalized to any finite number of positive integers. By the definition of $exp(G)$ there is a $g_i \in G$ with $o(g_i) = p_i^{e_i} r_i$ with p_i and r_i relatively prime. Let $h_i = g_i^{r_i}$. Then from Theorem 6.4.6, we get $o(h_i) = p_i^{e_i}$. Now let $g = h_1 h_2 \cdots h_k$. From Corollary 6.4.2 (extended to any finite set of numbers) we have $o(g) = p_1^{e_1} \cdots p_k^{e_k} = exp(G)$.

$\square$

If K is a field then the multiplicative subgroup of nonzero elements of K is an abelian group $K^\star$. The above results lead to the fact that a finite subgroup of $K^\star$ must actually be cyclic. For this proof we need the fact that a polynomial of degree m over a field can have at most m zeros. We will prove this fact in Chapter 12. We also use the usual notation $K[x]$ for the set of all polynomials in the variable x with coefficients in K.

Theorem 6.4.7. *Let K be a field. Then any finite subgroup of $K^\star$ is cyclic.*

Proof. Let $A \subset K^\star$ be a finite subgroup of $K^\star$ with $|A| = n$. Suppose that $m = exp(A)$. Consider the polynomial $f(x) = x^m - 1 \in K[x]$. Since the order of each element in A divides m it follows that $a^m = 1$ for all $a \in A$ and hence each $a \in A$ is a zero of the polynomial $f(x)$. Hence $f(x)$ has at least n zeros. Since a polynomial of degree m over a field can have at most m zeros it follows that $n \leq m$. From Lemma 6.4.3 there is an element $a \in A$ with $o(a) = m$. Since $|A| = n$ it follows from Lagrange's theorem that $m|n$ and hence $m \leq n$. Therefore $m = n$ and hence $A = \langle a \rangle$ showing that A is cyclic. $\qquad\square$

We close this section and end this chapter with two other results concerning cyclic groups. The first proves, using group theory, a very interesting number theoretic result concerning the Euler phi function.

Theorem 6.4.8. *For $n > 1$ and for $d \geq 1$*

$$\sum_{d|n} \phi(d) = n.$$

Proof. Consider a cyclic group G of order n. For each $d|n$, $d \geq 1$ there is a unique cyclic subgroup H of order d according to Theorem 6.4.3. Then H has $\phi(d)$ generators by Corollary 6.4.1. Each element in G generates its own cyclic subgroup H_1, say of order d where d must divide n by Lagrange's Theorem and hence must be included in the $\phi(d)$ generators of H_1. Therefore

$$\sum_{d|n} \phi(d) = \text{ sum of the numbers of generators of the cyclic subgroups of } G$$

But since we have done this for each element of G, this must be the whole group and hence this sum is n. $\qquad\square$

We shall make use of the above theorem directly in the following theorem.

Theorem 6.4.9. *If $|G| = n$ and if for each positive d such that $d|n$, G has at most one cyclic subgroup of order d, then G is cyclic (and consequently has exactly one cyclic subgroup of order d).*

Proof. For each $d|n$, $d > 0$, let $\Psi(d) = $ the number of elements in G of order d. Since by Lagrange's theorem, the order of any element must divide the order of the group if we compute $\Psi(d)$ for every positive d such that $d \mid n$ and then add all these numbers up we must get the number of elements in the whole group, G. Thus

$$\sum_{d|n} \Psi(d) = n.$$

Now suppose that $\Psi(d) \neq 0$ for a given $d|n$. Then there exists an $a \in G$ of order d which generates a cyclic subgroup, $\langle a \rangle$, of order d of G. We claim that all elements of G of order d are in $\langle a \rangle$. Indeed, if $b \in G$ with $o(b) = d$ and $b \notin \langle a \rangle$, then $\langle b \rangle$ is a second cyclic subgroup of order d, distinct from $\langle a \rangle$. This contradicts the hypothesis, so the claim is proven. Thus, if $\Psi(d) \neq 0$, then $\Psi(d) = \phi(d)$. In general, we have $\Psi(d) \leq \phi(d)$, for all positive $d|n$. But

$$n = \sum_{d|n} \Psi(d) \leq \sum_{d|n} \phi(d) = n,$$

by the previous Theorem 6.4.8. It follows, clearly, from this that $\Psi(d) = \phi(d)$ for all $d|n$. In particular, $\Psi(n) = \phi(n) \geq 1$. Hence, there exists at least one element of G of order n; hence G is cyclic. This completes the proof. □

Corollary 6.4.3. *If in a group G of order n, for each $d|n$, the equation $x^d = 1$ has at most d solutions in G, then G is cyclic.*

Proof. The hypothesis clearly implies that G can have at most one cyclic subgroup of order d since all elements of such a subgroup satisfy the equation. So Theorem 6.4.9, applies to give our result. □

6.5 Exercises

EXERCISES FOR SECTION 6.1

6.1.1. Let G be a group and $H \subset G$. Prove H is a subgroup of G if and only if

(1) $H \neq \emptyset$

(2) H is closed with respect to the operation in G and inverses, i.e., $a, b \in H \implies ab \in H$ and $a^{-1}, b^{-1} \in H$.

6.1.2. Let G be a group and $H \subset G$. and assume H is nonempty. Prove that H is a subgroup of G if and only if $a, b \in H \implies ab^{-1} \in H$. (This is sometimes called the one-step subgroup test.)

6.1.3. (Finite Subgroup Test) Let H be a nonempty finite subset of a group G such that $a, b \in H \implies ab \in H$. Then show that H is a subgroup of G. Give an example to show this does not work if $|H| = \infty$

(HINT: Use a proof like the one used to show that a finite integral domain is a field.)

6.1.4. Suppose that G is a group, $g \in G$ with $g^m = 1$ for some positive integer m. Then let n be the smallest positive integer such that $g^n = 1$. Here n is called the order of g denoted $o(g) = n$. If no such m exists then $o(g) = \infty$.

Prove:

(a) The set of elements $\{1, g, g^2, \ldots, g^{n-1}\}$ are all distinct, i.e., for $i \neq j \implies g^i \neq g^j$ for $0 \leq i, j \leq n - 1$.

(HINT: Suppose without loss of generality that $i > j$. To deduce a contradiction assume that $g^i = g^j$. But this implies that $g^{i-j} = 1$ (why?). Since $i - j < n$, this contradicts the definition of what n was.)

(b) For any other power g^k we have $g^k = g^t$ for some $t = 0, 1, \ldots, n - 1$.

(HINT:Use the division algorithm (see Chapter 4) to divide k by n and then use the claimed properties of exponentiation.)

6.1.5. Let G be a group and $g \in G$. Prove that $o(g^{-1}) = o(g)$.

(HINT: See the previous problem for the definition of $o(g)$. Consider two cases: (1) $o(g) < \infty$. (2) $o(g) = \infty$.)

6.1.6. Let $a, b \in G$, a group. Suppose $o(a) = o(b) = o(ab) = 2$. Then show that $ab = ba$.

6.1.7. Let G be a group and let $a \in G$. Let

$$C_G(a) = \{x \in G : ax = xa\}.$$

Sometimes we omit the subscript G and just write $C(a)$. Prove that $C(a) \subset G$ is a subgroup of G. This subgroup is called the **centralizer of A in G**.

6.1.8. Suppose G is a group which has only one element $a \in G$ of order two, i.e., a is the only element such that $o(a) = 2$. Prove that $ax = xa$ for all $x \in G$.

(HINT: Consider $(x^{-1}ax)^2$.)

6.1.9. (a) Show that the intersection of any nonempty collection of subgroups of a group G is a subgroup of G.

(b) Show that if H, K are subgroups of G and the union $H \cup K$ is also a subgroup of G then either $H \subset K$ or $K \subset H$.

6.1.10. Let G be a group. Referring to exercises 6.1.7 and 6.1.9, the subgroup

$$Z(G) = \bigcap_{a \in G} C_G(a)$$

is called the **center of** G. Describe in words $Z(G)$, i.e., without using intersections.

6.1.11. Show that if G is a finite group, its multiplication table (or group table) is a Latin square, i.e., each element of the group appears once and only once in each row and in each column of the table.

EXERCISES FOR SECTION 6.2

6.2.1. Prove that for F a field, $SL(n, F)$ is a subgroup of $GL(n, F)$.

6.2.2. In the proof that $|A| = |B| \implies S_A \cong S_B$, prove that the map $F : S_A \to S_B$ give by $f \longmapsto \sigma f \sigma^{-1}$ is onto. This would complete the proof that $S_A \cong S_B$.

(HINT:For any $h \in S_B$ consider $\sigma^{-1} h \sigma$.)

6.2.3. If $A_1 \subset A$ prove that $stab(A_1)$ is a subgroup of S_A.

6.2.4. Explain why the "trick" given in the text to find the inverse of a permutation given in two-row form works. The trick is, if f is a permutation on a finite set given in two-row form, then f^{-1} is found by interchanging the first and second rows.

6.2.5. For the permutations a, b, c, d, e of S_3 as defined in Example 6.2.1 of the text, show by actual permutation mutiplication that the following relations hold:

(a) $a^2 = b$.
(b) $a^3 = 1$.
(c) $c^2 = 1$.
(d) $ac = d$.

(e) $a^2c = e$.

(f) $ac = ca^2$.

(g) Why are the 6 elements of S_3 just the following

$$1, a, a^2, c, ac, a^2c?$$

Note these are the elements on the guide row and guide column of the multiplication table of this group given in the text. This verifies that the Cayley table given in the text is actually one for the symmetric group S_3.

6.2.6. Verify using the relations $a^3 = c^2 = 1, ac = ca^2$ all the products in the multiplication table of this group given in the text.

6.2.7. Without using the text multiplication table for S_3 go back to the permutations in S_3 given by their two-row forms and write out the Cayley table for S_3 just from them.

6.2.8. Complete the proof of Cayley's theorem by showing the following:

(a) π_g is a permutation on the set G. (Here you must show that $\pi_g : G \to G$ is a bijection.)

(b) The map $\varphi : G \to S_G$ given by $\varphi(g) = \pi_g$ is a 1-1 group homomorphism. Here you must show that φ is 1-1 and φ preserves the operation.

6.2.9. If G is a group and $a, b \in G$, prove the equations $ax = b$ and $ya = b$ always have unique solutions for x and y in G.

*6.2.10. Show that the following is an alternate definition for a group: Suppose G is a set with a binary operation on it which is associative and for any $a, b \in G$ there always exist $x, y \in G$ such that $ax = b$ and $ya = b$, then G is a group. In other words the equations $ax = b$ and $ya = b$ are always solvable in G. (NOTE: Problem 6.2.9 shows that the conditions given here are always true in a group, but here you must show if these conditions hold then they imply that there is a group structure on G.)

6.2.11. Show that the following is an alternate definition for a *finite* group: Suppose G is a finite set with a binary operation on it which is associative and the cancellation laws hold (both left and right), then G is a group. If the word "finite" is deleted does this condition still work? Why or why not? If not, you must give a counterexample.

(HINT: Use the same kind of proof as was used to show that a finite integral domain is a field.)

6.2.12. Prove that a planar Euclidean motion is completely determined by its action on three noncollinear points — that is, on a single triangle.

Hence if T, U agree on three points then $T \equiv U$. In particular if T fixes three points then $T = I$.

(HINT: If an isometry fixes two points it has to fix the perpendicular bisector.)

6.2.13. Prove that a planar Euclidean motion that has two fixed points is either a reflection or the identity.

(HINT: If T fixes A and B then it fixes the line AB. Now consider a point C not on this line and consider the action on the triangle ABC.)

6.2.14. Prove that if a planar Euclidean motion has exactly one fixed point then it is a rotation.

(HINT: Let T be the motion and let O be the fixed point and consider the triangle $T(O)OT^2(O)$. Compare what a rotation about O would do to this triangle.)

6.2.15. Prove that a planar Euclidean motion with no fixed points and preserving orientation is a translation and if it reverses orientation it's a glide reflection.

EXERCISES FOR SECTION 6.3

For Problems 6.3.1–6.3.3, let G be a group, H is a subgroup of G, and $a, b \in G$.

6.3.1. Show that $Ha = Hb$ if and only if $ba^{-1} \in H$.

(NOTE: Here you may either use the equivalence relation which gives rise to right cosets or the definition of right cosets — as was done for left cosets in the text.)

6.3.2. Let H be a subgroup of the group G. Prove the following properties of cosets:

    (a) $hH = H$ if h is any element in H.
    (b) $x \in Hx$ for any $x \in G$.
    (c) $H^2 = H \cdot H = H$.

6.3.3. Show that the mapping $f : aH \to bH$ defined by $f(ah) = bh$ for any $h \in H$ is 1-1 and onto.

6.3.4. Prove the map $F : \{$all left cosets of H in $G\} \to \{$all right cosets of H in $G\}$ given by $F(aH) = Ha^{-1}$ is a bijection.

6.3.5. Find all the cosets of $H = \langle 4 \rangle = 4\mathbb{Z}$ in $\mathbb{Z}$. Recall that the operation here is $+$. So the cosets should be written additively, e.g., $0 + \mathbb{Z} = \mathbb{Z}$. Does it make any difference whether we use right or left cosets here? Tell what the index $[\mathbb{Z} : 4\mathbb{Z}]$ is.

6.3.6. Let our group be S_3 and our subgroup H be $\langle \begin{pmatrix} 1 & 2 & 3 \\ 2 & 1 & 3 \end{pmatrix} \rangle$. Find both the left and right coset decomposition, (partition), of S_3 with respect to H. Are these partitions the same?

6.3.7. Let G be an abelian group of order 6. Show that there exists an element $a \in G$ such that $G = \{e, a, a^2, a^3, a^4, a^5\} = \langle a \rangle$, i.e., $o(a) = 6$. This means G is cyclic.

(HINT: Use the Corollary 6.3.1 of Lagrange's theorem first to determine the possible orders of elements in G. Next show that if G has more than one element of order 2, then G must have a subgroup of order 4. This is a contradiction (why?). Thus G can only have at most one element of order 2. Similarly show that G can only have at most one subgroup of order 3. Use a counting argument to show that either you have already found an element of order 6 (i.e., an element a that was to be shown exists) or there must exist an x and y in G such that $o(x) = 2$ and $o(y) = 3$ and $y \notin \langle x \rangle = \{1, x\}$. Show this implies that G must have an element of order 6 by considering xy.)

6.3.8. Prove that any group of prime order is cyclic.

(HINT: Use Lagrange's theorem.)

6.3.9. Let G be a group of order pq where p and q are primes such that $p < q$. Prove that G does not contain two distinct subgroups of order q.

(HINT: Use a proof by contradiction using Theorem 6.3.3 (the product formula) and Problem 6.3.8 above.)

6.3.10. If U and V are subgroups of a group G such that $UV = V$ then prove that $U \subset V$.

6.3.11 For r, s, S and U as in Theorem 6.3.4, prove that

$$r(\bigcup_{s \in S} sU) = \bigcup_{s \in S} rsU.$$

6.3.12 Show that if $p, q, m \in \mathbb{N}$ where $p|m, q|m$ and $\gcd(p, q) = 1$ then $pq|m$.

6.3.13. In the proof of the Theorem of Poincaré, we used the fact that UV is a subgroup. Tell where it was used for the first time.

6.3.14. In the corollary to Poincare's theorem, tell where the hypothesis that G is a finite group was used for the **last time**. Also tell where the hypothesis that UV is a subgroup was used in this proof.

6.3.15. Find 8^{103} (mod 13). Note that this must be a number between 0 and 12 inclusive.

(HINT: Use Fermat's little theorem twice. The first time divide 103 by 12 and use Fermat's theorem to reduce. Then use Fermat's theorem again.)

EXERCISES FOR SECTION 6.4

6.4.1. Let G be a group. Explain why if $\{U_\alpha\}$ is the collection of all subgroups of G containing S, then $\cap_\alpha U_\alpha$ is the smallest subgroup of G containing S.

6.4.2. Prove that if G is a group and $g \in G$ has $o(g) = n$. Then $g^k = g^l$ if and only if $k \equiv l \bmod n$.

6.4.3. Let G be a group and $S \subset G$ be any complex. Let E be the set of all finite products of elements of S and their inverses (this includes single elements of S). Prove that E is a subgroup of G.

6.4.4. Let our group be $(\mathbb{Z}, +)$. Find
(a) $\langle 3 \rangle \cap \langle 4 \rangle$
(b) $\langle 6 \rangle + \langle 9 \rangle$

6.4.5. Let $G = \langle a \rangle$ with $|G| = n$. Let $s_1, s_2 > 0$ such that $s_i \mid n$ $(i = 1, 2)$. If $H_1 = \langle a^{s_1} \rangle$ and $H_2 = \langle a^{s_2} \rangle$ then prove that,

$$H_1 \cap H_2 = \langle a^{lcm(s_1, s_2)} \rangle$$

and

$$H_1 H_2 = \langle a^{\gcd(s_1, s_2)} \rangle.$$

(HINT: First explain why it is clear that $H_1 \cap H_2$ and $H_1 \cdot H_2$ are both subgroups here. For the latter, use the fact that the gcd is a linear combination of s_1 and s_2, i.e., Theorem 1.3.)

6.4.6. Let G be a group and $a \in G$ such that $o(a) = mn$ where $\gcd(m, n) = 1$. Show that one can write $a = bc$ where $o(b) = m$, $o(c) = n$, and $bc = cb$. Moreover, prove the uniqueness of such a representation.

(HINT: Write $mx + ny = 1$, let $b = a^{ny}, c = a^{mx}$ and use Theorem 6.4.10. For uniqueness, show using $mx + ny = 1$, that if b and c satisfy the stated conditions in the problem, then they must be given as stated in this hint.)

6.4.7. Prove that a group of order p^m, where p is a prime and $m \in \mathbb{N}$, must contain a subgroup of order p.

(HINT: Use Theorem 6.4.6.)

6.4.8. If in a group G of order n, for each positive $d \mid n$, the equation $x^d = 1$ has less than $d + \phi(d)$ solutions, then show G is cyclic.

(HINT: Use Corollary 6.4.3 and Theorem 6.4.9.)

6.4.9. Prove that a finite group G is cyclic if and only if G has no more than k, kth roots of 1 for every $k \in \mathbb{N}$.

(HINT: A kth root of 1 is a solution to $x^k = 1$ — here we assume 1 is the identity of G. Use Theorem 6.4.4 for the only if part and Theorem 6.4.9 for the if part.)

6.4.10. Suppose that A is a finite subset of nonzero integers.

(a) Extend the definition of GCD of two nonzero integers to $\gcd(A)$. Make sure to include the fact that the GCD is positive. Why does one want that?

(b) Extend the definition of LCM of two nonzero integers to $\operatorname{lcm}(A)$. Make sure to include the fact that the lcm is positive. Why does one want that?

(c) Prove that $\gcd(A)$ is a linear combination of the elements of A. That is, if $A = \{a_1, \ldots, a_n\}$ then

$$\gcd(A) = \sum_i x_i a_i$$

for some integers x_i.

(HINT: Return to the proof for two nonzero integers and extend it to any finite number.)

6.4.11. Using the definition given in Problem 6.4.10 above we can say that a finite set of nonzero integers is **relatively prime** if its GCD is 1. Extend the corollary which says that two elements of a group G commute and have finite relatively prime orders then the order of their product is the product of their orders to the case where one has any finite number of elements that commute in pairs and have relatively prime finite orders in pairs.

(HINT: Use mathematical induction.)

6.4.12. (a) Prove that $x_i, k \in \mathbb{Z}$ for all $i = 1, \ldots, n$ and if $\mathrm{GCD}(x_1, \ldots, x_n)$ $= 1$ and $x_i | k$ for all $i = 1, \ldots, n$ then $x_1 \cdots x_n | k$.

(b) Prove the following fact used in the text to prove the lemma about the exponent of a finite abelian group: Given any finite set of positive integers, their LCM can be found by writing the standard prime factorization of each of these integers and then taking the product of all the primes with each raised to the maximum power to which it occurs in any of the given integers. Note we can always write each integer in this finite set with same prime factors by allowing exponents to be 0 if necessary in the standard prime decomposition.

6.4.13. Let

$$G = SL(2, \mathbb{Z}) = \{ \begin{pmatrix} a & b \\ c & d \end{pmatrix} : a, b, c, d \in \mathbb{Z}, ad - bc = 1 \}.$$

Show that G is generated by

$$A = \begin{pmatrix} 0 & -1 \\ 1 & 0 \end{pmatrix} \text{ and } T = \begin{pmatrix} 1 & 1 \\ 0 & 1 \end{pmatrix}.$$

(HINT: Let $U = \begin{pmatrix} a & b \\ c & d \end{pmatrix} \in G$. Then

$$AU = \begin{pmatrix} -c & -d \\ a & b \end{pmatrix}$$

and

$$T^k U = \begin{pmatrix} a + kc & b + kd \\ c & d \end{pmatrix}.$$

If $|a| < |c|$ then apply A from the left. If $|a| > |c|$ then use the Euclidean algorithm to get a k such that $a = -kc + R$ with $0 < |r| < |c|$ and apply T^k from the left. If $a = c$ then consider $ATAU$ and if $a = -c$ consider $A^{-1}TU$. Repeating this we eventually get a matrix $\begin{pmatrix} \alpha & \beta \\ 0 & \alpha \end{pmatrix}$ with $\alpha = \pm 1$. If $\alpha = -1$ then apply A^2.)

Chapter 7

Factor Groups and the Group Isomorphism Theorems

In this chapter, we consider a construction that is fundamental to all of group theory. This is the factor group or quotient group construction. To construct factor groups, we need to introduce a new class of subgroups which are called normal subgroups. We shall see that normal subgroups enable us to construct factor groups using as our set for the factor group the set of all cosets relative to the normal subgroup.

It turns out that normal subgroups, factor groups and homomorphisms are intimately related. This is the content of the first isomorphism theorem. After this, we also consider the second and third isomorphism theorems for groups.

7.1 Normal Subgroups

We saw in Section 6.3 that given a group G and a subgroup H there is a set of equivalence classes called cosets. In this section we will look at a special type of subgroups called **normal subgroups** for which we can construct a group structure on the set of cosets. This new group will be called the **quotient group** or **factor group** of G modulo H. In Chapter 11 we will introduce and study the analogous construction for rings. The special subrings that correspond to normal subgroups will be called **ideals** and from them we can build quotient rings.

Definition 7.1.1. *Let G be an arbitrary group and suppose that H_1 and H_2*

are subgroups of G. We say that H_2 is **conjugate** to H_1 in G if there exists an element $a \in G$ such that $H_2 = aH_1a^{-1}$. H_1, H_2 are the called **conjugate subgroups** of G.

Lemma 7.1.1. *Let G be an arbitrary group. Then the relation of conjugacy is an equivalence relation on the set of subgroups of G.*

Proof. We must show that conjugacy is reflexive, symmetric and transitive. If H is a subgroup of G then $1(H)1^{-1} = H$ and hence H is conjugate to itself and therefore the relation is reflexive. (We note that we would get the same result if instead of using 1 we used any $h \in H$ because $hHh^{-1} = H$ for any such h.) Suppose that H_1 is conjugate to H_2. Then there exists a $g \in G$ with $H_1 = gH_2g^{-1}$. This implies that $g^{-1}H_1g = H_2$. However, $(g^{-1})^{-1} = g$ and hence letting $g^{-1} = g_1$ we have $g_1H_1g_1^{-1} = H_2$. Therefore H_2 is conjugate to H_1 and conjugacy is symmetric. Finally suppose that H_1 is conjugate to H_2 and H_2 is conjugate to H_3. Then there exist $g_2, g_3 \in G$ with $H_1 = g_2H_2g_2^{-1}$ and $H_2 = g_3H_3g_3^{-1}$. Then

$$H_1 = g_2g_3H_3g_3^{-1}g_2^{-1} = (g_2g_3)H_3(g_2g_3)^{-1}.$$

Therefore H_1 is conjugate to H_3 and conjugacy is transitive.   □

Analogous to the definition of the centralizer of an element (see problem 6.1.7), we define the **normalizer of the subgroup** H in G, denoted by $N_G(H)$ as follows:

$$N_G(H) = \{a \in G : aHa^{-1} = H.\}$$

We leave it to the exercises to show that $N_G(H)$ is a subgroup of G. But it is also clear that $H \subset N_G(H)$. When it is clear which group we are working in, we may write $N(H)$ for $N_G(H)$.

Theorem 7.1.1. *Let H be a subgroup of the finite group G. Then the number of distinct subgroups of G conjugate to H is $[G : N_G(H)]$. In other words, this is the number of conjugates of H in G.*

Proof. In this proof we will write $N(H)$ instead of $N_G(H)$. Let

$$G = \bigcup_{i=1}^{n} g_iN(H) \text{ (disjoint)}.$$

Here $n = [G : N(H)]$. Now if we can show that any two elements of the same coset of $N(H)$, i.e. of $g_i N(H)$ for some i, yield the same conjugate of H while elements from different cosets of $N(H)$ yield different conjugates of H, then we will be done because then the number of distinct conjugates of H will be equal to the number of distinct left cosets of $N(H)$ in G, that number being $[G : N(H)]$.

We first note that if $x \in N(H)$, then $g_i x H x^{-1} g_i^{-1} = g_i H g_i^{-1}$. (Why?) This implies that if you multiply any g_i by an element of $N(H)$ on the right it does not change the conjugate of H which the element gives. Consider $g_i H g_i^{-1}$ $= g_j H g_j^{-1}$. Then this is equivalent to $H = g_i^{-1} g_j H g_j^{-1} g_i = g_i^{-1} g_j H (g_i^{-1} g_j)^{-1}$. This in turn is equivalent to $g_i^{-1} g_j \in N(H)$. Lemma 6.3.2 implies that this is equivalent to $g_i N(H) = g_j N(H)$. This says that two conjugates of H are equal if and only if the two corresponding left cosets are equal. (Note that multiplying an element g_i on the right by an element of $N(H)$ does not change the conjugate of H it gives — we established this first.) But elements of different left cosets will yield different conjugates. This means that the number of different conjugates of H in G is equal to the number of left cosets of $N(H)$ in G, i.e., $[G : N(H)]$. $\square$

We observe that the above proof also shows that the order of an equivalence class under the equivalence relation given in Lemma 7.1.1 is equal to the order of a left coset of $N(H)$. This follows because what the above proof implies is that the decomposition of G with respect to the left cosets of $N(H)$ is the same as the decomposition of G with respect to the equivalence relation of conjugacy on H.

We next consider conjugacy of elements. Let G be any group. Recall that an automorphism of G is an isomorphism of G onto itself.

Lemma 7.1.2. *Let G be an arbitrary group. Then for $g \in G$ the map*

$$T_g : a \mapsto g^{-1} a g$$

is an automorphism on G.

Proof. For a fixed $g \in G$ define the map $T_g : G \to G$ by $T_g(a) = g^{-1} a g$ for $a \in G$. We must show that this is a homomorphism and that it is one-to-one and onto.

Let $a_1, a_2 \in G$. Then

$$T_g(a_1 a_2) = g^{-1} a_1 a_2 g = (g^{-1} a_1 g)(g^{-1} a_2 g) = T_g(a_1) T_g(a_2).$$

Hence T_g is a homomorphism.

If $T_g(a_1) = T_g(a_2)$ then $g^{-1}a_1g = g^{-1}a_2g$. Clearly by the cancellation law which holds in G we then have $a_1 = a_2$ and hence T_g is one-to-one.

Finally let $a \in G$ and let $a_1 = gag^{-1}$ which clearly is in G. Then $a = g^{-1}a_1g$ and hence $T_g(a_1) = a$. It follows that T_g is onto and therefore T_g is an automorphism on G. $\qquad\square$

We note that since the automorphism $T_g(a) = g^{-1}ag$ in the above lemma maps the subgroup H onto $a^{-1}Ha = T_g(H)$, it is clear from Lemma 2.7.6 of Chapter 2 or from Theorem 7.3.1 of this chapter that $a^{-1}Ha$ is a subgroup of G.

For a given group G we denote by $\text{Aut}(G)$ the set of all automorphisms of G.

Lemma 7.1.3. *For any group G the set $\text{Aut}(G)$ is a group with respect to the binary operation of composition of mappings.*

Proof. Clearly $I_G \in \text{Aut}(G)$, i.e., $I_G(x) = x$ for all $x \in G$ is the identity element or 1. The associative law is true with respect to composition for mappings and if $f \in \text{Aut}(G)$, then f^{-1} exists (as a map) and further $f^{-1} \in \text{Aut}(G)$ since for any $a, b \in G$,

$$f^{-1}(ab) = f^{-1}(ff^{-1}(a)ff^{-1}(b))$$

$$= f^{-1}(f(f^{-1}(a)f^{-1}(b)))$$

since f preserves the operation

$$= f^{-1}(a)f^{-1}(b).$$

Is the above sufficient to show that $\text{Aut}(G)$ is a group under composition of functions? Why or why not? (See the exercises.) $\qquad\square$

Now we consider two special kinds of automorphisms of a group. Let $a \in G$ and consider the map $T_a : G \to G$ defined by $T_a(g) = a^{-1}ga$ for all $g \in G$. The above lemma (Lemma 7.1.2) shows that T_a belongs to $\text{Aut}(G)$. The automorphism T_a is called the **inner automorphism of** G **determined by** a.

If we set

$$\text{Inn}(G) = \{T \in \text{Aut}(G) : T \text{ is an inner automorphism of } G\},$$

then it is not hard to show that $\mathrm{Inn}(G)$ is a subgroup of $\mathrm{Aut}(G)$. (See the exercises.)

We note that here we have written the inner automorphism $T_a(g) = a^{-1}ga$. Sometimes we also call $F_a(g) = aga^{-1}$ the inner automorphism determined by a. Even though $T_a \neq F_a$ (necessarily — they could be equal for example if G is abelian), it is not hard to show that the set $\mathrm{Inn}(G)$ remains the same no matter which definition we take (see the exercises). All elements of $\mathrm{Aut}(G)$ which are not inner automorphisms, if there are any, are called **outer automorphisms** of G.

The following example will be very important when we discuss Galois theory in Chapter 15.

EXAMPLE 7.1.1. Let K be an arbitrary field and let $F \subset K$ be a subfield of K. We say that K is an extension field of F. Just as for groups, we introduce the set

$$\mathrm{Aut}(K) = \{f : K \to K : f \text{ is an automorphism of } K\}.$$

Here an automorphism is a bijection from K onto K that is a ring homomorphism, i.e., preserves both addition and multiplication of the field K. $\mathrm{Aut}(K)$ forms a group with respect to composition of functions. This fact is shown more or less the same way as for groups — except that now there are two operations $+$ and $\cdot$. We leave the proof that $\mathrm{Aut}(K)$ is a group as an exercise. Since an automorphism of K is a bijection of K onto K and the operation is composition, $\mathrm{Aut}(K)$ is a subgroup of the group of all permutations of K, i.e., the symmetric group S_K. So we have $\mathrm{Aut}(K) \subset S_K$. If we denote by $G(K/F)$ the set of all automorphisms of K leaving F fixed elementwise, i.e., $f \in G(K/F)$ means $f \in \mathrm{Aut}(K)$ and $f(x) = x$ for all $x \in F$ then $G(K/F)$ is a subgroup of $\mathrm{Aut}(K)$. This is a straightforward verification and so will also be left to the exercises.

In general a subgroup H of a group G may have many different conjugates. As we have just seen in Theorem 7.1.1, the number of conjugates of a subgroup of a finite group is the index of its normalizer. However, in certain situations the only conjugate of a subgroup H is H itself. If this is the case we say that H is a **normal subgroup**.

Definition 7.1.2. *Let G be an arbitrary group. A subgroup H is a **normal subgroup** of G, which we denote by $H \lhd G$ if $g^{-1}Hg = H$ for all $g \in G$.*

We note that since $g^{-1}Hg = H$ for all $g \in G$, that this is equivalent to $gHg^{-1} = H$ for all $g \in G$. (See the exercises.) We also note that $H \lhd G$ if and only if $N_G(H) = G$. Thus $[G : N_G(H)] = 1$ which as we commented before the definition, confirms that normality is equivalent to the subgroup having just one conjugate, i.e., itself.

Since the conjugation map is an isomorphism it follows that if $g^{-1}Hg \subset H$ for all $g \in G$ then $g^{-1}Hg = H$ (by Lemma 7.1.2 and see Problem 2.2.7 of Section 2.2). Hence in order to show that a subgroup is normal we need only show inclusion. But this can be done without that reference as follows.

Lemma 7.1.4. *Let N be a subgroup of a group G. Then if $aNa^{-1} \subset N$ for all $a \in G$, then $aNa^{-1} = N$. In particular, $aNa^{-1} \subset N$ for all $a \in G$ implies that N is a normal subgroup. (We note showing $a^{-1}Na \subset N$ $\forall a \in G$ also implies that $N \lhd G$.)*

Proof. Suppose that $aNa^{-1} \subset N$ for all $a \in G$. We need to show that $aNa^{-1} = N$ for all $a \in G$. Since $aNa^{-1} \subset N$ for all $a \in G$, it must hold for a^{-1}. This means $a^{-1}Na \subset N$. But if you multiply both sides of a set inclusion by the same element of a group, then the set inclusion remains valid. (See the exercises.) So if we multiply on the left by a and then on the right by a^{-1} we get $N \subset aNa^{-1}$. This gives the necessary opposite inclusion and so $N = aNa^{-1}$. Similarly, we can start with $a^{-1}Na \subset N$ and show that $N = aNa^{-1}$. (See the exercises.) $\square$

Notice that if $g^{-1}Hg = H$ then $Hg = gH$. That is, *as sets* the left coset gH is equal to the right coset Hg. This does *not* say that g commutes with every element of H, but rather for each $h_1 \in H$ there is an $h_2 \in H$ with $gh_1 = h_2g$ and similarly for each element h_2 of H_2. If $H \lhd G$ this is true for all $g \in G$. Further if H is normal in G then for the product of two cosets g_1H and g_2H considered as the product of complexes as defined in Section 6.3, we have

$$(g_1H)(g_2H) = g_1(Hg_2)H = g_1g_2(HH) = g_1g_2H.$$

If $(g_1H)(g_2H) = (g_1g_2)H$ for all $g_1, g_2 \in G$ we necessarily have $gHg^{-1} = H$ for all $g \in G$. To see that note that if we set $g_2 = g_1^{-1}$, we get $g_1Hg_1^{-1}H = H$. Now consider any element of $g_1Hg_1^{-1}$ say, $g_1h_1g_1^{-1}$ where $h_1 \in H$. Since $g_1Hg_1^{-1}H = H$, $g_1h_1g_1^{-1} \cdot h_2 = h_3$ where h_2 and h_3 both $\in H$. But this implies that $g_1h_1g_1^{-1} = h_3h_2^{-1} \in H$. Thus we have shown that for any g_1, $g_1Hg_1^{-1} \subset H$ which by Lemma 7.1.4 is sufficient for $g_1Hg_1^{-1} = H$.

Hence we have proved (together with the exercises).

Lemma 7.1.5. *Let H be a subgroup of a group G. Then the following are equivalent:*

(1) *H is a normal subgroup of G.*
(2) *$gHg^{-1} = H$ for all $g \in G$.*
(3) *$gH = Hg$ for all $g \in G$.*
(4) *$(g_1 H)(g_2 H) = (g_1 g_2)H$ for all $g_1, g_2 \in G$.*

This is precisely the condition needed to construct factor groups. First we give some examples of normal subgroups.

Lemma 7.1.6. *Every subgroup of an abelian group is normal.*

Proof. Let G be abelian and H a subgroup of G. Suppose $g \in G$ then $gh = hg$ for all $h \in H$ since G is abelian. It follows that $gH = Hg$. Since this is true for every $g \in G$ it follows that H is normal. $\square$

Lemma 7.1.7. *Let $H \subset G$ be a subgroup of index 2, that is, $[G : H] = 2$. Then H is normal in G.*

Proof. Suppose that $[G : H] = 2$. We must show that $gH = Hg$ for all $g \in G$. If $g \in H$ then clearly $H = gH = Hg$. Therefore we may assume that g is not in H. Then there are only two left cosets and two right cosets. That is,

$$G = H \cup gH = H \cup Hg.$$

Since the union is a disjoint union we must have $gH = Hg$ and hence H is normal. $\square$

Lemma 7.1.8. *Let K be any field. Then the group $SL(n, K)$ is a normal subgroup of $GL(n, K)$ for any positive integer $n \geq 2$.*

Proof. Recall that $GL(n, K)$ is the group of $n \times n$ matrices over the field K with nonzero determinant while $SL(n, K)$ is the subgroup of $n \times n$ matrices over the field K with determinant equal to 1. Let $U \in SL(n, K)$ and $T \in GL(n, K)$. Now using the properties of determinants assumed in the proof of Lemma 6.2.2, we consider $T^{-1}UT$. Then

$$det(T^{-1}UT) = det(T^{-1})det(U)det(T) = det(U)det(T^{-1}T)$$

$$= det(U)det(I) = det(U) = 1.$$

Hence $T^{-1}UT \in SL(n, K)$ for any $U \in SL(n, K)$ and any $T \in GL(n, K)$. It follows that $T^{-1}SL(n, K)T \subset SL(n, K)$ and therefore $SL(n, K)$ is normal in $GL(n, K)$. □

The intersection of normal subgroups is again normal and the product of normal subgroups is a normal subgroup.

Lemma 7.1.9. *Let N_1, N_2 be normal subgroups of the group G. Then:*
(1) $N_1 \cap N_2$ is a normal subgroup of G.
(2) $N_1 N_2$ is a normal subgroup of G.
(3) If H is any subgroup of G then $N_1 \cap H$ is a normal subgroup of H and $N_1 H = H N_1$ and hence $N_1 H$ is a subgroup of G.

Proof. (1) Let $n \in N_1 \cap N_2$ and $g \in G$. Then $g^{-1}ng \in N_1$ since N_1 is normal. Similarly $g^{-1}ng \in N_2$ since N_2 is normal. Hence $g^{-1}ng \in N_1 \cap N_2$. It follows that $g^{-1}(N_1 \cap N_2)g \subset N_1 \cap N_2$ and therefore $N_1 \cap N_2$ is normal.
(2) Let $n_1 \in N_1, n_2 \in N_2$. Since N_1, N_2 are both normal $N_1 N_2 = N_2 N_1$ as sets and the complex $N_1 N_2$ forms a subgroup of G by Theorem 6.3.2. Let $g \in G$ and $n_1 n_2 \in N_1 N_2$. Then

$$g^{-1}(n_1 n_2)g = (g^{-1}n_1 g)(g^{-1}n_2 g) \in N_1 N_2$$

since $g^{-1}n_1 g \in N_1$ and $g^{-1}n_2 g \in N_2$. Therefore $N_1 N_2$ is normal in G.
(3) Let $h \in H$ and $n \in N_1 \cap H$. Then $h^{-1}nh \in N_1$ because $N_1 \lhd G$ and $h^{-1}nh \in H$ because h, n both do and therefore $h^{-1}nh \in N_1 \cap H$. Hence $N_1 \cap H$ is a normal subgroup of H. But since $N_1 \lhd G$ it follows that $N_1 H = H N_1$. Again by Theorem 6.3.2, this implies that $N_1 H$ is a subgroup of G. □

For a group G we have introduced the group $\text{Aut}(G)$ and it subgroup $\text{Inn}(G)$. We now show the following.

Lemma 7.1.10. *For any group G, $\text{Inn}(G)$ is a normal subgroup of $\text{Aut}(G)$.*

Proof. We have already seen that $\text{Inn}(G)$ is a subgroup. Therefore to show that it is normal we must show that any conjugate of any element in $\text{Inn}(G)$ is still in $\text{Inn}(G)$.

For an element $g \in G$ we let T_g denote the corresponding inner automorphism, i.e., $T_g \in \text{Inn}(G)$. Here we take $T_g(x) = gxg^{-1}$ for all $x \in G$. Let

$\alpha \in \text{Aut}(G)$ and consider the element $\alpha T_g \alpha^{-1} \in \text{Aut}(G)$. This is a mapping and we consider its action on any element $x \in G$. Thus we have

$$\alpha T_g \alpha^{-1}(x) = \alpha(T_g(\alpha^{-1}(x)) = \alpha(g\alpha^{-1}(x)g^{-1}) = \alpha(g)\alpha(\alpha^{-1}(x))\alpha(g^{-1})$$

$$= \alpha(g)x(\alpha(g))^{-1} = T_{\alpha(g)}(x)$$

since α is a homomorphism.

Therefore conjugating the inner automorphism T_g by an arbitrary automorphism α is the same mapping as the inner automorphism determined by $\alpha(g)$ and hence the conjugate of T_g by α is still in $\text{Inn}(G)$ proving that it is normal. □

7.2 Factor Groups

We now construct **factor groups** or **quotient groups** of a group modulo a normal subgroup.

Definition 7.2.1. *Let G be an arbitrary group and H a normal subgroup of G. Let G/H denote the set of distinct left (and hence also right) cosets of H in G. On G/H define the multiplication*

$$(g_1 H)(g_2 H) = g_1 g_2 H$$

for any elements $g_1 H, g_2 H$ in G/H.

Theorem 7.2.1. *Let G be a group and H a normal subgroup of G. Then G/H under the operation defined above forms a group. This group is called the **factor group** or **quotient group** of G modulo H. The identity element is the coset $1H = H$ and the inverse of a coset gH is $g^{-1}H$.*

Proof. We first show that the operation on G/H is well-defined. Suppose that $a'H = aH$ and $b'H = bH$. Then $a' = ah_1$ where $h_1 \in H$ and similarly $b' = bh_2$ where $h_2 \in H$. Therefore

$$a'Hb'H = (ah_1 H)(bh_2 H) = (aH)(bH) = a(Hb)H = ab(HH) = abH.$$

These follow since H is normal which implies that $b^{-1}Hb = H$ so that $Hb = bH$. Similarly $aH = Ha$.

Thus we have shown that if $H \triangleleft G$ then since $a'Hb'H = a'b'H$ by definition and similarly $(aH)(bH) = (ab)H$ and we have shown above that $a'Hb'H = abH$, we have shown that the operation on G/H, is indeed, well-defined.

The associative law is true because coset multiplication as defined above uses the ordinary group operation which is by definition associative.

The coset H serves as the identity element of G/H. Notice that

$$aH \cdot H = aH^2 = aH$$

and

$$H \cdot aH = aH^2 = aH.$$

The inverse of aH is $a^{-1}H$ since

$$aHa^{-1}H = aa^{-1}H^2 = H$$

and

$$a^{-1}HaH = a^{-1}aH^2 = H.$$

$\square$

We emphasize that the elements of G/H are cosets and thus subsets (complexes) of G. If $|G| < \infty$, then $|G/H| = [G : H] = \frac{|G|}{|H|}$, the number of cosets of H in G. It is also to be emphasized that in order for G/H to be a group H must be a normal subgroup of G.

In some cases properties of G are preserved in factor groups.

Lemma 7.2.1. *If G is abelian then any factor group of G is also abelian. If G is cyclic then any factor group of G is also cyclic.*

Proof. Suppose that G is abelian and H is a subgroup of G. H is necessarily normal from Lemma 7.1.6 so that we can form the factor group G/H. Let $g_1H, g_2H \in G/H$. Since G is abelian we have $g_1g_2 = g_2g_1$. Then in G/H,

$$(g_1H)(g_2H) = (g_1g_2)H = (g_2g_1)H = (g_2H)(g_1H).$$

Therefore G/H is abelian.

We leave the proof of the second part to the exercises. $\square$

An extremely important concept is when a group contains no proper normal subgroups other than the identity subgroup $\{1\}$.

Definition 7.2.2. *A group $G \neq \{1\}$ is* **simple** *provided that $N \triangleleft G$ implies $N = G$ or $N = \{1\}$.*

One of the most outstanding problems in group theory has been to give a complete classification of all finite simple groups. In other words, this is the program to discover all finite simple groups and to prove that there are no more to be found. This was accomplished through the efforts of many mathematicians. The proof of this magnificent result took thousands of pages. We refer the reader to [G] for a complete discussion of this. We give one elementary example.

Lemma 7.2.2. *Any finite group of prime order is simple and cyclic.*

Proof. Suppose that G is a finite group and $|G| = p$ where p is a prime. Let $g \in G$ with $g \neq 1$. Then $\langle g \rangle$ is a nontrivial subgroup of G so its order divides the order of G by Lagrange's theorem. Since $g \neq 1$ and p is a prime we must have $|\langle g \rangle| = p$. Therefore $\langle g \rangle$ is all of G, that is, $G = \langle g \rangle$ and hence G is cyclic. The argument above shows that G has no nontrivial proper subgroups and therefore no nontrivial normal subgroups. Therefore G is simple. $\square$

In Chapter 9 we will mention certain other finite simple groups.

We conclude this section with a nice application of factor groups. In particular, we will show that any group of order p^2 where p is a prime is abelian. To do this we need a few facts. Recall from Problem 6.1.10, that the **center**, $Z(G)$, of a group G is the following subgroup

$$Z(G) = \{z \in G : zx = xz \; \forall x \in G\}.$$

It is not hard to show that $Z(G)$ is normal in G, i.e., $Z(G) \triangleleft G$ (see the exercises). Here we need the following fact.

Lemma 7.2.3. *If G is a group such that $G/Z(G)$ is cyclic, then G is abelian.*

Proof. We will write Z for $Z(G)$. We first remark that G/Z is defined because $Z \triangleleft G$ as commented before the statement of our lemma. Now since G/Z is cyclic, let us write $G/Z = \langle aZ \rangle$ for $a \in G$. Now let us decompose G via the left cosets of Z, realizing that each such left coset is a power of aZ. This gives

$$G = \bigcup_{n \in \mathbb{Z}} a^n Z,$$

where one must understand that we are not assuming G is finite nor are these cosets necessarily disjoint. But what is certainly true is if we take any g and h in G then $g \in a^k Z$ and $h \in a^m Z$ for some integers k, m. This gives that $g = a^k z_1$ and $h = a^m z_2$ where $z_1, z_2 \in Z$. Thus

$$gh = (a^k z_1)(a^m z_2) = a^k a^m z_1 z_2,$$

since z_1 and z_2 are in the center and so commute with everything. Hence

$$gh = (a^m z_2)(a^k z_1) = hg.$$

So that G is abelian. $\square$

We need one last fact which we will prove in Chapter 9. This is that the center of a group of order p^n where p is a prime must be nontrivial. We can now obtain our desired application.

Theorem 7.2.2. *A group of order p^2 where p is a prime is abelian.*

Proof. Let G be a group such that $|G| = p^2$ and let $Z = Z(G)$ be the center of G. Since Z is nontrivial from Lagrange's theorem $|Z| = p$ or p^2. If $|Z| = p^2$ then $G = Z$ and so G is abelian and we are done. On the other hand if $|Z| = p$, we consider G/Z. Again this makes sense because Z is normal in G. Now $|G/Z| = \frac{|G|}{|Z|} = \frac{p^2}{p} = p$ by Lagrange's theorem. Thus by Lemma 7.2.2, G/Z is cyclic. Then G is again abelian by the above lemma and actually $|Z| = p^2$. $\square$

Actually one can prove more and also show that there are precisely two nonisomorphic groups of order p^2, but we will consider this in the next chapter.

7.2.1 Examples of Factor Groups

We now consider some examples of factor groups. Before going to the examples, however, we emphasize that if G is a group and N is a normal subgroup of G, i.e., $N \triangleleft G$, then the elements of G/N are complexes (cosets which are nonempty subsets) of G. If $|G| < \infty$, then from Lagrange's theorem

$$|G/N| = \text{the number of left cosets of } N \text{ in } G = [G : N] = \frac{|G|}{|N|}.$$

Since N is normal in G, left and right cosets are the same. As our examples will show, it is possible to have G infinite, N infinite, but G/N finite.

EXAMPLE 7.2.1. Let $6\mathbb{Z} = \{6n : n \in \mathbb{Z}\} = \{\ldots, -12, -6, 0, 6, 12, \ldots\}$. So $6\mathbb{Z}$ is the cyclic subgroup $\langle 6 \rangle$ of the group $(\mathbb{Z}, +)$, of integers under addition. Since $\mathbb{Z}$ is abelian by Lemma 7.1.6, $6\mathbb{Z} \lhd \mathbb{Z}$.

To construct $\mathbb{Z}/6\mathbb{Z}$, we first find all (left) cosets of $6\mathbb{Z}$ in $\mathbb{Z}$. That is, we find the partition of $\mathbb{Z}$ which $6\mathbb{Z}$ gives. Here the left cosets are written additively because the group operation is $+$.

Consider the 6 cosets:

$$
\begin{aligned}
0 + 6\mathbb{Z} = 6\mathbb{Z} &= \{\ldots, -12, -6, 0, 6, 12, \ldots\}. \\
1 + 6\mathbb{Z} \quad &= \{\ldots, -11, -5, 1, 7, 13, \ldots\}. \\
2 + 6\mathbb{Z} \quad &= \{\ldots, -10, -4, 2, 8, 14, \ldots\}. \\
3 + 6\mathbb{Z} \quad &= \{\ldots, -9, -3, 3, 9, 15, \ldots\}. \\
4 + 6\mathbb{Z} \quad &= \{\ldots, -8, -2, 4, 10, 16, \ldots\}. \\
5 + 6\mathbb{Z} \quad &= \{\ldots, -7, -1, 5, 11, 17, \ldots\}.
\end{aligned}
$$

From the above, it is evident that $\mathbb{Z} = 6\mathbb{Z} \cup (1 + 6\mathbb{Z}) \cup \cdots \cup (5 + 6\mathbb{Z})$ and that this union is disjoint. This shows that these are all the cosets of $6\mathbb{Z}$ in $\mathbb{Z}$. This is also clear from the division algorithm. For if $n \in \mathbb{Z}$, then $n = 6q + r$ where $0 \leq r < 6$. Thus

$$
n + 6\mathbb{Z} = 6q + r + 6\mathbb{Z} = r + 6\mathbb{Z}.
$$

We also note that this shows that if $[n] \in \mathbb{Z}_6$ is the equivalence class of n under the equivalence relation of congruence modulo 6, then $[n] = [r] = r + 6\mathbb{Z}$.

Now that we know the elements of the factor group $\mathbb{Z}/6\mathbb{Z}$, we write its Cayley table.

	$0 + 6\mathbb{Z}$	$1 + 6\mathbb{Z}$	$2 + 6\mathbb{Z}$	$3 + 6\mathbb{Z}$	$4 + 6\mathbb{Z}$	$5 + 6\mathbb{Z}$
$0 + 6\mathbb{Z}$	$0 + 6\mathbb{Z}$	$1 + 6\mathbb{Z}$	$2 + 6\mathbb{Z}$	$3 + 6\mathbb{Z}$	$4 + 6\mathbb{Z}$	$5 + 6\mathbb{Z}$
$1 + 6\mathbb{Z}$	$1 + 6\mathbb{Z}$	$2 + 6\mathbb{Z}$	$3 + 6\mathbb{Z}$	$4 + 6\mathbb{Z}$	$5 + 6\mathbb{Z}$	$0 + 6\mathbb{Z}$
$2 + 6\mathbb{Z}$	$2 + 6\mathbb{Z}$	$3 + 6\mathbb{Z}$	$4 + 6\mathbb{Z}$	$5 + 6\mathbb{Z}$	$0 + 6\mathbb{Z}$	$1 + 6\mathbb{Z}$
$3 + 6\mathbb{Z}$	$3 + 6\mathbb{Z}$	$4 + 6\mathbb{Z}$	$5 + 6\mathbb{Z}$	$0 + 6\mathbb{Z}$	$1 + 6\mathbb{Z}$	$2 + 6\mathbb{Z}$
$4 + 6\mathbb{Z}$	$4 + 6\mathbb{Z}$	$5 + 6\mathbb{Z}$	$0 + 6\mathbb{Z}$	$1 + 6\mathbb{Z}$	$2 + 6\mathbb{Z}$	$3 + 6\mathbb{Z}$
$5 + 6\mathbb{Z}$	$5 + 6\mathbb{Z}$	$0 + 6\mathbb{Z}$	$1 + 6\mathbb{Z}$	$2 + 6\mathbb{Z}$	$3 + 6\mathbb{Z}$	$4 + 6\mathbb{Z}$

Here we note that since the group operation is $+$, the operation on cosets becomes addition rather than multiplication and so we get

$$
(a + 6\mathbb{Z}) + (b + 6\mathbb{Z}) = (a + b) + 6\mathbb{Z}.
$$

It is easy to verify from the above that

$$\mathbb{Z}/6\mathbb{Z} \cong \mathbb{Z}_6 \quad (n + 6\mathbb{Z}) \mapsto [n].$$

More generally if $n \in \mathbb{N}$ and we let $n\mathbb{Z} = \langle n \rangle = \{0, \pm n, \pm 2n, \ldots\}$, then $\mathbb{Z}/n\mathbb{Z} \cong \mathbb{Z}_n$.

We note that in the above example $\mathbb{Z}$ and $6\mathbb{Z}$ were both infinite, but $\mathbb{Z}/6\mathbb{Z}$ was not; it has order 6.

EXAMPLE 7.2.2. In Chapter 6 (see Example 6.2.1) we gave a Cayley table for S_3 in terms of $1, a, a^2, c, ac, a^2c$ where $a = \begin{pmatrix} 1 & 2 & 3 \\ 2 & 3 & 1 \end{pmatrix}$ and $c = \begin{pmatrix} 1 & 2 & 3 \\ 2 & 1 & 3 \end{pmatrix}$. If we consider $N = \langle a \rangle$, we see it has order 3 because $a^3 = 1$. Thus it is normal by Lemma 7.1.7. You should note that $\langle c \rangle$ is *not* normal. (Can you see why?) Then $|G/N| = 2$. We see this immediately from the following using the Cayley table given for S_3 :

$$\begin{aligned} 1N &= N = \{1, a, a^2\}, \\ cN &= \{c, ca, ca^2\} = \{c, a^2c, ac\}. \end{aligned}$$

So the whole group $G = S_3 = N \cup cN$ and this union is disjoint. This again shows that N is of index 2. Thus $G/N = \{N, cN\}$ is a group of order 2 where the coset N is the identity. Here is its Cayley table.

	N	cN
N	N	cN
cN	cN	N

We note $c^2 = 1$, so $cN \cdot cN = c^2 N = N$.

EXAMPLE 7.2.3. For a less trivial example of factor groups using permutations let $G = S_4$. So now $|G| = 4! = 24$. Let us consider the following four of the 24 permutation:

$$f_1 = 1 = \begin{pmatrix} 1 & 2 & 3 & 4 \\ 1 & 2 & 3 & 4 \end{pmatrix}, \quad f_2 = \begin{pmatrix} 1 & 2 & 3 & 4 \\ 2 & 1 & 4 & 3 \end{pmatrix}, \quad f_3 = \begin{pmatrix} 1 & 2 & 3 & 4 \\ 3 & 4 & 1 & 2 \end{pmatrix},$$
$$f_4 = \begin{pmatrix} 1 & 2 & 3 & 4 \\ 4 & 3 & 2 & 1 \end{pmatrix}.$$

It is not hard to show just by mutiplying these permutations out using the above two-row form that $f_2^2 = f_3^2 = f_4^2 = \dot{1}$. (See the exercises.) Further we claim for i, j, k any permutation of $2, 3, 4$ that

$$f_i f_j = f_j f_i = f_k.$$

For example,

$$f_2 f_3 = \begin{pmatrix} 1 & 2 & 3 & 4 \\ 2 & 1 & 4 & 3 \end{pmatrix} \begin{pmatrix} 1 & 2 & 3 & 4 \\ 3 & 4 & 1 & 2 \end{pmatrix} = \begin{pmatrix} 1 & 2 & 3 & 4 \\ 4 & 3 & 2 & 1 \end{pmatrix} = f_4$$

but also

$$f_3 f_2 = \begin{pmatrix} 1 & 2 & 3 & 4 \\ 3 & 4 & 1 & 2 \end{pmatrix} \begin{pmatrix} 1 & 2 & 3 & 4 \\ 2 & 1 & 4 & 3 \end{pmatrix} = \begin{pmatrix} 1 & 2 & 3 & 4 \\ 4 & 3 & 2 & 1 \end{pmatrix} = f_4$$

There are four more multiplications like this which we leave to the exercises to verify that a Cayley table for the set $V_4 = \{f_1, f_2, f_3, f_3\}$ is as follows.

	f_1	f_2	f_3	f_4
f_1	f_1	f_2	f_3	f_4
f_2	f_2	f_1	f_4	f_3
f_3	f_3	f_4	f_1	f_2
f_4	f_4	f_3	f_2	f_1

This is sufficient to show that V_4 is a subgroup of $G = S_4$. V_4 is called the **Klein 4-group**. You should note that V_4 is abelian but not cyclic. (Why?) Here we state that V_4 is a normal subgroup of S_4. To show this from the definition of normality is quite tedious because we would need to show that for all $g \in S_4$ that $g V_4 g^{-1} \subset V_4$. There are other ways of showing that $V_4 \triangleleft S_4$. We will not give a proof of this here. (Later in the text we will present a very nice way to prove this [see Example 9.1.3]). But just to give an example of what one would have to do to show it from the definition of normality, we take an element in S_4 not in V_4 so that the containment we want to show is not trivially true. Let, for example, $g = \begin{pmatrix} 1 & 2 & 3 & 4 \\ 3 & 2 & 1 & 4 \end{pmatrix} \in S_4 - V_4$. Then

$$g^{-1} = \begin{pmatrix} 3 & 2 & 1 & 4 \\ 1 & 2 & 3 & 4 \end{pmatrix} = \begin{pmatrix} 1 & 2 & 3 & 4 \\ 3 & 2 & 1 & 4 \end{pmatrix} = g.$$

Thus if we compute

$$g f_2 g^{-1} = \begin{pmatrix} 1 & 2 & 3 & 4 \\ 3 & 2 & 1 & 4 \end{pmatrix} \begin{pmatrix} 1 & 2 & 3 & 4 \\ 2 & 1 & 4 & 3 \end{pmatrix} \begin{pmatrix} 1 & 2 & 3 & 4 \\ 3 & 2 & 1 & 4 \end{pmatrix}$$

$$= \begin{pmatrix} 1 & 2 & 3 & 4 \\ 4 & 3 & 2 & 1 \end{pmatrix} = f_4 \in V_4.$$

The fact that $g f_1 g^{-1} \in V_4$ is trivial (Why?). But just for this one element g we still have to show that both $g f_3 g^{-1}$ and $g f_4 g^{-1}$ are in V_4. We will leave this for the exercises. This, by the way, still does not complete the verification that V_4 is normal in S_4. If we put $G = S_4$ and $N = V_4$ then we have that the factor group G/N has order $6 = |S_4/V_4|$. Since we do not have as yet efficient ways of mutiplying permutations (such ways do exist), we will not continue with writing down the Cayley table for this factor group. It turns out that G/N is a nonabelian group of order 6 and what is more that $G/N \cong S_3$. We just state these facts without proof here.

EXAMPLE 7.2.4. Let $G = GL(n, \mathbb{R})$ and $N = SL(n, \mathbb{R})$. So now our group consists of $n \times n$ invertible matrix with real entries. We first note that by Lemma 7.1.8 $N \triangleleft G$. Then

$$G/N = \{AN : A \in G\}.$$

Here you must understand that A is an $n \times n$ invertible matrix but N is a subgroup, so that AN is a left coset of N in G. We note that $X \in AN$ if and only if $X = AB$ where $B \in N = SL(n, \mathbb{R})$. We claim that the element AN of G/N consists of all matrices with the same determinant as A. Let us set D_A equal to this set, that is,

$$D_A = \{X \in GL(n, \mathbb{R}) : \det(X) = \det(A)\}.$$

We need to show that the two sets AN and D_A are equal. If $X \in AN$, then $\det(X) = \det(A) \det(B) = \det(A)$ because $B \in SL(n, \mathbb{R})$. Thus we have that $X \in D_A$. So $AN \subset D_A$. Suppose that $Y \in D_A$. Then $\det(Y) = \det(A)$. Let $C = A^{-1}Y$. Then $\det(C) = 1$ using the properties of determinants noted in the proof of Lemma 6.2.2. So this implies that $C \in SL(n, \mathbb{R}) = N$. Thus $Y = AC \in AN$. This shows that $D_A \subset AN$ and hence $D_A = AN$. Thus every coset consists of matrices with the same nonzero determinant.

But since the cosets partition the group, we must have

$$G = GL(n, \mathbb{R}) = \cup \, AN \quad (\text{disjoint})$$

where this union is then taken over matrices A with different real nonzero determinants. If we choose for each nonzero real number α an $A_\alpha \in G$ such that $\det(A_\alpha) = \alpha$, then we put $\Lambda = \mathbb{R} - \{0\}$. We note that given a nonzero real number α we always can find an $n \times n$ invertible matrix, A, in $GL(n, \mathbb{R})$ such that $\det(A) = \alpha$. (Can you see why?) Hence we have

$$G = \bigcup_{\alpha \in \Lambda} A_\alpha N.$$

Moreover if $\alpha, \beta \in \Lambda$ with $\alpha \neq \beta$, then in G/N, $(A_\alpha N)(A_\beta N) = A_{\alpha\beta} N$. If we think of A_α and A_β as coset representatives, then we can suppress N and think of this mutiplication in G/N as being given by $A_\alpha \cdot A_\beta = A_{\alpha\beta}$. Thus from this, one can show that G/N is isomorphic to the group of nonzero reals under multiplication. This tells us a fact which is not immediately evident otherwise. In particular, G/N is an abelian group. Note that neither G nor N are abelian.

EXAMPLE 7.2.5. In Lemma 7.1.10 we saw that for any group G the inner automorphism group $\mathrm{Inn}(G)$ is a normal subgroup of the automorphism group $\mathrm{Aut}(G)$. The resulting factor group $\mathrm{Aut}(G)/\mathrm{Inn}(G)$ is called the **outer automorphism group** which is denoted $\mathrm{Out}(G)$.

When we create the factor group G/N, it is important to understand that we are really defining every element of N to be the identity. This is apparent from the previous example where at the end we just suppressed the N. In the example using $\mathbb{Z}/6\mathbb{Z}$, we are saying that any multiple of 6 is 0. That is why $8 + 6\mathbb{Z} = 2 + 6 + 6\mathbb{Z} = 2 + 6\mathbb{Z}$, etc. In the example where we put $G = S_3, N = \langle a \rangle$, we have $\begin{pmatrix} 1 & 2 & 3 \\ 2 & 1 & 3 \end{pmatrix} N = \begin{pmatrix} 1 & 2 & 3 \\ 1 & 3 & 2 \end{pmatrix} N$ because $\begin{pmatrix} 1 & 2 & 3 \\ 1 & 3 & 2 \end{pmatrix} = \begin{pmatrix} 1 & 2 & 3 \\ 2 & 1 & 3 \end{pmatrix} \begin{pmatrix} 1 & 2 & 3 \\ 2 & 3 & 1 \end{pmatrix}$. But $a = \begin{pmatrix} 1 & 2 & 3 \\ 2 & 3 & 1 \end{pmatrix} \in N$. So going to the factor group makes a the identity and so in the factor group $\begin{pmatrix} 1 & 2 & 3 \\ 2 & 1 & 3 \end{pmatrix} = \begin{pmatrix} 1 & 2 & 3 \\ 1 & 3 & 2 \end{pmatrix}$ just as their cosets are equal. For this reason, group theorists often refer to the process of creating the factor group G/N as "killing" N.

7.3   The Group Isomorphism Theorems

In this section we show that there is a close relationship between group homomorphisms and factor groups. In particular, for each normal subgroup, and consequently for each factor group, there is a group homomorphism that has that normal subgroup as its kernel. Conversely, for each group homomorphism its kernel is a normal subgroup and the corresponding factor group is isomorphic to the image of the homomorphism. This is called the **group isomorphism theorem**. In Chapter 11, we will look at the analogous results in rings. We will then examine some consequences of this result that will be crucial in the Galois theory of fields.

Definition 7.3.1. *If G_1 and G_2 are groups and $f : G_1 \to G_2$ is a group homomorphism then the **kernel** of f, denoted $ker(f)$, is defined as*

$$ker(f) = \{g \in G_1 : f(g) = 1\}.$$

*That is, the kernel is the set of the elements of G_1 that map onto the identity of G_2. (Using previous concepts, $ker(f) = f^{-1}(\{1\})$.) The **image** of f, denoted $im(f)$, also called the **range** of f, is the set of elements of G_2 mapped onto by f from elements of G_1. That is,*

$$im(f) = \{g \in G_2 : f(g_1) = g_2 \text{ for some } g_1 \in G_1\} = \{f(g) : g \in G\} = f(G).$$

Note that if f is a surjection then $im(f) = G_2$.

The kernel measures how far a homomorphism is from being an injection, that is, a one-to-one mapping.

Lemma 7.3.1. *Let G_1 and G_2 be groups and $f : G_1 \to G_2$ be a group homomorphism. Then f is injective if and only if $ker(f) = \{1\}$.*

Proof. Suppose that f is injective. Since $f(1) = 1$ we always have $1 \in ker(f)$. Suppose that $g \in ker(f)$. Then $f(g) = f(1)$. Since f is injective this implies that $g = 1$ and hence $ker(f) = \{1\}$. Conversely suppose that $ker(f) = \{1\}$ and $f(g_1) = f(g_2)$. Then

$$f(g_1)(f(g_2))^{-1} = 1 \implies f(g_1 g_2^{-1}) = 1 \implies g_1 g_2^{-1} \in ker(f).$$

Then since $ker(f) = \{1\}$ we have $g_1 g_2^{-1} = 1$ and hence $g_1 = g_2$. Therefore f is injective. $\qquad\square$

We now state the **group isomorphism theorem**. We note that this is sometimes called the first group isomorphism theorem or the fundamental homomorphism theorem.

Theorem 7.3.1. *(Group Isomorphism Theorem)*
    (a) Let G_1 and G_2 be groups and $f : G_1 \to G_2$ a group homomorphism. Then $\ker(f)$ is a normal subgroup of G_1, $im(f)$ is a subgroup of G_2 and

$$G/\ker(f) \cong im(f).$$

    *(b) Conversely, suppose that N is a normal subgroup of a group G. Then there exists a group H and a homomorphism $f : G \to H$ such that $\ker(f) = N$ and $im(f) = H$. In particular $H = G/N$ and the map $f : G \to G/N$ is called the **canonical homomorphism**. This is given by $f(g) = gN$.*

Proof. (a) Since $1 \in \ker(f)$ the kernel is nonempty. Suppose that $g_1, g_2 \in \ker(f)$. Then $f(g_1) = f(g_2) = 1$. It follows that $f(g_1 g_2^{-1}) = f(g_1)(f(g_2))^{-1} = 1$. Hence $g_1 g_2^{-1} \in \ker(f)$ and therefore $\ker(f)$ is a subgroup of G_1. Further for any $g \in G_1$ we have

$$f(g^{-1}g_1 g) = (f(g))^{-1} f(g_1) f(g) = (f(g))^{-1} \cdot 1 \cdot f(g) = f(g^{-1}g) = f(1) = 1.$$

Hence $g^{-1}g_1 g \in \ker(f)$. This implies that $g^{-1}(\ker(g))g \subset \ker(f)$ and thus $\ker(f)$ is a normal subgroup.

It is straightforward to show that $im(f)$ is a subgroup of G_2.

Consider the map $\widehat{f} : G/\ker(f) \to im(f)$ defined by

$$\widehat{f}(g\ker(f)) = f(g).$$

We show that this is an isomorphism.

We must first show that this $\widehat{f}$ is well-defined. Suppose that $g_1\ker(f) = g_2\ker(f)$ then $g_1^{-1}g_2 \in \ker(f)$ by Lemma 6.3.2, so that $f(g_1^{-1}g_2) = 1$. This implies that $f(g_1) = f(g_2)$ and hence the map is $\widehat{f}$ well-defined. Now

$$\widehat{f}(g_1\ker(f)g_2\ker(f)) = \widehat{f}(g_1 g_2\ker(f)) = f(g_1 g_2) =$$

$$= f(g_1)f(g_2) = \widehat{f}(g_1\ker(f))\widehat{f}(g_2\ker(f))$$

where we have used the fact that f is a homomorphism to go from the first to the second line and therefore $\widehat{f}$ is a homomorphism. Suppose that

$$\widehat{f}(g_1\ker(f)) = \widehat{f}(g_2\ker(f))$$

then $f(g_1) = f(g_2)$ so as above $g_1^{-1}g_2 \in \ker(f)$ and hence $g_1\ker(f) = g_2\ker(f)$ again by Lemma 6.3.2. It follows that $\widehat{f}$ is injective. Finally suppose that $h \in \operatorname{im}(f)$. Then there exists a $g \in G_1$ with $f(g) = h$. Then $\widehat{f}(g\ker(f)) = h$ and $\widehat{f}$ is a surjection onto $\operatorname{im}(f)$. Therefore $\widehat{f}$ is an isomorphism completing the proof of part (a).

(b) Conversely suppose that N is a normal subgroup of G. Define the map $f : G \to G/N$ by $f(g) = gN$ for $g \in G$. By the definition of the product in the quotient group G/N it is clear that f is a homomorphism with $\operatorname{im}(f) = G/N$. If $g \in N$, then $f(g) = gN = N$ which is the identity in G/N. This shows that $N \subset \ker(f)$. If $g \in \ker(f)$ then $f(g) = gN = N$ since N is the identity in G/N. However, this implies that $g \in N$ so that $\ker(f) \subset N$, and hence it follows that $\ker(f) = N$. Taking $H = G/N$ completes the proof. $\qquad\square$

At this point the reader should look back at Theorem 2.3.2.

EXAMPLE 7.3.1. Let $f : \mathbb{Z}_6 \to \mathbb{Z}_3$ defined by $f = \begin{pmatrix} 0 & 1 & 2 & 3 & 4 & 5 \\ 0 & 1 & 2 & 0 & 1 & 2 \end{pmatrix}$. It is not hard to verify that f is a homomorphism of $\mathbb{Z}_6$ onto $\mathbb{Z}_3$. $\ker(f) = \{0, 3\}$ is the subgroup of $\mathbb{Z}_6$ generated by 3, i.e., $\langle 3 \rangle$. The group isomorphism theorem implies $\mathbb{Z}_6/\langle 3 \rangle \cong \mathbb{Z}_3$. We note that this example could also have been given as follows: Let $f : \mathbb{Z}/6\mathbb{Z} \to \mathbb{Z}/3\mathbb{Z}$ defined by $f(i + 6\mathbb{Z}) = i + 3\mathbb{Z}$ for any $i \in \mathbb{Z}$. The reader should verify that this is a homomorphism, find its kernel, and state the conclusion of the Group Isomorphism Theorem in this case.

EXAMPLE 7.3.2. Applying the group isomorphism theorem to the Example at the end of Section 3.4, i.e., $f : \mathbb{Z} \to \mathbb{Z}_n$ defined by $f(z) = [z]$ gives $\mathbb{Z}/n\mathbb{Z} \cong \mathbb{Z}_n$.

EXAMPLE 7.3.3. We have seen that the set of all inner automorphisms $\operatorname{Inn}(G)$ of a group G forms a normal subgroup of the automorphism group $\operatorname{Aut}(G)$ of all automorphisms of G. We mentioned that the resulting factor group $\operatorname{Aut}(G)/\operatorname{Inn}(G)$ is called the outer automorphism group denoted $Out(G)$. Here we show that $\operatorname{Inn}(G) \cong G/Z(G)$ where $Z(G)$ is the center of G, that is,

$$Z(G) = \{x \in G : xg = gx \text{ for all } g \in G\}.$$

It is straightforward to show that $Z(G)$ is a normal subgroup of G.

Consider the map $\varphi : G \to Inn(G)$ given by $\varphi(g) = T_g$ where $T_g(x) = gxg^{-1}$. This is clearly surjective. Now $\varphi(ab) = T_{ab}$ for $a, b \in G$. But

$$T_a T_b(x) = T_a(bxb^{-1}) = a(bxb^{-1})a^{-1} = (ab)x(ab)^{-1} = T_{ab}(x).$$

Thus $T_{ab} = T_a T_b$. This shows that the map φ is a homormorphism.

Consider the kernel K of this map. K consists of those and only those $x \in G$ such that $\varphi(x) = I_G$, i.e., $\varphi(x) = T_x = I_G$. But $T_x(g) = xgx^{-1}$ and $I_G(g) = g$. So $xgx^{-1} = g$ for all $g \in G$. This is equivalent to $xg = gx$ for all $g \in G$ and hence $x \in Z(G)$. It follows that the kernel is precisely the center $K = Z(G)$. Now applying the group isomorphism theorem we get the result that

$$G/Z(G) \cong Inn(G).$$

There are two related theorems that are called the second isomorphism theorem and the third isomorphism theorems.

Theorem 7.3.2. *(Second Isomorphism Theorem) Let N be a normal subgroup of a group G and U a subgroup of G. Then $U \cap N$ is normal in U and*

$$(UN)/N \cong U/(U \cap N).$$

Proof. From Lemma 7.1.9 we know that $U \cap N$ is normal in U and that UN is a subgroup of G. Define the map

$$\alpha : UN \to U/U \cap N$$

by $\alpha(un) = u(U \cap N)$. To show that α is well-defined consider $un = u'n'$ where $u, u' \in U$ and $n, n' \in N$. Then $u'^{-1}u = n'n^{-1} \in U \cap N$. Therefore $u'(U \cap N) = u(U \cap N)$ by Lemma 6.3.2 and hence the map α is well-defined.

Suppose that $un, u'n' \in UN$ where u, u', n, n' are as before. Since N is normal in G we have that, $nu' = u'n''$ for $n'' \in N$. Thus $unu'n' = u(nu')n' = u\,u'n''n' \in uu'N$. Hence $unu'n' = uu'n'''$ with $n''' \in N$. Then

$$\alpha(unu'n') = \alpha(uu'N) = uu'(U \cap N).$$

However, $U \cap N$ is normal in U so

$$uu'(U \cap N) = u(U \cap N)u'(U \cap N) = \alpha(un)\alpha(u'n').$$

Therefore α is a homomorphism.

We have $\text{im}(\alpha) = U/(U \cap N)$ by definition. Consider $n \in N$. Then $n = 1 \cdot n \in UN$ and $\alpha(n) = \alpha(1 \cdot n) = 1 \cdot U \cap N = U \cap N$ which is the identity in $U/U \cap N$. This shows that $N \subset \ker(\alpha)$. Suppose that $un \in \ker(\alpha)$. Then $\alpha(un) = U \cap N \subset N$ which implies $u \in N$, so that $un \in N$. This shows that $\ker(\alpha) \subset N$. Therefore $\ker(f) = N$. From the group isomorphism theorem we then have

$$UN/N \cong U/(U \cap N)$$

proving the theorem. □

The Second Isomorphism Theorem probably can be best remembered by the following mnemomic device

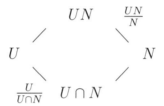

Label the vertices of the figure as indicated, it being immaterial which side one writes U or N on. One then reads the isomorphism by reading "modulo the opposite sides." Should U also be normal in G, then we obtain, symmetrically, that

$$\frac{UN}{U} \cong \frac{N}{U \cap N},$$

which may be read by reading "modulo" the other pair of opposite sides of the figure.

Theorem 7.3.3. *(Third Isomorphism Theorem) Let N and M be normal subgroups of a group G with N a subgroup of M. Then M/N is a normal subgroup in G/N and*

$$(G/N)/(M/N) \cong G/M.$$

Proof. Define the map $\beta : G/N \to G/M$ by

$$\beta(gN) = gM.$$

It is straightforward that β is well-defined, onto, and a homomorphism. (See the exercises.) If $m \in M$, $\beta(m) = mM = M$ which is the identity in G/M. This shows that $M/N \subset \ker(\beta)$. If $gN \in \ker(\beta)$ then $\beta(gN) = gM = M$

and hence $g \in M$. So that $\ker(\beta) \subset M/N$. It follows that $\ker(\beta) = M/N$. In particular this also shows that M/N is normal in G/N since a kernel of a homomorphism is always a normal subgroup by group isomorphism theorem. From the group isomorphism theorem, again, we have that

$$(G/N)/(M/N) \cong G/M.$$

$\square$

For a normal subgroup N in G the canonical homomorphism $f : G \to G/N$ provides a one-to-one correspondence between subgroups of G containing N and the subgroups of G/N. This correspondence will play a fundamental role in the study of subfields of a field.

Theorem 7.3.4. *(Correspondence Theorem) Let N be a normal subgroup of a group G and let f be the corresponding homomorphism $f : G \to G/N$. Then the mapping*

$$\phi : H \to f(H)$$

where H is a subgroup of G containing N provides a one-to-one correspondence between all the subgroups of G/N and the subgroups of G containing N.

Proof. We first show that the mapping ϕ is surjective. Let H_1 be a subgroup of G/N and let

$$H = \{g \in G : f(g) \in H_1\}.$$

We show that H is a subgroup of G and that $N \subset H$. We note that $H = f^{-1}(H_1)$. We also note that f is a surjection.

First we note that $1 \in H$ since $f(1) = N$, the identity in G/N. But H_1 is a subgroup so $f(1) = N \in H_1$ which shows that $1 \in H$. Thus $H \neq \emptyset$.

If $g_1, g_2 \in H$ then $f(g_1) \in H_1$ and $f(g_2) \in H_1$. Therefore $f(g_1)f(g_2) \in H_1$ and hence $f(g_1 g_2) \in H_1$. Therefore $g_1 g_2 \in H$. In an identical fashion $g_1^{-1} \in H$. Therefore H is a subgroup of G.

If $n \in N$ then $f(n) = 1 \in H_1$ (note: here $1 = N$ since H_1 is in the factor group) and hence $n \in H$. Therefore $N \subset H$. So H is a subgroup of G containing N. But then $\phi(H) = f(f^{-1}(H_1)) = H_1$ since f is onto. (See Problem 2.2.8 (b)). This shows that the map ϕ is surjective.

Suppose that $\phi(H_1) = \phi(H_2)$ where H_1 and H_2 are subgroups of G containing N. This implies that $f(H_1) = f(H_2)$. Let $g_1 \in H_1$. Then $f(g_1) =$

$f(g_2)$ for some $g_2 \in H_2$. Then $f(g_1 g_2^{-1}) = 1$ so $g_1 g_2^{-1} \in \ker(f) = N \subset H_2$. Since $g_2 \in H_2$ and since we have shown that $g_1 g_2^{-1} \in H_2$, this gives that $g_1 \in H_2$. Hence $H_1 \subset H_2$. In a similar fashion $H_2 \subset H_1$ (we leave this verification to the exercises) and therefore $H_1 = H_2$. It follows that ϕ is injective.

□

Corollary 7.3.1. *Let N be a normal subgroup of a group G and let f be the canonical homomorphism $f : G \to G/N$. Then in the correspondence theorem for any subgroup H of G containing N, H is normal in G if and only $f(H)$ is normal in G/N.*

Proof. Suppose that H is as in the statement of the corollary and H is normal in G. Consider $f(H)$ in G/N. Let $g \in G$. Then $f(g) = gN$ and also

$$(gN)(f(H))(gN)^{-1} = f(g)f(H)(f(g))^{-1} = f(gHg^{-1})$$

since f is a homomorphism. But since H is normal in G

$$f(gHg^{-1}) = f(H).$$

This shows that $f(H)$ is normal in G/N.

Conversely suppose that $f(H)$ is normal in G/N. Then for any $g \in G$ consider gHg^{-1}. Since f is a homomorphism

$$f(gHg^{-1}) = f(g)f(H)(f(g))^{-1} = f(H)$$

since $f(H)$ is normal is G/N. But the map ϕ in the correspondence theorem was shown to be one-to-one and since

$$\phi(gHg^{-1}) = f(gHg^{-1}) = f(H) = \phi(H)$$

we must have that $gHg^{-1} = H$. It follows that H is normal in G.

We note that if $N \subset H$ then $N \subset gHg^{-1}$ because $N = gNg^{-1} \subset gHg^{-1}$.

□

We end this section with a corollary to the above theorem which tells us what the form of any subgroup of a factor looks like.

Corollary 7.3.2. *Let G be a group and $N \triangleleft G$. Any subgroup of G/N is of the form H/N where H is a subgroup of G containing N.*

Proof. This follows directly from the correspondence theorem, above, because the map ϕ is onto. So if H_1 is any subgroup of G/N then there must exist a subgroup H of G with $N \subset H$ such that $\phi(H) = H_1$. But $\phi(H) = f(H) = H/N$. Thus

$$H_1 = H/N.$$

□

7.4 Exercises

EXERCISES FOR SECTION 7.1

7.1.1. If G is a group and H is a subgroup, then prove that $N_G(H)$, i.e., the normalizer of H in G, is also a subgroup of G.

7.1.2. Show that if H is a subgroup of a group G then $a^{-1}Ha = H \ \forall a \in G$ is equivalent to $aHa^{-1} = H \ \forall a \in G$.

(HINT: Think of inverses.)

7.1.3. Prove that if N is a subgroup such that $a^{-1}Na \subset N$ for all $a \in G$, then $N = a^{-1}Na$ for all $a \in G$, i.e., N is normal in G.

7.1.4. In the text it was shown that the number of distinct conjugates of a subgroup H in a group G is equal to the index in G of the normalizer in G, that is, $[G : N_G(H)]$. Use the proof to define a map from the class of all conjugates of H in G to the collection of left cosets of $N_G(H)$ in G. Show that your mapping is a bijection.

7.1.5. In the text right before Lemma 7.1.4, the analogous result to the above problem 7.1.2, i.e., if $aNa^{-1} \subset N$ for all $a \in G$, then N is normal in G, it says that "Since the conjugation map is an isomorphism it follows that if $g^{-1}Hg \subset H$ then $g^{-1}Hg = H$." It then gives the appropriate items one needs to prove this. Explain the proof that if this condition holds then since conjugation is an isomorphism we are done.

7.1.6. Prove that if G is a cyclic group, then any factor group of G is cyclic.

7.1.7. Let G be a group. Recall that the subgroup defined in Problem 6.1.10 was the **center of** G,

$$Z(G) = \{z \in G : xz = zx \ \ \forall x \in G\}.$$

Prove that $Z(G) \lhd G$.

7.1.8. Show that $H = \{1 = \begin{pmatrix} 1 & 2 & 3 \\ 1 & 2 & 3 \end{pmatrix}, \begin{pmatrix} 1 & 2 & 3 \\ 2 & 1 & 3 \end{pmatrix}\}$ is not a normal subgroup of S_3.

7.1.9. (a) Prove that if A, B are both normal subgroups of G then their intersection $A \cap B$ is also a normal subgroup of G.

(b) Extend part (a) to any collection of normal subgroups of G.

7.1.10. (a) If A is a subset of G, $A \subset G$ then explain why the previous Problem 7.1.9 shows that it makes sense to talk about the smallest normal subgroup of G containing A. We denote this by A^G and call it the **normal closure** of A in G.

(NOTE: If $A = \emptyset$ we put $A^G = \{1\}$.)
        (b) Prove that $A^G = gp\{g^{-1}ag : a \in A, g \in G\}$.

7.1.11. Let G be any group. Let $\text{Aut}(G)$ be its automorphism group. In the text you were asked if the given argument was sufficient to show that $\text{Aut}(G)$ is a group under composition. Was it? Why or why not? If not, then provide the missing steps.

7.1.12. Let $F \subset K$ be fields and $\text{Aut}(K)$ and $G(K/F)$ be defined as in Example 7.1.1. Prove that $\text{Aut}(K)$ is a group and $G(K/F)$ is a subgroup under the binary operation of composition.

(HINT: You may follow the proof for groups to show that $\text{Aut}(K)$ is a group but to show that if $f \in \text{Aut}(K)$ then $f^{-1} \in \text{Aut}(K)$ not only do you have to show that f^{-1} preserves multiplication, you also have to show that f^{-1} preserves addition. Also consider Problem 7.1.11 above.)

7.1.13. Let G be any group. In the text we defined the set $\text{Inn}(G) = \{T_g : g \in G\}$. You will note that sometimes we define $T_g(x) = gxg^{-1}$ for $x \in G$ and other times we may say that $T_g(x) = g^{-1}xg$ for $x \in G$. Prove that the set of all inner automorphisms $\text{Inn}(G)$ remains the same no matter which of these two one takes as the definition of an inner automorphism.

7.1.14 Let G be a group and $\text{Inn}(G)$ be the set of all inner automorphisms. Prove that $\text{Inn}(G)$ is a normal subgroup of $\text{Aut}(G)$.

EXERCISES FOR SECTION 7.2

7.2.1. (a) Find the cosets of $4\mathbb{Z}$ in $\mathbb{Z}$.

(b) Use these to make a Cayley table for $\mathbb{Z}/4\mathbb{Z}$ — as was done in Example 7.1 of this section.

(c) We note that $|\mathbb{Z}/4\mathbb{Z}| = 4$ even though $\mathbb{Z}$ and $4\mathbb{Z}$ are infinite. In Example 7.3 of this section another group of order four is given, V_4. Is $\mathbb{Z}/4\mathbb{Z}$ isomorphic to V_4? Why or why not?

7.2.2. (a) For V_4 as in Example 7.3, show that for g as given in that example show that gf_3g^{-1} and gf_4g^{-1} both belong to V_4.

(b) Take any other element $h \in S_4 - V_4$ where $h \neq g$ above. Show that for the element h you chose that $hV_4h^{-1} \subset V_4$. Is this sufficient to show that $V_4 \lhd S_4$?

7.2.3. Let $G = GL(n, \mathbb{R})$ and $N = SL(n, \mathbb{R})$ as in Example 7.4.

(a) Let $A \in G$. Then show that the left coset AN is the set of all matrices in $GL(n, \mathbb{R})$ with the same determinant as A, i.e.,

$$AN = \{X \in GL(n, \mathbb{R}) : \det(X) = \det(A)\}.$$

(b) Here we note that in contrast to Problem 7.2.1 G, N are infinite and also so is G/N. Prove that the $G/N \cong \mathbb{R}^*$, i.e., the group of nonzero reals under multiplication.

(c) Is $G/N \cong G$? Why or why not?

(NOTE: Without saying whether or not it is true here we note that it is possible for G/N to be isomorphic to G. If this is true for some nontrivial N we say the group G is **non-hopfian**.)

7.2.4. Suppose that G is a group and $H \lhd G$. Let $a \in G$ and suppose that $o(a) = n$. Show that $o(aH) \mid n$.

7.2.5. Let $G = \langle a \rangle$ be a cyclic group. Let H be any subgroup of G.

(a) Show that $H = \langle a^s \rangle$ where s is an arbitrary positive integer if $|G| = \infty$ while if $|G| < \infty$, then $s \mid |G|$.

(HINT: Go back to the section on cyclic groups, Section 6.4.)

(b) Explain why $H \lhd G$ whether G is finite or infinite — it makes no difference. Again without reference to the size of G, show that $G/H = \{H, aH, \ldots, a^{s-1}H\} = \langle aH \rangle$, i.e., the factor group of a cyclic group is finite cyclic no matter whether the original group is finite or infinite.

EXERCISES FOR SECTION 7.3

7.3.1. Let G_1, G_2, G_3 be groups. Suppose that $f_1 : G_1 \to G_2$ and $f_2 : G_2 \to G_3$ are both group homomorphisms. Then show that $f_2 \circ f_1 : G_1 \to G_3$ is also a homomorphism.

7.3.2. Let $f : G_1 \to G_2$ be a group homomorphism. Show that $\operatorname{im}(f) = f(G_1) = \{f(g) : g \in G_1\}$ is a subgroup of G_2.

7.3.3. Show that the map $\beta : G/N \to G/M$ defined in the third isomorphism theorem is
(a) well defined;
(b) onto;
(c) a homomorphism.

7.3.4. In the proof of the correspondence theorem to show that

$$\phi : \{\text{set of all subgroups of } G \text{ containing } N\} \to \{\text{set of all subgroups of } G/N\}$$

was an injection, we needed to show that if $\phi(H_1) = \phi(H_2)$ then $H_1 = H_2$. We showed in the text that $H_1 \subset H_2$. Complete the proof by showing that $H_2 \subset H_1$.

7.3.5. Verify that $f : \mathbb{R} \to \mathbb{C}$ defined by $f(x) = \cos(2\pi x) + i\sin(2\pi x) = e^{2\pi i x}$ is a homomorphism of the additive group of reals $\mathbb{R}$ onto the multiplicative group of all complex numbers of absolute value 1, i.e., $f : \mathbb{R} \to \{z \in \mathbb{C} : |z| = 1\}$. (Recall $z = a + b\mathbf{i}$, $|z| = \sqrt{a^2 + b^2}$.) What is ker(f)? State the conclusion of the group isomorphism theorem for this map.

7.3.6. Prove that if H_1 and H_2 are subgroups of a group G with $H_2 \lhd G$ then the map ϕ given by $\phi(h_1) = h_1 H_2$ for any $h_1 \in H_1$ is a homomorphism from H_1 into $H_1 H_2 / H_2$.

(NOTE: Here you must first show why $H_1 H_2$ is a subgroup.)

7.3.7. If $N \lhd G$, a finite group, and if $[G : N]$ and $|N|$ are relatively prime, then prove that N contains every subgroup of G whose order is a divisor of $|N|$.

(HINT: Let H be any subgroup of G whose order is a divisor of the order of N, i.e., $|H| \mid |N|$. Let $h \in H$. Consider $o(h)$ for h an element of H and $o(hN)$ for an element of G/N. Use Problem 7.2.4 and Lagrange's theorem. Use this to prove that $h \in N$. We note that this result would not be true necessarily if $[G : N]$ and $|N|$ were **not** relatively prime.)

7.3.8. Let H_1 and H_2 be subgroups of a group G such that $H_2 \triangleleft H_1$; also let H be any subgroup of G. Finally let $\overline{H} = H_2 \cap H$ and $\widehat{H} = H_1 \cap H$.

(a) Prove that $\overline{H} \triangleleft \widehat{H}$.

(b) Use the second isomorphism theorem to show that $\widehat{H} / \overline{H}$ is isomorphic to a subgroup of H_1/H_2.

(HINT: Use the mnemomic device given for remembering the second isomorphism theorem on $H_1 \cap H$ and H_2.)

Chapter 8

Direct Products and Abelian Groups

8.1 Direct Products of Groups

In this section we look at a very important construction, the direct product, which allows us to build new groups out of existing groups. Given a finite number of groups we can use this construction (the external direct product) to get new groups. Actually the process of constructing the direct product can be extended to the case where an infinite number of groups are given, but we shall not go into those matters here.

This construction is the analog for groups of the direct sum of rings which we discussed briefly in Chapter 3. Here we shall also consider the related situation of decomposing a given group in a certain fashion, the internal direct product, into a product (with the usual meaning) of a finite number of subgroups. We will investigate the relationship between this situation, the internal direct product, and the first, external direct product, of our considerations. We shall see that up to isomorphism the two concepts of direct product are indistinguishable.

In order to make the material more palpable, we first consider the simpler situation of the direct product of two groups. We will then consider the case of the direct product of any finite number of groups. As an application of this construction in the next section, we present a theorem which completely describes the structure of all finite abelian groups.

8.1.1   Direct Products of Two Groups

Let G_1 and G_2 be groups and let G be the cartesian product of G_1 and G_2. That is,

$$G = G_1 \times G_2 = \{(a,b) : a \in G_1,\ b_2 \in G_2\}.$$

On G define

$$(a_1, b_1) \cdot (a_2, b_2) = (a_1 a_2, b_1 b_2).$$

With this operation it is straightforward to verify the group axioms for G (see the exercises) and hence G becomes a group.

Theorem 8.1.1. *Let G_1 and G_2 be groups and G the cartesian product $G_1 \times G_2$ with the operation defined above. Then G forms a group called the (external)* **direct product** *of G_1 and G_2. The identity element is $(1,1)$ where the first 1 is the identity of G_1 and the second 1 is the identity of G_2 — even though we have used the same notation for them here they can be quite different. The inverse of an arbitrary element is $(g, h)^{-1} = (g^{-1}, h^{-1})$.*

Theorem 8.1.2. *For groups G_1 and G_2, we have $G_1 \times G_2 \cong G_2 \times G_1$ and $G_1 \times G_2$ is abelian if and only if each $G_i, i = 1, 2$, is abelian.*

Proof. The map $(a, b) \mapsto (b, a)$ where $a \in G_1, b \in G_2$ provides an isomorphism from $G_1 \times G_2 \to G_2 \times G_1$. (See the exercises for verification of the details.)

Suppose that both G_1 and G_2 are abelian. Then if (a_1, b_1) and (a_2, b_2) are arbitrary elements of $G_1 \times G_2$, we have

$$(a_1, b_1)(a_2, b_2) = (a_1 a_2, b_1 b_2).$$

But since G_1 and G_2 are abelian $a_1 a_2 = a_2 a_1$ and $b_1 b_2 = b_2 b_1$, where by definition of the direct product the a's are in G_1 and the b's are in G_2. Thus

$$(a_1 a_2, b_1 b_2) = (a_2 a_1, b_2 b_1) = (a_2, b_2)(a_1, b_2)$$

and hence $G_1 \times G_2$ is abelian.

Conversely suppose $G_1 \times G_2$ is abelian and suppose that $a_1, a_2 \in G_1$ are arbitrary. Then writing 1 for the identity G_2, we have

$$(a_1 a_2, 1) = (a_1, 1)(a_2, 1) = (a_2, 1)(a_1, 1) = (a_2 a_1, 1).$$

Therefore $a_1 a_2 = a_2 a_1$ and G_1 is abelian. Identically using the second components rather than the first we find G_2 is abelian. □

We show next that in $G_1 \times G_2$ there are normal subgroups H_1, H_2 with

$$H_1 \cong G_1 \text{ and } H_2 \cong G_2.$$

Theorem 8.1.3. *Let* $G = G_1 \times G_2$. *Let* $H_1 = \{(a, 1) : a \in G_1\}$ *and* $H_2 = \{(1, b) : b \in G_2\}$. *Then both* H_1 *and* H_2 *are normal subgroups of* G *with* $G = H_1 H_2$ *(the usual product of subgroups). Every element of* H_1 *commutes with each element of* H_2 *and* $H_1 \cap H_2 = \{(1, 1)\}$. *(Note that* $(1, 1)$ *is the identity or 1 for* G.) *Further* $H_1 \cong G_1, H_2 \cong G_2$ *and* $G/H_1 \cong G_2$ *and* $G/H_2 \cong G_1$.

Proof. Map $G_1 \times G_2$ onto G_2 by $(a, b) \mapsto b$. It is not hard to show that this map is a homomorphism from G onto G_2 (see the exercises). Its kernel is $\{(a, b) : (a, b) \mapsto 1\} = \{(a, 1) : a \in G_1\} = H_1$. This establishes that H_1 is a normal subgroup of G and that $G/H_1 \cong G_2$ by the group isomorphism theorem (Theorem 7.3.1). In an identical fashion using the map $(a, b) \mapsto a$, it can be shown that H_2 is a normal subgroup of G and $G/H_2 \cong G_1$. (See the exercises.) The map $(a, 1) \mapsto a$ provides the isomorphism from H_1 onto G_1 and the map $(1, b) \mapsto b$ provides the isomorphism from H_2 onto G_2. (See the exercises for the details.)

Since H_1, H_2 are normal subgroups of $G = G_1 \times G_2$, then it follows from Lemma 7.1.9 that $H_1 H_2 \subset G$ is a subgroup. Let (a, b) be any element of $G = G_1 \times G_2$. Then $(a, b) = (a, 1)(1, b) \in H_1 H_2$. This shows $G \subset H_1 H_2$. Hence $G = H_1 H_2$.

Consider arbitrary elements $(a, 1)$ and $(1, b)$ of H_1 and H_2, respectively. Now $(a, 1)(1, b) = (a, b) = (1, b)(a, 1)$. Thus H_1 and H_2 commute element-wise.

Finally, consider $H_1 \cap H_2$. If $(x, y) \in H_1 \cap H_2$ then $(x, y) \in H_1$ and $(x, y) \in H_2$. But if $(x, y) \in H_1$ then $y = 1$ and if $(x, y) \in H_2$ then $x = 1$. This implies that $(x, y) = (1, 1)$. This completes our proof. □

If the factors are finite it is easy to find the order of $G_1 \times G_2$. The size of the cartesian product is just the product of the sizes of the factors.

Lemma 8.1.1. *If* $|G_1|$ *and* $|G_2|$ *are finite then* $|G_1 \times G_2| = |G_1| |G_2|$.

We have just considered the external direct product of two groups. We now consider the internal situation. Suppose that G is a group such that

(a) G_1 and G_2 are normal subgroups of G.
(b) $G = G_1 G_2$ (the usual product).
(c) $G_1 \cap G_2 = \{1\}$.

Then we will show that G is isomorphic to the direct product $G_1 \times G_2$. In this case we say that G is the **internal direct product** of its subgroups and that G_1 and G_2 are **direct factors** of G.

Lemma 8.1.2. *Given that G is the internal direct product of its subgroups G_1 and G_2, then every element of G_1 commutes with each element of G_2 and every element of $g \in G$ can be written **uniquely** as a product of an element of G_1 times an element of G_2, i.e., $g = g_1 g_2$ where $g_1 \in G_1, g_2 \in G_2$ and this representation is unique.*

Proof. Let a be an arbitrary element of G_1 and b be an arbitrary element of G_2. Consider the element $a^{-1}b^{-1}ab \in G$. Since $G_1 \triangleleft G$ then $b^{-1}ab \in G_1$. So $a^{-1}b^{-1}ab \in G_1$ since G_1 is a subgroup. Similarly $a^{-1}b^{-1}a \in G_2$ since G_2 is a normal subgroup of G. But then $a^{-1}b^{-1}ab \in G_2$ since G_2 is a subgroup. Thus we have shown that $a^{-1}b^{-1}ab \in G_1 \cap G_2 = \{1\}$. So $a^{-1}b^{-1}ab = 1$. This in turn implies that $ab = ba$. So we have shown that G_1 commutes elementwise with G_2.

Now since $G = G_1 G_2$ every element of G can be written as a product $g_1 g_2$ where $g_1 \in G_1, g_2 \in G_2$. But here we claim that this is the only way it can be so written. For suppose that $g_1 g_2 = h_1 h_2$ where $g_1, h_1 \in G_1, g_2, h_2 \in G_2$. Then

$$g_1 g_2 = h_1 h_2 \implies h_1^{-1} g_1 = h_2 g_2^{-1}.$$

But $h_1^{-1} g_1 \in G_1$ and $h_2 g_2^{-1} \in G_2$ since G_1 and G_2 are subgroups. Thus the element $h_1^{-1} g_1 = h_2 g_2^{-1} \in G_1 \cap G_2 = \{1\}$. This implies that $g_1 = h_1$ and $g_2 = h_2$. So the representation is unique. $\square$

Theorem 8.1.4. *Suppose that G is a group with normal subgroups G_1 and G_2 such that $G = G_1 G_2$ and $G_1 \cap G_2 = \{1\}$. In other words, G is the internal direct product of its subgroups G_1, G_2. Then G is isomorphic to the direct product $G_1 \times G_2$.*

Proof. Since $G = G_1G_2$ each element of G has the form ab with $a \in G_1$, $b \in G_2$. Moreover according to Lemma 8.1.2, this representation is unique and each $a \in G_1$ commutes with each $b \in G_2$.

Now consider the map

$$f : G \rightarrow G_1 \times G_2$$

given by $f(ab) = (a, b)$ where $a \in G_1$ and $b \in G_2$. We claim that this is an isomorphism. It is clearly onto. Now

$$f((a_1b_1)(a_2b_2)) = f(a_1a_2b_1b_2) = (a_1a_2, b_1b_2)$$

since the a's are in G_1 and the b's are in G_2 they commute so $b_1a_2 = a_2b_1$ which gives the above together with associativity. But

$$(a_1a_2, b_1b_2) = (a_1, b_1)(a_2, b_2) = f(a_1b_1)f(a_2b_2)$$

so that f is a homomorphism. The kernel of f is $\ker(f) = \{g \in G : f(g) = (1, 1)\}$. But from the definition of f, this means that $g = 1 \cdot 1 = 1$ and from Lemma 8.1.2 this representation is unique. So the $\ker(f) = \{1\}$. This implies that f is an isomorphism. □

Although the final resulting groups are isomorphic we call $G_1 \times G_2$ an external direct product if we started with the groups G_1, G_2 and constructed $G_1 \times G_2$ and call $G_1 \times G_2$ an internal direct product if we started with a group G having normal subgroups and decompose G into G_1G_2 as in the theorem. To distinguish between these two, if necessary, we write for the first $G_1 \times G_2$ (external) and for the second $G_1 \times G_2$ (internal).

We consider two examples of direct products before going to the next subsection. In both cases, the operation is addition and is commutative. We will still, however, use multiplicative notation in our general discussions.

EXAMPLE 8.1.1. $\mathbb{Z}_2 \times \mathbb{Z}_3$ is a cyclic group.

$$\mathbb{Z}_2 \times \mathbb{Z}_3 = \{(a, b) : a \in \mathbb{Z}_2, b \in \mathbb{Z}_3\}.$$

By Lemma 8.1.1, $|\mathbb{Z}_2 \times \mathbb{Z}_3| = |\mathbb{Z}_2||\mathbb{Z}_3| = 2 \cdot 3 = 6$. We could also just enumerate the elements of $\mathbb{Z}_2 \times \mathbb{Z}_3$ since there are so few of them. We can now use Corollary 6.4.2 together with Theorem 8.1.2 to note that the elements $(1, 0)$ and $(0, 1)$ in $\mathbb{Z}_2 \times \mathbb{Z}_3$ commute and have relatively prime orders, i.e., 2 and 3, respectively. Thus $(1, 0) + (0, 1) = (1, 1)$ has order $2 \cdot 3 = 6$. This shows that $\mathbb{Z}_2 \times \mathbb{Z}_3$ is cyclic. Rather than relying on this theory, we could also just compute the order of $(1,1) \in \mathbb{Z}_2 \times \mathbb{Z}_3$ as follows:

$$
\begin{aligned}
2(1,1) &= (1,1) + (1,1) = (0,2), \\
3(1,1) &= 2(1,1) + (1,1) = (0,2) + (1,1) = (1,0), \\
4(1,1) &= 3(1,1) + (1,1) = (1,0) + (1,1) = (0,1), \\
5(1,1) &= 4(1,1) + (1,1) = (0,1) + (1,1) = (1,2), \\
6(1,1) &= 5(1,1) + (1,1) = (1,2) + (1,1) = (0,0).
\end{aligned}
$$

This shows that the least positive multiple of $(1,1)$ which is the identity is $6(1,1)$. Thus $o(1,1) = 6$ and so $(1,1)$ generates all of $\mathbb{Z}_2 \times \mathbb{Z}_3$. Note that even though $\mathbb{Z}_2 \times \mathbb{Z}_3$ has order 6 as does S_3, they are not isomorphic. (Why?)

EXAMPLE 8.1.2. $\mathbb{Z}_2 \times \mathbb{Z}_2$ is not a cyclic group,

$$\mathbb{Z}_2 \times \mathbb{Z}_2 = \{(a,b) : a \in \mathbb{Z}_2, b \in \mathbb{Z}_2\}.$$

By Lemma 8.1.1, $|\mathbb{Z}_2 \times \mathbb{Z}_2| = |\mathbb{Z}_2| \, |\mathbb{Z}_2| = 2 \cdot 2 = 4$. But we claim that $\mathbb{Z}_2 \times \mathbb{Z}_2$ has no element of order 4. The easiest way to see this is just to write out the four elements of $\mathbb{Z}_2 \times \mathbb{Z}_2$ and compute their orders directly.

$$\mathbb{Z}_2 \times \mathbb{Z}_2 = \{(0,0), (1,0), (0,1), (1,1)\}.$$

Then, for example, $2(1,1) = (1,1) + (1,1) = (0,0)$. Similarly for the other nonidentity elements. Thus each nonidentity element of $\mathbb{Z}_2 \times \mathbb{Z}_2$ has order 2. This shows that $\mathbb{Z}_2 \times \mathbb{Z}_2$ is not cyclic. We note that $\mathbb{Z}_2 \times \mathbb{Z}_2$ is not isomorphic to $\mathbb{Z}_4$ which is cyclic, but $\mathbb{Z}_2 \times \mathbb{Z}_2$ is isomorphic to a group of order 4 which we did bring up in Chapter 7. In particular, $\mathbb{Z}_2 \times \mathbb{Z}_2 \cong V_4$, the Klein 4-group.

8.1.2   Direct Products of Any Finite Number of Groups

We now generalize everything we did in the previous subsection to the case where we have a product of any finite number of groups.

Definition 8.1.1. *Let $G_1, G_2, \ldots, G_n$ be a finite collection of groups. We form the set*

$$G = G_1 \times G_2 \times \cdots \times G_n,$$

the cartesian product of the sets $G_1, G_2, \ldots, G_n$. Thus G consists of all ordered n-tuples of the form $(a_1, a_2, \ldots, a_n)$ where $a_i \in G_i$, for $i = 1, 2, \ldots, n$.

We introduce the componentwise operation which will make G into a group as follows. For two such n-tuples, we define

$$(a_1, a_2, \ldots, a_n)(b_1, b_2, \ldots, b_n) = (a_1 b_1, a_2 b_2, \ldots, a_n b_n).$$

The group G so constructed is called the (external) **direct product** *of the given groups. We denote this by $G_1 \times G_2 \times \cdots \times G_n$ (external).*

It is understood that each product $a_i b_i$ is performed with the operation of G_i. It is now straightforward to show that for this operation, the associative law is satisfied. Also the element $(1, 1, \ldots, 1)$ where 1 is the identity element of G_i — and each such identity could be very different from the others — functions as the identity element of G. Finally, the inverse of the element $(a_1, a_2, \ldots, a_n)$ is the element $(a_1^{-1}, a_2^{-1}, \ldots, a_n^{-1})$; hence G is a group with respect to the given operation. It is also clear that the following is true.

Lemma 8.1.3. *Let $G = G_1 \times G_2 \times \cdots \times G_n$.*
(a) If each G_i is finite and $|G_i| = n_i$ then

$$|G| = \prod_i n_i.$$

(b) If each G_i is abelian, then G is abelian.

Now we consider the following situation which will subsequently be shown to be related. Before doing so, let us just state that if $S_1, S_2, \ldots, S_n$ are complexes of a group G, then their product is

$$S_1 S_2 \cdots S_n = \{s_1 s_2 \cdots s_n : s_i \in S_i\}.$$

Definition 8.1.2. *Let G be a given group and let $G_1, G_2, \ldots, G_n$ be normal subgroups of G such that*

$$G = G_1 G_2 \cdots G_n$$

(where this is the usual product of sets in a group as defined above) and $G_i \cap G_1 \cdots G_{i-1} G_{i+1} \cdots G_n = \{1\}$ for every $i = 1, 2, \ldots, n$. In this situation, we say that G is decomposed into the (internal) **direct product** *of the subgroups $G_1, G_2, \ldots, G_n$, and we shall write $G = G_1 \times G_2 \times \cdots \times G_n$ (internal).*

For the time being, we shall write in parentheses after an expression of the form $G_1 \times G_2 \times \cdots \times G_n$ either external or internal to distinguish which of the two situations above actually prevails. After we have seen the interconnection beteen these two concepts, it will be clear that we can drop this accompanying label without fear of confusion.

Our first theorem related to these direct products is the generalization of Lemma 8.1.2 to any finite number of terms and so is concerned with when the group G is the internal direct product of its subgroups $G_1, G_2, \ldots, G_n$.

Theorem 8.1.5. $G = G_1 \times G_2 \times \cdots \times G_n$ *(internal) if and only if both of the following hold.*

(1) $a_i a_j = a_j a_i$ *for any* $a_i \in G_i$ *and any* $a_j \in G_j$ *where* $i \neq j$.

(2) Every element of G can be written uniquely in the form $a_1 a_2 \cdots a_n$ *where* $a_i \in G_i$

Proof. Suppose first that $G = G_1 \times G_2 \times \cdots \times G_n$ (internal) and let $a_i \in G_i$ and $a_j \in G_j$ where $i \neq j$. Then conisider the commutator $[a_i, a_j]$ which is defined as $[a_i, a_j] = a_i a_j a_i^{-1} a_j^{-1} \in G_j$ because $G_j \lhd G$ and $a_j \in G_j$. However, $a_i^{-1} \in G_i$ and $a_j a_i^{-1} a_j^{-1} \in G_i$ since $G_i \lhd G$; therefore $[a_i, a_j] \in G_i$, but $G_i \cap G_j = \{1\}$ (Why?). Hence we have that

$$[a_i, a_j] = a_i a_j a_i^{-1} a_j^{-1} = 1$$

which implies that $a_i a_j = a_j a_i$ and proves (1). Next since $G = G_1 G_2 \cdots G_n$, any $a \in G$ can be written in the form $a = a_1 a_2 \cdots a_n$ where $a_i \in G_i$. If also $a = b_1 b_2 \cdots b_n$ where $b_i \in G_i$, then

$$a = a_1 a_2 \cdots a_n = b_1 b_2 \cdots b_n$$

using the commutativity of elements in different G_i's, we get

$$b_i a_i^{-1} = b_1^{-1} a_1 \cdots b_{i-1}^{-1} a_{i-1} b_{i+1}^{-1} a_{i+1} \cdots b_n^{-1} a_n.$$

Since G is the internal direct product of the G_i, this in turn implies that

$$b_i a_i^{-1} = 1$$

or that $b_i = a_i$. But this can be done for every $i = 1, 2, \ldots, n$. This establishes (2).

Conversely suppose that (1) and (2) hold. Then each $G_i \lhd G$ for if $g_i \in G_i$ and $a = a_1 a_2 \cdots a_n$, where $a_j \in G_j$, is an arbitrary element of G, then

$$
\begin{aligned}
a g_i a^{-1} &= a_1 a_2 \cdots a_n g_i a_n^{-1} \cdots a_2^{-1} a_1^{-1} \\
&= a_i g_i a_i^{-1} \in G_i
\end{aligned}
$$

since by (1), a_j commutes with g_i for all $j > i$ and a_j commutes with $a_i g_i a_i^{-1}$ for all $j < i$. This is sufficient to prove that each G_i is normal in G. From (2), we have that $G = G_1 G_2 \cdots G_n$. Finally we note that a typical element of $G_1 \cdots G_{i-1} G_{i+1} \cdots G_n$ is of the form $a_1 a_2 \cdots a_{i-1} a_{i+1} \cdots a_n$ where $a_j \in G_j$. Suppose such an element is also in G_i and, hence, is equal to some $a_i \in G_i$. Then we have

$$
a_i = a_1 a_2 \cdots a_{i-1} a_{i+1} \cdots a_n
$$

or

$$
1 \cdots 1 a_i 1 \cdots 1 = a_1 a_2 \cdots a_{i-1} 1 a_{i+1} \cdots a_n
$$

where 1 is the identity of G. By the uniqueness part of 2, we now have that $a_i = 1$, so

$$
G_i \cap G_1 \cdots G_{i-1} G_{i+1} \cdots G_n = \{1\}.
$$

$\square$

We now proceed to establish two theorems which will show that in the future, we need not distinguish between internal and external direct products.

Theorem 8.1.6. *Let* $G = G_1 \times G_2 \times \cdots \times G_n$ *(internal) and let* $G_i \cong G_i^*, i = 1, 2, \ldots, n$. *Form* $G^* = G_1^* \times G_2^* \times \cdots \times G_n^*$ *(external); then* $G \cong G^*$.

Proof. Let $f_i : G_i^* \to G_i$ $(1 \le i \le n)$ be an isomorphism of G_i^* onto G_i which we are given exists. We now define a mapping

$$
f : G^* \to G
$$

by

$$
f((a_1, a_2, \ldots, a_n)) = f_1(a_1) f_2(a_2) \cdots f_n(a_n).
$$

We claim that f is an isomorphism of G^* onto G. First we observe that f is onto G: For any element g of G, we know is of the form $g = b_1 b_2 \cdots b_n$, where

$b_i \in G_i$. There since each f_i is onto there exist $a_i \in G_i^*$ so that $b_i = f_i(a_i)$, and so

$$
\begin{aligned}
g &= b_1 b_2 \cdots b_n = f_1(a_1) f_2(a_2) \cdots f_n(a_n) \\
&= f((a_1, a_2, \ldots, a_n))
\end{aligned}
$$

where $(a_1, a_2, \ldots, a_n) \in G^*$. Thus f is onto G. In the following for ease of notation, we will at times drop the double parentheses around the n-tuple, i.e., we may write $f((a_1, a_2, \ldots, a_n))$ as $f(a_1, a_2, \ldots, a_n)$.

Next, we note that f is a homomorphism, i.e., f preserves the operation; for

$$
f((a_1, a_2, \ldots, a_n)(b_1, b_2, \ldots, b_n)) =
$$

$$
f(a_1 b_1, a_2 b_2, \ldots, a_n b_n) = f_1(a_1 b_1) f_2(a_2 b_2) \cdots f_n(a_n b_n)
$$

and since each f_i is a homomorphism

$$
\begin{aligned}
&= f_1(a_1) f_1(b_1) f_2(a_2) f_2(b_2) \cdots f_n(a_n) f_n(b_n) \\
&= (f_1(a_1) f_2(a_2) \cdots f_n(a_n)) \cdot (f_1(b_1) f_2(b_2) \cdots f_n(b_n))
\end{aligned}
$$

since elements from different factors of an internal direct product commute from Theorem 8.1.5. Finally, we have from the definition of f that the above is equal to

$$
f(a_1, a_2, \ldots, a_n) f(b_1, b_2, \ldots, b_n)
$$

which shows that f preserves the operation.

Furthermore, f is an injection. To see this suppose that $f(a_1, a_2, \ldots, a_n) = 1$, i.e., $(a_1, a_2, \ldots, a_n) \in \ker(f)$. Then $1 \cdot 1 \cdots 1 = f_1(a_1) f_2(a_2) \cdots f_n(a_n)$, which by the uniqueness of representation, again by Theorem 8.1.5, implies that each $f_i(a_i) = 1$ for all i, $1 \leq i \leq n$. Since each f_i is an isomorphism, Lemma 7.3.1 implies that $a_i = 1$ for all i, $1 \leq i \leq n$, where it must be understood that each 1 is the identity of G_i^* and these can be very different. Therefore $(a_1, a_2, \ldots, a_n) = (1, 1, \ldots, 1)$ is the identity of G^*, i.e., $\ker(f)$ is trivial. Thus Lemma 7.3.1 implies that f is 1-1.   □

We note, in particular, taking $G_i^* = G_i, i = 1, 2, \ldots, n$, in Theorem 8.1.6 that if $G = G_1 \times G_2 \times \cdots \times G_n$ (internal), then forming $G^* = G_1 \times G_2 \times \cdots \times G_n$ (external) gives a group isomorphic to G.

Theorem 8.1.7. *Let* $G = G_1 \times G_2 \times \cdots \times G_n$ *(external); also let*

$$H_i = \{(1, \ldots, 1, a_i, 1, \ldots, 1) : a_i \in G_i\}.$$

Then each $H_i \lhd G$ *and*

$$G = H_1 \times H_2 \times \cdots \times H_n \quad (internal),$$

and $H_i \cong G_i$ *for every* i, $i = 1, 2, \ldots, n$.

We note that this is the analog of the theorem for two direct factors, i.e., Theorem 8.1.3.

Proof. Clearly, each H_i is a subgroup of G (see the exercises) and the mapping $G_i \to H_i$ given by $a_i \mapsto (1, \ldots, 1, a_i, 1, \ldots, 1)$ is also clearly an isomorphism of G_i onto H_i. Moreover,

$$(b_1, \ldots, b_n)(1, \ldots, 1, a_i, 1, \ldots, 1)(b_1^{-1}, \ldots, b_n^{-1}) \quad =$$
$$(1, \ldots, 1, b_i a_i b_i^{-1}, 1, \ldots, 1)$$

which shows that each $H_i \lhd G$. Now let $(a_1, a_2, \ldots, a_n)$ be an arbitrary element of G, we can write this in the form

$$(a_1, a_2, \ldots, a_n) = a_1' a_2' \cdots a_n'$$

where $a_i' = (1, \ldots, 1, a_i, 1, \ldots, 1) \in H_i$. Hence $G = H_1 H_2 \cdots H_n$. Finally any element of $H_1 \cdots H_{i-1} H_{i+1} \cdots H_n$ is of the form $(a_1, \ldots, a_{i-1} 1, a_{i+1}, \ldots, a_n)$ where $a_j \in H_j$. Thus it follows immediately that

$$H_i \cap (H_1 \cdots H_{i-1} H_{i+1} \cdots H_n) = \{1\}$$

for all $i = 1, \ldots, n$ where the 1 above is the identity of G, i.e., $1 = (1, \ldots, 1)$.
$$\square$$

From now on, we may drop writing in parentheses after an expression $G_1 \times G_2 \times \cdots \times G_n$ either external or internal. It should be clear from the context what is meant.

8.2   Abelian Groups

If A and B are abelian groups then we have seen that their direct product $A \times B$ is also abelian. It follows that the direct product of finitely many cyclic groups is still abelian. In the remainder of this chapter we will show that the structure of any finitely generated abelian groups is completely determined up to isomorphism as a direct product of cyclic groups. We first look at finite abelian groups. If G is a finite abelian group it has only a finite number of elements and hence it must also be finitely generated.

8.2.1   Finite Abelian Groups

The following theorem completely provides the structure of finite abelian groups.

Theorem 8.2.1. *(Basis Theorem for Finite Abelian Groups or Fundamental Theorem for Finite Abelian Groups) Let G be a finite abelian group. Then G is a direct product of cyclic groups of prime power order. Furthermore, this decomposition is unique up to the order of the direct factors.*

Before giving the proof we give two examples showing how this theorem leads to the classification of finite abelian groups. Since all cyclic groups of order n are isomorphic to $(\mathbb{Z}_n, +)$ we will denote a cyclic group of order n by $\mathbb{Z}_n$.

We need a few more facts before going to the examples.

Lemma 8.2.1. *If m and n are relatively prime natural numbers, that is, $GCD(m, n) = 1$, then $\mathbb{Z}_m \times \mathbb{Z}_n \cong \mathbb{Z}_{mn}$.*

Proof. To show this isomorphism it is sufficient to show that $\mathbb{Z}_m \times \mathbb{Z}_n$ is cyclic. For it is of order mn and if cyclic then it must be isomorphic to $\mathbb{Z}_{mn}$. Thus if we can find an element of $\mathbb{Z}_m \times \mathbb{Z}_n$ of order mn, then we will be done. Consider then element $(1, 1) \in \mathbb{Z}_m \times \mathbb{Z}_n$. As we know the order is the smallest "power" of $(1, 1)$ which gives the identity $(0, 0)$. Since the operation in $\mathbb{Z}_m$ and $\mathbb{Z}_n$ is written additively, taking a "power" of $(1, 1)$ in our additive notation will involve adding $(1, 1)$ to itself repeatedly or taking a natural number multiple of $(1, 1)$. Under componentwise addition, the first component $1 \in \mathbb{Z}_m$ yields 0 only after m summands, or $2m$ summands, etc., while the second component $1 \in \mathbb{Z}_n$ yields 0 only after n summands, or $2n$ summands, etc. Thus for them to give 0 simultaneously, the number of summands must be a multiple of both m and n. Since $\text{GCD}(m,n) = 1$, the smallest such number which is $o(1,1) = \text{lcm}(m,n) = mn$. Thus $\mathbb{Z}_m \times \mathbb{Z}_n = \langle (1,1) \rangle$ and so is isomorphic to $\mathbb{Z}_{mn}$. $\qquad\square$

This lemma clearly can be extended to a direct product of more than two factors (using an induction argument). Here we prove that a cyclic group of order n can be written as an (internal) direct product of cyclic groups of prime power order.

Theorem 8.2.2. *If G is a cyclic group of order $n = p_1^{e_1} p_2^{e_2} \cdots p_m^{e_m}$ where the p_i are all distinct primes (this is what we called the standard prime decomposition), then G is a direct product of cyclic groups of orders $p_i^{e_i}$, $i = 1, 2, \ldots, m$.*

Proof. We designate by G_i the unique cyclic subgroup of order $p_i^{e_i}$ of G (see Theorem 6.4.3(a)) and let $H = G_1 G_2 \cdots G_m$. H is, of course, a subgroup by Theorem 6.3.2 since the G_i commute because G is cyclic (and thus, of course, abelian). Also $H \supseteq G_i$ for every $i = 1, 2, \ldots, m$; therefore $p_i^{e_i} \mid |H|$ for every $i = 1, 2, \ldots, m$. Thus $n = \text{lcm}(p_i^{e_i})$ must also divide $|H|$. Since $H \subset G$, $|H| \mid |G| = n$. Since all the numbers involved are positive, this implies that $n = |H|$. This proves $G = H = G_1 G_2 \cdots G_m$.

Next, we designate by H_i the cyclic subgroup of G of order $n_i = n/(p_i^{e_i})$, $i = 1, 2, \ldots, m$, and let $W_i = H_i \cap G_i$. Then W_i is a subgroup of both H_i and G_i (by Lemma 6.3.4). Thus $|W_i| \mid |H_i|$ and $|W_i| \mid |G_i|$, but $\text{GCD}(n_i, p_i^{e_i}) = 1$, so $|W_i| = 1$. Hence $W_i = H_i \cap G_i = \{1\}$. However, $p_j^{e_j} \mid |H_i|$ for all $j \neq i$. Thus $H_i \supseteq G_j$ for $j \neq i$ since H_i being cyclic contains a subgroup of order $p_j^{e_j}$ and that subgroup must be $G_j \subset H_i \subset G$ by uniqueness (Theorem 6.4.3(a)), so $H_i \supseteq G_1 \cdots G_{i-1} G_{i+1} \cdots G_m$ and therefore

$$(G_1 \cdots G_{i-1} G_{i+1} \cdots G_m) \cap G_i = \{1\}$$

since the bigger group H_i has trivial intersection with G_i and this must hold for all $i = 1, 2, \ldots, m$. This proves that G is the (internal) direct product of the G_i for $i = 1, 2, \ldots, m$ where G_i is cyclic of order $p_i^{e_i}$. $\qquad\square$

Recalling that all cyclic groups of order n are isomorphic to $\mathbb{Z}_n$, we see the G_i in the above proof are such that $G_i \cong \mathbb{Z}_{p_i^{e_i}}$. Thus we can replace the G_i in the above proof by $\mathbb{Z}_{p_i^{e_i}}$. So this says if n is any integer with $n > 1$ and if n has the standard prime decomposition $n = p_1^{e_1} p_2^{e_2} \cdots p_m^{e_m}$ where the p_i are all distinct, then

$$\mathbb{Z}_n \cong \mathbb{Z}_{p_1^{e_1}} \times \mathbb{Z}_{p_2^{e_2}} \times \cdots \times \mathbb{Z}_{p_m^{e_m}}.$$

For example, $\mathbb{Z}_{60} \cong \mathbb{Z}_3 \times \mathbb{Z}_4 \times \mathbb{Z}_5$. We do note, however, $60 = 6 \cdot 10$ but 6 and 10 are not relatively prime. If we take $\mathbb{Z}_6 \times \mathbb{Z}_{10}$, we get an abelian group of order 60, but it is no longer isomorphic to $\mathbb{Z}_{60}$ because the largest order of any element is 30. (Can you see why?)

Finally getting back to the fundamental theorem if G is any finite abelian group, then

$$G \cong \mathbb{Z}_{p_1^{e_1}} \times \mathbb{Z}_{p_2^{e_2}} \times \cdots \times \mathbb{Z}_{p_m^{e_m}}$$

where we consider the direct product on the right-hand side to be external because G may not even consist of integers and now it must be understood that the primes p_i are not necessarily distinct but the prime powers $p_1^{e_1}$, $p_2^{e_2}$, ..., $p_m^{e_m}$ are uniquely determined by G. The fact that the primes may not be distinct is unavoidable; for example, $\mathbb{Z}_5 \times \mathbb{Z}_5 \times \mathbb{Z}_6$ is not isomorphic to $\mathbb{Z}_{25} \times \mathbb{Z}_6$ for the former group has elements at most of order 30 while the latter is cyclic of order 150 (see Lemma 8.2.1). Writing down a group in this direct product form is called **determining the isomorphism class of the group.**

We will use our basis theorem or fundamental theorem to get an algorithm for constructing all nonisomorphic abelian groups of any order. To get some appreciation of how powerful this theorem is, we can contrast it with finding all the nonabelian groups of a given order. Even for orders as small as 8, of which there are only two, to prove it at this point in our discussions of group theory would not be easy. To find all the nonabelian groups of order 12 (there are three of them) is already beyond the scope of this text.

In a moment, we will see that there are three nonisomorphic abelian groups of order 8. Thus to start our program for finding all abelian groups of a given order, we first consider abelian groups of prime power order. For simplicity sake, we will just consider here groups of order p^n where p is a prime and $n \leq 4$.

We list these in the following table.

$\|G\|$	Direct products for G
p	$\mathbb{Z}_p$
p^2	$\mathbb{Z}_{p^2}$
	$\mathbb{Z}_p \times \mathbb{Z}_p$
p^3	$\mathbb{Z}_{p^3}$
	$\mathbb{Z}_{p^2} \times \mathbb{Z}_p$
	$\mathbb{Z}_p \times \mathbb{Z}_p \times \mathbb{Z}_p$
p^4	$\mathbb{Z}_{p^4}$
	$\mathbb{Z}_{p^3} \times \mathbb{Z}_p$
	$\mathbb{Z}_{p^2} \times \mathbb{Z}_{p^2}$
	$\mathbb{Z}_{p^2} \times \mathbb{Z}_p \times \mathbb{Z}_p$
	$\mathbb{Z}_p \times \mathbb{Z}_p \times \mathbb{Z}_p \times \mathbb{Z}_p$

The table shows as mentioned above that there are three nonisomorphic abelian groups of order 8. More generally, there is one group of order p^n for each set of positive integers whose sum is n. Such a set of positive integers is called a **partition** of n. So if n can be written as

$$n = n_1 + n_2 + \cdots + n_k$$

where each n_i is a positive integer, then

$$\mathbb{Z}_{p^{n_1}} \times \mathbb{Z}_{p^{n_2}} \times \cdots \times \mathbb{Z}_{p^{n_k}}$$

is an abelian group of order p^n. Moreover, the uniqueness part of the basis theorem guarantees that distinct partitions of n give nonisomorphic groups. Thus $\mathbb{Z}_8 \times \mathbb{Z}_2$ is not isomorphic to $\mathbb{Z}_4 \times \mathbb{Z}_2 \times \mathbb{Z}_2$ since the first group arises from the partition of $4 = 3 + 1$ while the second one arises from the partition of $4 = 2 + 1 + 1$. We note, of course, that both groups have order $2^4 = 16$. If you require more justification than this that these two groups are not isomorphic, you may observe that the first group has elements of order 8 while the largest order of any element in the seond group is 4. A nice theorem for comparing whether or not external direct products are isomorphic is the cancellation theorem (due to Ronald Hirshon [RH]) which states:

Theorem 8.2.3. *(Cancellation in Direct Products) If G is a finite group, then for any groups H and K, $G \times H \cong G \times K$ if and only if $H \cong K$.*

So, for example, $\mathbb{Z}_{25} \times \mathbb{Z}_5$ is not isomorphic to $\mathbb{Z}_5 \times \mathbb{Z}_5 \times \mathbb{Z}_5$ because $\mathbb{Z}_{25}$ is not isomorphic to $\mathbb{Z}_5 \times \mathbb{Z}_5$.

Now we know how to construct all abelian groups of prime power order. Let us consider the problem of constructing all abelian groups of a certain order n where n has two or more distinct prime divisors. We start by writing the standard prime decomposition

$$n = p_1^{e_1} \, p_2^{e_2} \cdots \, p_m^{e_m}.$$

Next, form all abelian groups of order $p_1^{e_1}$, then of order $p_2^{e_2}$, etc. as just described. Finally form all possible external direct products of these and you are done!

EXAMPLE 8.2.1. Classify all abelian groups of order 60. Let G be an abelian group of order 60. Now $60 = 2^2 \cdot 3 \cdot 5$ so the only primes involved are $2, 3$, and 5. Since 2^2 and $2 \cdot 2$ correspond to the two partitions of the exponent 2, the complete list of distinct isomorphism classes of abelian groups of order 60 consists of only two groups

$$
\begin{aligned}
G &\cong \mathbb{Z}_4 \times \mathbb{Z}_3 \times \mathbb{Z}_5, \\
G &\cong \mathbb{Z}_2 \times \mathbb{Z}_2 \times \mathbb{Z}_3 \times \mathbb{Z}_5.
\end{aligned}
$$

EXAMPLE 8.2.2. Find all abelian groups (up to isomorphism) of order 360. Let G be an abelian group of order 360. Write the standard prime decomposition form for $360 = 2^3 \cdot 3^2 \cdot 5$ so the primes involved again are $2, 3$, and 5. But now there are three partitions of 3. So we get $2^3, 2^2 \cdot 2$, and $2 \cdot 2 \cdot 2$. There are two partitions again of 2, i.e., giving 3^2 and $3 \cdot 3$. Hence up to isomorphism there are six abelian groups of order 360:

$$
\begin{aligned}
G &\cong \mathbb{Z}_8 \times \mathbb{Z}_9 \times \mathbb{Z}_5, \\
G &\cong \mathbb{Z}_4 \times \mathbb{Z}_2 \times \mathbb{Z}_9 \times \mathbb{Z}_5, \\
G &\cong \mathbb{Z}_2 \times \mathbb{Z}_2 \times \mathbb{Z}_2 \times \mathbb{Z}_9 \times \mathbb{Z}_5, \\
G &\cong \mathbb{Z}_8 \times \mathbb{Z}_3 \times \mathbb{Z}_3 \times \mathbb{Z}_5, \\
G &\cong \mathbb{Z}_4 \times \mathbb{Z}_2 \times \mathbb{Z}_3 \times \mathbb{Z}_3 \times \mathbb{Z}_5, \\
G &\cong \mathbb{Z}_2 \times \mathbb{Z}_2 \times \mathbb{Z}_2 \times \mathbb{Z}_3 \times \mathbb{Z}_3 \times \mathbb{Z}_5.
\end{aligned}
$$

Given a particular abelian group G of order, let's say 360, the question as to which of the above six isomorphism classes represents G can be answered by comparing the orders of elements. We note it can be shown that two abelian groups are isomorphic if and only if they have the same number of elements of each order. For example, if we could determine that G has an element of order 8, then G must be isomorphic to either the first or fourth group above. To narrow G down to a single choice, we now need only determine whether or not G has an element of order 9. Since the first product above has such an element while the fourth does not.

The proof of Theorem 8.2.1. involves the following lemmas. Here we will not prove the uniqueness of the direct decomposition of any finite abelian group but only the existence of such a decomposition.

Lemma 8.2.2. *Let G be a finite abelian group and let p divide $|G|$ where p is a prime. Then the set of all the elements of G whose orders are a power of p form a normal subgroup of G. This subgroup is called the p-**primary component** of G, which we will denote by G_p.*

Proof. Let p be a prime with $p \mid |G|$. Since $1 \in G_p$, it is certainly true that G_p is nonempty. (As a matter of fact shortly we will show there has to exist a nontrivial element in G_p but this will suffice for now). Since it is clear that the set consisting of 1, i.e., $\{1\}$ is a subgroup, we now assume that $|G_p| > p^0 = 1$. Let a and b be two elements of G of order a power of p. Since G is abelian the order of ab divides the LCM of the orders which is again a power of p. (See Chapter 6, Lemma 6.4.2.) Therefore $ab \in G_p$. The order of a^{-1} is the same as the order of a so $a^{-1} \in G_p$ and therefore G_p is a subgroup. Since G is abelian, it is a normal subgroup. (See Chapter 7, Lemma 7.1.6.) □

Lemma 8.2.3. *Let G be a finite abelian group of order n. Suppose that $n = p_1^{e_1} \cdots p_k^{e_k}$ with $p_1, \ldots, p_k$ distinct primes. Then*

$$G = G_{p_1} \times \cdots \times G_{p_k}$$

where G_{p_i} is the p_i-primary component of G.

Proof. Each G_{p_i} is normal since G is abelian and since distinct primes are relatively prime, the intersection G_{p_i} with $\prod_{j \neq i} G_{p_j}$ is the identity. (See the exercises to fill in the details here.) Also this is true for every $i = 1, \ldots, k$. Therefore the result will follow by showing that each element of G is a product of elements in the G_{p_i}, that is, the collection of the p-primary components generates G.

It is somewhat easier to complete the proof if we consider the group using additive notation. That is, the operation is conisdered to be $+$, the identity is 0, and powers are given by multiples. Hence if an element $g \in G$ has order p^k then in additive notation $p^k g = 0$.

Now let $g \in G$. Then the order of g is $n = p_1^{f_1} \cdots p_k^{f_k}$ where each $f_i \leq e_i$ and each f_i is a nonnegative integer. Hence $ng = 0$. For each $i = 1, \ldots, k$ let $n_i = \frac{n}{p_i^{f_i}}$. Since the prime divisors of n_i are all primes, in the factorization of n, $p_j \neq p_i$, it follows that the set of the n_i cannot have a common divisor other than 1 and hence $\mathrm{GCD}(n_1, \ldots, n_k) = 1$. From Problem 6.4.10, it follows that there are integers $x_1, \ldots, x_k$ with $\sum_{i=1}^{k} (x_i n_i) = 1$. Therefore

$$\sum_{i=1}^{k} (x_i n_i) g = g.$$

Consider the term $h_i = (x_i n_i) g$. Then

$$p_i^{f_i} h_i = p_i^{f_i} x_i n_i g = x_i n g = 0.$$

It follows that for each i the term $h_i \in G_{p_i}$. Hence the element g is expressed as a sum with each term in a p-primary component. This shows that the set of p-primary components generates the whole group G which completes the proof of this lemma. $\square$

In general if p is a prime then a **finite p-group** is a group where every element has order a power of p. Clearly, from Lagrange's Theorem if a group has order a power of p then it is a p-group. We note that the p-primary components considered in Lemma 8.2.2 are finite abelian p-groups by this definition. We next show that any finite abelian p-group has order p^n for some integer n. This is a consequence of the following lemma which says that if G is a finite abelian group and $p \mid |G|$ with p a prime, then G has an element of order p. In the next chapter, we will prove that this result holds in any finite group, abelian or not. This is called Cauchy's theorem but for now we only need it in abelian groups. Recall from chapter 6 that if G is a cyclic group of order n and d is a positive divisor of n, then G has a unique subgroup, hence a cyclic subgroup, of order d, i.e., it has an element of order d.

Lemma 8.2.4. *Let G be a finite abelian group and let p be a prime such that p divides the order of G. Then G contains at least one element of order p.*

Proof. The proof will proceed by induction on $|G|$. If $|G| = p$, a prime, then G is cyclic and so the result is true. Thus suppose $|G| = n$ where n is composite and also suppose that the result has been proven for all groups whose order is less than n. Suppose p is a prime such that $p \mid n$. We need to show that G has an element of order p. If G is cyclic, then we are done by the remarks above. So suppose that G is not cyclic and let $a \in G, a \neq 1$; then $\langle a \rangle$ is a proper subgroup of G. (Why?) Since G has a proper subgroup, i.e., $\langle a \rangle$, there is a proper subgroup, let's call it H, of maximal order. If $p \mid |H|$, then since $|H| < |G|$, we have by the induction hypothesis that there exists an $h \in H$ such that $o(h) = p$, but clearly, $h \in G$, also, so we are done. If, however, $p \nmid |H|$, then since H is a proper subgroup of G, let $b \in G - H$. Let $K = \langle b \rangle$. Now the product HK is a subgroup of G by Theorem 6.3.2 since G is abelian. Also $H \subset HK$ properly, i.e., $H \neq HK$, because $b \in HK$ but $b \notin H$. However, H was a maximal proper subgroup. Thus it must be that $HK = G$. Then by Theorerm 6.3.3,

$$|G| = \frac{|H|\,|K|}{|H \cap K|} = \frac{|H| \cdot o(b)}{d}$$

where $d = |H \cap K|$. Thus

$$d\,|G| = |H| \cdot o(b),$$

and since $p \mid |G|$, we must have that $p \mid |H| \cdot o(b)$. However, we have assumed $p \nmid |H|$. Thus $\mathrm{GCD}(p, |H|) = 1$ and so $p \mid o(b)$. Thus we are done again because then the cyclic subgroup $K = \langle b \rangle$ of G has an element of order p. $\square$

Corollary 8.2.1. *Let G be a finite abelian p-group. Then $|G| = p^m$ for some positive integer m.*

Proof. By definition every element of G has order a power of p. Suppose q is a prime with $q \neq p$ and $q \mid |G|$. Then from the previous lemma, G has an element of order q which is a contradiction. Therefore the only prime that can divide the order of G is p and hence $|G| = p^m$ for some m. $\square$

Thus according to this corollary, the p_i-primary components occuring in Lemma 8.2.3 each have order a power of p_i. We next need the concept of a **basis.** Let G be a finitely generated abelian group (finite or infinite) and let $\{g_1, \ldots, g_n\}$ be a set of generators for G. The generators $g_1, \ldots, g_n$ form a **basis** for G if

$$G = \langle g_1 \rangle \times \cdots \times \langle g_n \rangle,$$

that is, G is the direct product of the cyclic subgroups generated by the g_i. The basis theorem for finite abelian groups says that any finite abelian group has a basis.

Suppose that G is a finite abelian group with a basis $g_1, \ldots, g_k$ so that $G = \langle g_1 \rangle \times \cdots \times \langle g_k \rangle$. Since G is finite each g_i has finite order say m_i. It follows then, from the fact that G is a direct product, that each $g \in G$ can be expressed as

$$g = g_1^{n_i} \cdots g_k^{n_k}$$

and further, the integers n_i are unique modulo the order of g_i for $i = 1, \cdots, k$. Hence each integer n_i can be chosen in the range $0, 1, \ldots, m_i - 1$ and within this range for the element g the integer n_i is unique.

From Lemma 8.2.3, each finite abelian group splits into a direct product of its p-primary components for different primes p. From Corollary 8.2.1 each p-primary component has order p^m for some m. Hence to complete the proof of the basis theorem we must show that any finite abelian group of order p^m for some prime p has a basis. From previously introduced terminology an abelian group of order p^m is an abelian p-group.

Consider an abelian group G of order p^m for a prime p. As above it is somewhat easier to complete the proof if we consider the group using additive notation. As explained this means that the operation is considered to be $+$, the identity is 0 and powers are given by multiples. Hence if an element $g \in G$ has order p^k then in additive notation $p^k g = 0$. A set of elements $\{g_1, \ldots, g_k\}$ is then a basis for G if each $g \in G$ can be expressed uniquely as $g = m_1 g_1 + \cdots + m_k g_k$ where the m_i are unique modulo the order of g_i. We say that the $g_1, \ldots, g_k$ are **independent** and this is equivalent to the fact that whenever $m_1 g_1 + \cdots + m_k g_k = 0$ then $m_i \equiv 0$ modulo the order of g_i. We now prove that any finite abelian p-group has a basis. We note that since the trivial group is a p-group for every prime p, we make the following convention. The empty set is a basis for the trivial group.

Lemma 8.2.5. *Let G be a finite abelian group of prime power order p^n for some prime p. Then G has a basis, that is, G is a direct product of cyclic groups.*

Proof. Let G be a finite abelian group of prime power order p^n for some prime p. Notice that in the group G we have $p^n g = 0$ for all $g \in G$ as a consequence of Lagrange's theorem. Further every element has order a power of p. The smallest power of p say p^r such that $p^r g = 0$ for all $g \in G$ is called the exponent of G. Any finite abelian p-group must have some exponent p^r. The proof of this lemma is by induction on the exponent.

The lowest possible exponent is p, so suppose first that $pg = 0$ for all $g \in G$. Since G is finite, it has a finite system of generators. Let $S = \{g_1, \ldots, g_k\}$ be a minimal set of generators for G. This exists by LWO. We claim that this is a basis. Since this is a set of generators, to show that it forms a basis we must show that they are independent. Hence suppose that we have

$$m_1 g_1 + \cdots + m_k g_k = 0 \tag{1}$$

for some set of integers m_i. Since the order of each g_i is p, as explained above, we may assume that $0 \le m_i < p$ for $i = 1, \ldots, k$. Suppose that one $m_i \ne 0$. Then $(m_i, p) = 1$ and hence there exists an x_i with $m_i x_i \equiv 1 \pmod{p}$ (see Chapter 4). Multiplying the equation (1) by x_i we get mod p,

$$m_1 x_i g_1 + \cdots + g_i + \cdots + m_k x_i g_k = 0,$$

and rearranging

$$g_i = -m_1 x_i g_1 - \cdots - m_k x_i g_k.$$

But then g_i can be expressed in terms of the other g_j and therefore the set $\{g_1, \ldots, g_k\}$ is not minimal. It follows that $g_1, \ldots, g_k$ constitute a basis and the lemma is true for the exponent p.

Now suppose that any finite abelian group of exponent less than or equal to p^{n-1} has basis and assume that G has exponent p^n. Consider the set $\overline{G} = pG = \{pg : g \in G\}$. It is straightforward that this forms a subgroup (see the exercises). Since $p^n g = 0$ for all $g \in G$ it follows that $p^{n-1} g = 0$ for all $g \in \overline{G}$ and so the exponent of $\overline{G} \le p^{n-1}$. By the inductive hypothesis, $\overline{G}$ has a basis

$$S = \{pg_1, \ldots, pg_k\}.$$

Consider the set $\{g_1, \ldots, g_k\}$ and adjoin to this set the set of all elements $h \in G$ satisfying $ph = 0$. Call this set S_1 so that we have

$$S_1 = \{g_1, \ldots, g_k, h_1, \ldots, h_t\}.$$

We claim that S_1 is a set of generators for G. Let $g \in G$. Then $pg \in \overline{G}$ which has the basis $pg_1, \ldots, pg_k$ so that

$$pg = m_1 pg_1 + \cdots + m_k pg_k.$$

This implies that

$$p(g - m_1 g_1 - \cdots - m_k g_k) = 0$$

so that $g - m_1 g_1 - \cdots - m_k g_k$ must be one of the h_i. Hence

$$g - m_1 g_1 - \cdots - m_k g_k = h_i$$

so that

$$g = m_1 g_1 + \cdots + m_k g_k + h_i$$

proving the claim.

Now since S_1 is finite there is a minimal subset of S_1 that is still a generating system for G. Note this can not be all of S_1 because by construction S_1 contains $0 \in G$ since $p \cdot 0 = 0$ and we adjoined all elements $h \in G$ such that $ph = 0$. But no generating system containing 0 can be minimal. Call this new subset S_0 and suppose that S_0, renumbering if necessary, is

$$S_0 = \{g_1, \ldots, g_r, h_1, \ldots, h_s\}.$$

We claim now that $g_1, \ldots, g_r, h_1, \ldots, h_s$ are independent and hence form a basis for G.

Suppose that

$$m_1 g_1 + \cdots + m_r g_r + n_1 h_1 + \cdots + n_s h_s = 0 \tag{2}$$

for some integers $m_1, \ldots, m_r, n_1, \ldots, n_s$. Each m_i and n_i must be divisible by p. Suppose for example that some m_i is not. Then $(m_i, p) = 1$ and then $(m_i, p^n) = 1$. This implies that there exists an x_i with $m_i x_i \equiv 1 \pmod{p^n}$. Multiplying through by x_i and rearranging we then obtain

$$g_i = -m_1 x_i g_1 - \cdots - n_s x_i h_s.$$

Therefore g_i can be expressed in terms of the remaining elements of S_0 contradicting the minimality of S_0. An identical argument works if some n_i is not divisible by p. Therefore the relation (2) takes the form

$$a_1pg_1 + \cdots + a_kpg_k + b_1ph_1 + \cdots + b_sph_s = 0. \tag{3}$$

Each of the terms $ph_i = 0$ so that (3) becomes

$$a_1pg_1 + \cdots + a_kpg_k = 0.$$

The $pg_1, \ldots, pg_k$ are independent and hence $a_i = 0$ for each i and hence $m_i = 0$ for all i. Now (2) becomes

$$n_1h_1 + \cdots + n_sh_s = 0.$$

However, we have seen that $p \mid n_i$ so each $n_i \equiv 0 \pmod{p}$. Since each h_i has order p this completes the claim. Therefore the whole group G has a basis proving the lemma by induction. $\qquad\square$

The proof of the existence part of the basis theorem for finite abelian groups (Theorem 8.2.1) now follows directly from the lemmas. Let G be a finite abelian group. From Lemma 8.2.3, G is a direct product of its p-primary components. Each p-primary component is a finite abelian p-group and from Lemma 8.2.5 is then a direct product of cyclic groups. Therefore the whole group G is a direct product of cyclic groups.

8.2.2 Free Abelian Groups

The basis theorem for finite abelian groups completely determined the structure of any finite abelian group. We now consider infinite but finitely generated abelian groups. Recall from the previous section the concept of a **basis**. Let G be any finitely generated abelian group and let $\{g_1, \ldots, g_n\}$ be a set of generators for G. The generators $g_1, \ldots, g_n$ form a **basis** for G if

$$G = \langle g_1 \rangle \times \cdots \times \langle g_n \rangle,$$

that is, G is the direct product of the cyclic subgroups generated by the g_i. From the basis theorem for finite abelian groups it follows that any finite abelian group has a basis. The main result on finitely generated infinite abelian groups is that they also have bases. We next need the concept of a free abelian group.

Definition 8.2.1. *A **free abelian group of rank** n is a direct product of n copies of the infinite cylic group $\mathbb{Z}$, that is, $G = \mathbb{Z} \times \mathbb{Z} \times \cdots \times \mathbb{Z}$. If $g_1, \ldots, g_n$ are generators of the respective factors then $g_1, \ldots, g_n$ is a **free basis** for G. We usually denote a free abelian group of rank n by $\mathbb{Z}^n$.*

We note that there are free abelian groups of infinite rank but we will not consider them here.

In Chapter 13 we will see that the dimension of a vector space is unique. Using this fact we will be able to prove the following lemma.

Lemma 8.2.6. *A free abelian group of rank n is isomorphic to a free abelian group of rank m if and only if $n = m$. It follows that the rank of a free abelian group is unique.*

The following is analogous to a result in linear algebra on bases for vector spaces. We will see this in Chapter 13 also and at that time prove the lemma.

Lemma 8.2.7. *Let F be free abelian of rank n with free basis $g_1, \ldots, g_n$. Then every $g \in G$ can be expressed uniquely as a product $g = g_1^{m_1} g_2^{m_2} \cdots g_n^{m_n}$ with each m_i an integer.*

We say that a group G is **torsion-free** if every element except the identity has infinite order. Let G be an abelian group which is torsion-free and finitely generated. If G has a basis $g_1, \ldots, g_n$ then each g_i must have infinite order and hence generate an infinite cyclic group isomorphic to $\mathbb{Z}$. It follows that G must be free abelian. Hence we have the result.

Lemma 8.2.8. *A torsion-free abelian group with a basis must be free abelian.*

Our main result is that any torsion-free finitely generated abelian group must be free abelian. We need one more preliminary result. Let $\mathbb{Q}^*$ be the multiplicative group of the nonzero rationals.

Lemma 8.2.9. *Any torsion-free finitely generated subgroup of the additive group of the rationals $\mathbb{Q}$ is infinite cyclic.*

Proof. Let H be a finitely generated subgroup of the additive group of $\mathbb{Q}$. Suppose that $\{\frac{a_1}{b_1}, \ldots, \frac{a_n}{b_n}\}$ is a set of generators for H. Let $b = b_1 b_2 \cdots b_n$. Let $h \in H$ then $h = m_1 \frac{a_1}{b_1} + \cdots + m_n \frac{a_n}{b_n}$ for some integers $m_1, \ldots, m_n$. It follows that bh is then an integer. Consider the map $f : H \to \mathbb{Z}$ defined by $f(h) = bh$ where $\mathbb{Z}$ is considered as the additive group of the integers $\mathbb{Z}$ and is hence an infinite cyclic group. Since H is abelian, $b(h_1 + h_2) = bh_1 + bh_2$ so that this map is a homomorphism. It is also easy to see that f is one-to-one. (See the exercises.) Hence this map injects H as a subgroup of the infinite cyclic subgroup $\mathbb{Z}$. In other words, H is isomorphic to a subgroup of $\mathbb{Z}$. Since subgroups of infinite cyclic groups are also infinite cyclic (see Chapter 6) it follows that H is infinite cyclic. □

The above lemma shows that any finitely generated subgroup of the additive group $\mathbb{Q}$ is cyclic; the terminology for this is that $(\mathbb{Q},+)$ is **locally cyclic**. To prove the above lemma we used the map $f : H \to \mathbb{Z}$. Can you see how to prove the same thing, i.e., the rationals under addition form a locally cyclic group without using this map? (See the exercises.)

Theorem 8.2.4. *Let G be a finitely generated torsion-free abelian group. Then G has a basis. It follows that G is a free abelian group.*

Proof. Let G be a finitely generated torsion-free abelian group. If G has a basis, then it follows from Lemma 8.2.8 that G is a free abelian group. So we prove that G has a basis by induction on the number of generators for G. Suppose that G is nontrivial and has one generator so that $G = \langle g \rangle$ where $g \neq 1$. (Note if G is trivial we can take the empty set as its basis.) Since G is torsion-free it is then infinite cyclic and so $o(g) = \infty$ and therefore $\{g\}$ is a basis. Note here $\langle g \rangle \cong \mathbb{Z}$.

We now suppose that any torsion-free abelian group with $n-1$ generators or less has a basis and suppose that $G = \langle g_1, \ldots, g_n \rangle$. Note that we are assuming here that G cannot be generated by fewer than n elements. If it could be so generated it would already have a basis by our assumption. So in particular g_n does not lie in $\langle g_1, \ldots, g_{n-1} \rangle$. The subgroup generated by $g_1, \ldots, g_{n-1}$ has a basis by the inductive hypothesis so without loss of generality we may assume that $g_1, \ldots, g_{n-1}$ are a basis for the subgroup they generate. Let $\langle g_n \rangle$ be the cyclic subgroup generated by g_n and define the subset

$$\overline{\langle g_n \rangle} = \{g \in G : g^m \in \langle g_n \rangle \text{ for some nonzero integer } m\}.$$

It is straightfoward that $\overline{\langle g_n \rangle}$ is a subgroup of G. (See the exercises for a proof.) Since G is abelian this subgroup is normal in G, so we may consider the factor group $G/\overline{\langle g_n \rangle}$. Since $g_1, \ldots, g_n$ generate G and $g_n \in \overline{\langle g_n \rangle}$ it follows that the images of $g_1, \ldots, g_{n-1}$ generate $G/\overline{\langle g_n \rangle}$. Further, $G/\overline{\langle g_n \rangle}$ is also torsion-free. To see this, suppose that $\overline{h} = h\langle g_n \rangle$ has finite order in $G/\overline{\langle g_n \rangle}$. Then $\overline{h}^m$ must be trival in $G/\overline{\langle g_n \rangle}$ for some integer m. That means $\overline{h}^m = h^m \langle g_n \rangle = \langle g_n \rangle$. This in turn means that $h^m \in \langle g_n \rangle$. Therefore by definition of $\overline{\langle g_n \rangle}$, $h \in \overline{\langle g_n \rangle}$. Then $\overline{h}$ is the identity in the factor group, $G/\overline{\langle g_n \rangle}$. This shows that the only way an element of $G/\overline{\langle g_n \rangle}$ can have finite order is if the element itself is the identity and hence $G/\overline{\langle g_n \rangle}$ is torsion-free.

Therefore $G/\overline{\langle g_n \rangle}$ is a torsion-free abelian group generated by $n - 1$ elements so by the inductive hypothesis it has a basis. It follows that the original group G is the direct product of $\overline{\langle g_n \rangle}$ and a free abelian group of finite rank. To see this suppose that a basis for $G/\overline{\langle g_n \rangle}$ is $\overline{a_1}, \ldots, \overline{a_r}$. Let $f : G \to G/\overline{\langle g_n \rangle}$ be the canonical homomorphism and choose an $a_i \in f^{-1}(\{\overline{a_i}\})$ for $i = 1, \ldots, r$. We can certainly do this since f is surjective. We claim that

$$G = \langle a_1 \rangle \times \cdots \times \langle a_r \rangle \times \overline{\langle g_n \rangle}.$$

Let us first note that if this is true, then since each $a_i \in G$ for all $i = 1, \ldots, r$ and G is torsion-free each a_i has infinite order. So that each group $\langle a_i \rangle \cong \mathbb{Z}$ for all i. Thus $\langle a_1 \rangle \times \cdots \times \langle a_r \rangle$ is a free abelian group of finite rank.

To prove our claim we use Theorem 8.1.5. We first need that G is generated by $a_1, \ldots, a_r, \overline{\langle g_n \rangle}$. Let $g \in G - \{1\}$. Put $\overline{g} = f(g) = g\langle g_n \rangle$. Then since $G/\overline{\langle g_n \rangle}$ has as a basis $\overline{a_1}, \ldots, \overline{a_r}$. It must be that

$$\overline{g} = \prod_{i=1}^{r} \overline{a_i}^{n_i}.$$

But this implies that $g = a_1^{n_1} \cdots a_r^{n_r} h$ where $h \in \overline{\langle g_n \rangle}$. (See the exercises to supply the details here.) This shows that $a_1, \ldots, a_r, \overline{\langle g_n \rangle}$ generate G. The commutativity requirement in Theorem 8.1.5 is automatically satisfied because G is abelian. So we will be finished with the proof of the claim if we can show that every nontrivial element of G has a **unique** representation in terms of $a_1, \ldots, a_r, \overline{\langle g_n \rangle}$. So suppose that $g = a_1^{n_1} \cdots a_r^{n_r} h$ and $g = a_1^{m_1} \cdots a_r^{m_r} h_1$ where $h_1 \in \overline{\langle g_n \rangle}$. Using f to map $g \mapsto \overline{g}$ gives that

$$\overline{a_1}^{n_1} \cdots \overline{a_r}^{n_r} = \overline{a_1}^{m_1} \cdots \overline{a_r}^{m_r}.$$

But since $\overline{a_1}, \ldots, \overline{a_r}$ is a basis for $G/\overline{\langle g_n \rangle}$, we get $n_i = m_i$ for all $i = 1, \ldots, r$. Finally since $a_1^{n_1} \cdots a_r^{n_r} h = a_1^{m_1} \cdots a_r^{m_r} h_1$ and each n_i equals the correrponding m_i, we can cancel all a's from both sides of this equality to get that $h = h_1$. This completes the proof of the claim.

Since we have now shown that G is the direct product of $\overline{\langle g_n \rangle}$ and a free abelian group of finite rank, the proof of the theorem is completed if we can show that $\overline{\langle g_n \rangle}$ is infinite cyclic.

Suppose that $h \in \overline{\langle g_n \rangle}$. Then $h^m = g_n^k$ for some $m \neq 0$. Since the group G is torsion-free, the integers m and k are uniquely determined by h. This allows us to define a map φ from $\overline{\langle g_n \rangle}$ to the additive group of the rationals $\mathbb{Q}$, i.e. $\langle \mathbb{Q}, + \rangle$ by

$$\phi(h) = \frac{k}{m}$$

with k and m as defined above. If $h_1, h_2 \in \overline{\langle g_n \rangle}$ so that $h_1^{m_1} = g_n^{k_1}$ and $h_2^{m_2} = g_n^{k_2}$, then it is not hard to show that $(h_1 h_2)^{m_1 m_2} = g_n^{m_2 k_1 + m_1 k_2}$ and hence the map ϕ is a homomorphism from $\overline{\langle g_n \rangle}$ to the additive group of the rationals. (See the exercises to fill in the details.) Further, this map is one-to-one (see the exercises) and therefore embeds $\overline{\langle g_n \rangle}$ as a torsion-free subgroup of the additive group of the rationals. We also note that $G/(\langle a_1 \rangle \times \cdots \times \langle a_r \rangle) \cong \overline{\langle g_n \rangle}$. This follows from Theorem 8.1.3 and Problem 8.1.10. But a quotient group of a finitely generated group is itself finitely generated or put another way a homomorphic image of a finitely generated group is finitely generated. (See the exercises for a proof of this.) Thus since G is finitely generated so is $G/(\langle a_1 \rangle \times \cdots \times \langle a_r \rangle)$. This in turn implies that $\overline{\langle g_n \rangle}$ is also finitely generated since these groups are isomorphic. From the previous Lemma 8.2.9, $\overline{\langle g_n \rangle}$ is then an infinite cyclic group, completing the proof of Theorem 8.2.4. $\qquad \square$

We close this section by showing that subgroups of free abelian groups of finite rank n are still free abelian and of rank $k \leq n$.

Theorem 8.2.5. *Let F be a free abelian group of rank n and suppose that $H \subset F$ is a subgroup of F. Then H is also free abelian of rank k with $k \leq n$.*

Proof. Let F be free abelian of rank n and suppose that $H \subset F$ is a subgroup. The proof is by induction on the rank n. If $n = 1$ then F is an infinite cyclic group. From Theorem 6.4.3, we have that any nontrivial subgroup of an infinite cyclic group is again an infinite cyclic group and hence H is free abelian of rank ≤ 1. (By convention, we say that the trivial group has rank 0.)

We suppose that the result is true for any free abelian group of rank $< n$. Let F be free abelian of rank n with a basis $g_1, \ldots, g_n$. Let H be a subgroup of F and let $F_n = \langle g_1, \ldots, g_{n-1} \rangle$. Then $H_n = H \cap F_n$ is a subgroup of a free abelian group of rank $< n$ and hence H_n is free abelian of rank $\leq n - 1$. By the second isomorphism theorem (see Chapter 7) we have that

$$H/H_n = H/(H \cap F_n) \cong (H \cdot F_n)/F_n \subset F/F_n \cong \mathbb{Z}.$$

It follows just as above from Theorem 6.4.3, that H/H_n is infinite cyclic as it is isomorphic to a subgroup of an infinite cyclic group. Then Problem 8.1.10 implies

$$H \cong H_n \times H/H_n.$$

Now H_n is free abelian of rank $\leq n - 1$ and H/H_n is free abelian of rank 1. Therefore from above, we have that H is free abelian of rank $\leq n$. □

8.2.3   The Basis Theorem for F.G. Abelian Groups

In this section we show that any finitely generated abelian group has a basis. In any group G, an element g of finite order is called a **torsion element**. We have already defined a group to be torsion-free if it has no nontrivial elements of finite order. We first show that any finitely generated abelian group can be decomposed as a direct product of a finite abelian group and a torsion-free abelian group.

Let G be an abelian group. We let $T(G)$ denote the set of torsion elements in G, that is, the set of elements in G that have finite order.

Lemma 8.2.10. *For any abelian group G the set $T(G)$ forms a subgroup called the **torsion subgroup** of G. Further, $G/T(G)$ is torsion-free.*

Proof. Let G be an abelian group and T(G) its set of elements of finite order. Notice first that the identity 1 always has finite order so $1 \in T(G)$ and hence $T(G) \neq \phi$.

Next suppose that $g_1, g_2 \in T(G)$. Since G is abelian and g_1, g_2 are of finite order it follows from Lemma 6.4.2 that $g_1 g_2$ also has finite order and hence $g_1 g_2 \in T(G)$. Further, $o(g_1^{-1}) = o(g_1)$ so $g_1^{-1} \in T(G)$. Therefore $T(G)$ is a subgroup of G.

Since G is abelian, $T(G)$ is then a normal subgroup and we can form the factor group $G/T(G)$. Suppose that $\bar{g} = gT(G) \in G/T(G)$. Suppose that $\overline{g}$ has finite order m. This implies that $\overline{g}^m = 1$ in $G/T(G)$ and hence $g^m \in T(G)$. This implies that g^m has finite order in G so there exists a natural number k with $(g^m)^k = g^{mk} = 1$. Hence g has finite order in G and so $g \in T(G)$ which implies that $\overline{g} = gT(G)$ is the identity in $G/T(G)$. Therefore the only element of finite order in the factor group $G/T(G)$ is the identity and hence $G/T(G)$ is torsion-free. $\qquad\qquad\qquad\qquad\square$

If G is an abelian group and A is a subgroup of G then we say that A has **complement** B in G if $G = A \times B$.

Lemma 8.2.11. *If G is a finitely generated abelian group then the torsion subgroup has a complement in G. Further, the complement is free abelian.*

Proof. Let G be a finitely generated abelian group and $T(G)$ its torsion subgroup. From Lemma 8.2.10, it follows that $G/T(G)$ is then a torsion-free finitely generated abelian group (see the exercises for the fact that it is finitely generated). From Theorem 8.2.4, we have that $G/T(G)$ is finitely generated free abelian. Let $\{\overline{g_1}, \ldots, \overline{g_n}\}$ be a basis for $G/T(G)$ and let $\{g_1, \ldots, g_n\}$ be a set of preimages of these in G. That is, if $f : G \to G/T(G)$ is the canonical homomorphism (see Chapter 7) then we choose an element in each $f^{-1}(\{\overline{g_i}\})$ and call it g_i for all $i = 1, \ldots, n$. Let

$$G' = \langle g_1, \ldots, g_n \rangle.$$

We claim that $G = T(G) \times G'$, i.e., G' is a complement of $T(G)$, and that G' is free abelian.

Let $g \in G$ and $f(g) = \overline{g} = gT(G)$ be its image in $G/T(G)$. Since $\{\overline{g_1}, \ldots, \overline{g_n}\}$ is a basis for $G/T(G)$ we have that $\overline{g} = \overline{g_1}^{m_1} \cdots \overline{g_n}^{m_n}$ for some integers $m_1, \ldots, m_n$. It follows that $\overline{g} = g_1^{m_1} \cdots g_n^{m_n} h$ for some $h \in T(G)$. Hence $g_1, \ldots, g_n$ together with $T(G)$ generate G. Since G is abelian each of $\langle g_1, \ldots, g_n \rangle$ and $T(G)$ are normal subgroups and therefore to show that $G = T(G) \times G'$ all that is required is to show that $G' \cap T(G) = \{1\}$.

Let $g \in G' \cap T(G)$. Then since $g \in T(G)$, $\overline{g} = gT(G)$ is the identity in $G/T(G)$. Since $\{\overline{g_1}, \ldots, \overline{g_n}\}$ is a basis for $G/T(G)$ the element $\overline{g}$ can be expressed uniquely in terms of these. If $\overline{g} = \overline{g_1}^{m_1} \cdots \overline{g_n}^{m_n}$ this would imply that $m_1 = m_2 = \ldots = m_n = 0$. But then if $g \in G'$, then $g = g_1^{k_1} \cdots g_n^{k_n}$ since G' is generated by $g_1, \ldots, g_n$. Applying f to g gives $\overline{g} = \overline{g_1}^{k_1} \cdots \overline{g_n}^{k_n}$. Thus each $k_i = m_i = 0$ and so $g = 1$. It follows that $G' \cap T(G) = \{1\}$ and from this we have $G = T(G) \times G'$.

We claim that the above argument also shows the subgroup G' is torsion-free. For suppose that some nontrivial element $g \in G'$ is such that $g^m = 1$ for some nonnegative m. We need to show that $m = 0$. But $g \in G'$ implies that $g = g_1^{k_1} \cdots g_n^{k_n}$ which in turn implies that $\overline{g} = \overline{g_1}^{k_1} \cdots \overline{g_n}^{k_n}$. Now if $g^m = 1$, then

$$1 = \prod_i \overline{g_i}^{mk_i}.$$

Since $\{\overline{g_1}, \ldots, \overline{g_n}\}$ is a basis, this implies that $mk_i = 0$ for all $i = 1, \ldots, n$. But since g is itself nontrivial, there must exist at least one k_j such that $k_j \neq 0$, then since $k_j m = 0$, we get that $m = 0$. Thus G' is a finitely generated torsion free abelian group, so by Theorem 8.2.4 it is free abelian. $\square$

We note here that being finitely generated is crucial. There exist infinitely generated abelian groups G where $T(G)$ does not have a complement.

Lemma 8.2.12. *Let G be a finitely generated abelian group. Then the torsion-subgroup $T(G)$ is a finite group.*

Proof. From the previous lemma since G is finitely generated it follows that $T(G)$ has a complement in G, that is, $G = T(G) \times G'$. By Theorem 8.1.3, $G/G' \cong T(G)$. Note that G/G' is finitely generated since G is (see the exercises). Thus $T(G)$ is also finitely generated. Suppose that $\{g_1, \ldots, g_n\}$ is a set of generators for $T(G)$. Since each $g_i \in T(G)$, each has finite order say $m_i = o(g_i)$. If $g \in T(G)$ it follows that $g = g_1^{k_1} \cdots g_n^{k_n}$ for some integers $k_1, \ldots, k_n$ since $g_1, \ldots, g_n$ generate $T(G)$. However, since the order of g_i is m_i it follows that there are m_i choices $0, 1, 2, \ldots, m_i - 1$ for each k_i and therefore at most $m_1 m_2 \cdots m_n$ choices for g. It follows that the order of $T(G)$ is at most $m_1 m_2 \cdots m_n$ and hence $T(G)$ is a finite group. $\square$

We now get to the main result.

Theorem 8.2.6. *Let G be a finitely generated abelian group. Then G has a basis. It follows that G is a direct product of cyclic groups of prime power order and a free abelian group of rank n. Further, G is completely determined by the rank n and the structure of its torsion subgroup.*

Proof. Let G be a finitely generated abelian group. From Lemma 8.2.11, we have that $G = T(G) \times G'$ where $T(G)$ is a finite abelian group from Lemma 8.2.12 and G' is free abelian of finite rank, say n. G' clearly has a basis $\{g_1, \ldots, g_n\}$. From the basis theorem for finite abelian groups (Theorem 8.2.1) $T(G)$ has a basis, say $\{h_1, \ldots, h_m\}$, where each h_i generates a cyclic group of prime power order. The set

$$\{h_1, \ldots, h_m, g_1, \ldots, g_n\}$$

determines a basis for G. If $G = T(G) \times G''$ then it follows that G'' is free abelian from Lemma 8.2.11. But Theorem 8.1.3 implies that $G/T(G) \cong G'$ and $G/T(G) \cong G''$. Then since isomorphism is an equivalence relation (see Chapter 2), we have that $G'' \cong G'$. Finally Lemma 8.2.6 implies then that G'' must have the same rank, n, as G'. This completes the proof. $\square$

We remark first that in the above proof the fact that $G'' \cong G'$ can also be established using the cancellation theorem 8.2.3 since $T(G)$ is a finite group. From applications in topology the rank n of the free abelian part of a finitely generated abelian group, G, is called the **Betti number** of G.

8.3 Exercises

EXERCISES FOR SECTION 8.1

8.1.1. Prove that if G and H are groups both with operation denoted by multiplication or juxtaposition, then $G \times H$ is also a group with the operation defined componentwise.

8.1.2. If G_1 and G_2 are groups, prove that $G_1 \times G_2 \cong G_2 \times G_1$.

8.1.3. Let $\mathbb{Z}_2 = \{0, 1\}$ with the usual operation of addition mod 2 and let S_3 be the symmetric group on $\{1, 2, 3\}$.
(a) Write the multiplication table out for $\mathbb{Z}_2 \times S_3$.
(Note: This group has order 12.)
(b) Is this group, $\mathbb{Z}_2 \times S_3$, abelian or not? Why?
(c) Find a subgroup of $\mathbb{Z}_2 \times S_3$ isomorphic to $\mathbb{Z}_2$ and one isomorphic to S_3.

8.1.4. Find all proper subgroups of $\mathbb{Z}_2 \times \mathbb{Z}_2$.

(HINT: Write out the Cayley table for this group.)

8.1.5. Let $G = G_1 \times G_2$. Consider the map $\pi_1 : G \to G_1$ defined by $\pi_1(a, b) = a$. (We can also define $\pi_2(a, b) = b$.) This is usually called the projection onto the first component.

(a) Prove that π_1 is an onto (surjection) homomorphism. (Of course, π_2 is also an onto homomorphism by the same kind of argument.)

(b) Let

$$H_1 = \{(a, 1) : a \in G_1 \text{ and } 1 \in G_2\}$$

and

$$H_2 = \{(1, b) : 1 \in G_1 \text{ and } b \in G_2\}.$$

Show that $\ker(\pi_1) = H_2$ and so $G/H_2 \cong G_1$.

(c) Show that $H_1 \cong G_1$ and $H_2 \cong G_2$.

8.1.6. Prove Lemma 8.1.3.

8.1.7. Verify the first two statements in Theorem 8.1.7; i.e.,

(a) Each H_i is a subgroup of G

(b) The map $a_i \mapsto (1, \ldots, 1, a_i, 1, \ldots, 1)$ is an isomorphism of G_i onto H_i.

*8.1.8. Let $H_1, H_2 \lhd G$ such that the canonical homomorphism $G \to G/H_2$ when restricted to H_1 gives an isomorphism of H_1 onto G/H_2. Then prove $G = H_1 \times H_2$ (internal).

8.1.9. Let A be a finitely generated abelian group, in particular let $A = \langle a_1, \ldots, a_n \rangle$. Prove $A = \langle a_1 \rangle \times \cdots \times \langle a_n \rangle$ if and only if $\prod_{i=1}^{n} a_i^{e_i} = 1$ always implies that $a_i^{e_i} = 1$ for all $i = 1, 2, \ldots, n$.

8.1.10. Let G be an abelian group and H a subgroup of G such that G/H is an infinite cyclic group. Then prove that $G \cong H \times G/H$.

(HINT: Use exercise 8.1.8.)

8.1.11. Let $G = G_1 \times \cdots \times G_n$. Here we prove that extension of Problem 8.1.5(b) to this direct product which we will have use for in a later section of this chapter. Let $H_i = \{(1, \ldots, h_i, \ldots, 1) : h_i \in G_i\}$. This means that there is an identity element in every component except possibly in the i-th component. Here $i = 1, \ldots, n$.

(a) Prove $H_i \cong G_i$ and $H_i \lhd G$ for every i.

(b) $G/G_i \cong G_1 \times \cdots \times G_{i-1} \times G_{i+1} \times \cdots \times G_n$. Here we are identifying G_i with H_i.

(HINT: For (b), although this can be proven directly using the group isomorphism theorem as in 5(b), it is probably easier to use the result of 5(b) in the following way. $G \cong (G_1 \times \cdots \times G_{i-1} \times G_{i+1} \times \cdots \times G_n) \times G_i$. Now put $H = G_1 \times \cdots \times G_{i-1} \times G_{i+1} \times \cdots \times G_n$. So we get $G \cong H \times G_i$. Identifying G with $H \times G_i$, we use the result of 5(b) to show that $G/G_i \cong H$. This gives the required result. We also note that this argument gives that $G/(G_1 \times \cdots \times G_{i-1} \times G_{i+1} \times \cdots \times G_n) \cong G_i$ by using the corresponding result to 5(b) proved in the text.)

EXERCISES FOR SECTION 8.2

8.2.1. Prove that if $m_1, \ldots, m_k$ are pairwise relatively prime positive integers (i.e., $\mathrm{GCD}(m_i, m_j) = 1$ for $i \neq j$), then $\mathbb{Z}_{m_1} \times \mathbb{Z}_{m_2} \times \cdots \times \mathbb{Z}_{m_k}$ where $k > 2$ is cyclic.

(HINT: Use Lemma 8.2.1 and induction on k.)

8.2.2. Explain why the largest order of any element in $\mathbb{Z}_6 \times \mathbb{Z}_{10}$ is 30 even though the group has order 60.

8.2.3. Find all abelian groups up to isomorphism (i.e., determine all possible isomorphism classes) and write these out as direct products of cyclic groups for the following orders:

(a) 720

(b) 1089

8.2.4. Determine how many isomorphism classes of abelian groups of each order there are

(a) 24

(b) 25

(c) (24)(25)

(d) 10^5

8.2.5. In each case decide whether or not the given groups are isomorphic or not. You must give a reason why they are or why they are not isomorphic.

(a) $\mathbb{Z}_{60}$ and $\mathbb{Z}_3 \times \mathbb{Z}_4 \times \mathbb{Z}_5$.

(b) $\mathbb{Z}_{60}$ and $\mathbb{Z}_6 \times \mathbb{Z}_{10}$.

(c) $\mathbb{Z}_2 \times \mathbb{Z}_2 \times \mathbb{Z}_{15}$ and $\mathbb{Z}_4 \times \mathbb{Z}_{15}$.

8.2.6. Prove that if G is an abelian group of order n and m is any positive divisor of n, then G has a subgroup $H \subset G$ such that $|H| = m$.

What theorem is this almost a converse to? Do you think that this holds for an arbitrary finite group?

(HINT: Use the fundamental theorem for finite abelian groups applied to G.)

8.2.7. Give an example to show that a finite abelian group may not be the internal direct product of two of its proper subgroups.

8.2.8. Consider the following subgroups of S_3:

$$N = \left\{ \begin{pmatrix} 1 & 2 & 3 \\ 1 & 2 & 3 \end{pmatrix}, \begin{pmatrix} 1 & 2 & 3 \\ 2 & 3 & 1 \end{pmatrix}, \begin{pmatrix} 1 & 2 & 3 \\ 3 & 1 & 2 \end{pmatrix} \right\}$$

and

$$K = \left\{ \begin{pmatrix} 1 & 2 & 3 \\ 1 & 2 & 3 \end{pmatrix}, \begin{pmatrix} 1 & 2 & 3 \\ 1 & 3 & 2 \end{pmatrix} \right\}.$$

Show that S_3 is not the internal direct product of these.

8.2.9. Use the fundamental theorem for finite abelian groups (or the basis theorem) to show that if n is a **square free** integer, i.e., n is not divisible by the square of any prime, then every abelian group of order n is cyclic.

8.2.10. Let G be a finite abelian group with the property that it contains a subgroup $H_0 \neq \{1\}$ which lies in every subgroup $H \neq \{1\}$
(a) Use Lemma 8.2.4 to determine the form of $|G|$.
(b) Use induction to show that G must be cyclic.

8.2.11. Find the order of the torsion subgroup, $T(G)$, where G is each of the following groups:
(a) $G = \mathbb{Z} \times \mathbb{Z}_3 \times \mathbb{Z}_4$;
(b) $G = \mathbb{Z} \times \mathbb{Z}_8 \times \mathbb{Z} \times \mathbb{Z} \times \mathbb{Z}_3$;
(c) $G = \mathbb{Z}_{12} \times \mathbb{Z} \times \mathbb{Z}_{12}$.

8.2.12. Find the Betti number of the group

$$\mathbb{Z} \times \mathbb{Z}_6 \times \mathbb{Z} \times \mathbb{Z} \times \mathbb{Z}_{12} \times \mathbb{Z}_{10} \times \mathbb{Z}.$$

8.2.13. (Some knowledge of the complex numbers, $\mathbb{C}$ (see Chapter 5), is needed here.) Find the torsion subgroup, $T(\mathbb{C}^*)$, of the multiplicative group $\mathbb{C}^*$ of nonzero complex numbers.
(HINT: Use DeMoivre's theorem.)

8.2.14. (This problem gives an alternate proof to the one in the text for the fact that the rationals $\mathbb{Q}$ under addition are locally cyclic, i.e., any finitely generated subgroup is cyclic — taken from [MH].) Let $H \subset \mathbb{Q}$ be a subgroup and let $H = \langle \frac{a_1}{b_1}, \ldots, \frac{a_n}{b_n} \rangle$ where it is assumed that all fractions are in lowest terms, i.e., $\mathrm{GCD}(a_i, b_i) = 1$ for all $i = 1, \ldots, n$. Consider any element $h \in H$. It is of the form

$$h = c_1 \frac{a_1}{b_1} + \cdots + c_n \frac{a_n}{b_n}$$

for some integers $c_1, \ldots, c_n$. (Why?) But this can be written as

$$h = \frac{c_1 a_1 b_2 \cdots b_n + c_2 a_2 b_1 b_3 \cdots b_n + \cdots + c_n a_n b_1 b_2 \cdots b_{n-1}}{b_1 b_2 \cdots b_n}.$$

WHY? Prove that the set

$$H_1 = \left\{ \begin{array}{c} c_1 a_1 b_2 \cdots b_n + c_2 a_2 b_1 b_3 \cdots b_n + \cdots + c_n a_n b_1 b_2 \cdots b_{n-1} : c_i \\ \text{are arbitrary integers for } 1 \leq i \leq n \end{array} \right\}$$

is a subgroup of $\mathbb{Z}$. But since $\mathbb{Z}$ is a cyclic group, so is H_1. (Why?) Let $H_1 = \langle d \rangle$. But then any element of H is of the form $x \cdot \frac{d}{b_1 b_2 \cdots b_n}$ where $x \in \mathbb{Z}$. (Why?) Thus we are done. (Why?)

The next five problems deal with free abelian groups. Here we show that any finitely generated abelian group is the homomorphic image of a free abelian group.

8.2.15. Let $G = G_1 \times G_2$ be an abelian group and suppose that H is any abelian group. Suppose that there exist homomorphisms θ and ϕ of G_1 into H and G_2 into H, respectively. Prove there exists a homomorphism $\gamma : G \to H$ such that $\gamma \mid_{G_1} = \theta$ and $\gamma \mid_{G_2} = \phi$.

(HINT: For any $g \in G$ we have $g = g_1 g_2$ uniquely where $g_i \in G_i$ for $i = 1, 2$. (Why?) Define $\gamma(g) = \theta(g_1)\phi(g_2)$. Explain why this is a valid map from G into H and prove it is a homomorphism with the required properties.)

8.2.16. Now let $G = G_1 \times \cdots \times G_n$ be an abelian group and for each i there exists a homomorphism $\theta_i : G_i \to H$ where H is some abelian group. Prove that there exists a homomorphism $\theta : G \to H$ such that $\theta \mid_{G_i} = \theta_i$ for all $i = 1, \ldots, n$. We say that θ **extends** the mappings θ_i or that θ is an **extension** of the maps θ_i.

(HINT: Proceed just like in Problem 8.2.15.)

8.2.17. Let $J = \langle c \rangle$ be an infinite cyclic group and let H be any abelian group. Prove that if φ is a mapping from $\{c\}$ into H, then there exists a homomorphism $\theta : J \to H$ such that $\theta \mid_{\{c\}} = \varphi$.

(HINT: Define $\theta(c^n) = (\varphi(c))^n$ and prove that θ is a homomorphism from J into H.)

8.2.18. Now let $G = G_1 \times \cdots \times G_n$ where each G_i is infinite cyclic ($G_i \cong \mathbb{Z}$) and put $G_i = \langle c_i \rangle$ for all i. Thus G is a free abelian group of rank n. Let $X = \{c_1, \ldots, c_n\}$ be a basis for G. If there is a mapping $\theta : X \to H$ where H is any abelian group, then prove that there is a homomorphism $\theta^* : G \to H$ such that $\theta^* \mid_X = \theta$.

(HINT: Use Problem 8.2.17 to get a map from each G_i into H and then use Problem 8.2.16 to get the homomorphism θ^*.)

8.2.19. Prove that every finitely generated abelian group is the homomorphic image of some free abelian group.

(HINT: If G is any finitely generated abelian group and $G = \langle g_1, \ldots, g_n \rangle$, then let $G_i = \langle x_i \rangle$ be an infinite cyclic group for all $i = 1, \ldots, n$. Then define the map $\theta : x_i \mapsto g_i$. Then the group $F = G_1 \times \cdots \times G_n$ is a free abelian group of rank n. (Why?) Then use problem 8.2.18 to extend this θ to θ^*. Also you must prove that θ^* is a surjection.)

We note that actually more is true. Every abelian group is the homomorphic image of some free abelian group. But to prove this would require direct products with infinitely many factors which we have not discussed.

8.2.20. Let G be a finitely generated torsion-free abelian group with $G = \langle g_1, \ldots, g_n \rangle$. Let $\overline{\langle g_n \rangle} = \{g \in G : g^m \in \langle g_n \rangle$ for some nonzero integer $m\}$. Prove that $\overline{\langle g_n \rangle}$ is a subgroup of G.

8.2.21. Suppose that G and $\overline{\langle g_n \rangle}$ are as in Problem 8.2.20.
(a) Why is $\overline{\langle g_n \rangle} \triangleleft G$?
(b) $G/\overline{\langle g_n \rangle}$ has as a basis $\overline{a_1}, \ldots, \overline{a_r}$. Thus each $\overline{a_i} = a_i \overline{\langle g_n \rangle}$. Why?
(c) It must be that if $\overline{g}$ is any element of $G/\overline{\langle g_n \rangle}$ then $\overline{g} = \prod_{i=1}^{r} \overline{a_i}^{n_i}$. Show this implies that $g = a_1^{n_1} \cdots a_r^{n_r} h$ where $h \in \overline{\langle g_n \rangle}$.

8.2.22. Again suppose that G and $\overline{\langle g_n \rangle}$ are as in Problem 8.2.20. Let $h \in \overline{\langle g_n \rangle}$. Then $h^m = g_n^k$ for some $m \neq 0$. Define a map φ from $\overline{\langle g_n \rangle}$ to the additive group of the rationals $\mathbb{Q}$, i.e., $\langle \mathbb{Q}, + \rangle$ by

$$\phi(h) = \frac{k}{m}.$$

(a) Use the fact that G is torsion-free to show that the integers m and k are uniquely determined by h so that this makes sense as a map.
(b) Show that if $h_1, h_2 \in \overline{\langle g_n \rangle}$ so that $h_1^{m_1} = g_n^{k_1}$ and $h_2^{m_2} = g_n^{k_2}$, then $(h_1 h_2)^{m_1 m_2} = g_n^{m_2 k_1 + m_1 k_2}$. Use this to show that the map defined above, ϕ, is a homomorphism from $\overline{\langle g_n \rangle}$ into $\mathbb{Q}$ under addition.
(c) Finally show that ϕ is injective.

8.2.23. Let G and H be groups. Also $f : G \to H$ is a surjective homomorphism. Show that if G is finitely generated, then so is H. In other words, the homomorphic image of a finitely generated group is itself finitely generated.

(HINT: Let $G = \langle g_1, \ldots, g_n \rangle$. Show that $H = \langle f(g_1), \ldots, f(g_n) \rangle$.)

8.2.24. Let G be a group and $N \lhd G$. If G is finitely generated, then so is G/N.

(HINT: Use Problem 8.2.23)

8.2.25. In the last theorem in Section 8.2.2 use the cancellation theorem for direct products, Theorem 8.2.3, to show that $G' \cong G''$ rather than the argument given in the text.

Chapter 9

Symmetric and Alternating Groups

9.1 Symmetric Groups and Cycle Structure

Recall that if A is a nonempty set, the set S_A of all permutations (bijections from A to A) on A forms a group under composition called the **symmetric group** on A. If $|A| > 2$ then S_A is nonabelian. Further, if A, B have the same cardinality or size, then $S_A \cong S_B$ (see Theorem 6.2.1). In general a **permutation group** is any subgroup of S_A for a set A.

For this section we will only consider finite symmetric groups S_n and always consider the set A as $A = \{1, 2, 3, \ldots, n\}$. There is no loss of generality in doing this because from the above comments, if B is any other set with $|B| = n$ then $S_B \cong S_A$. Also the reader should keep in mind that the binary operation here is always composition of functions.

Definition 9.1.1. *Suppose that f is a permutation of $A = \{1, 2, \ldots, n\}$, which has the following effect on the elements of A: There exists an element $a_1 \in A$ such that $f(a_1) = a_2$, $f(a_2) = a_3, \ldots$, $f(a_{k-1}) = a_k$, $f(a_k) = a_1$, and f leaves all other elements (if there are any) of A fixed, i.e., $f(a_j) = a_j$ for $a_j \neq a_i$, $i = 1, 2, \ldots, k$. Such a permutation f is called a **cycle** or a **k-cycle**.*

We use the following notation for a k-cycle, f, as given above

$$f = (a_1, a_2, \ldots, a_k).$$

The cycle notation is read from left to right, it says f takes a_1 into a_2, a_2 into a_3, etc., and finally a_k, the last symbol, into a_1, the first symbol.

Moreover, f leaves all the other elements not appearing in the representation above fixed.

Note that one can write the same cycle in many ways using this type of notation; e.g., $f = (a_2, a_3, \ldots, a_k, a_1)$. In fact any cyclic rearrangement of the symbols gives the same cycle. The integer k is the **length** of the cycle. Note that we allow a cycle to have length 1, i.e., $f = (a_1)$, for instance, this is just the identity map. For this reason, we will usually designate the identity of S_n by (1) or just 1. (Of course, it also could be written as (a_i) where $a_i \in A$.)

If f and g are two cycles, they are called **disjoint cycles** if the elements moved by one are left fixed by the other, that is, their representations (see the display above) contain different elements of the set A (their representations are disjoint as sets).

Lemma 9.1.1. *If f and g are disjoint cycles, then they must commute, that is, $fg = gf$.*

Proof. Since the cycles f and g are disjoint, each element moved by f is fixed by g and vice versa. First suppose $f(a_i) \neq a_i$. This implies that $g(a_i) = a_i$ and $f^2(a_i) \neq f(a_i)$. Can you see why? But since $f^2(a_i) \neq f(a_i)$, $g(f(a_i)) = f(a_i)$. Now $(fg)(a_i) = f(g(a_i)) = f(a_i)$. Thus if $f(a_i) \neq a_i$, we have $(gf)(a_i) = (fg)(a_i)$. Similarly if $g(a_j) \neq a_j$, then $(fg)(a_j) = (gf)(a_j)$. Finally, if $f(a_k) = a_k$ and $g(a_k) = a_k$ then clearly $(fg)(a_k) = a_k = (gf)(a_k)$. Thus $gf = fg$. $\qquad\square$

Before proceeding further with the theory, let us consider a specific example. Let $A = \{1, 2, \ldots, 8\} \in S_8$ and let

$$f = \begin{pmatrix} 1 & 2 & 3 & 4 & 5 & 6 & 7 & 8 \\ 2 & 4 & 6 & 5 & 1 & 7 & 3 & 8 \end{pmatrix} \in S_8.$$

We pick an arbitrary number from the set A, say 1. Then $f(1) = 2$, $f(2) = 4$, $f(4) = 5$, $f(5) = 1$. Now select an element from A not in the set $\{1, 2, 4, 5\}$, say 3. Then $f(3) = 6$, $f(6) = 7$, $f(7) = 3$. Next select any element of A not occurring in the set $\{1, 2, 4, 5\} \cup \{3, 6, 7\}$. The only element left is 8, and $f(8) = 8$. It is clear that we can now write the permutation f as a product of cycles:

$$f = (1, 2, 4, 5)(3, 6, 7)(8)$$

where the order of the cycles is immaterial since they are disjoint and there-fore commute. It is customary to omit such cycles as (8) and write f simply as

$$f = (1, 2, 4, 5)(3, 6, 7)$$

with the understanding that the elements of A not appearing are left fixed by f.

It is not difficult to generalize what was done here for a specific example, and show that any permutation f can be written uniquely, except for order, as a product of disjoint cycles. Thus let f be a nonidentity permutation on the set $A = \{1, 2, \ldots, n\}$, and let a_1 be the smallest element of A such that $f(a_1) \neq a_1$. Let $f(a_1) = a_2$, $f^2(a_1) = f(a_2) = a_3$, etc., and continue until a repetition is obtained. We claim that this first occurs for a_1, that is, the first repetition is say $f^k(a_1) = f(a_k) = a_{k+1} = a_1$. For suppose the first repetition occurs at the kth iterate of f and

$$f^k(a_1) = f(a_k) = a_{k+1},$$

and $a_{k+1} = a_j$, where $j < k$ since the first repetition occurs at f^k. Then

$$f^k(a_1) = f^{j-1}(a_1),$$

and so $f^{k-j+1}(a_1) = a_1$. However, $k - j + 1 < k$ if $j \neq 1$, and we assumed that the first repetition occurred for k. Thus, $j = 1$ and so f does cyclically permute the set $\{a_1, a_2, \ldots, a_k\}$. If $k < n$, then there exists $b_1 \in A$ such that $b_1 \notin \{a_1, a_2, \ldots, a_k\}$ and we may proceed similarly with b_1. We continue in this manner until all the elements of A are accounted for. It is then seen that f can be written in the form

$$f = (a_1, \ldots, a_k)(b_1, \ldots, b_\ell)(c_1, \ldots, c_m) \cdots (h_1, \ldots, h_t).$$

Note that all powers $f^i(a_1)$ belong to the set $\{a_1 = f^0(a_1) = f^k(a_1), a_2 = f^1(a_1), \ldots, a_k = f^{k-1}(a_1)\}$, all powers $f^i(b_1)$ belong to the set $\{b_1 = f^0(b_1) = f^\ell(b_1), b_2 = f^1(b_1), \ldots, b_\ell = f^{\ell-1}(b_1)\}$, Here, we define b_1 as the smallest element in $\{1, 2, \ldots, n\}$ which does not belong to $\{a_1 = f^0(a_1) = f^k(a_1), a_2 = f^1(a_1), \ldots, a_k = f^{k-1}(a_1)\}$, and c_1 is the smallest element in $\{1, 2, \ldots, n\}$ which does not belong to

$$\{a_1 = f^0(a_1) = f^k(a_1), a_2 = f^1(a_1), \ldots, a_k = f^{k-1}(a_1)\}$$

$$\cup \{b_1 = f^0(b_1) = f^\ell(b_1), b_2 = f^1(b_1), \ldots, b_\ell = f^{\ell-1}(b_1)\} \cup \cdots \text{ etc.}.$$

Therefore by construction, all the cycles are disjoint. To see that the cycles are all disjoint it is sufficient to show that $(a_1, \ldots, a_k)$ and $(b_1, \ldots, b_\ell)$ are disjoint. Any two others can be shown in the same manner. Suppose $f^t(a_1) = f^s(b_1)$. Then $f^{t-s}(a_1) = b_1$ which can't be true by the definition of b_1. From this it follows that $k + \ell + m + \cdots + t = n$ assuming we have written all cycles including the identity 1-cycles for elements which f does not move. It is clear that this factorization is unique except for the order of the factors since it tells explicitly what effect f has on each element of A.

In summary we have proven the following result.

Theorem 9.1.1. *Every permutation of S_n can be written uniquely as a product of disjoint cycles (up to order).*

If $(a_1, a_2, \ldots, a_k)$ is a k-cycle it is easy to see that $(a_1, a_2, \ldots, a_k)^{-1} = (a_k, a_{k-1}, \ldots, a_1)$ (see the exercises). The following is really a corollary of the proof of Theorem 9.1.1.

Corollary 9.1.1. *Let $(a_1, a_2, \ldots, a_k)$ be a k-cycle in S_n. Then the order of $(a_1, a_2, \ldots, a_k)$ is k. That is, $o((a_1, a_2, \ldots, a_k)) = k$.*

Proof. It is easy to see that $(a_1, a_2, \ldots, a_k)^k = 1$ (see the exercises). We note that $f = (a_1, \ldots, a_k)$ is a permutation on the set $A = \{1, 2, \ldots, n\}$. Then from the first part of the proof of Theorem 9.1.1 if we start taking powers of f we have $f^2(a_1) = a_3$ etc. and continue until a repetition is obtained then the first time this occurs is for a_1, i.e, $f^j(a_1) = f(a_j) = a_{j+1} = a_1$. Moreover since f is a cycle, the first time that occurs is for $j = k$. Also since f is a cycle once we get that $f^k(a_1) = a_1$ then it must be also true that $f^k(a_i) = a_i$ for all i. This shows that the smallest positive power of f which is the identity is k and hence $o(f) = k$. $\qquad\square$

Corollary 9.1.2. *Let $f \in S_n$. Suppose that the disjoint cycle representation of f is*

$$f = g_1 g_2 \cdots g_r$$

where each g_i is an n_i-cycle with $i = 1, 2, \ldots, r$. Then

$$o(f) = LCM(n_1, n_2, \ldots, n_r).$$

Proof. For the definition of the LCM for more than two nonzero integers see Problem 6.4.10. Let $m = LCM(n_1, n_2, \ldots, n_r)$. Here we need to show that $f^m = 1$, the identity permutation, and that m is the least positive integer with this property. By Lemma 9.1.1 all the g_i commute. Thus $f^m = g_1^m g_2^m \cdots g_r^m$. Our previous corollary shows that $o(g_i) = n_i$ for all $i = 1, 2, \ldots, r$. Hence $m = LCM(o(g_1), \ldots, o(g_r))$ and hence $f^m = 1$.

We claim that if $f^k = 1$ for some positive integer k then $n_i | k$ for all $i, 1 \leq i \leq r$ (see the exercises for a proof of this). However, if this is true then $m = \mathrm{LCM}(n_1, \ldots, n_r)$ also divides k. This implies that $m \leq k$ and proves that m is the least positive integer for which $f^k = 1$. $\square$

EXAMPLE 9.1.1. The elements of S_3 can be written in cycle notation as $1 = (1), (1, 2), (1, 3), (2, 3), (1, 2, 3), (1, 3, 2)$. This is the largest symmetric group which consists entirely of cycles.

In S_4, for example, the element $(1, 2)(3, 4)$ is not a cycle.

Suppose we multiply two elements of S_3 say $(1, 2)$ and $(1, 3)$. In forming the product or composition here, we read from right to left. Thus to compute $(1, 2)(1, 3)$: We note the permutation $(1, 3)$ takes 1 into 3 and then the permutation $(1, 2)$ takes 3 into 3 so the composite $(1, 2)(1, 3)$ takes 1 into 3. Continuing the permutation $(1, 3)$ takes 3 into 1 and then the permutation $(1, 2)$ takes 1 into 2, so the composite $(1, 2)(1, 3)$ takes 3 into 2. Finally $(1, 3)$ takes 2 into 2 and then $(1, 2)$ takes 2 into 1 so $(1, 2)(1, 3)$ takes 2 into 1. Thus we see that

$$(1, 2)(1, 3) = (1, 3, 2).$$

EXAMPLE 9.1.2. As another example of this **cycle multiplication** consider the product in S_5,

$$(1, 2)(2, 4, 5)(1, 3)(1, 2, 5).$$

Reading from right to left

$$1 \mapsto 2 \mapsto 2 \mapsto 4 \mapsto 4$$

so altogether

$$1 \mapsto 4.$$

Now

$$4 \mapsto 4 \mapsto 4 \mapsto 5 \mapsto 5$$

so altogether

$$4 \mapsto 5.$$

Next

$$5 \mapsto 1 \mapsto 3 \mapsto 3 \mapsto 3$$

so
$$5 \mapsto 3.$$

Then
$$3 \mapsto 3 \mapsto 1 \mapsto 1 \mapsto 2$$

so
$$3 \mapsto 2.$$

Finally
$$2 \mapsto 5 \mapsto 5 \mapsto 2 \mapsto 1$$

so
$$2 \mapsto 1.$$

Since all the elements of $A = \{1, 2, 3, 4, 5\}$ have been accounted for, we have

$$(1, 2)(2, 4, 5)(1, 3)(1, 2, 5) = (1, 4, 5, 3, 2).$$

Let $f \in S_n$. If f is a cycle of length 2, i.e., $f = (a_1, a_2)$ where $a_1, a_2 \in A$, then f is called a **transposition**. We have the following easy lemma.

Lemma 9.1.2. *Any cycle can be written as a product of transpositions.*

Proof. Consider the k-cycle $(a_1, \ldots, a_k)$. Then by straightforward computation we get

$$(a_1, \ldots, a_k) = (a_1, a_k)(a_1, a_{k-1}) \cdots (a_1, a_2).$$

This proves the lemma. $\qquad\qquad\qquad\qquad\qquad\qquad\qquad\qquad\qquad\square$

From Theorem 9.1.1 any permutation can be written in terms of cycles, but from the above any cycle can be written as a product of transpositions. Thus we have the following result.

Theorem 9.1.2. *Let $f \in S_n$ be any permutation of degree n. Then f can be written as a product of transpositions.*

9.1.1 The Alternating Groups

From the end of the previous section we can see that any k-cycle can be written as a product of transpositions as follows:

$$(a_1, \ldots, a_k) = (a_1, a_k)(a_1, a_{k-1}) \cdots (a_1, a_2).$$

Note that there are $k - 1$ transpositions in this product.

Hence if f is a permutation with a disjoint cycle decomposition

$$(a_1, \ldots, a_k)(b_1, \ldots, b_j) \cdots (m_1, \ldots, m_t)$$

then f can be written as a product of

$$W(f) = (k - 1) + (j - 1) + \cdots + (t - 1)$$

transpositions. The number $W(f)$ is uniquely associated with the permutation f since f is uniquely represented (up to order) as a product of disjoint cycles. However, there is nothing unique about the number of transpositions occurring in an arbitrary representation of f as a product of transpositions. For example, in S_3

$$(1, 3, 2) = (1, 2)(1, 3) = (1, 2)(1, 3)(1, 2)(1, 2),$$

since $(1, 2)(1, 2) = (1)$, the identity permutation of S_3.

Although the number of transpositions is not unique in the representation of a permutation, f, as a product of transpositions, we shall show, however, that the parity (even or oddness) of that number is unique. Moreover, this depends solely on the number $W(f)$ uniquely associated with the representation of f. More explicitly, we have the following result.

Theorem 9.1.3. *(Uniqueness of Parity) If f is a permutation written as a product of disjoint cycles and if $W(f)$ is the associated integer given above, then if $W(f)$ is even or odd any representation of f as a product of transpositions must contain respectively an even or odd number of transpositions.*

Proof. We first observe the following:

$$(a, b)(b, c_1, \ldots, c_t)(a, b_1, \ldots, b_k) = (a, b_1, \ldots, b_k, b, c_1, \ldots, c_t),$$

$$(a, b)(a, b_1, \ldots, b_k, b, c_1, \ldots, c_t) = (a, b_1, \ldots, b_k)(b, c_1, \ldots, c_t).$$

We leave the verification of these identities to the exercises. Suppose now that f is represented as a product of disjoint cycles, where we include all the 1-cycles of elements of A which f fixes, if any. Let a and b be any distinct elements moved by f. If a and b occur in the same cycle in this representation for f,

$$f = \cdots (a, b_1, \ldots, b_k, b, c_1, \ldots, c_t) \cdots$$

then in the computation of $W(f)$ this cycle contributes $k + t + 1$. Now consider $(a, b)f$. Since the cycles are disjoint and disjoint cycles commute,

$$(a, b)f = \cdots (a, b)(a, b_1, \ldots, b_k, b, c_1, \ldots, c_t)\ldots$$

since neither a nor b can occur in any factor of f other than $(a, b_1, \ldots, b_k, b, c_1, \ldots, c_t)$. Using the first of the above identities we see that (a, b) cancels out and we find that $(a, b)f = \ldots(b, c_1, \ldots, c_t)(a, b_1, \ldots, b_k)\ldots$. Since $W((b, c_1, \ldots, c_t)(a, b_1, \ldots, b_k)) = k + t$ but $W(a, b_1, \ldots, b_k, b, c_1, \ldots, c_t) = k + t + 1$, we have $W((a, b)f) = W(f) - 1$.

A similar analysis shows that in the case where a and b occur in different cycles in the representation of f, then $W((a, b)f) = W(f) + 1$. (See the exercises for verification.) Combining both cases, we have

$$W((a, b)f) = W(f) \pm 1.$$

Now let f be written as a product of m transpositions, say

$$f = (a_1, b_1)(a_2, b_2) \cdots (a_m, b_m).$$

Then

$$(a_m, b_m) \cdots (a_2, b_2)(a_1, b_1)f = 1.$$

Iterating this, together with the fact that $W(1) = 0$, shows that

$$W(f) + (\pm 1) + (\pm 1) + (\pm) \cdots + (\pm 1) = 0,$$

where there are m terms of the form ± 1. Thus

$$W(f) = (\mp 1) + (\mp 1) + \cdots + (\mp 1),$$

with m terms. Note that, if exactly p are $+1$ and $q = m - p$ are -1 then $m = p + q$ and $W(f) = p - q$. Hence $m \equiv W(f) \pmod 2$. Thus, $W(f)$ is even if and only if m is even and this completes the proof.   □

It now makes sense to state the following definition since we know that the parity is indeed unique.

Definition 9.1.2. *A permutation $f \in S_n$ is said to be* **even** *if it can be written as a product of an even number of transpositions. Similarly, f is called* **odd** *if it can be written as a product of an odd number of transpositions.*

Definition 9.1.3. *On the group S_n we define the* **sign function**

$$sgn : S_n \to (\mathbb{Z}_2, +)$$

by $sgn(\pi) = 0$ if π is an even permutation and $sgn(\pi) = 1$ if π is an odd permutation.

We note that if f and g are even permutations then so are fg and f^{-1} and also the identity permutation is even. Further, if f is even and g is odd it is clear that fg is odd. From this it is straightforward to establish the following.

Lemma 9.1.3. *The sign function sgn is a homomorphism from S_n onto $(\mathbb{Z}_2, +)$*

We now let

$$A_n = \{\pi \in S_n : \operatorname{sgn}(\pi) = 0\}.$$

That is, A_n is precisely the set of even permutations in S_n.

Theorem 9.1.4. *For each $n \in \mathbb{N}$ the set A_n forms a normal subgroup of index 2 in S_n called the* **alternating group** *on n symbols. Further, $|A_n| = \frac{n!}{2}$.*

Proof. From Lemma 9.1.3 the mapping sgn $: S_n \to (\mathbb{Z}_2, +)$ is a homomorphism. Then $\ker(\operatorname{sgn}) = A_n$ and therefore A_n is a normal subgroup of S_n from Theorem 7.3.1. Since $\operatorname{im}(\operatorname{sgn}) = \mathbb{Z}_2$ we have the size of the image is 2, that is, $|\operatorname{im}(\operatorname{sgn})| = 2$ and hence $|S_n/A_n| = 2$. Therefore $[S_n : A_n] = 2$. Since $|S_n| = n!$ then $|A_n| = \frac{n!}{2}$ follows from Lagrange's theorem. $\square$

9.1.2 Conjugation in S_n

Recall that in a group G two elements $x, y \in G$ are **conjugates** in G if there exists a $g \in G$ with $gxg^{-1} = y$. Conjugacy is an equivalence relation on G. In the symmetric groups S_n it is easy to determine if two elements are conjugates. We say that two permutations in S_n have the **same cycle structure** if, in their disjoint cycle decomposition they have the same number of cycles and the lengths are the same. Hence for example in S_8 the permutations

$$\pi_1 = (1,3,6,7)(2,5) \text{ and } \pi_2 = (2,3,5,6)(1,8)$$

have the same cycle structure.

We will prove that if π_1, π_2 are two permutations in S_n then π_1, π_2 are conjugates in S_n if and only if they have the same cycle structure. Therefore in S_8 the permutations

$$\pi_1 = (1,3,6,7)(2,5) \text{ and } \pi_2 = (2,3,5,6)(1,8)$$

are conjugates.

Lemma 9.1.4. *Let*

$$\pi = (a_{11}, a_{12}, \ldots, a_{1k_1}) \cdots (a_{s1}, a_{s2}, \ldots, a_{sk_s})$$

be the disjoint cycle decomposition of $\pi \in S_n$. Let $\tau \in S_n$ and denote by a_{ij}^τ the image of a_{ij} under τ, that is, $a_{ij}^\tau = \tau(a_{ij})$. Then

$$\tau\pi\tau^{-1} = (a_{11}^\tau, a_{12}^\tau, \ldots, a_{1k_1}^\tau)\ldots(a_{s1}^\tau, a_{s2}^\tau, \ldots, a_{sk_s}^\tau)$$

and this is the disjoint cycle decomposition of $\tau\pi\tau^{-1}$.

Proof. We first note that since τ is a bijection and the given decomposition of π is disjoint, then the above decomposition of $\tau\pi\tau^{-1}$ must also be disjoint. (See the exercises for verification.)

Now consider a_{11}. Then operating on the left as functions we have

$$\tau\pi\tau^{-1}(a_{11}^\tau) = \tau\pi(a_{11}) = \tau(a_{12}) = a_{12}^\tau.$$

The same computation then follows for all the symbols, a_{ij}, proving the lemma. □

Theorem 9.1.5. *Two permutations $\pi_1, \pi_2 \in S_n$ are conjugates in S_n if and only if they have the same cycle structure.*

Proof. Suppose that $\pi_2 = \tau \pi_1 \tau^{-1}$. Then from Lemma 9.1.4 we have that π_1 and π_2 have the same cycle structure.

Conversely suppose that π_1 and π_2 have the same cycle structure. Let

$$\pi_1 = (a_{11}, a_{12}, \ldots, a_{1k_1}) \cdots (a_{s1}, a_{s2}, \ldots, a_{sk_s})$$

$$\pi_2 = (b_{11}, b_{12}, \ldots, b_{1k_1}) \cdots (b_{s1}, b_{s2}, \ldots, b_{sk_s})$$

where we place the cycles of the same length under each other. Let τ be the permutation in S_n that maps each symbol in π_1 to the digit below it in π_2. Then from Lemma 9.1.4 we have $\tau \pi_1 \tau^{-1} = \pi_2$ and hence π_1 and π_2 are conjugate. $\square$

EXAMPLE 9.1.3. We can now easily prove a statement made in Example 7.2.3 of Chapter 7. There it was noted that to show the Klein 4-group $V_4 = \{f_1, f_2, f_3, f_4\}$ is a normal subgroup of S_4 was a tedious computation. Here though we note that $f_1 = (1), f_2 = (1,2)(3,4), f_3 = (1,3)(2,4), f_4 = (1,4)(2,3)$. Since these all have the same cycle structure and by Theorem 9.1.5 conjugacy preserves cycle structure and these are all the permutations in S_4 with this cycle structure (see Problem 9.1.5), $gV_4g^{-1} \subset V_4$ for all $g \in S_4$. This is sufficient to show that V_4 is normal in S_4.

9.2 The Simplicity of A_n

A **simple group** is a group G with no nontrivial proper normal subgroups. Up to this point the only examples we have of simple groups are cyclic groups of prime order. In this section we prove that if $n \geq 5$ each alternating group A_n is a simple group.

Theorem 9.2.1. *For each $n \geq 3$ each $\pi \in A_n$ is a product of cycles of length 3.*

Proof. Let $\pi \in A_n$. Since π is a product of an even number of transpositions to prove the theorem it suffices to show that if τ_1, τ_2 are transpositions then $\tau_1 \tau_2$ is a product of 3-cycles.

Suppose that a, b, c, d are different digits in $\{1, \ldots, n\}$. There are three cases to consider. First:

Case (1): $(a, b)(a, b) = 1 = (1, 2, 3)^0$ and hence it is true here.

Next:

Case (2): $(a, b)(b, c) = (c, a, b)$ and hence it is true here also.

Finally:

Case (3): $(a, b)(c, d) = (a, b)(b, c)(b, c)(c, d) = (c, a, b)(d, b, c)$ since $(b, c)(b, c) = 1$. Therefore it is also true here, proving the theorem.

$\square$

Now our main result:

Theorem 9.2.2. *For $n \geq 5$ the alternating group A_n is a simple nonabelian group.*

Proof. Suppose that N is a nontrivial normal subgroup of A_n with $n \geq 5$. We show that $N = A_n$ and hence that A_n is simple.

We claim first that N must contain a 3-cycle. Let $1 \neq \pi \in N$ then π is not a transposition since $\pi \in A_n$. Therefore π moves at least three digits. If π moves exactly three digits then it is a 3-cycle and we are done. Suppose then, that π moves at least four digits. Let $\pi = \tau_1 \cdots \tau_r$ with τ_i disjoint cycles.

Case (1): There is a $\tau_i = (\ldots, a, b, c, d)$. Set $\sigma = (a, b, c) \in A_n$. Then

$$\pi \sigma \pi^{-1} = \tau_i \sigma \tau_i^{-1} = (b, c, d),$$

where the first equality comes from the fact that the only cycle in π which does not commute with σ is τ_i.

Further, since $\pi \in N$ and N is normal we have

$$\pi(\sigma \pi^{-1} \sigma^{-1}) = (b, c, d)(a, c, b) = (a, d, b).$$

Therefore in this case N contains a 3-cycle $(a, d, b) \in N$.

Case (2): There is a τ_i which is a 3-cycle. Then

$$\pi = (a, b, c)(d, e, \ldots).$$

Now set $\sigma = (a, b, d) \in A_n$ and then

$$\pi \sigma \pi^{-1} = (b, c, e) = (a^\pi, b^\pi, d^\pi)$$

and again since $\pi \in N$ and N is normal in A_4,

$$\sigma^{-1}\pi\sigma\pi^{-1} = (d, b, a)(b, c, e) = (b, c, e, a, d) \in N.$$

Now N contains a cycle which moves at least four elements and thus we can now use Case (1). Therefore in this case N has a 3-cycle.

In the final case π is a disjoint product of transpositions.

Case (3): $\pi = (a, b)(c, d) \cdots$. Since $n \geq 5$ there is an $e \neq a, b, c, d$. Let $\sigma = (a, c, e) \in A_n$. Then

$$\pi\sigma\pi^{-1} = (b, d, e_1) \text{ with } e_1 = e^\pi \neq b, d.$$

So $(a^\pi, c^\pi, e^\pi) = (b, d, e_1)$. Let $\gamma = (\sigma^{-1}\pi\sigma)\pi^{-1}$. This is in N since N is normal and $\pi \in N$. If $e = e_1$ then $\gamma = (e, c, a)(b, d, e) = (a, e, b, d, c)$ and this says that N again contains a cycle which moves at least four elements so we can use Case (1) to get that N contains a 3-cycle. If $e \neq e_1$ then $\gamma = (e, c, a)(b, d, e_1) \in N$ and then we can use Case (2) to obtain that N contains a 3-cycle.

These three cases show that N must contain a 3-cycle.

If N is normal in A_n then from the argument above N contains a 3-cycle τ. However, from Theorem 9.1.5 any two 3-cycles in S_n are conjugate. Hence τ is conjugate to any other 3-cycle in S_n. Since N is normal and $\tau \in N$ each of these conjugates must also be in N. Therefore N contains all 3-cycles in S_n. From Theorem 9.2.1 each element of A_n is a product of 3-cycles. It follows then that each element of A_n is in N. However, since $N \subset A_n$ this is only possible if $N = A_n$, completing the proof.

$\square$

The theorem shows that A_n is simple if $n \geq 5$. This is not true if $n < 5$. A_4 is not simple because it contains V_4 and as pointed out in Example 9.1.3 V_4 is normal in S_4 and hence certainly it is normal in A_4. A_3 however has size 3 so it is cyclic of order 3 and hence simple.

We need the following result for the final theorem in this chapter.

Lemma 9.2.1. *Let p be a prime and $\alpha \in S_p$ be a p-cycle. Then each of the powers $\alpha^i, 1 \leq i \leq p - 1$ is also a p-cycle. Further, no such α^i has a fixed point.*

Proof. By Corollary 9.1.1 the order of α is p, that is, $o(\alpha) = p$. Then by Theorem 6.4.6

$$o(\alpha^i) = \frac{p}{\gcd(p, i)}.$$

For $1 \leq i \leq p-1$ we have $\gcd(p, i) = 1$ and so $o(\alpha^i) = p$ for all $i, 1 \leq i \leq p-1$.

Now consider any α^i for $1 \leq i \leq p - 1$. It is a permutation of S_p so it must have a representation as a product of disjoint cycles. Suppose that

$$\alpha^i = g_1 g_2 \cdots g_r$$

is the disjoint cycle representation of α^i. By Corollary 9.1.2, $p = o(\alpha^i) = lcm(o(g_1), \ldots, o(g_r))$. Since we assume that the disjoint cycle representation only contains nontrivial cycles, each of the cycles $g_1, g_2, \ldots, g_r$ must have order p. Thus each g_i is a p-cycle for $1 \leq i \leq r$. But S_p only involves p elements (that is, $A = \{1, 2, \ldots, p\}$) being permuted. The fact that these cycles are disjoint then forces $r = 1$ and α^i to be a p-cycle also. Again since the permutations in S_p only permute the p elements in A and α^i is a p-cycle we must have that α^i has no fixed points for $1 \leq i \leq p - 1$. $\square$

Theorem 9.2.3. *Let p be a prime and $U \subset S_p$ a subgroup. Let τ be a transposition and α a p-cycle with $\alpha, \tau \in U$. Then $U = S_p$.*

Proof. Suppose without loss of generality that $\tau = (1, 2)$. This can be done because if $\tau = (a_1, a_2)$ where $a_1, a_2 \in A = \{1, 2, \ldots, p\}$ then there is a $\gamma \in S_p$ such that $\gamma(a_1) = 1, \gamma(a_2) = 2$. By Lemma 9.1.4 we then have $\gamma\tau\gamma^{-1} = (1, 2)$. Thus if we consider the subgroup $\gamma U \gamma^{-1}$ then $(1, 2) \in \gamma U \gamma^{-1}$. But if we can show that $\gamma U \gamma^{-1} = S_p$ then $U = \gamma^{-1} S_p \gamma = S_p$ also.

By the above Lemma 9.2.1, $\alpha, \alpha^2, \ldots, \alpha^{p-1}$ are all p-cycles with no fixed points. Thus some power of α takes any element in A into any other one. So there exists an i with $\alpha^i(1) = 2$. Without loss of generality we may assume that

$$\alpha = (1, 2, a_3, \ldots, a_p).$$

The reason we may do this is that if $\alpha \in U$ then certainly $\alpha^i \in U$. But as already pointed out α^i is a p-cycle and so $\alpha^i = (1, 2, a_3, \ldots, a_p)$. So if α is replaced by α^i which it can be we get the form of α above.

Now let

$$\pi = \begin{pmatrix} 1 & 2 & a_3 & \cdots & a_p \\ 1 & 2 & 3 & \cdots & p \end{pmatrix}.$$

Then from Lemma 9.1.4 we have

$$\pi\alpha\pi^{-1} = (1, 2, \ldots, p).$$

Further, $\pi(1, 2)\pi^{-1} = (1, 2)$. Hence $U_1 = \pi U \pi^{-1}$ contains $(1, 2)$ and $(1, 2., \ldots, p)$.

Now we have

$$(1, 2, \ldots, p)(1, 2)(1, 2, \ldots, p)^{-1} = (2, 3) \in U_1.$$

Analogously

$$(1, 2, \ldots, p)(2, 3)(1, 2, \ldots, p)^{-1} = (3, 4) \in U_1,$$

and so on until

$$(1, 2, \ldots, p)(p - 2, p - 1)(1, 2, \ldots, p)^{-1} = (p - 1, p) \in U_1.$$

Hence the transpositions $(1, 2), (2, 3), \ldots, (p - 1, p) \in U_1$.
    Moreover
$$(1, 2)(2, 3)(1, 2) = (1, 3) \in U_1.$$
In an identical fashion each $(1, k) \in U_1$. Then for any digits s, t we have

$$(1, s)(1, t)(1, s) = (s, t) \in U_1.$$

Therefore U_1 contains all the transpositions of S_p and hence $U_1 = S_p$. Since $U = \pi U_1 \pi^{-1}$ we must have $U = S_p$ also.

$\square$

9.3   Exercises

EXERCISES FOR SECTION 9.1

9.1.1. Consider the following three permutations in S_5.

$$a = \begin{pmatrix} 1 & 2 & 3 & 4 & 5 \\ 2 & 3 & 4 & 5 & 1 \end{pmatrix}, b = \begin{pmatrix} 1 & 2 & 3 & 4 & 5 \\ 2 & 1 & 4 & 5 & 3 \end{pmatrix}, c = \begin{pmatrix} 1 & 2 & 3 & 4 & 5 \\ 3 & 2 & 5 & 4 & 1 \end{pmatrix}.$$

Determine the following permutations
    (a) $a^2 b$.
    (b) a^{-1}.
    (c) $ab^2 c^{-1}$.
9.1.2. In S_5 perform the indicated operations and then write the result of each product in the 2-row form for a permutation in S_5.

(a) $(2,3)(1,3)(1,4,5)(1,2)$.

(b) $(3,4)^{-1}(1,2)(1,5,3,2)$.

(c) $(1,3)^{-1}(1,2,4,5)^2(1,3)$.

9.1.3. Find the disjoint cycle decomposition for the following permutations

(a) $\begin{pmatrix} 1 & 2 & 3 & 4 & 5 & 6 & 7 & 8 & 9 \\ 7 & 4 & 9 & 2 & 3 & 8 & 1 & 6 & 5 \end{pmatrix}$.

(b) $\begin{pmatrix} 1 & 2 & 3 & 4 & 5 \\ 2 & 3 & 4 & 5 & 1 \end{pmatrix}$.

9.1.4. Find the orders of each of the permutations in Problem 9.1.2.

9.1.5. Express each of the permutations in Problem 9.1.2 as a product of transpositions.

9.1.6. Determine all subgroups of S_3.

9.1.7. Show that if f is a permutation then if $f(a_i) \neq a_i$, $f^2(a_i) \neq f(a_i)$. (This is used in the proof of Lemma 9.1.1.)

9.1.8. Show that in the proof of Lemma 9.1.1, if $g(a_j) \neq a_j$ then $(fg)(a_j) = (gf)(a_j)$.

9.1.9 Prove that if $(a_1, a_2, \ldots, a_k)$ is a k-cycle in S_n then $(a_1, a_2, \ldots, a_k)^{-1} = (a_k, a_{k-1}, \ldots, a_1)$.

9.1.10. (a) Determine all the elements of S_4.

(b) Write each permutation in S_4 as a product of disjoint cycles.

9.1.11. Find the parities of each of the permutations in Problem 9.2.1. Which are in the corresponding alternating groups?

9.1.12. Verify the following identity.

$$(a_1, a_2, \ldots, a_k) = (a_1, a_k)(a_1, a_{k-1}) \cdots (a_1, a_2).$$

This shows that any k-cycle can be written as a product of transpositions and hence any permutation can be written as a product of transpositions.

9.1.13. Let

$$a = \begin{pmatrix} 1 & 2 & 3 & 4 & 5 \\ 2 & 3 & 4 & 5 & 1 \end{pmatrix}, \quad b = \begin{pmatrix} 1 & 2 & 3 & 4 & 5 \\ 2 & 1 & 4 & 5 & 3 \end{pmatrix}.$$

Determine the conjugate $b^{-1}ab$.

9.1.14. Show that S_n is nonabelian if $n > 2$.

9.1.15. Show that the order of any k-cycle is k. (In Corollary 9.1.1 where it is shown that the order of a k-cycle $(a_1, a_2, \ldots, a_k)$ is k, prove that $(a_1, a_2, \ldots, a_k)^k = 1$ where 1 is the identity permutation.)

9.1.16. A permutation π on A **fixes** a point $P \in A$ if $\pi(P) = P$. Show that if $P \in A$ then the permutations that fix P form a subgroup of A.

9.1.17. Let A be an infinite set. Let H be the set of all permutations on A that fix all but a finite number of elements of A. Show that H is a subgroup of S_A.

9.1.18. In the proof of Corollary 9.1.2 show that if $f^k = 1$ then $n_i | k$ for all $i = 1, \ldots, r$.

(HINT: First explain why it is true that if g_1 is an n_1-cycle then the elements in the cycle g_1 moved by any power of f can only come from those moved by powers of g_1. We claim that if $f^k = 1$ then $g_i^k = 1$ where here 1 is the identity permutation. Explain why this claim is sufficient to solve the problem.

In order to establish the claim use a proof by contradiction. So supposing the claim is not true means that there exists some i such that $g_i^k \neq 1$. In the equation $1 = g_1^k \cdots g_i^k \cdots g_r^k$ solve for g_i^k and use the first part of this hint to show that the equation which you then get is impossible.)

9.1.19. Show that if $\sigma \in S_n$ and $(a_1, \ldots, a_k)$ and $(b_1, \ldots, b_r)$ are disjoint cycles then $(\sigma(a_1), \ldots, \sigma(a_k))$ and $(\sigma(b_1), \ldots, \sigma(b_r))$ are also disjoint cycles.

9.1.20. Which of the following pairs of permutations, written in cycle notation, are conjugate and which are not.
    (a) $(1,3)(4,5,9)(2,7)$ and $(1,4)(3,6,7)(8,9)$.
    (b) $(1,3)(4,5,8,9)(2,7)$ and $(1,4)(3,6,7)(8,9)$.
    (c) $(1,3)(6,8)(4,5,9)(2,7)$ and $(1,4)(8,9)(2,5)(3,6,7)$.

9.1.21. In the proof of Theorem 9.1.3, prove the first two identities stated.

EXERCISES FOR SECTION 9.2

9.2.1. Let H be a nontrivial proper normal subgroup of S_n with $n \geq 5$. Show that $H = A_n$.

9.2.2. Prove that S_n is generated by $(1,2)$ and $(1,2,3,\ldots,n)$.

(HINT: Use an argument just like that in the proof of Theorem 9.2.3 to show that any transposition $(s,t) \in \langle (1,2), (1,2,\ldots,n) \rangle$ for all $s, t \in \{1, 2, \ldots, n\}$.)

9.2.3. In the proof of Theorem 9.2.1 show the following:

(a) In case (2) show that if $(bcead) \in N$ then N contains a 3-cycle.

(b) In case (3) show that if $(aebdc) \in N$ then N contains a 3-cycle. Here we are assuming that $e_1 \neq e$. Explain why you can't just use case (2) here when $e \neq e_1$.

9.2.4. Show that there is no subgroup of S_n strictly between A_n and S_n. (HINT: Recall that $[S_n : A_n] = 2$.)

9.2.5. Here is an example to show that the converse of Lagrange's theorem is not true in general: Prove that A_5 cannot have a subgroup of order 30.

(NOTE: This is not a minimal example. We shall see another example of this later.)

9.2.6. Write down all the elements of A_3 and A_4.

9.2.7. Show that A_3 is simple but A_4 is not simple.

Chapter 10

Group Actions and Topics in Group Theory

10.1 Group Actions

Here we discuss a very important tool in group theory called a group action.

Definition 10.1.1. *A **group action** of a group G on a set A is a homomorphism from G into S_A, the symmetric group on A. We say that G **acts** on A. Hence G acts on A if to each $g \in G$ corresponds a permutation*

$$\pi_g : A \to A$$

such that

(1) $\pi_{g_1}(\pi_{g_2}(a)) = \pi_{g_1 g_2}(a)$ for all $g_1, g_2 \in G$ and for all $a \in A$
(2) $\pi_1(a) = a$ for all $a \in A$, that is, $\pi_1 = 1$.

In (2) above we have used 1 for the identity in G and 1 for the identity permutation in S_A. For the remainder of this chapter if $g \in G$ and $a \in A$ we will write ga for $\pi_g(a)$.

Group actions are an extremely important idea and we use this idea in the present chapter to prove several fundamental results in group theory.

If G acts on the set A then we say that two elements $a_1, a_2 \in A$ are **congruent** under G if there exists a $g \in G$ with $ga_1 = a_2$. The set

$$G_a = \{a_1 \in A : a_1 = ga \text{ for some } g \in G\}$$

is called the **orbit** of a. It consists of elements of A congruent to a under G.

Lemma 10.1.1. *If G acts on A then congruence under G is an equivalence relation on A.*

Proof. Any element $a \in A$ is congruent to itself via the identity map and hence the relation is reflexive. If $a_1 \sim a_2$ so that $ga_1 = a_2$ for some $g \in G$ then $g^{-1}a_2 = a_1$ and so $a_2 \sim a_1$ and the relation is symmetric. Finally is $g_1a_1 = a_2$ and $g_2a_2 = a_3$ then $g_2g_1a_1 = a_3$ and the relation is transitive. $\qquad\square$

Recall that the equivalence classes under an equivalence relation partition a set. For a given $a \in A$ its equivalence class under this relation is precisely its orbit as defined above.

Corollary 10.1.1. *If G acts on the set A then the orbits under G partition the set A.*

We say that G **acts transitively** on A if any two elements of A are congruent under G. That is, the action is transitive if for any $a_1, a_2 \in A$ there is some $g \in G$ such that $ga_1 = a_2$.

If $a \in A$ the **stabilizer** of a consists of those $g \in G$ that **fix** a. Hence

$$\text{Stab}_G(a) = \{g \in G : ga = a\}.$$

The following is easily proved and left to the exercises.

Lemma 10.1.2. *If G acts on A then for any $a \in A$ the stabilizer $Stab_G(a)$ is a subgroup of G.*

We now prove the crucial theorem concerning group actions.

Theorem 10.1.1. *Suppose that G acts on A and $a \in A$. Let G_a be the orbit of a under G and $Stab_G(a)$ its stabilizer. Then*

$$[G : Stab_G(a)] = |G_a|.$$

That is, the size of the orbit of a is the index of its stabilizer in G.

Proof. Let $\mathcal{L}$ be the set of all left cosets of $\text{Stab}_G(a)$ in G. We define a map $f : \mathcal{L} \to G_a$ and show that it is a bijection. For $g \in G$ define $f(g\text{Stab}_G(a)) = ga$. We must first show that f is well-defined. Suppose that $g_1, g_2 \in G$ and $g_1\text{Stab}_G(a) = g_2\text{Stab}_g(a)$, that is, g_1 and g_2 define the same left coset of the stabilizer. Then $g_2^{-1}g_1 \in \text{Stab}_G(a)$. This implies that $g_2^{-1}g_1a = a$

so that $g_1a = g_2a$. Hence any two elements in the same left coset of the stabilizer produce the same image of a in G_a. Thus the map f is well-defined. It is clear that f is onto. Finally to show that f is injective suppose that $f(g_1) = f(g_2)$. That is, $g_1a = g_2a$. Then $g_2^{-1}g_1a = a$ and hence $g_2^{-1}g_1 \in \mathrm{Stab}_G(a)$. Therefore g_1, g_2 define the same left coset of the stabilizer, that is, $g_1\mathrm{Stab}(a) = g_2\mathrm{Stab}(a)$. Therefore f is injective and hence a bijection. It follows that the size of G_a is precisely the index in G of the stabilizer, $\mathrm{Stab}_G(a)$. $\qquad\square$

We will use this theorem repeatedly with different group actions to obtain important group theoretic results.

10.2 Conjugacy Classes and the Class Equation

Recall that if G is a group then two elements $g_1, g_2 \in G$ are **conjugate** in G if there exists a $g \in G$ with $gg_1g^{-1} = g_2$. We saw that conjugacy is an equivalence relation on G. For $g \in G$ its equivalence class is called its **conjugacy class** that we will denote by $Cl(g)$. Thus

$$Cl(g) = \{g_1 \in G : g_1 \text{ is conjugate to } g\}.$$

If $g \in G$ then its **centralizer** $C_G(g)$ is the set of elements in G that commute with g;

$$C_G(g) = \{g_1 \in G : gg_1 = g_1g\}.$$

Theorem 10.2.1. *Let G be a finite group and $g \in G$. Then the centralizer of g is a subgroup of G and*

$$[G : C_G(g)] = |Cl(g)|.$$

That is, the index of the centralizer of g is the size of its conjugacy class.

In particular for a finite group the size of each conjugacy class divides the order of the group.

Proof. Let the group G act on itself by conjugation. That is, $g(g_1) = gg_1g^{-1}$. It is easy to show that this is an action on the set G (see the exercises). For $g \in G$ its orbit under this action is precisely its conjugacy class $Cl(g)$ and the stabilizer is its centralizer $C_G(g)$. The statements in the theorem then follow directly from Theorem 10.1.1 and Lemma 10.1.2.

$\square$

For any group G, since conjugacy is an equivalence relation, the conjugacy classes partition G. Hence

$$G = \bigcup_{g \in \Lambda} Cl(g)$$

where Λ can be chosen so that this union is a disjoint union. It follows that if G is a finite group

$$|G| = \sum_{g \in \Lambda} |Cl(g)|.$$

If $Cl(g) = \{g\}$, that is, the conjugacy class of g is g alone then $C_G(g) = G$ so that g commutes with all of G. Therefore in this case $g \in Z(G)$ where $Z(G)$ is the center of G. This is true for every element of the center and therefore

$$G = Z(G) \cup \bigcup_{g \notin Z(G)} Cl(g)$$

where again the second union is a disjoint union. The size of G is then the sum of these disjoint pieces so

$$|G| = |Z(G)| + \sum_{g \notin Z(G)} |Cl(g)|.$$

However, from Theorem 10.2.1 $|Cl(g)| = [G : C_G(g)]$ so the equation above becomes

$$|G| = |Z(G)| + \sum_{g \notin Z(G)} [G : C_G(g)].$$

This is known as the **class equation**.

Theorem 10.2.2. *(The Class Equation) Let G be a finite group. Then*

$$|G| = |Z(G)| + \sum_{g \notin Z(G)} [G : C_G(g)]$$

where the sum is taken over the distinct centralizers.

As a first application we prove the result that certain finite groups have nontrivial center.

Theorem 10.2.3. *Let G be a finite group of order p^n where p is a prime. Then G has a nontrivial center.*

Proof. Let G be a finite group with $|G| = p^n$ for equation

$$|G| = |Z(G)| + \sum_{g \notin Z(G)} [G : C_G(g)].$$

Since $[G : C_G(g)]$ divides $|G|$ for each $g \in G$ we must have that $p|[G : C_G(g)]$ for each $g \in G$. Further, $p||G|$. Therefore p must divide $|Z(G)|$ and hence $|Z(G)| = p^m$ for some $m \geq 1$ since $|Z(G)|$ divides p^n and therefore $Z(G)$ is nontrivial.

$\square$

The idea of conjugacy and the centralizer of an element can be extended to subgroups. If H_1, H_2 are subgroups of a group G then H_1, H_2 are conjugate if there exists a $g \in G$ such that $gH_1g^{-1} = H_2$. As it is for elements, conjugacy is an equivalence relation on the set of subgroups of G. We have seen this in Chapter 7.

If $H \subset G$ is a subgroup then its **conjugacy class** consists of all the subgroups of G conjugate to it. The **normalizer** of H is

$$N_G(H) = \{g \in G : gHg^{-1} = H\}.$$

As for elements, let G act on the set of subgroups of G by conjugation. That is, for $g \in G$ the map is $g : H \to gHg^{-1}$ where H is a subgroup. For $H \subset G$ the stabilizer under this action is precisely the normalizer. Hence exactly as for elements we obtain the following theorem. See Thoerem 7.1.1 for a different proof.

Theorem 10.2.4. *Let G be a group and $H \subset G$ a subgroup. Then the normalizer $N_G(H)$ of H is a subgroup of G, H is normal in $N_G(H)$ and*

$$[G : N_G(H)] = \text{ number of conjugates of } H \text{ in } G.$$

10.3 The Sylow Theorems

If G is a finite group and $H \subset G$ is a subgroup then Lagrange's theorem guarantees that the order of H divides the order of G. However, the converse of Lagrange's theorem is false. That is, if G is a finite group of order n and if $d|n$, then G need not contain a subgroup of order d. If d is a prime p or a power of a prime p^e, however, then we shall see that G must contain subgroups of that order. In particular, we shall see that if p^d is the highest power of p that divides n, then all subgroups of that order are actually conjugate, and we shall finally get a formula concerning the number of such subgroups. These theorems constitute the Sylow theorems which we will examine in this section. First we give an example where the converse of Lagrange's theorem is false.

Lemma 10.3.1. *The alternating group on four symbols A_4 has order 12 but has no subgroup of order 6.*

Proof. Suppose that there exists a subgroup $U \subset A_4$ with $|U| = 6$. Then $[A_4 : U] = 2$ since $|A_4| = 12$ and hence U is normal in A_4.

Now 1, $(1,2)(3,4)$, $(1,3)(2,4)$ and $(1,4)(2,3)$ are in A_4. These each have order 2 and commute so they form a subgroup $V \subset A_4$ of order 4. This subgroup $V \cong \mathbb{Z}_2 \times \mathbb{Z}_2$. Then by Theorem 6.3.3

$$12 = |A_4| \geq |VU| = \frac{|V||U|}{|V \cap U|} = \frac{4 \cdot 6}{|V \cap U|}.$$

It follows that $V \cap U \neq \{1\}$ and since U is normal and V is normal in A_4 (see Example 9.1.3) we have that $V \cap U$ is also normal in A_4 by Lemma 7.1.9(i).

Now $(1,2)(3,4) \in V$ and by renaming the entries in V if necessary we may assume that it is also in U so that $(1,2)(3,4) \in V \cap U$. This is true because $U \cap V \neq 1$ so $U \cap V$ must contain some nontrivial element of V. Since $(1,3,2) \in A_4$ and $(1,3,2)^{-1} = (1,2,3)$ we have

$$(3,2,1)(1,2)(3,4)(1,2,3) = (1,3)(2,4) \in V \cap U$$

But then

$$(1,3,2)(1,3)(2,4)(1,2,3) = (1,4)(2,3) \in U \cap V.$$

But then $V \subset V \cap U$ and so $V \subset U$. But this is impossible since $|V| = 4$ which doesn't divide $|U| = 6$.

$\square$

Definition 10.3.1. *Let G be a finite group with $|G| = n$ and let p be a prime such that $p^a | n$ but no higher power of p divides n. A subgroup of G of order p^a is called a p-**Sylow** subgroup.*

It is not clear that a p-Sylow subgroup must exist. We will prove that for each $p|n$ a p-Sylow subgroup exists.

We first consider a very special case that was already proven as Lemma 8.2.4. Here we just restate it as

Theorem 10.3.1. *Let G be a finite abelian group and let p be a prime such that $p||G|$. Then G contains at least one element of order p.*

Therefore if G is an abelian group and if $p|n$, then G contains a subgroup of order p, the cyclic subgroup of order p generated by an element $a \in G$ of order p whose existence is guaranteed by the above theorem. We now present the first Sylow theorem.

Theorem 10.3.2. *(The First Sylow Theorem) Let G be a finite group and let $p||G|$, then G contains a p-Sylow subgroup, that is, a p-Sylow subgroup exists.*

Proof. Let G be a finite group of order pn. We do induction on n. If $n = 1$ then G is cyclic and G is its own maximal p-subgroup and hence all of G is a p-Sylow subgroup. We assume then that if $|G| = pm$ with $m < n$ then G has a p-Sylow subgroup.

Assume that $|G| = p^t m$ with $(m, p) = 1$. If $m = 1$ then G is its own p-Sylow subgroup. Now let $m \geq 2$. We must show that G contains a subgroup of order p^t. If H is a proper subgroup whose index is prime to p then $|H| = p^t m_1$ with $m_1 < m$. Therefore by the inductive hypothesis H has a p-Sylow subgroup of order p^t. This will also be a subgroup of G and hence a p-Sylow subgroup of G.

Therefore we may assume that the index of any proper subgroup H of G must have index divisible by p. Now consider the class equation for G,

$$|G| = |Z(G)| + \sum_{g \notin Z(G)} [G : C_G(g)].$$

By assumption each of the indices are divisible by p and also $p||G|$. Therefore $p||Z(G)|$. It follows that $Z(G)$ is a finite abelian group whose order is divisible by p. From Theorem 10.3.1 there exists an element $g \in Z(G) \subset G$ of order p. Since $g \in Z(G)$ we must have $\langle g \rangle$ normal in G. The factor group $G/\langle g \rangle$ then has order $p^{t-1}m$ and by the inductive hypothesis must have a p-Sylow subgroup $\overline{K}$ of order p^{t-1} and hence of index m. By Corollary 7.3.2 since $\overline{K}$ is a subgroup of $G/\langle g \rangle$ we have $\overline{K} = K/\langle g \rangle$ where K is a subgroup of

G containing g. But then $|K| = p^t$ by Lagrange's theorem and so K is a p-Sylow subgroup of G.   □

On the basis of this theorem, we can now strengthen the result obtained in Theorem 10.3.1.

Theorem 10.3.3. *(Cauchy) If G is a finite group and if p is a prime such that $p||G|$, then G contains at least one element of order p.*

Proof. Let P be a p-Sylow subgroup of G, and let $|P| = p^t$. If $g \in P$, $g \neq 1$, then the order of g is p^{t_1}. Then the cyclic group $\langle g \rangle$ must have a unique subgroup of order p, that is, $\langle g^t \rangle$ by Theorem 6.4.3. Thus $g^t \in G$ and $o(g^t) = p$.   □

Any group, finite or infinite, is called p-**group** for some prime p if every element of the group has order a power of p. It is not hard to see from the above theorem (Cauchy's theorem) the following characterization of finite p-groups.

Corollary 10.3.1. *Let G be a finite nontrivial group and let p be a prime. Then G is a p-group if and only if $|G| = p^n$ for some positive integer n.*

For the proof see the exercises. We have seen that p-Sylow subgroups exist. We now wish to show that any two p-Sylow subgroups are conjugate. This is the content of the second Sylow theorem.

Theorem 10.3.4. *(The Second Sylow Theorem) Let G be a finite group and p a prime such that $p||G|$. Then any p-subgroup H of G is contained in a p-Sylow subgroup. Further, all p-Sylow subgroups of G are conjugate in G. That is, if P_1 and P_2 are any two p-Sylow subgroups of G then there exists an $a \in G$ such that $P_1 = aP_2a^{-1}$.*

Proof. Let Ω be the set of p-Sylow subgroups of G and let G act on Ω by conjugation. Since conjugation is an automorphism (see Lemma 7.1.2) this action certainly takes a p-Sylow subgroup to a p-Sylow subgroup. This action will of course partition Ω into disjoint orbits. Let P be a fixed p-Sylow subgroup and Ω_P be its orbit under the conjugation action. The size of the orbit, by Theorem 10.2.4, is the index of its stabilizer under this action, that is, $|\Omega_P| = [G : \text{Stab}_G(P)]$. Now $P \subset \text{Stab}_G(P)$ and P is a maximal p-subgroup of G. It follows that the index of $\text{Stab}_G(P)$ must be prime to p

and so the number of p-Sylow subgroups conjugate to P is prime to p. Since the action is conjugation the stabilizer is the normalizer.

Now let H be a p-subgroup of G and let H act on Ω_P by conjugation. We note from Corollary 10.3.1 above that the order of H must be a power of p. Ω_P will itself decompose into disjoint orbits under this action. Further, the size of each orbit is an index of a subgroup of H and hence must be a power of p. On the other hand the size of the whole orbit is prime to p. Therefore there must be one orbit that has size exactly 1. This orbit contains a p-Sylow subgroup P' and P' is fixed by H under conjugation, that is, H normalizes P'. It follows that HP' is a subgroup of G by Theorem 6.3.2 and P' is normal in HP'. From the second isomorphism theorem, Theorem 7.3.2, we then obtain

$$HP'/P' \cong H/(H \cap P').$$

Since H is a p-group the size of $H/(H \cap P')$ is a power of p and therefore so is the size of HP'/P'. But P' is also a p-group so it follows that HP' also has order a power of p. Now $P' \subset HP'$ but P' is a maximal p-subgroup of G. Hence $HP' = P'$. This is possible only if $H \subset P'$ (see Problem 6.3.10) proving the first assertion in the theorem. Therefore any p-subgroup of G is contained in a p-Sylow subgroup.

Now let H be a p-Sylow subgroup P_1 and let P_1 act on Ω_P. Exactly as in the argument above, $P_1 \subset P'$ where P' is a conjugate of P. Since P_1 and P' are both p-Sylow subgroups they have the same size and hence $P_1 = P'$. This implies that P_1 is a conjugate of P. Since P_1 and P are arbitrary p-Sylow subgroups it follows that all p-Sylow subgroups are conjugate. □

We come now to the last of the three Sylow theorems. This one gives us information concerning the number of p-Sylow subgroups.

Theorem 10.3.5. *(The Third Sylow Theorem) Let G be a finite group and p a prime such that $p||G|$. Then the number of p-Sylow subgroups of G is of the form $1 + pk$ and divides the order of $|G|$. It follows that if $|G| = p^a m$ with $(p, m) = 1$ then the number of p-Sylow subgroups divides m.*

Proof. Let P be a p-Sylow subgroup and let P act on Ω, the set of all p-Sylow subgroups, by conjugation. Now P normalizes itself so there is one orbit, namely P that has size exactly 1. Every other orbit has size a power of p since the size is the index of a nontrivial subgroup of P and therefore must be divisible by p. Hence the size of the Ω is $1 + pk$.

□

10.3.1   Some Applications of the Sylow Theorems

We now give some applications of the Sylow theorems. First we show that the converse of Lagrange's theorem is true for both finite p-groups and for finite abelian groups. For finite abelian groups, this also follows from the fundamental theorem of finite abelian groups (see Problem 8.2.6).

Theorem 10.3.6. *Let G be a group of order p^n. Then G contains at least one normal subgroup of order p^m, for each m such that $0 \leq m \leq n$.*

Proof. We use induction on n. For $n = 1$ the theorem is trivial. By Theorem 7.2.2 any group of order p^2 is abelian. This together with Theorem 10.3.1 establishes the truth of the theorem for $n = 2$.

We now assume the theorem is true for all groups G of order p^k where $1 \leq k < n$, where $n > 2$. Let G be a group of order p^n. From Theorem 10.2.3 G has a nontrivial center. Thus by Lagrange's theorem, $|Z(G)| \geq p$ and hence there is an element $g \in Z(G)$ of order p. Let $N = \langle g \rangle$. Since $g \in Z(G)$ it follows that N is normal subgroup of order p. Then G/N is of order p^{n-1}, and therefore, contains (by the induction hypothesis) normal subgroups of orders p^{m-1}, for $0 \leq m-1 \leq n-1$. By Corollaries 7.3.1 and 7.3.2 these groups are of the form H/N, where H is a normal subgroup in G and H contains N. H is of order p^m, $1 \leq m \leq n$, because $|H| = |N|[H : N] = |N| \cdot |H/N|$. $\square$

On the basis of the first Sylow theorem together with the above result we see that if G is a finite group and if $p^k || G|$, then G must contain a subgroup of order p^k. One can actually show that, as in the case of Sylow p-groups, the number of such subgroups is of the form $1 + pt$, but we shall not prove this here.

Theorem 10.3.7. *Let G be a finite abelian group of order n. Suppose that $d|n$. Then G contains a subgroup of order d.*

Proof. Suppose that $n = p_1^{e_1} \cdots p_k^{e_k}$ is the prime factorization of n. Then $d = p_1^{f_1} \cdots p_k^{f_k}$ for some nonnegative $f_1, \ldots, f_k$ with $0 \leq f_i < e_i$ for all $i = 1, \ldots, k$. Now G has p_1-Sylow subgroup H_1 of order $p_1^{e_1}$. Hence from Theorem 10.3.6 H_1 has a subgroup K_1 of order $p_1^{f_1}$. Similarly there are subgroups $K_2, \ldots, K_k$ of G of respective orders $p_2^{f_2}, \ldots, p_k^{f_k}$. Further, since the orders are relatively prime $K_i \cap K_j = \{1\}$ if $i \neq j$. It is not hard to show that the product $K_1 K_2 \cdots K_k$ is the direct product of the K_i and thus by Lemma 8.1.3 it has order $|K_1||K_2| \cdots |K_k| = p_1^{f_1} \cdots p_k^{f_k} = d$. $\square$

Theorem 10.3.8. *Let p, q be distinct primes with $p < q$ and q not congruent to 1 mod p. Then any group of order pq is cyclic. For example any group of order 15 must be cyclic.*

Proof. Suppose that $|G| = pq$ with $p < q$ and q not congruent to 1 mod p. The number of q-Sylow subgroups is of the form $1 + qk$ and divides p. Since q is greater than p this implies that there can be only one and hence there is a normal q-Sylow subgroup H. Since q is a prime, H is cyclic of order q and therefore there is an element g of order q.

The number of p-Sylow subgroups is of the form $1 + pk$ and divides q. Since q is not congruent to 1 mod p and since q is a prime, $q = (1 + pk)n$ for some integer n implies that $k = 0$. That is, there also can be only one p-Sylow subgroup and hence there is a normal p-Sylow subgroup K. Since p is a prime K is cyclic of order p and therefore there is an element h of order p. Since p, q are distinct primes $H \cap K = \{1\}$. Consider the element $g^{-1}h^{-1}gh$. Since K is normal $g^{-1}hg \in K$. Then $g^{-1}h^{-1}gh = (g^{-1}h^{-1}g)h \in K$. But H is also normal so $h^{-1}gh \in H$. This then implies that $g^{-1}h^{-1}gh = g^{-1}(h^{-1}gh) \in H$ and therefore $g^{-1}h^{-1}gh \in K \cap H$. It follows then that $g^{-1}h^{-1}gh = 1$ or $gh = hg$. Since g, h commute and $\langle g \rangle \cap \langle h \rangle = \{1\}$, then by Lemma 6.4.2, the order of gh, $o(gh)$, is the LCM of the orders of g and h which is pq. Therefore G has an element of order pq. Since $|G| = pq$ this implies that G is cyclic. $\qquad\square$

In the above theorem since we assumed that q is not congruent to 1 mod p and hence $p \neq 2$. In the case when $p = 2$ we get another possibility.

Theorem 10.3.9. *Let p be an odd prime and G a finite group of order $2p$. Then either G is cyclic or G is isomorphic to the dihedral group of order $2p$, that is, the group of symmetries of a regular p-gon. In this latter case G is generated by two elements g and h which satisfy the relations $g^p = h^2 = (gh)^2 = 1$.*

Proof. As in the proof of Theorem 10.3.8, G must have a normal cyclic subgroup of order p, say $\langle g \rangle$. Since $|G|$ is even, the group G must have an element of order 2, say h, by Cauchy's theorem 10.3.3. Consider the order of $(gh) = o(gh)$. By Lagrange's theorem this element can have order $1, 2, p, 2p$. If the order is 1 then $gh = 1$ or $g = h^{-1} = h$. This is impossible since g has order p and h has order 2. If the order of gh is p since there is only one p-Sylow subgroup $\langle g \rangle$, then from the second Sylow theorem $gh \in \langle g \rangle$. But this implies that $h \in \langle g \rangle$ which is impossible since every nontrivial element of $\langle g \rangle$ has order p. Therefore the order of gh is either 2 or $2p$.

If the order of gh is $2p$ then since G has order $2p$ it must be cyclic.

If the order of gh is 2 then within G we have the relations $g^p = h^2 = (gh)^2 = 1$. Let $H = \langle g, h \rangle$ be the subgroup of G generated by g and h. The relations $g^p = h^2 = (gh)^2 = 1$ imply that H has order $2p$. To see this notice that the relation $(gh)^2 = 1$ gives us $ghgh = 1 \implies hg = g^{-1}h^{-1} = g^{p-1}h$. This implies that any element of G can be written in the form $g^t h^s$ where $t = 0, 1, \ldots, p-1$ and $s = 0, 1$. Hence there are $2p$ such elements. Since $|G| = 2p$ we get that $H = G$. G is isomorphic to the dihedral group D_p of order $2p$ (see the exercises).

In the above description g represents a rotation of $\frac{2\pi}{p}$ of a regular p-gon about its center while h represents any reflection across a line of symmetry of the regular p-gon.

$\square$

We have looked at the finite fields $\mathbb{Z}_p$. We give an example of a p-Sylow subgroup of a matrix group over $\mathbb{Z}_p$. This example requires some linear algebra.

EXAMPLE 10.3.1. Consider $GL(n, p)$, the group of $n \times n$ invertible matrices over $\mathbb{Z}_p$. If $\{v_1, \ldots, v_n\}$ is a basis for the vector space $(\mathbb{Z}_p)^n$ over $\mathbb{Z}_p$ then we recall that an $n \times n$ matrix can be thought of as a linear transformation of this vector space to itself. Moreover if the matrix of a linear transformation is invertible then so is the linear transformation. Finally such an invertible linear transformation must take a linearly independent set to a linearly independent set. It follows that the size of $GL(n, p)$ is the number of independent images $\{w_1, \ldots, w_n\}$ of $\{v_1, \ldots, v_n\}$. Since w_1 can be anything except the 0 vector then for w_1 there are $p^n - 1$ choices. Having chosen w_1 by independence it follows that w_2 can't be the zero vector or any nontrivial multiple of w_1. Hence there are there are $p^n - p$ choices for w_2. Similarly there $p^n - p^2$ choices for w_3 and so on. (See the exercises for a verification of this.) It follows that

$$|GL(n, p)| = (p^n - 1)(p^n - p) \cdots (p^n - p^{n-1}).$$

Now if we multiply this out we obtain

$$|GL(n, p)| = p^{1 + 2 + \cdots + (n-1)}((p^n - 1)(p^{n-1} - 1) \ldots (p - 1))$$

where the second term is ± 1 plus a sum of terms of the form $\pm p^k$ where $k = 1, 2, \ldots, (n + (n-1) + \cdots + 1)$. It follows that

$$|GL(n,p)| = p^{1+2+\cdots+(n-1)}m = p^{\frac{n(n-1)}{2}}m$$

where m is relatively prime to p Thus a p-Sylow subgroup must have size $p^{\frac{n(n-1)}{2}}$.

Let P be the subgroup of upper triangular matrices with 1's on the diagonal. See the exercises to show that this is a subgroup. Then P has size $p^{1+2+\cdots+(n-1)} = p^{\frac{n(n-1)}{2}}$ and is therefore a p-Sylow subgroup of $GL(n,p)$. See the exercises for a verification of the size of P.

The final example is a bit more difficult. We mentioned that a major result on finite groups is the classification of the finite simple groups. This classification showed that any finite simple group is either cyclic of prime order, in one of several classes of groups such as the A_n or one of a number of special examples called sporadic groups. For a nice discussion of these, the interested reader can see the article by D. Gorenstein, entitled "The Enormous Theorem" [G] in the *Scientific American*. One of the major tools in this classification is the following famous result called the Feit-Thompson theorem that showed that any finite group G of odd order is solvable (we will explain what solvable means later on in the chapter) and, in addition, if G is not cyclic then G is nonsimple.

Theorem 10.3.10. *(Feit-Thompson Theorem) Any finite group of odd order is solvable.*

The proof of this theorem, one of the major results in algebra in the twentieth century is way beyond the scope of this book — the proof is actually hundreds of pages in length when one counts the results used. However, here we discuss the smallest nonabelian simple group.

Theorem 10.3.11. *Suppose that G is a simple group of order 60. Then G is isomorphic to A_5. Further, A_5 is the smallest nonabelian finite simple group.*

In order to do this we first require the following fact.

Lemma 10.3.2. *A permutation group G (recall that this is any subgroup of S_n) which contains an odd permutation has a normal subgroup of index 2.*

Proof. (of lemma) Since G contains an odd permutation, G is not contained in A_n. Also since A_n is normal in S_n the set A_nG is a subgroup of S_n by Lemma 7.1.9. Thus $A_n \subset A_nG$ properly. But $[S_n : A_n] = 2$ and thus $A_nG = S_n$. By the second isomorphism theorem (Theorem 7.3.2) we have

$$2 = [A_nG : A_n] = [G : A_n \cap G].$$

Thus $A_n \cap G$ is subgroup of index 2 in G and hence a normal subgroup of G completing the proof of the lemma. $\qquad\square$

We can now prove the theorem.

Proof. (of the theorem) Suppose that G is a simple group of order $60 = 2^2 \cdot 3 \cdot 5$. The number of 5-Sylow subgroups is of the form $1 + 5k$ and divides 12. Hence there is 1 or 6. Since G is assumed simple and all 5-Sylow subgroups are conjugate there cannot be only one and hence there are 6.

Let G act on the set of 5-Sylow subgroups by conjugation. Since an action is a permutation this gives a homomorphism f from G into S_6. By the first isomorphism theorem $G/\ker(f) \cong \text{im}(f)$. However, since G is simple the kernel must be trivial and this implies that G would imbed into S_6. So without loss of generality we may assume that $G \subset S_6$. Since G is simple the previous lemma implies that $G \subset A_6$. But $[A_6 : G] = 6$. Now let G act on the six left cosets of G in A_6 by left multiplication by an element of G. Then just as above $G \subset S_6$. Under this action, however, G itself is a fixed point. Thus $G \subset S_5$. Again using the simplicity of G and the previous lemma it follows that $G \subset A_5$. Since $|G| = 60$ we must have $G = A_5$ or $G \cong A_5$.

The proof that A_5 is the smallest nonabelian simple group is actually brute force. We show that any group G of order less than 60 either has prime order or is nonsimple. There are strong tools that we can use. By the Feit-Thompson theorem we must only consider groups of even order. From Theorem 10.3.9 we don't have to consider orders $2p$, since a dihedral group always has a subgroup of index 2. The rest can be done by an analysis using Sylow theory. For example we show that any group of order 20 is nonsimple. Since $20 = 2^2 \cdot 5$ the number of 5-Sylow subgroups is $1 + 5k$ and divides 4. Hence there is only one and therefore it must be normal and so G is nonsimple. There is a strong theorem, whose proof is usually done with representation theory, which says that any group whose order is divisible by only two primes is solvable. Therefore for $|G| = 60$ we only have to show that groups of order $30 = 2 \cdot 3 \cdot 5$ and $42 = 2 \cdot 3 \cdot 7$ are nonsimple.

Suppose $|G| = 30$. The number of 5-Sylow subgroups is of the form $1 + 5k$ and divides 6; hence there are 1 or 6. Since G is assumed to be simple there are 6. Since each 5-Sylow subgroup is cyclic of order 5 they intersect only in the identity and hence these cover 24 distinct elements. The number of 3-Sylow subgroups is of the form $1 + 3k$ and divides 10; hence there are 1 or 10. If there were 10 these would cover an additional 20 distinct elements which is impossible since we already have 24 and G has order 30. Therefore there is only one and hence a normal 3-Sylow subgroup. It follows that G cannot be simple. The case $|G| = 42$ is even simpler. In this case, there must be a normal 7-Sylow subgroup. (See the exercises.) $\square$

10.4 Groups of Small Order

Classification is an extremely important concept in algebra. A large part of the theory is devoted to classifying all structures of a given type, for example all integral domains. In most cases this is not possible. Since for a given finite n there are only finitely many group tables it is theoretically possible to classify all groups of order n. However, even for small n this becomes impractical. In this section we look at some further results on finite groups and then use these to classify all the finite groups up to order 10.

Before stating the classification we give some further examples of groups that are needed.

EXAMPLE 10.4.1. In Chapter 6 we saw that the symmetry group of an equilateral triangle had six elements and is generated by elements r and f which satisfy the relations $r^3 = f^2 = 1$, $f^{-1}rf = r^{-1}$, where r is a rotation of 120° about the center of the triangle and f is a reflection through an altitude. This was called the dihedral group D_3 of order 6.

This can be generalized to any regular n-gon. If D is a regular n-gon, then the symmetry group D_n has $2n$ elements and is called the **dihedral group** of order $2n$. It is generated by elements r and f which satisfy the relations $r^n = f^2 = 1$, $f^{-1}rf = r^{n-1}$, where r is a rotation of $\frac{2\pi}{n}$ about the center of the n-gon and f is a reflection across a line of symmetry of the regular n-gon.

Hence, D_4, the symmetries of a square, has order 8 and D_5, the symmetries of a regular pentagon, has order 10.

EXAMPLE 10.4.2. Let $\mathbf{i}, \mathbf{j}, \mathbf{k}$ be the generators of the quaternions (see

Chapter 5). Then we have

$$\mathbf{i}^2 = \mathbf{j}^2 = \mathbf{k}^2 = -1, \ (-1)^2 = 1 \text{ and } \mathbf{ijk} = -1.$$

The elements $\{\pm 1, \pm \mathbf{i}, \pm \mathbf{j}, \pm \mathbf{k}\}$ then form a group of order 8 called the **quaternion group** denoted by Q. Since $\mathbf{ijk} = -1$ we have $\mathbf{ij} = -\mathbf{ji}$, and the generators $\mathbf{i}$ and $\mathbf{j}$ satisfy the relations $\mathbf{i}^4 = \mathbf{j}^4 = 1$, $\mathbf{i}^2 = \mathbf{j}^2$, $\mathbf{ij} = \mathbf{i}^2 \mathbf{ji}$ (see the exercises).

We now state the main classification and then prove it in a series of lemmas.

Theorem 10.4.1. *Let G be a finite group.*
    (a) If $|G| = 2$ then $G \cong \mathbb{Z}_2$.
    (b) If $|G| = 3$ then $G \cong \mathbb{Z}_3$.
    (c) If $|G| = 4$ then $G \cong \mathbb{Z}_4$ or $G \cong \mathbb{Z}_2 \times \mathbb{Z}_2$.
    (d) If $|G| = 5$ then $G \cong \mathbb{Z}_5$.
    (e) If $|G| = 6$ then $G \cong \mathbb{Z}_6 \cong \mathbb{Z}_2 \times \mathbb{Z}_3$ or $G \cong D_3$, the dihedral group with six elements. (Note $D_3 \cong S_3$ the symmetric group on 3 symbols.)
    (f) If $|G| = 7$ then $G \cong \mathbb{Z}_7$.
    (g) If $|G| = 8$ then $G \cong \mathbb{Z}_8$ or $G \cong \mathbb{Z}_4 \times \mathbb{Z}_2$ or $G \cong \mathbb{Z}_2 \times \mathbb{Z}_2 \times \mathbb{Z}_2$ or $G \cong D_4$, the dihedral group of order 8 or $G \cong Q$ the quaternion group.
    (h) If $|G| = 9$ then $G \cong \mathbb{Z}_9$ or $G \cong \mathbb{Z}_3 \times \mathbb{Z}_3$.
    (i) If $|G| = 10$ then $G \cong \mathbb{Z}_{10} \cong \mathbb{Z}_2 \times \mathbb{Z}_5$ or $G \cong D_5$, the dihedral group with ten elements.

Recall from Lemma 7.2.2 that a finite group of prime order must be cyclic. Hence in the theorem the cases $|G| = 2, 3, 5, 7$ are handled. We saw in Chapter 7 that if a group G has order p^2 where p is a prime then G is abelian. Here we give a proof that uses the fundamental theorem of finite abelian groups.

Definition 10.4.1. *If G is a group then its **center** denoted $Z(G)$ is the set of elements in G which commute with everything in G. That is*

$$Z(G) = \{g \in G : gh = hg \text{ for any } h \in G\}.$$

Lemma 10.4.1. *For any group G:*
    (a) the center $Z(G)$ is a normal subgroup.
    (b) $G = Z(G)$ if and only if G is abelian.
    (c) if $G/Z(G)$ is cyclic then G is abelian.

Proof. (a) and (b) are direct and we leave them to the exercises. Consider the case where $G/Z(G)$ is cyclic. Then each coset of $Z(G)$ has the form $g^m Z(G)$ where $g \in G$. Let $a, b \in G$. Then since a, b are in cosets of the center we have $a = g^m u$ and $b = g^n v$ with $u, v \in Z(G)$. Then

$$ab = (g^m u)(g^n v) = (g^m g^n)(uv) = (g^n g^m)(vu) = (g^n v)(g^m u) = ba$$

since u, v commute with everything. Therefore G is abelian. $\qquad\square$

Recall that a finite p-**group** has prime power order (see Corollary 10.3.1). Combining Theorem 10.2.3 with Corollary 10.3.1 gives:

Lemma 10.4.2. *A finite p-group has a nontrivial center of order at least p.*

Lemma 10.4.3. *If $|G| = p^2$ with p a prime then G is abelian and hence $G \cong \mathbb{Z}_{p^2}$ or $G \cong \mathbb{Z}_p \times \mathbb{Z}_p$.*

Proof. Suppose that $|G| = p^2$. Then from the previous lemma G has a nontrivial center and hence $|Z(G)| = p$ or $|Z(G)| = p^2$. If $|Z(G)| = p^2$ then $G = Z(G)$ and G is abelian. If $|Z(G)| = p$ then $|G/Z(G)| = p$. Since p is a prime this implies that $G/Z(G)$ is cyclic and hence from Lemma 10.4.1 G is abelian. The fundamental theorem of finite abelian groups (Theorem 8.2.1) implies that $G \cong \mathbb{Z}_{p^2}$ or $G \cong \mathbb{Z}_p \times \mathbb{Z}_p$. $\qquad\square$

The last lemma handles the cases $n = 4$ and $n = 9$. Therefore if $|G| = 4$ we must have $G \cong \mathbb{Z}_4$ or $G \cong \mathbb{Z}_2 \times \mathbb{Z}_2$ and if $|G| = 9$ we must have $G \cong \mathbb{Z}_9$ or $G \cong \mathbb{Z}_3 \times \mathbb{Z}_3$.

This leaves $n = 6, 8, 10$. We next handle 6 and 10.

Lemma 10.4.4. *If G is any group where every nontrivial element has order 2 then G is abelian.*

Proof. Suppose that $g^2 = 1$ for all $g \in G$. This implies that $g = g^{-1}$ for all $g \in G$. Let a, b be arbitrary elements of G. Then

$$(ab)^2 = 1 \implies abab = 1 \implies ab = b^{-1}a^{-1} = ba.$$

Therefore a, b commute and G is abelian. $\qquad\square$

Lemma 10.4.5. *If $|G| = 6$ then $G \cong \mathbb{Z}_6$ or $G \cong D_3$.*

Proof. Since $6 = 2 \cdot 3$ if G was abelian then $G \cong \mathbb{Z}_2 \times \mathbb{Z}_3$. Notice that if an abelian group has an element of order m and an element of order n with $(n, m) = 1$ then it has an element of order mn (see Lemma 8.2.1). Therefore for 6 if G is abelian there is an element of order 6 and hence $G \cong \mathbb{Z}_2 \times \mathbb{Z}_3 \cong \mathbb{Z}_6$.

Now suppose that G is nonabelian. The nontrivial elements of G have orders 2, 3 or 6. If there is an element of order 6 then G is cyclic and hence abelian. If every element has order 2 then G is abelian. Therefore there is an element of order 3 say $g \in G$. The cyclic subgroup $\langle g \rangle = \{1, g, g^2\}$ then has index 2 in G and is therefore normal. Let $h \in G$ with $h \notin \langle g \rangle$. Since g, g^2 both generate $\langle g \rangle$ we must have $\langle g \rangle \cap \langle h \rangle = \{1\}$. If h also had order 3 then by Theorem 6.3.3, $|\langle g, h \rangle| = \frac{|\langle g \rangle||\langle h \rangle|}{|\langle g \rangle \cap \langle h \rangle|} = 9$ which is impossible. Therefore h must have order 2. Since $\langle g \rangle$ is normal we have $h^{-1}gh = g^t$ for $t = 1, 2$. If $h^{-1}gh = g$ then g, h commute and the group G is abelian. Therefore $h^{-1}gh = g^2 = g^{-1}$. It follows that g, h generate a subgroup of G satisfying

$$g^3 = h^2 = 1, h^{-1}gh = g^{-1}.$$

This defines a subgroup of order 6 isomorphic to D_3 and hence must be all of G. □

Lemma 10.4.6. *If* $|G| = 10$ *then* $G \cong \mathbb{Z}_{10}$ *or* $G \cong D_5$.

Proof. The proof is almost identical to that for $n = 6$. Since $10 = 2 \cdot 5$ if G were abelian $G \cong \mathbb{Z}_2 \times \mathbb{Z}_5 = \mathbb{Z}_{10}$.

Now suppose that G is nonabelian. As for $n = 6$, G must contain a normal cyclic subgroup of order 5 say $\langle g \rangle = \{1, g, g^2, g^3, g^4\}$. If $h \notin \langle g \rangle$ then exactly as for $n = 6$ it follows that h must have order 2 and $h^{-1}gh = g^t$ for $t = 1, 2, 3, 4$. If $h^{-1}gh = g$ then g, h commute and G is abelian. Notice that $h^{-1} = h$. Suppose that $h^{-1}gh = hgh = g^2$. Then since $h^2 = 1$ we have

$$g = h^2gh^2 = h(hgh)h = hg^2h.$$

This implies that

$$g^2 = hgh \implies g^4 = hg^2h = g \implies g = 1$$

which is a contradiction. Similarly $hgh = g^3$ leads to a contradiction (see the exercises). Therefore $h^{-1}gh = g^4 = g^{-1}$ and g, h generate a subgroup of order 10 satisfying

$$g^5 = h^2 = 1; h^{-1}gh = g^{-1}.$$

Therefore this is all of G and by Example 10.4.1 is isomorphic to D_5. □

This leaves the case $n = 8$, that is the most difficult. If $|G| = 8$ and G is abelian then by the fundamental theorem of finite abelian groups $G \cong \mathbb{Z}_8$ or $G \cong \mathbb{Z}_4 \times \mathbb{Z}_2$ or $G \cong \mathbb{Z}_2 \times \mathbb{Z}_2 \times \mathbb{Z}_2$. The proof of Theorem 10.4.1 is then completed with the following.

Lemma 10.4.7. *If G is a nonabelian group of order 8 then $G \cong D_4$ or $G \cong Q$.*

Proof. The nontrivial elements of G have orders 2, 4 or 8. If there is an element of order 8 then G is cyclic and hence abelian while if every element has order 2 then G is abelian. Hence we may assume that G has an element of order 4, say g. Then $\langle g \rangle$ has index 2 and is a normal subgroup. Suppose first that G has an element $h \notin \langle g \rangle$ of order 2. Then

$$h^{-1}gh = g^t \text{ for some } t = 1, 2, 3.$$

If $h^{-1}gh = g$ then as in the cases 6 and 10, $\langle g, h \rangle$ defines an abelian subgroup of order 8 and hence G is abelian. If $h^{-1}gh = g^2$ then

$$(h^{-1}gh)^2 = (g^2)^2 = g^4 = 1 \implies g = h^{-2}gh^2 = h^{-1}g^2h = g^4 \implies g^3 = 1$$

contradicting the fact that g has order 4. Therefore $h^{-1}gh = g^3 = g^{-1}$. It follows that g, h as before define a subgroup of order 8 isomorphic to D_4. Since $|G| = 8$ this must be all of G and $G \cong D_4$.

Therefore we may now assume that every element $h \in G$ with $h \notin \langle g \rangle$ has order 4. Let h be such an element. Then h^2 has order 2 so $h^2 \in \langle g \rangle$ which implies that $h^2 = g^2$. This further implies that g^2 is central, that is, commutes with everything. To see this notice that g^2 commutes with both g and h which generate the whole group. Identifying g with $\mathbf{i}$, h with $\mathbf{j}$ and g^2 with -1 we get that G is isomorphic to Q, completing the lemma and the proof of Theorem 10.4.1. $\qquad\square$

In principle this type of analysis can be used to determine the structure of any finite group although it quickly becomes impractical.

10.5   Solvability and Solvable Groups

The original motivation for Galois theory grew out of a famous problem in the theory of equations. This problem was to to determine the solvability or insolvability of a polynomial equation of degree 5 or higher in terms of a formula involving the coefficents of the polynomial and only using algebraic operations and radicals. This question arose out of the well-known quadratic formula.

The ability to solve quadratic equations and in essence the quadratic formula was known to the Babylonians some 3600 years ago. With the discovery of imaginary numbers, the quadratic formula then says that any degree 2 polynomial over $\mathbf{C}$ can be solved by radicals in terms of the coefficients. In the sixteenth century the Italian mathematician Niccolo Tartaglia discovered a similar formula in terms of radicals to solve cubic equations. This **cubic formula** is now known erroneously as **Cardano's formula** in honor of Cardano, who first published it in 1545. An earlier special version of this formula was discovered by Scipione del Ferro. Cardano's student Ferrari extended the formula to solutions by radicals for fourth-degree polynomials. The combination of these formulas says that polynomial equations of degree 4 or less over the complex numbers can be solved by radicals.

From Cardano's work until the very early nineteenth century, attempts were made to find similar formulas for degree 5 polynomials. In 1805 Ruffini proved that fifth-degree polynomial equations are insolvable by radicals in general. Therefore there exists no comparable formula for degree 5. Abel in 1825–1826 and Galois in 1831 extended Ruffini's result and proved the insolubility by radicals for all degrees 5 or greater. In doing this, Galois developed a general theory of field extensions and its relationship to group theory. This has come to be known as **Galois theory**. We will examine this theory in more detail in Chapters 14 and 15.

The solution of the insolvability of the quintic and higher order equations involved a translation of the problem into a group theory setting. For a polynomial equation to be solvable by radicals its corresponding Galois group (a concept we will introduce in Chapter 14) must be a **solvable group**. This is a group with a certain defined structure. In the next section we introduce and discuss this class of groups.

10.5.1   Solvable Groups

A **normal series** for a group G is a finite chain of subgroups beginning with G and ending with the identity subgroup $\{1\}$

$$G = G_0 \supset G_1 \supset G_2 \supset \ldots \supset G_n \supset G_{n+1} = \{1\}$$

in which each G_{i+1} is a proper normal subgroup of G_i. The factor groups G_i/G_{i+1} are called the **factors** of the series and n is the length of the series.

Definition 10.5.1. *A group G is* **solvable** *if it has a normal series with abelian factors, that is, G_i/G_{i+1} is abelian for all $i = 0, 1, \ldots, n$. Such a normal series is called a* **solvable series**.

If G is an abelian group then $G = G_0 \supset \{1\}$ provides a solvable series. Hence any abelian group is solvable. Further, the symmetric group S_3 on 3-symbols is also solvable, however nonabelian. Consider the series

$$S_3 \supset A_3 \supset \{1\}.$$

Since $|S_3| = 6$ we have $|A_3| = 3$ and hence A_3 is cyclic and therefore abelian. Further, $|S_3/A_3| = 2$ and hence the factor group S_3/A_3 is also cyclic and hence abelian. Therefore the series above gives a solvable series for S_3.

We also note that there is nothing unique about a normal series. For example the symmetric group S_4 has the following normal series among others:

$$S_4 \supset A_4 \supset V_4 \supset \mathbb{Z}_2 \supset \{1\},$$

$$S_4 \supset A_4 \supset V_4 \supset \{1\}$$
$$S_4 \supset V_4 \supset \mathbb{Z}_2 \supset \{1\}.$$

where V_4 is the Klein 4-group (see Example 9.1.3) and $\mathbb{Z}_2$ is the cyclic group of order 2. Here we make $\mathbb{Z}_2 = \langle (1,2)(3,4) \rangle$.

Lemma 10.5.1. *If G is a finite solvable group then G has a normal series with cyclic factors.*

Proof. If G is a finite solvable group then by definition it has a normal series with abelian factors. Hence to prove the lemma it suffices to show that a finite abelian group has a normal series with cyclic factors. To see that note that

$$G = G_0 \supset G_2 \supset \cdots \supset G_n \supset G_{n+1} = \{1\}$$

where for $i = 0, 1, \ldots, n$ each factor G_i/G_{i+1} is abelian. Thus each of the groups G_i/G_{i+1} is a finite abelian group.

We claim that if a finite abelian group has a normal series with noncyclic factors then we can fit groups between G_i and G_{i+1} so as to make the factors in this new longer series cyclic. In particular we show how to do this for

$G = G_0$ and $G_1 \subset G_0$. However, if we do it there, then the same procedure can be used between any two G_i and G_{i+1}.

Since G_0/G_1 is a finite abelian group, we suppose there exist $A_1 \supset A_2 \supset \dots A_k = \{1\}$ with A_j/A_{j+1} cyclic for $j = 1, 2, \dots, k-1$. Thus each A_j can be thought of as a subgroup of the abelian factor group G_0/G_1. By Corollary 7.3.2 of the correspondence theorem there must exist groups $G_{0,j}$ with $1 \le j \le k$ all containing G_1, with $G_{0,1} = G_0$ and $G_{0,k} = G_1$ such that $A_j = G_{0,j}/G_1$. But then by Theorem 7.3.3 (the third isomorphism theorem)

$$A_j/A_{j+1} = (G_{0,j}/G_1)/(G_{0,j+1}/G_1) \cong G_{0,j}/G_{o,j+1}.$$

Note that in that theorem the subgroups are assumed to be normal. Why is that clear here?

Thus the factors in the series

$$G = G_0 \supset G_{0,1} \supset G_{0,2} \supset \cdots \supset G_{0,k} = G_1$$

are all, establishing the claim.

Now let A be a nontrivial finite abelian group. We do an induction on the order of A. If $|A| = 2$ then A itself is cyclic and the result follows. Suppose that $|A| > 2$. Choose a nontrivial $a \in A$. Let $N = \langle a \rangle$ so that N is cyclic. Then we have the normal series $A \supset N \supset \{1\}$ with $N/\{1\}$ cyclic. Further, A/N has order less than A so A/N has a normal series with cyclic factors and we can use the same procedure as above to get a normal series between A and N with cyclic factors. The whole result then follows.   □

Solvability is preserved under subgroups and factor groups.

Theorem 10.5.1. *Let G be a solvable group then*
    (1) any subgroup H of G is also solvable.
    (2) any factor group G/N of G is also solvable.

Proof. (1) Let G be a solvable group and suppose that

$$G = G_0 \supset G_1 \supset \cdots \supset G_r \supset \{1\}$$

is a solvable series for G. Hence G_{i+1} is a normal subgroup of G_i for each i and the factor group G_i/G_{i+1} is abelian.

Now let H be a subgroup of G and consider the chain of subgroups

$$H = H \cap G_0 \supset H \cap G_1 \supset \cdots \supset H \cap G_r \supset \{1\}.$$

Since G_{i+1} is normal in G_i we know that $H \cap G_{i+1}$ is normal in $H \cap G_i$ (see the exercises) and hence this gives a finite normal series for H. Further, from the second isomorphism theorem (Theorem 7.3.2) we have for each i,

$$(H \cap G_i)/(H \cap G_{i+1}) = (H \cap G_i)/((H \cap G_i) \cap G_{i+1})$$

$$\cong (H \cap G_i)G_{i+1}/G_{i+1} \subset G_i/G_{i+1}.$$

However, G_i/G_{i+1} is abelian so each factor in the normal series for H is abelian. Therefore the above series is a solvable series for H and hence H is also solvable.

(2) Let N be a normal subgroup of G. Then from (1) N is also solvable. As above let

$$G = G_0 \supset G_1 \supset \cdots \supset G_r \supset \{1\}$$

be a solvable series for G. Consider the chain of subgroups

$$G/N = G_0N/N \supset G_1N/N \supset \cdots \supset G_rN/N \supset N/N = \{1\}.$$

Let $m \in G_i, n \in N$. Then since N is normal in G,

$$(mn)^{-1}G_{i+1}N(mn) = n^{-1}m^{-1}G_{i+1}mnN = n^{-1}G_{i+1}nN$$

$$= n^{-1}NG_{i+1} = NG_{i+1} = G_{i+1}N.$$

It follows that $G_{i+1}N$ is normal in G_iN for each i. Since $G_{i+1}N, N$ are normal subgroups of G_iN, the third isomorphism theorem (Theorem 7.3.3) implies that $G_{i+1}N/N$ is normal in G_iN/N and therefore the series for G/N is a normal series.

Further, again from the third isomorphism theorem we have

$$(G_iN/N)/(G_{i+1}N/N) \cong \frac{G_iN}{G_{i+1}N} = \frac{G_i(G_{i+1}N)}{G_{i+1}N} \qquad (*)$$

The above is true because N is normal in G and G_{i+1} is a subgroup of G_i implies that $G_iG_{i+1} = G_i$ (see the exercises). Now using the second isomorphism theorem (Theorem 7.3.2) we get

$$\frac{G_i(G_{i+1}N)}{G_{i+1}N} \cong \frac{G_i}{G_i \cap (G_{i+1}N)}. \qquad (**)$$

Finally the third isomorphism theorem can be applied again because G_{i+1} is normal in $G_i \cap G_{i+1}N$ (see the exercises) to obtain

$$\frac{G_i}{G_i \cap (G_{i+1}N)} \cong \frac{G_i/G_{i+1}}{(G_i \cap (G_{i+1}N))/G_{i+1}}. \qquad (\ast\ast\ast)$$

Thus $(\ast), (\ast\ast)$, and $(\ast\ast\ast)$ imply the factors

$$\frac{(G_iN/N)}{(G_{i+1}N/N)} \cong \frac{(G_i/G_{i+1})}{(G_i \cap G_{i+1}N)/G_{i+1}}.$$

However, the last group $(G_i/G_{i+1})/((G_i \cap G_{i+1}N)/G_{i+1})$ is a factor group of the group G_i/G_{i+1} which is abelian. Hence this last group is also abelian and therefore each factor in the normal series for G/N is abelian. Hence this series is a solvable series and G/N is solvable.

$\square$

The following is a type of converse of the above theorem.

Theorem 10.5.2. *Let G be a group and H a normal subgroup of G. If both H and G/H are solvable then G is solvable.*

Proof. Suppose that

$$H = H_0 \supset H_1 \supset \cdots \supset H_r \supset \{1\}$$

$$G/H = G_o/H \supset G_1/H \supset \cdots \supset G_s/H \supset H/H = \{1\}$$

are solvable series for H and G/H respectively. Then

$$G = G_0 \supset G_1 \supset \cdots \supset G_s = H \supset H_1 \supset \cdots \supset \{1\}$$

gives a normal series for G. Further, from the third isomorphism theorem

$$G_i/G_{i+1} \cong (G_i/H)/(G_{i+1}/H)$$

and hence each factor is abelian. Therefore this is a solvable series for G and hence G is solvable.

$\square$

This theorem allows us to prove that solvability is preserved under direct products.

Corollary 10.5.1. *Let G and H be solvable groups. Then their direct product $G \times H$ is also solvable.*

Proof. Suppose that G and H are solvable groups and $K = G \times H$. Here we are using the isomorphism between internal and external direct products. Recall from Theorem 8.1.3 that G can be considered as a normal subgroup of K with $K/G \cong H$. Therefore G is a solvable subgroup of K and G/K is a solvable quotient. It follows then from Theorem 10.5.2 that K is solvable.

$\square$

We saw that the symmetric group S_3 is solvable. However, the following theorem shows that the symmetric group S_n is not solvable for $n \geq 5$. This result will be crucial to the proof of the insolvability of the quintic and higher order equations.

Theorem 10.5.3. *For $n \geq 5$ the symmetric group S_n is not solvable.*

Proof. For $n \geq 5$ we saw that the alternating group A_n is simple. Further, A_n is nonabelian. Hence A_n cannot have a nontrivial normal series and so no solvable series. Therefore A_n is not solvable. If S_n were solvable for $n \geq 5$ then from Theorem 10.5.1 A_n would also be solvable. Therefore S_n must also be nonsolvable for $n \geq 5$.

$\square$

In general for a simple, solvable group we have the following.

Lemma 10.5.2. *If a group G is both simple and solvable then G is cyclic of prime order.*

Proof. Suppose that G is a nontrivial simple, solvable group. Since G is simple the only normal series for G is $G = G_0 \supset \{1\}$. Since G is solvable the factors are abelian and hence G is abelian. Again since G is simple G must be cyclic. If G were infinite then $G \cong (\mathbb{Z}, +)$. However, then $2\mathbb{Z}$ is a proper normal subgroup, a contradiction. Therefore G must be finite cyclic. If the order were not prime then for each proper divisor of the order there would be a nontrivial proper normal subgroup. Therefore G must be of prime order.

$\square$

In general a finite p-group is solvable.

Theorem 10.5.4. *A finite p-group G is solvable.*

Proof. Suppose that $|G| = p^n$. We do this by induction on n. If $n = 1$ then $|G| = p$ and G is cyclic, hence abelian and therefore solvable. Suppose that $n > 1$. Then as used previously G has a nontrivial center $Z(G)$ (see Lemma 10.4.2). If $Z(G) = G$ then G is abelian and hence solvable. If $Z(G) \neq G$ then $Z(G)$ is a finite p-group of order less than p^n. From our inductive hypothesis $Z(G)$ must be solvable. Further, $G/Z(G)$ is then also a finite p-group of order less than p^n so it is also solvable. Hence $Z(G)$ and $G/Z(G)$ are both solvable so from Theorem 10.5.2, G is solvable.

$\square$

10.5.2   The Derived Series

Let G be a group and let $a, b \in G$. The product $aba^{-1}b^{-1}$ is called the **commutator** of a and b. We write $[a,b] = aba^{-1}b^{-1}$.

Clearly $[a,b] = 1$ if and only if a and b commute.

Definition 10.5.2. *Let G' be the subgroup of G which is generated by the set of all commutators*

$$G' = gp(\{[x,y] : x, y \in G\}).$$

G' is called the **commutator** *or* (**derived**) **subgroup** *of G. We sometimes write $G' = [G,G]$.*

Theorem 10.5.5. *For any group G the commutator subgroup G' is a normal subgroup of G and G/G' is abelian. Further, if H is a normal subgroup of G then G/H is abelian if and only if $G' \subset H$.*

Proof. The commutator subgroup G' consists of all finite products of commutators and inverses of commutators. However,

$$[a,b]^{-1} = (aba^{-1}b^{-1})^{-1} = bab^{-1}a^{-1} = [b,a]$$

and so the inverse of a commutator is once again a commutator. It then follows that G' is precisely the set of all finite products of commutators, i.e., G' is the set of all elements of the form

$$h_1 h_2 \cdots h_n$$

where each h_i is a commutator of elements of G.

If $h = [a, b]$ for $a, b \in G$, and $x \in G$, $xhx^{-1} = [xax^{-1}, xbx^{-1}]$ is again a commutator of elements of G. Now from our previous comments, an arbitrary element of G' has the form $h_1 h_2 \cdots h_n$, where each h_i is a commutator. Thus $x(h_1 h_2 \cdots h_n)x^{-1} = (xh_1 x^{-1})(xh_2 x^{-1}) \cdots (xh_n x^{-1})$ and, since by the above each $xh_i x^{-1}$ is a commutator, $x(h_1 h_2 \cdots h_n)x^{-1} \in G'$. It follows that G' is a normal subgroup of G.

Here we first use the fact mentioned earlier that $[a, b] = 1$ if and only if a and b commute. We use this in the quotient group G/G'. Let aG' and bG' be any two elements of G/G'. Then

$$[aG', bG'] = aG' \cdot bG' \cdot (aG')^{-1} \cdot (bG')^{-1}$$

$$= aG' \cdot bG' \cdot a^{-1}G' \cdot b^{-1}G' = aba^{-1}b^{-1}G' = G'$$

since $[a, b] \in G'$. Now $G' = \{1\}$ in G/G'. In other words, any two elements of G/G' commute and therefore G/G' is abelian.

Now let N be a normal subgroup of G with G/N abelian. Let $a, b \in G$ then aN and bN commute since G/N is abelian. Therefore

$$[aN, bN] = aNbNa^{-1}Nb^{-1}N = aba^{-1}b^{-1}N = N.$$

It follows that $[a, b] \in N$. Therefore all commutators of elements in G lie in N and therefore the commutator subgroup $G' \subset N$.

$\square$

From the second part of Theorem 10.5.5 we see that G' is the minimal normal subgroup of G such that G/N is abelian. We call $G/G' = G_{ab}$ the **abelianization** of G.

We consider next the following inductively defined sequence of subgroups of an arbitrary group G called the **derived series**.

Definition 10.5.3. *For an arbitrary group G define $G^{(0)} = G$ and $G^{(1)} = G'$ and then inductively $G^{(n+1)} = (G^{(n)})'$. That is, $G^{(n+1)}$ is the commutator subgroup or derived group of $G^{(n)}$. The chain of subgroups*

$$G = G^{(0)} \supset G^{(1)} \supset \cdots \supset G^{(n)} \supset \cdots$$

*is called the **derived series** for G.*

Notice that since $G^{(i+1)}$ is the commutator subgroup of $G^{(i)}$ we have that $G^{(i)}/G^{(i+1)}$ is abelian. If the derived series ends in the identity in a finite number of steps then G would have a normal series with abelian factors and hence be solvable. We denote this by saying the **derived series is finite**. The converse is also true and characterizes solvable groups in terms of the derived series.

Theorem 10.5.6. *A group G is solvable if and only if its derived series is finite. That is, there exists an n such that $G^{(n)} = \{1\}$.*

Proof. If $G^{(n)} = \{1\}$ for some n then as explained above the derived series provides a solvable series for G and hence G is solvable.

Conversely suppose that G is solvable and let

$$G = G_0 \supset G_1 \supset \cdots \supset G_r = \{1\}$$

be a solvable series for G. We claim first that $G_i \supset G^{(i)}$ for all i. We do this by induction on the number of terms in the solvable series. Since G_0/G_1 is abelian it follows that $G^{(1)} = G' \subset G_1$ by Theorem 10.5.5, which starts the induction. Suppose that $G_i \supset G^{(i)}$. Then $G_i' \supset (G^{(i)})' = G^{(i+1)}$ where we have used the fact that $A \subset B$ and then $A' \subset B'$. A proof of this fact is left to the exercises. Since G_i/G_{i+1} is abelian it follows from Theorem 10.5.5 that $G_{i+1} \supset G_i'$. Therefore $G_{i+1} \supset G^{(i+1)}$ establishing the claim.

Now if G is solvable from the claim we have that $G_r \supset G^{(r)}$. However, $G_r = \{1\}$ and therefore $G^{(r)} = \{1\}$, proving the theorem.

$\square$

The length of the derived series is called the **solvability length** of a solvable group G. The set of solvable groups of **class c** consists of those solvable groups of solvability length c or less.

10.6   Composition Series and the Jordan-Hölder Theorem

The concept of a normal series is extremely important in the structure theory of groups. This is especially true for finite groups. If

$$G = G_0 \supset G_1 \supset \cdots \supset G_n \supset \{1\}$$

$$G = H_0 \supset H_1 \supset \cdots \supset H_m \supset \{1\}$$

are two normal series for the group G then the first is a **refinement** of the second if all the terms of the second occur in the first series. Further, two normal series are called **equivalent** or (**isomorphic**) if there exists a 1-1 correspondence between the factors (hence the length must be the same), that is, $n = m$, of the two series such that the corresponding factors are isomorphic. As an example consider the three normal series in S_4 already given:

$$S_4 \supset A_4 \supset V_4 \supset \mathbb{Z}_2 \supset \{1\},$$

$$S_4 \supset A_4 \supset V_4 \supset \{1\},$$

$$S_4 \supset V_4 \supset \mathbb{Z}_2 \supset \{1\}.$$

Then the first is a refinement of the second and the third but the second is not a refinement of the third.

Note that in the above examples of normal series in S_4 the only possibility for equivalent normal series is the second and third (why?). But they are not equivalent (see the exercises).

The following theorem, known as Schreier's theorem, is crucial to proving the Jordan-Hölder theorem which is a type of unique prime decomposition theorem for finite groups.

Theorem 10.6.1. *(Schreier's Theorem) Any two normal series for a group G have equivalent refinements.*

The proof of this result, while within the scope of this book in difficulty, is long and detailed, so we refer to [R] or [AMG] for a complete proof. We note, however, that the proof depends on the additional result known as Zassenhaus's lemma which we also just state.

Theorem 10.6.2. *(Zassenhaus's Lemma) Let G be a group with subgroups G_1, G_2, H_1, H_2 such that H_1 is a normal subgroup of G_1 and H_2 is a normal subgroup of G_2. Then*

$$(G_1 \cap H_2)H_1 \text{ is a normal subgroup of } (G_1 \cap G_2)H_1,$$

$$(H_1 \cap G_2)H_2 \text{ is a normal subgroup of } (G_1 \cap G_2)H_2,$$

and further

$$\frac{(G_1 \cap G_2)H_1}{(G_1 \cap H_2)H_1} \cong \frac{(G_1 \cap G_2)H_2}{(H_1 \cap G_2)H_2}.$$

A proper normal subgroup N of a group G is called **maximal** in G if there does not exist any normal subgroup $N \subset M \subset G$ with all inclusions proper. This is the group theoretic analog of a maximal ideal which we discuss in Chapter 11. An alternative characterization is the following: N is a maximal normal subgroup of G if and only if G/N is simple. (See the exercises for a proof.)

A normal series where each factor is simple can have no refinements (see the exercises).

Definition 10.6.1. *A* **composition series** *for a group G is a normal series where all the inclusions are proper and such that G_{i+1} is maximal in G_i. Equivalently a composition series is a normal series where each factor is simple.*

For example, in the case of the previously given normal series for S_4 only the first

$$S_4 \supset A_4 \supset V_4 \supset \mathbb{Z}_2 \supset \{1\}$$

is a composition series S_4. A composition series for A_4 would be

$$A_4 \supset V_4 \supset \mathbb{Z}_2 \supset \{1\}.$$

Note that $A_4 \supset V_4 \supset \{1\}$ would not be a composition series for A_4. (Why?)

It is possible that an arbitrary group does not have a composition series or, even if it does have one, a subgroup of it may not have one. Of course, a finite group does have a composition series.

In the case in which a group, G, does have a composition series the following important theorem, called the Jordan-Hölder theorem, provides a type of unique factorization.

Theorem 10.6.3. *(Jordan-Hölder Theorem) If a group G has a composition series, then any two composition series are equivalent, that is, the composition factors are unique.*

Proof. Suppose we are given two composition series. Applying Theorem 10.6.1 we get that the two composition series have equivalent refinements. But the only refinement of a composition series is one obtained by introducing repetitions. If in the 1-1 correspondence between the factors of these refinements, the paired factors equal to $\{e\}$ are disregarded, that is, if we drop the repetitions, we clearly get that the original composition series are equivalent. $\square$

The simple groups are important because they play a role in finite group theory somewhat analogous to that of the primes in number theory. In particular, an arbitrary finite group, G, can be broken down into simple components. These uniquely determined simple components are, according to the Jordan-Hölder theorem, the factors of a composition series for G.

We close by giving an application of the Jordan-Hölder theorem. In particular we use it to prove the uniqueness part of the fundamental theorem of arithmetic (see Theorem 4.4.3). The theorem states that every positive integer > 1 can be factored uniquely as a product of primes with uniqueness up to ordering. The existence of such a factorization is easy to see and can be proven as in Chapter 4.

Using the Jordan-Hölder theorem we can easily show the uniqueness part of the above result. To do this suppose that

$$n = p_1 p_2 \cdots p_s$$

and

$$n = q_1 q_2 \cdots q_t$$

where the p_i and q_j are primes. Then denoting as usual the cyclic group of order k by $\mathbb{Z}_k$ we have two composition series for $\mathbb{Z}_n$:

$$\mathbb{Z}_n \supset \mathbb{Z}_{p_2 \cdots p_s} \supset \mathbb{Z}_{p_3 \cdots p_s} \supset \ldots \subset \mathbb{Z}_{p_s} \supset \{0\}$$

$$\mathbb{Z}_n \supset \mathbb{Z}_{q_2 \cdots q_t} \supset \mathbb{Z}_{q_3 \cdots q_t} \supset \ldots \subset \mathbb{Z}_{q_t} \supset \{0\}$$

The Jordan-Hölder theorem implies that these must be equivalent; hence we must have $s = t$ and by suitably arranging $p_i = q_i; i = 1, \ldots, s$. Thus we have established the uniqueness of prime factorization for positive integers as an application of the Jordan-Hölder theorem

10.7 Exercises

EXERCISES FOR SECTION 10.1

10.1.1. Let G be a group and for each $a \in G$ define the map $a : g \mapsto a^{-1}ga$. Show that this is a homomorphism on G. In the language of this chapter this says that conjugation is an action on a group G. Explain why this is true according to Definition 10.1.1. Here you must say what the set A is and what the permutation $\pi_g \in S_A$ is for any $g \in G$.

10.1.2. Let G act on the set A and let $A_1 \subset A$ be nonempty. Show that the stabilizer

$$\mathrm{Stab}_G(A_1) = \{g \in G : g : A_1 \to A_1\}$$

is a subgroup of G.

10.1.3. Let G act on a set A. On A define $a_1 \sim a_2$ if there exists a $g \in G$ with $g(a_1) = g(a_2)$. Show that this is an equivalence relation on A. (We say that a_1 and a_2 are congruent modulo G.) The equivalence classes under this equivalence relation then partition the set A. These subsets are called the **orbits** under G. If $a_1 \in A$ the orbit of a_1 under the action of G is denoted G_{a_1}.

10.1.4. Show that if G is a group then conjugation is an action on the set of subgroups of G.

10.1.5. Let G be the subgroup of S_5 given by $G = \langle (1,2,3)(4,5) \rangle$. G acts on $\{1,2,3,4,5\}$.

    (a) What is $|G|$? Write out the elements of G.
    (b) Determine the orbits: G_1, G_2, G_4.
    (c) Find $\mathrm{Stab}_G(1)$.
    (d) Use the results of (b) and (c) to illustrate Theorem 10.1.1.

10.1.6. Let G be a group and H a subgroup of G. Let A be the set of left cosets of H in G and for each $g \in G$ define the map $\pi_g : A \to A$ by $\pi_g : g_1 H \mapsto gg_1 H$. Show that this is an action of G on A, that is, the map $\pi_g : A \to A$, is a permutation on A and the map $g \to \pi_g$ is a homomorphism from G into S_A.

EXERCISES FOR SECTION 10.2

10.2.1. Let G be a group and $Z(G)$ its center. Show that $Z(G)$ is a normal subgroup of G.

10.2.2. Let G be a group and $Z(G)$ its center. Show that G is abelian if and only if $G = Z(G)$.

10.2.3. Let $G = S_3$ the symmetric group on three symbols. Express G as $\{1, a, a^2, b, ab, a^2b\}$ where $a^3 = 1, b^2 = 1, ab = ba^2$. Find the conjugacy classes of G.

10.2.4. Let $G = S_3$ the symmetric group on three symbols. Express G as $\{1, a, a^2, b, ab, a^2b\}$ where $a^3 = 1, b^2 = 1, ab = ba^2$. Find the centralizer of the element a and the centralizer of the element b. Show that the center of G is just the identity.

10.2.5. Do Prolems 10.2.3 and 10.2.4 for the quaternion group

$$Q = \{1, \mathbf{i}, \mathbf{j}, \mathbf{ij}, -1, -\mathbf{i}, -\mathbf{j}, -\mathbf{ij}\}$$

where -1 is central, $(-1)^2 = 1$, $\mathbf{i}^2 = \mathbf{j}^2 = -1$ and $\mathbf{ij} = -\mathbf{ji}$. For Problem 10.2.4; find the centralizer of $\mathbf{i}$ and the centralizer if $\mathbf{j}$. Also find the center of the quaternions, that is, $Z(Q)$.

10.2.6. In the quaternion group Q find the normalizer of the cyclic subgroup $\langle \mathbf{ij} \rangle$.

10.2.7. Let G be a group and H a subgroup of G with $[G : H] = n$. The map $g \mapsto \pi_g$ of Problem 10.1.6 then defines a homomorphism of G into S_n. Show that the kernel of this homomorphism is

$$\{g \in G : g_1^{-1}gg_1 \in H \text{ for all } g_1 \in G\}.$$

10.2.8. Suppose that G is a finite group with precisely two conjugacy classes. Prove that $|G| = 2$.

(HINT: Decompose G into conjugacy classes where one of the classes is $Cl(1)$ where 1 is the identity of G. Write the class equation of G. What is $|Cl(1)|$? Use Lagrange's theorem to write the other terms in this equation in terms of $|G|$. Solve your equation for $|G|$ and use this to prove $|G| = 2$.)

EXERCISES FOR SECTION 10.3

10.3.1. If p is a prime then a p-group is a group where all elements have order a power of p. Show that if G is a finite p-group then $|G| = p^n$ for some natural number n.

(HINT: Use Cauchy's theorem.)

10.3.2. Prove the converse of Problem 10.3.1, that is, if $|G| = p^n$ then every element has order a power of p.

A consequence of Problems 10.3.1 and 10.3.2 is an if and only if statement. Write out that if and only if statement.

10.3.3. Show that any group G of order 12 must have a normal Sylow subgroup.

10.3.4. Show that any group of order 45 has a normal 3-Sylow subgroup.

10.3.5. Suppose that p is a prime and $n = 2p$. Let G be a group of order $2p$. Show that if G is nonabelian then G is isomorphic to D_p the dihedral group of order $2p$. Recall that this is the group of symmetries of a regular p-gon and is generated by r, f with $r^p = 1$, $f^2 = 1$ and $fr = r^{-1}f$.

Therefore as a consequence there are exactly two nonisomorphic groups of order $2p$ (see Theorem 10.3.9).

10.3.6. Let p, q be distinct odd primes with $p > q$. Let G be a group of order pq. Show that G has a normal p-Sylow subgroup.

10.3.7. Let p, q be distinct odd primes with $p > q$. Let G be a group of order $p^n q$ with $n \geq 1$. Show that G has a normal p-Sylow subgroup.

10.3.8. Let P be a p-Sylow subgroup of G which exists by Theorem 10.3.2. Show that every subgroup H of G which contains the normalizer of P is its own normalizer. The normalizer of P, $N_G(P) = \{x \in G : xPx^{-1} = P\}$. Here we write $N(P)$ for $N_G(P)$.

(HINT: Suppose that $N(P) \subset H$. P is a p-Sylow subgroup of H (why?). Let $x \in N(H)$. We need to show that $x \in H$. Note that xPx^{-1} is also a p-Sylow subgoup of H (why?). Now use Theorem 10.3.4 (The second Sylow theorem) applied to the above p-Sylow subgroups of H. From this and the fact that $N(P) \subset H$ establish the desired result.)

10.3.9. Let P be a p-Sylow subgroup of G. Let N be the normalizer of P in G and N_1 the normalizer of N. Show that $N = N_1$. (Put in another way $N(N(P)) = N(P)$.

(HINT: Use Problem 10.3.8.)

10.3.10. Prove that the subgroup $K_1 K_2 \cdots K_k$ in the proof of Theorem 10.3.7 is actually the direct product $K_1 \times K_2 \times \cdots \times K_k$ and thus its order is $|K_1||K_2| \ldots |K_k|$.

(HINT: Since each $K_i \subset G$ and G is abelian all we need to show is that

$$K_i \cap K_1 \ldots K_{i-1} \cap K_{i+1} \ldots K_k = \{1\}$$

for each $i = 1, 2, \ldots, k$. Use an argument on the orders of the elements in the K_i and K_j with $i \neq j$ to show this must be true.)

10.3.11. In Theorem 10.3.5 (the third Sylow theorem) show that the number of p-Sylow subgroups $1 + pk$ must divide the order of G. (Show that it is the index of a subgroup of G.) Now show that if $|G| = p^n m$ with $(p, m) = 1$ then the number of p-Sylow subgroups must divide m.

10.3.12. (a) In Example 10.3.1, show why there are $p^n - p^2$ choices for w_3. So finally $p^n - p^{n-1}$ choices for w_n. This explains why this gives the stated formula for $|GL(n, p)|$.

(b) Show that the set P of all $n \times n$ upper triangular matrices with 1's on the main diagnonal is a subgroup of $GL(n, p)$.

(c) Explain why $|P| = p^{n(n-1)/2}$. This makes P a p-Sylow subgroup of G.

10.3.13. Let G be a finite group with $p||G|$. If P is a p-Sylow subgroup of G show that any conjugate gPg^{-1} is also a p-Sylow subgroup of G.

10.3.14. Let G be a finite group and N a normal subgroup of G with $|N| = p^n$ for some n. Prove that P is contained in every p-Sylow subgroup of G. (Use the second Sylow theorem 10.3.4)

10.3.15. Let G be a finite group and P a p-Sylow subgroup of G. Prove that if $x \in N_G(P)$ and the order of x is a power of p then $x \in P$.

(HINT: Here again $N_G(P) = \{x \in G : xP = Px\}$. Use the fact that $x \in N_G(P)$ to conclude that $\langle x \rangle P$ is a subgroup. Then use the formula for the order of a product of two subgroups (Theorem 6.3.3) to show that if $x \notin P$ you get a contradiction of the maximality of P.)

10.3.16. Let G be a finite group and P a p-Sylow subgroup of G. Prove that P is the only p-Sylow subgroup contained in $N_G(P)$.

(HINT: See Problem 10.3.15 for the definition of $N_G(P)$ and use the second Sylow theorem.)

10.3.17. Using the Sylow theorems show that there are no simple groups of orders 18 and 204.

10.3.18. Find all the 3-Sylow and 2-Sylow subgroups of A_4.

(HINT: The result of Problem 10.3.9 may be helpful here. Also recall something about V_4 in A_4.)

10.3.19. It was shown in the text that if G is a finite abelian group of order n then for any $d|n$ with $d > 0$ there exists a subgroup of order d of G. Does this always imply that G has an element of order d?

(HINT: Consider $\mathbb{Z}_2 \times \mathbb{Z}_2 \times \mathbb{Z}_2$.)

10.3.20. Prove that if a group has exactly $1 + p$ Sylow subgroups of order p^n with p a prime then any two of these subgroups have exactly p^{n-1} elements in common.

(HINT: Suppose that P_1, P_2 are any 2 p-Sylow subgroups. Use an argument like that in the proof of Theorem 10.3.5. Let P_1 act by conjugation on Ω_p the set of all p-Sylow subgroups. Now P_1 normalizes itself so there is one orbit namely that of P_1 that has size 1. Every other orbit has size a power of p since the sizes are the index of a nontrivial subgroup of P_1. But there are only $1 + p$ conjugates and hence the other orbit (an orbit of P_2) must contain p elements. Now use the result of Problem 10.3.8 to note that an element $x \in P_1$ normalizes P_2 if and only if $x \in P_1 \cap P_2$. Now count the size of $|P_1 \cap P_2|$ using Lagrange's theorem applied to $\mathrm{Stab}_{P_1}(P_2)$.)

10.3.21. Prove that there is no simple group of order 42.

This completes the proof of Theorem 10.3.11.

EXERCISES FOR SECTION 10.4

10.4.1. Show that in the quaternion group the generators $\mathbf{i}, \mathbf{j}$ satisfy the relations

$$\mathbf{i}^4 = \mathbf{j}^4 = 1, \mathbf{i}^2 = \mathbf{j}^2 \text{ and } \mathbf{ij} = \mathbf{i}^2\mathbf{ji}.$$

10.4.2. Show in the proof of Lemma 10.4.6 that $hgh = g^3$ leads to a contradiction where $g, h \in G, |G| = 10$ and g, h have the properties given in that proof.

10.4.3. Show in the proof of Lemma 10.4.7 that if $|G| = 8$ and $g, h \in G$ with $o(g) = o(h) = 4$ and $h \notin \langle g \rangle$ then it must be that G is generated by g and h.

(HINT: Consider what the possible orders of the subgroup $\langle g \rangle \cdot \langle h \rangle \subset G$ are. Note that this must be a subgroup (why?). Then show that $\langle g \rangle \cdot \langle h \rangle = G$.)

10.4.4. Classify all groups with $|G| \leq 11$. We did not go to 12 because to find all groups of order 12 is beyond the scope of this book.

10.4.5. Suppose that the group G is generated by x and y and suppose the the the subgroup $\langle x \rangle$ is normal in G. Show that every element in G can be written in the form $x^n y^m$ for some integers m and n. Does this mean that the group G is finite?

10.4.6. Recall that a complex in a group G is any nonemepty subset of G not necessarily a subgroup. Now let G be a group, H a subgroup of G and S any complex in G. Prove that if $HS = H$ or $SH = H$ then $S \subset H$.

10.4.7. Find all subgroups of order 4 in the dihedral group D_4 of order 8.

(HINT: Make a multiplication table for D_4.)

10.4.8. Prove that if G is a group and H and K are finite subgroups whose orders are relatively prime then $H \cap K = \{1\}$.

10.4.9. If G is a nonabelian group of order 12 and there exists $x \in G$ with $o(x) = 6$ can $G \cong A_4$? Why or why not?

10.4.10. Suppose that G is a finite group and H is a subgroup of G. Also suppose that

$$G = \cup_{i=1}^n H x_i$$

is a decomposition of G into disjoint right cosets of H. Prove that a decomposition of G into disjoint left cosets of H is given by

$$G = \cup_{i=1}^n x_i^{-1} H.$$

EXERCISES FOR SECTION 10.5

10.5.1. In the proof of Lemma 10.5.1 it uses the third isomorphism theorem. In the text it asks why the appropriate subgroups must be normal. Explain why.

10.5.2. Let p be a prime. Recall that a finite p-group is any group G with $|G| = p^n$ for some n. Show that all finite p-groups are solvable.
(HINT: Do induction on n.)

10.5.3. Let p, q be distinct primes. Show that any group of order pq must be solvable.

10.5.4. Compute the commutator subgroups of S_3, A_4 and D_4.

10.5.5. If H is any subgroup of the group G and G_{i+1} is normal in G_i, then prove that $H \cap G_{i+1}$ is normal in $H \cap G_i$.

10.5.6. Show that in the proof of Theorem 10.5.1 (2):
  (a) Since N is normal in G and G_{i+1} is normal in G_i then $G_i N = G_i(G_{i+1}N)$ as indicated in the text.
  (b) Since G_{i+1} is normal in G_i then G_{i+1} is normal in $G_i \cap (G_{i+1}N)$.

10.5.7. If G is a group let G' denote its derived group or commutator subgroup. If A, B are subgroups of a group G and $A \subset B$ show that $A' \subset B'$.

10.5.8. Consider the three normal series for S_4 given in the text;

$$S_4 \supset A_4 \supset V_4 \supset \mathbb{Z}_2 \supset \{1\},$$

$$S_4 \supset A_4 \supset V_4 \supset \{1\},$$

$$S_4 \supset V_4 \supset \mathbb{Z}_2 \supset \{1\}.$$

(a) Explain why these are indeed normal series.

(b) Explain why only the second and third series above could be equivalent.

(c) Show that the second and third normal series as above are not equivalent.

10.5.9. Given the two series for $\mathbb{Z}_{15}$,

$$\mathbb{Z}_{15} \supset \langle 5 \rangle \supset \{0\},$$

$$\mathbb{Z}_{15} \supset \langle 3 \rangle \supset \{0\}.$$

(a) Explain why these are normal series.

(b) Show that they are equivalent.

10.5.10. Give a direct proof from the definition of solvability that a finite p-group is solvable. Note that this is unlike the proof given in Theorem 10.5.4.

(HINT: Use the fact that a finite p-group has order p^n for some n and Theorem 10.3.6.)

10.5.11. Use the alternative definition of solvability given in Theorem 10.5.6 to give a direct proof of Theorem 10.5.1. This is the definition of solvability using the derived series.

(HINT: Use the result of Problem 10.5.7 together with what the image of the commutator subgroup is under the canonical homomorphism.)

EXERCISES FOR SECTION 10.6

10.6.1 Show that the two normal series for S_4

$$S_4 \supset A_4 \supset V_4 \supset \{1\}$$

$$S_4 \supset V_4 \supset \mathbb{Z}_2 \supset \{1\}$$

are not equivalent.

10.6.2. Use the corollaries to the correspondence theorem (Corollary 7.3.1 and Corollary 7.3.2) to show that N is a maximal normal subgroup of G if and only if G/N is simple.

10.6.3. Prove that any group G has a normal series.

10.6.4. Prove that the only abelian simple groups are groups of prime order.

10.6.5. Prove that any infinite abelian group does not have a composition series.

(HINT: Suppose it does and arrive at a contradiction. You will also need the results of the previous exercise.)

10.6.6. Prove that a normal series where each factor is simple can have no refinements. (Here we assume that the inclusions in the series are all proper.)

10.6.7. Prove that a finite group is solvable if and only if the factors of a composition series are cyclic of prime orders.

10.6.8. Prove that any finite group has a composition series.

(HINT: Use induction on the order of the group.)

10.6.9. Prove that if G is a group which has a composition series then any normal subgroup of G and any factor group of G also have composition series with factors isomorphic to the composition factors of G.

(HINT: Mimic the proof of Theorem 10.5.1.)

Chapter 11

Topics in Ring Theory

11.1 Ideals in Rings

In Chapter 7 we looked at normal subgroups of a group G and showed how they can be used to construct factor groups or quotient groups. These played a large role in the structure theory of groups and led to the important group isomorphism theorems. In this chapter we look at the analogous ideas in rings. The special types of subrings that correspond to normal subgroups are called **ideals** and will be used to build **factor rings**. Further there is a tie between factor rings and ring homomorphisms. That is, the exact analog of the group isomorphism theorem. This is called the **ring isomorphism theorem**.

Definition 11.1.1. *Let R be a ring and I a subring of R. Then I is a*

(1) **left ideal** *if $a \in I$ and r is any element of R then $ra \in I$, that is, I is closed under left multiplication from the whole ring R;*

(2) **right ideal** *if $a \in I$ and r is any element of R then $ar \in I$, that is, I is closed under right multiplication from the whole ring R;*

(3) **two-sided ideal** *or just an* **ideal** *if $a \in I$ and r is any element of R then $ra \in I$ and $ar \in I$, that is, I is closed under both left and right multiplication from the whole ring R.*

We denote the fact that I forms an ideal in R by $I \triangleleft R$.

In a commutative ring R, the concepts of left and right ideals coincide so we only talk of ideals. For the rest of this chapter we will only be considering commutative rings unless otherwise indicated. The zero element alone, $\{0\}$, and the whole ring R are **trivial ideals** of R.

A very important example of ideals that always exist in a commutative ring R are principal ideals. These are analogous to cyclic subgroups of a group.

Lemma 11.1.1. *Let R be a commutative ring and $a \in R$. Then the set*

$$\langle a \rangle = aR = \{ar : r \in R\}$$

is an ideal of R. This ideal is called the **principal ideal generated by** *a.*

Proof. We must verify that aR is a subring and that it is closed under multiplication from R. Here, since R is assumed commutative, we must only show multiplication from either the left or the right. Since $a \in R$ we have $a \cdot 0 = 0 \in aR$ and therefore aR is nonempty. If $u = ar_1, v = ar_2$ are two elements of aR then

$$u \pm v = ar_1 \pm ar_2 = a(r_1 \pm r_2) \in aR,$$

so aR is closed under addition and additive inverses (to see that just take $r_1 = 0$ in the above). Further,

$$uv = (ar_1)(ar_2) = a(r_1 ar_2) \in aR$$

and so aR is closed under multiplication. It follows that aR forms a subring.
    Finally let $u = ar_1 \in aR$ and $r \in R$. Then

$$ru = rar_1 = a(rr_1) \in aR \text{ and } ur = ar_1 r = a(r_1 r) \in aR,$$

Hence aR is closed under multiplication from R and therefore aR forms an ideal. Recall that $a \in \langle a \rangle$ if the ring R has an identity.   □

Notice that if $n \in \mathbb{Z}$, then the principal ideal generated by n is precisely the ring $n\mathbb{Z}$, that we have already examined. Hence for each $n > 1$ the subring $n\mathbb{Z}$ is actually an ideal. We can show more.

Theorem 11.1.1. *Any subring of $\mathbb{Z}$ is of the form $n\mathbb{Z}$ for some n. Hence each subring of $\mathbb{Z}$ is actually a principal ideal.*

Proof. Let S be a subring of $\mathbb{Z}$. If $S = \{0\}$ then $S = 0\mathbb{Z}$ so we may assume that S has nonzero elements. Since S is a subring if it has nonzero elements it must have positive elements (since it has the additive inverse of any element in it).

Let S^+ be the set of positive elements in S. From the remarks above this is a nonempty set and so there must be a least positive element n. We claim that $S = n\mathbb{Z}$.

Let m be a positive element in S. By the division algorithm

$$m = qn + r,$$

where either $r = 0$ or $0 < r < n$. Suppose that $r \neq 0$. Then

$$r = m - qn.$$

Now $m \in S$ and $n \in S$. Since S is a subring it is closed under addition so that $qn \in S$. But S is a subring so $m - qn \in S$. It follows that $r \in S$. But this is a contradiction since n was the least positive element in S. Therefore $r = 0$ and $m = qn$. Hence each positive element in S is a multiple of n.

Now let m be a negative element of S. Then $-m \in S$ and $-m$ is positive. Hence $-m = qn$ and thus $m = (-q)n$. Therefore every element of S is a multiple of n and so $S = n\mathbb{Z}$.

It follows that every subring of $\mathbb{Z}$ is of this form and therefore every subring of $\mathbb{Z}$ is an ideal and in fact a principal ideal. $\qquad\square$

We mention that this result, every subring being an ideal, is true in $\mathbb{Z}$ but not always true. For example $\mathbb{Z}$ is a subring of $\mathbb{Q}$ but not an ideal.

The concept of a principal ideal can be extended within a commutative ring R to an ideal generated by finitely many elements.

Lemma 11.1.2. *Let R be a commutative ring and $a_1, \ldots, a_n \in R$ be a finite set of elements in R. Then the set*

$$\langle a_1, \ldots, a_n \rangle = \{r_1 a_1 + r_2 a_2 + \cdots + r_n a_n : r_i \in R\}$$

*is an ideal of R. This ideal is called the **ideal generated by** $a_1, \ldots, a_n$.*

Proof. The proof is a direct extension of the proof of Lemma 11.1.1 and we leave the details to the exercises. Notice further that if the ring R has an identity then the elements $a_1, \ldots, a_n$ are in the ideal $\langle a_1, \ldots, a_n \rangle$. $\qquad\square$

If R is a commutative ring with an identity, we recall that R is a field if every nonzero element has a multiplicative inverse. Fields can be further classified as precisely those commutative rings with identity that have only trivial ideals.

Theorem 11.1.2. *Let R be a commutative ring with an identity $1 \neq 0$. Then R is a field if and only if the only ideals in R are $\{0\}$ and R.*

Proof. Suppose that R is a field and $I \lhd R$ is an ideal. We must show that either $I = \{0\}$ or $I = R$. Suppose that $I \neq \{0\}$ then we must show that $I = R$.

Since $I \neq \{0\}$ there exists an element $a \in I$ with $a \neq 0$. Since R is a field this element a has an inverse a^{-1}. Since I is an ideal it follows that $a^{-1}a = 1 \in I$. Let $r \in R$ then, since $1 \in I$, we have $r \cdot 1 = r \in I$. Hence $R \subset I$ and thus $R = I$.

Conversely suppose that R is a commutative ring with an identity whose only ideals are $\{0\}$ and R. We must show that R is a field or equivalently that every nonzero element of R has a multiplicative inverse.

Let $a \in R$ with $a \neq 0$. Since R is a commutative ring and $a \neq 0$, the principal ideal aR is a nonzero ideal in R (why?). Hence $aR = R$. Therefore the multiplicative identity $1 \in aR$. It follows that there exists an $r \in R$ with $ar = 1$. Hence since R is commutative, a has a multiplicative inverse and R must be a field. $\square$

11.2   Factor Rings and the Ring Isomorphism Theorem

Given an ideal I in a ring R we can build a new ring called the **factor ring** or **quotient ring** of R modulo I. The special condition on the subring I that $rI \subset I$ and $Ir \subset I$ for all $r \in R$, that makes it an ideal, also guarantees that this construction is a ring. This is exactly analogous within the ring context to the construction of a factor group from a normal subgroup.

Definition 11.2.1. *Let I be an ideal in a ring R. Then a **coset** of I is a subset of R of the form*

$$r + I = \{r + i : i \in I\}$$

with r a fixed element of R.

Since an ideal is a subring it is an additive subgroup of the additive group of the ring. Hence a coset is just a group coset of the additive group of the ring. From the corresponding result in group theory we then get the following.

Lemma 11.2.1. *Let I be an ideal in a ring R. Then the cosets of I partition R, that is, any two cosets either coincide or are disjoint.*

Just as in groups we have the following.

Lemma 11.2.2. *Let I be an ideal in R and R any ring (not necessarily commutative). Then*

(1) $rI \subset I$ and $Ir \subset R$ for all $r \in R$.
(2) $i + I = I + i = I$ for all $i \in I$.
(3) $S + I = I + S = I$ for any set $S \subset I$.

Proof. Let I be an ideal in R, i.e., a two-sided ideal in the statement in (1). $rI \subset I$ and $Ir \subset I$ is just a restatement of the definition of an ideal, because for example $x \in rI$ means $x = ri$ for some $i \in I$. I being closed under left multiplication from the whole ring R means $ri \in I$ so $x \in I$ showing the inclusion.

Clearly (2) is a special case of (3). For if $S + I = I + S = I$ for any set S, then it is true for $S = \{i\}$ which would give (2). Hence we only have to show (3). But
$$S + I = \{s + i : s \in S, i \in I\}$$
and since $S \subset I$ and I is a subring $s + i \in I$ for all $s \in S$ and for all $i \in I$. Thus $S + I \subset I$. Now let $x \in I$. Then $x = s + (x - s)$ where $s \in S$. This show that $x \in S + I$ because I is a subring and $x - s \in I$. Therefore we have the reverse inclusion $I \subset S + I$. □

Now on the set of all cosets of an ideal we will build a new ring.

Theorem 11.2.1. *Let I be an ideal in a ring R with R not necessarily commutative. Let R/I be the set of all distinct cosets of I in R, that is*

$$R/I = \{r + I : r \in R\}.$$

We define addition and multiplication on R/I in the following manner

$$(r_1 + I) + (r_2 + I) = (r_1 + r_2) + I$$

$$(r_1 + I) \cdot (r_2 + I) = (r_1 \cdot r_2) + I.$$

*Then R/I forms a ring called the **factor ring** or **quotient ring** of R modulo I. The zero element of R/I is $0 + I$ and the additive inverse of $r + I$ is $-r + I$.*

Further if R is commutative then R/I is commutative and if R has an identity, 1, then R/I has an identity $1 + I$.

Proof. The proofs that R/I satisfies the ring axioms under the definitions above is straightforward. For example

$$(r_1 + I) + (r_2 + I) = (r_1 + r_2) + I = (r_2 + r_1) + I = (r_2 + I) + (r_1 + I)$$

and so addition is commutative. The rest are left to the exercises.

What must be shown is that both addition and multiplication are well-defined. That is, if

$$r_1 + I = r'_1 + I \text{ and } r_2 + I = r'_2 + I$$

then

$$(r_1 + I) + (r_2 + I) = (r'_1 + I) + (r'_2 + I)$$

and

$$(r_1 + I) \cdot (r_2 + I) = (r'_1 + I) \cdot (r'_2 + I).$$

Now if $r_1 + I = r'_1 + I$ then $r_1 \in r'_1 + I$ and so $r_1 = r'_1 + i_1$ for some $i_1 \in I$. Similarly if $r_2 + I = r'_2 + I$ then $r_2 \in r'_2 + I$ and so $r_2 = r'_2 + i_2$ for some $i_2 \in I$. Then by Lemma 11.2.2 we have

$$(r_1 + I) + (r_2 + I) = (r'_1 + i_1 + I) + (r'_2 + i_2 + I) = (r'_1 + I) + (r'_2 + I)$$

since $i_1 + I = I$ and $i_2 + I = I$. Similarly

$$(r_1 + I) \cdot (r_2 + I) = (r'_1 + i_1 + I) \cdot (r'_2 + i_2 + I) =$$

$$= (r'_1 \cdot r'_2) + I$$

since all the other products are in the ideal I and then Lemma 11.2.2 implies the last equality. Here we must be careful to keep the order multiplication in the correct order since we are not assuming that R is commutative. But by definition

$$r'_1 \cdot r'_2 + I = (r'_1 + I)(r'_2 + I).$$

This shows that addition and multiplication are well-defined. It also shows why the ideal property is necessary. $\square$

EXAMPLE 11.2.1. Let $R = \mathbb{Z}$ the integers. As we have seen each subring is an ideal and of the form $n\mathbb{Z}$ for some natural number n. The factor ring $\mathbb{Z}/n\mathbb{Z}$ is called the **residue class ring modulo** n denoted $\mathbb{Z}_n$. Notice that we can take as cosets

$$0 + n\mathbb{Z}, 1 + n\mathbb{Z}, \ldots, (n - 1) + n\mathbb{Z}.$$

Addition and multiplication of cosets is then just addition and multiplication modulo n, and as we can see, this is just a formalization of the ring $\mathbb{Z}_n$, that we have already looked at. Recall that $\mathbb{Z}_n$ is an integral domain if and only if n is prime and $\mathbb{Z}_n$ is a field for precisely the same n. If $n = 0$ then $\mathbb{Z}/n\mathbb{Z}$ is the same as $\mathbb{Z}$.

EXAMPLE 11.2.2. Consider the subring $\langle 3 \rangle = 3\mathbb{Z}$ and $\langle 9 \rangle = 9\mathbb{Z}$ of $\mathbb{Z}$. They are ideals and $9\mathbb{Z} \subset 3\mathbb{Z}$. Consider the factor ring $(3\mathbb{Z})/(9\mathbb{Z})$. We claim this factor ring has size 3 even though both $3\mathbb{Z}$ and $9\mathbb{Z}$ are infinite.

To see this note that:

$$3\mathbb{Z} = \{\ldots, -9, -6, -3, 0, 3, 6, 9, \ldots\},$$

$$9\mathbb{Z} = 0 + 9/Z = \{\ldots, -9, 0, 9, \ldots.\},$$

$$3 + 9\mathbb{Z} = \{\ldots, -6, 3, 12, \ldots\}.$$

$$6 + 9\mathbb{Z} = \{\ldots, -3, 6, 15, \ldots\}.$$

Hence $3\mathbb{Z} = 9\mathbb{Z} \cup (3 + 9\mathbb{Z}) + (6 + 9\mathbb{Z})$ and further this union is a disjoint union. Hence these represent the three cosets of $9\mathbb{Z}$ in $3\mathbb{Z}$ and therefore $|(3\mathbb{Z})/(9\mathbb{Z})| = 3$. Further as an abelian group $3\mathbb{Z}/9\mathbb{Z} \cong \mathbb{Z}_3$. However as a ring it has only zero multiplication.

EXAMPLE 11.2.3. Let $R = M_2(\mathbb{Z})$ the ring of two by two matrices with integral entries. Let

$$I = \{a \in M_2(\mathbb{Z}) : \text{ all entries of } A \text{ are even}\}$$

It is not hard to show that I is an ideal in $M_2(\mathbb{Z})$ and hence the factor ring is defined. We claim that the quotient ring $M_2(\mathbb{Z})/I$ has size 16 and leave the proof to the exercises.

We now show that ideals and factor rings are closely related to ring homomorphisms. Recall from Chapter 3 the following definitions and results concerning ring homomorphisms. In Chapter 3 we spoke about (1) — (3) in the following definition.

Definition 11.2.2. *Let R and S be rings. Then a mapping $f : R \to S$ is a* **ring homomorphism** *if*

$$f(r_1 + r_2) = f(r_1) + f(r_2) \text{ for any } r_1, r_2 \in R$$

$$f(r_1 \cdot r_2) = f(r_1) \cdot f(r_2) \text{ for any } r_1, r_2 \in R.$$

In addition:

    *(1) f is an **epimorphism** if it is surjective.*

    *(2) f is an **monomorphism** if it is injective.*

    *(3) f is an **isomorphism** if it is bijective, that is, both surjective and injective. In this case R and S are said to be **isomorphic rings** which we denote by $R \cong S$.*

    *(4) f is an **endomorphism** if $R = S$, that is, a ring homomorphism from a ring to itself.*

    *(5) f is an **automorphism** if $R = S$ and f is an isomorphism.*

Lemma 11.2.3. *Let R and S be rings and let $f : R \to S$ be a ring homomorphism. Then*

    (1) $f(0) = 0$ where the first 0 is the zero element of R and the second is the zero element of S.

    (2) $f(-r) = -f(r)$ for any $r \in R$.

Proof. We obtain $f(0) = 0$ from the equation $f(0) = f(0+0) = f(0) + f(0)$. Now add $-f(0)$ to both sides of this equation to obtain $f(0) = 0$. Hence $0 = f(0) = f(r - r) = f(r + (-r)) = f(r) + f(-r)$, that is, $f(-r) = -f(r)$. $\square$

Definition 11.2.3. *Let R and S be rings and let $f : R \to S$ be a ring homomorphism. Then the **kernel** of f, denoted $ker(f)$, is*

$$ker(f) = \{r \in R : f(r) = 0\}.$$

*The **image** of f, denoted $im(f)$, is the range of f within S. That is*

$$im(f) = \{s \in S : \text{ there exists } r \in R \text{ with } f(r) = s\}.$$

With these results we now state and prove the **ring isomorphism theorem**. The proof is almost identical to the group isomorphism theorem.

Theorem 11.2.2. *(Ring Isomorphism Theorem) Let R and S be rings and let*

$$f : R \to S$$

be a ring homomorphism. Then

    (1) $ker(f)$ is an ideal in R, $im(f)$ is a subring of S and

$$R/(ker(f)) \cong im(f).$$

    (2) Conversely suppose that I is an ideal in a ring R. Then the map $f : R \to R/I$ given by $f(r) = r + I$ for $r \in R$ is a ring homomorphism whose kernel is I and whose image is R/I.

The theorem says that the concepts of ideal of a ring and kernel of a ring homomorphism coincide, that is, each ideal is the kernel of a homomorphism and the kernel of each ring homomorphism is an ideal.

Proof. Let $f : R \to S$ be a ring homomorphism and let $I = \ker(f)$. We show first that I is an ideal. Now for any homomorphism $f(0) = 0$ and hence $0 \in I$ and therefore I is not empty. If $r_1, r_2 \in I$ then $f(r_1) = f(r_2) = 0$. It follows from the homomorphism property that

$$f(r_1 \pm r_2) = f(r_1) \pm f(r_2) = 0 + 0 = 0$$

$$f(r_1 \cdot r_2) = f(r_1) \cdot f(r_2) = 0 \cdot 0 = 0.$$

This shows that the kernel is closed under addition and multiplication. But if we take $r_1 = 0$ and $-$ in the first of the above two equations — here we get that I is also closed under additive inverses. Therefore I is a subring.

Now let $i \in I$ and $r \in R$. Then

$$f(r \cdot i) = f(r) \cdot f(i) = f(r) \cdot 0 = 0 \text{ and } f(i \cdot r) = f(i) \cdot f(r) = 0 \cdot f(r) = 0$$

and hence I is an ideal.

Consider the factor ring R/I. Let $f^* : R/I \to \operatorname{im}(f)$ by $f^*(r+I) = f(r)$. We show that f^* is an isomorphism.

First we show that this map f^* is well-defined. Suppose that $r_1 + I = r_2 + I$ then $r_1 - r_2 \in I = \ker(f)$. It follows that $f(r_1 - r_2) = 0$ so $f(r_1) = f(r_2)$. Hence $f^*(r_1 + I) = f^*(r_2 + I)$ and the map f^* is well-defined.

Now since f is a ring homomophism

$$f^*((r_1 + I) + (r_2 + I)) = f^*((r_1 + r_2) + I) = f(r_1 + r_2)$$

$$= f(r_1) + f(r_2) = f^*(r_1 + I) + f^*(r_2 + I)$$

and

$$f^*((r_1 + I) \cdot (r_2 + I)) = f^*((r_1 \cdot r_2) + I) = f(r_1 \cdot r_2)$$

$$= f(r_1) \cdot f(r_2) = f^*(r_1 + I) \cdot f^*(r_2 + I).$$

Hence f^* is a homomorphism. We must now show that it is injective and surjective.

Suppose that $f^*(r_1 + I) = f^*(r_2 + I)$. Then $f(r_1) = f(r_2)$ so that $f(r_1 - r_2) = 0$. Hence $r_1 - r_2 \in \ker(f) = I$. Therefore $r_1 \in r_2 + I$ and thus $r_1 + I = r_2 + I$ and the map f^* is injective.

Finally let $s \in \text{im}(f)$. Then there exists an $r \in R$ such that $f(r) = s$. Then $f^*(r + I) = s$ and the map f^* is surjective and hence an isomorphism. This proves the first part of the theorem.

To prove the second part let I be an ideal in R and R/I the factor ring. Consider the map $f : R \rightarrow R/I$ given by $f(r) = r + I$. From the definition of addition and multiplication in the factor ring R/I it is clear that this is a homomorphism. Consider the kernel of f. If $r \in \text{ker}(f)$ then $f(r) = r + I = 0 = 0 + I$. This implies that $r \in I$ and hence the kernel of this map is exactly the ideal I completing the theorem.

$\square$

EXAMPLE 11.2.4. Applying the ring isomorphism theorem to the example at the end of Section 3.4, that is the map $f : \mathbb{Z} \rightarrow \mathbb{Z}_n$ defined by $f(z) = [z]$, gives $\mathbb{Z}/n\mathbb{Z} \cong \mathbb{Z}_n$.

EXAMPLE 11.2.5. Let $R = \{A \in M_2(\mathbb{Z}) : A = \begin{pmatrix} x & 0 \\ y & x \end{pmatrix}\}$. It is not hard to show that R is a subring of $M_2(\mathbb{Z})$. Consider the map $f : R \rightarrow \mathbb{Z}$ given by $f(A) = x$. It is also not hard to show that f is a ring homomorphism with $\text{ker}(f) = \{\begin{pmatrix} 0 & 0 \\ y & 0 \end{pmatrix} : y \in \mathbb{Z}\}$. Thus the ring isomorphism theorem gives $R/\text{ker}(f) \equiv \mathbb{Z}$.

As a nontrivial application of the ring isomorphism theorem we prove here that given a ring R with identity then it essentially contains, up to isomorphism, one of the rings $\mathbb{Z}$ or $\mathbb{Z}_n$. Similarly given a field F it must contain either $\mathbb{Q}$ or $\mathbb{Z}_p$ for some prime p.

Suppose that R is a ring with identity 1. Recall that if $n \in \mathbb{N}$, $n \cdot 1$ means

$$n \cdot 1 = 1 + 1 + \cdots + 1$$

added to itself n times. If $n \in \mathbb{Z}$ but $n < 0$ then $n \cdot 1 = ((-n) \cdot 1)$. Finally if $n = 0$ then $n \cdot 1 = 0$. Recall from Chapter 3 that $\text{char}(R)$ is the least positive integer n such that $n \cdot 1 = 0$. If no such n exists then $\text{char}(R) = 0$. (See Lemma 3.6.2 and Definition 3.6.2.)

Lemma 11.2.4. *If R is any ring with an identity 1 then the mapping $\theta : \mathbb{Z} \rightarrow R$ defined by $\theta(n) = n \cdot 1$ is a ring homomorphism from $\mathbb{Z}$ into R.*

Proof. The fact that θ is a ring homomorphism follows from Lemma 3.2.5(1) and (5). $\square$

Corollary 11.2.1. *If R is any ring with identity and $char(R) = n > 0$ then R contains a subring isomorphic to $\mathbb{Z}_n$. If $char(R) = 0$ then R contains a subring isomorphic to $\mathbb{Z}$.*

Proof. The map $\theta : \mathbb{Z} \to R$ given by $\theta(m) = m \cdot 1$ is a ring homomorphism by Lemma 11.2.3. Then the kernel is an ideal in $\mathbb{Z}$ and the image is a subring of R. However all ideals in $\mathbb{Z}$ are of the form $t\mathbb{Z}$ for some $t \in \mathbb{Z}$ by Theorem 4.2.3. If $char(R) = n > 0$ then $\ker(\theta) = n\mathbb{Z}$. Then the image of $\theta = \operatorname{im}(\theta) = \theta(\mathbb{Z}) \subset R$ is isomorphic to $\mathbb{Z}/n\mathbb{Z} = \mathbb{Z}_n$ by the ring isomorphism theorem. Hence R contains an isomorphic copy of $\mathbb{Z}_n$. On the other hand if $char(R) = 0$ then $m \cdot 1 \neq 0$ for all $m \neq 0$ and hence $\ker(\theta) = \{0\}$. Thus θ is injective and the image of $\theta = \operatorname{im}(\theta) = \theta(\mathbb{Z}) \subset R$ is isomorphic to $\mathbb{Z}$. Therefore R contains an isomorphic copy of $\mathbb{Z}$. $\square$

Corollary 11.2.2. *Let F be any field. If $char(F) = p$ for some prime p, then F contains a subfield isomorphic to $\mathbb{Z}_p$. In the only other case where $char(F) = 0$, then F contains a subfield isomorphic to $\mathbb{Q}$.*

Proof. If $char(F) \neq 0$ then by the previous corollary, F contains a subring isomorphic to $\mathbb{Z}_n$ where $n = char(F)$. But for a field n must be a prime p. Thus in this case F contains a subfield isomorphic to $\mathbb{Z}_p$.

In the other case F must contain a subring isomorphic to $\mathbb{Z}$. Then by Corollary 5.2.1, F must contain a field of fractions for this subring and this field of fractions must be isomorphic to $\mathbb{Q}$. $\square$

We note that there is an analogous theorem for each algebraic structure. We have already seen the group isomorphism theorem. In Chapter 13 we will see the corresponding result for vector spaces.

11.3 Prime and Maximal Ideals

In the previous sections we defined ideals I in a ring R and then the factor ring R/I of R modulo the ideal I. We saw further that if R is commutative then R/I is also commutative and if R has an identity then so does R/I. This raises further questions concerning the structure of factor rings. In particular we can ask under what conditions does R/I form an integral domain and under what conditions does R/I form a field. These questions lead us to define certain special properties of ideals, called prime ideals and maximal ideals.

For motivation let us look back at the integers $\mathbb{Z}$. Recall that each proper ideal in $\mathbb{Z}$ has the form $n\mathbb{Z}$ for some $n > 1$ and the resulting factor ring $\mathbb{Z}/n\mathbb{Z}$ is isomorphic to $\mathbb{Z}_n$. We proved the following result.

Theorem 11.3.1. *$\mathbb{Z}_n = \mathbb{Z}/n\mathbb{Z}$ is an integral domain if and only if $n = p$ a prime. Further, $\mathbb{Z}_n$ is a field again if and only if $n = p$ is a prime.*

Hence for the integers $\mathbb{Z}$, a factor ring is a field if and only if it is an integral domain. We will see later that this is not true in general. However, what is clear is that the special ideals $n\mathbb{Z}$ leading to integral domains and fields are precisely when n is a prime. We look at the ideals $p\mathbb{Z}$ with p a prime in two different ways and then use these in subsequent sections to give the general definitions. We first recall Euclid's Lemma from Number Theory (see Chapter 4).

Lemma 11.3.1. *(Euclid) If p is a prime and $p|ab$ then $p|a$ or $p|b$.*

We now recast this lemma in two different ways in terms of the ideal $p\mathbb{Z}$. Notice that $p\mathbb{Z}$ consists precisely of all the multiples of p. Hence $p|ab$ is equivalent to $ab \in p\mathbb{Z}$. Therefore as a direct consequence of Euclid's Lemma we have.

Lemma 11.3.2. *If p is a prime and $ab \in p\mathbb{Z}$ then $a \in p\mathbb{Z}$ or $b \in p\mathbb{Z}$.*

This conclusion will be taken as the definition of a **prime ideal** in the next section. We next look at a different property of the ideal $p\mathbb{Z}$ when p is a prime.

Lemma 11.3.3. *If p is a prime and $p\mathbb{Z} \subset n\mathbb{Z}$ then $n = 1$ or $n = p$. That is, every ideal in $\mathbb{Z}$ containing $p\mathbb{Z}$ with p a prime is either all of $\mathbb{Z}$ or $p\mathbb{Z}$.*

Proof. Suppose that $p\mathbb{Z} \subset n\mathbb{Z}$. Then $p \in n\mathbb{Z}$ so p is a multiple of n. Since p is a prime it follows easily that either $n = 1$ or $n = p$.

$\square$

In Section 11.3.2 the conclusion of this lemma will be taken as the definition of a **maximal ideal**.

11.3.1 Prime Ideals and Integral Domains

Motivated by Lemma 11.3.2 we make the following general definition for commutative rings R with identity.

Definition 11.3.1. *Let R be a commutative ring with an identity $1 \neq 0$. An ideal P in R with $P \neq R$ is a **prime ideal** if whenever $ab \in P$ with $a, b \in R$ then either $a \in P$ or $b \in P$.*

This property of an ideal is precisely what is necessary and sufficient to make the factor ring R/I an integral domain.

Theorem 11.3.2. *Let R be a commutative ring with an identity $1 \neq 0$ and let P be a nontrivial ideal in R. Then P is a prime ideal if and only if the factor ring R/P is an integral domain.*

Proof. Let R be a commutative ring with an identity $1 \neq 0$ and let P be a prime ideal. We show that R/P is an integral domain. From Theorem 11.2.1 we have that R/P is again a commutative ring with an identity. Therefore we must show that there are no zero divisors in R/P. Suppose that $(a + I)(b + I) = 0$ in R/P. The zero element in R/P is $0 + P = P$ and hence

$$(a + P)(b + P) = 0 = 0 + P \implies ab + P = 0 + P \implies ab \in P.$$

However, P is a prime ideal so we must then have $a \in P$ or $b \in P$. If $a \in P$ then $a + P = P = 0 + P$ so $a + P = 0$ in R/P. The identical argument works if $b \in P$. Therefore there are no zero divisors in R/P and hence R/P is an integral domain.

Conversely suppose that R/P is an integral domain. We must show that P is a prime ideal. Suppose that $ab \in P$. Then $(a + P)(b + P) = ab + P = 0 + P$. Hence in R/P we have

$$(a + P)(b + P) = 0.$$

However, R/P is an integral domain so it has no zero divisors. It follows that either $a + P = 0$ and hence $a \in P$ or $b + P = 0$ and $b \in P$. Therefore either $a \in P$ or $b \in P$ so P is a prime ideal.

$\square$

In a commutative ring R we can define a multiplication of ideals. We then obtain an exact analog of Euclid's lemma. Since R is commutative each ideal is two-sided.

Definition 11.3.2. *Let R be a commutative ring with an identity $1 \neq 0$ and let A and B be ideals in R. Define*

$$AB = \{a_1 b_1 + \cdots + a_n b_n : a_i \in A, b_i \in B, n \in \mathbf{N}\}.$$

That is, AB is the set of finite sums of products ab with $a \in A$ and $b \in B$.

Lemma 11.3.4. *Let R be a commutative ring with an identity $1 \neq 0$ and let A and B be ideals in R. Then AB is an ideal. Moreover $AB \subset A \cap B$.*

Proof. We must verify that AB is a subring and that it is closed under multiplication from R. Le $r_1, r_2 \in AB$. Then

$$r_1 = a_1 b_1 + \cdots + a_n b_n \text{ for some } a_i \in A, b_i \in B$$

and

$$r_2 = a_1' b_1' + \cdots + a_m' b_m' \text{ for some } a_i' \in A, b_i' \in B$$

Then

$$r_1 \pm r_2 = a_1 b_1 + \cdots + a_n b_n \pm a_1' b_1' \pm \cdots \pm a_m' b_m'$$

where each time we have a $-$ sign we can write $-a_i' b_i' = +(-a_i') b_i'$. Hence $r_1 \pm r_2$ is in AB. Further,

$$r_1 \cdot r_2 = a_1 b_1 a_1' b_1' + \cdots + a_n b_n a_m' b_m'.$$

Consider for example the first term $a_1 b_1 a_1' b_1'$. Since R is commutative this is equal to

$$(a_1 a_1')(b_1 b_1').$$

Now $a_1 a_1' \in A$ since A is a subring and $b_1 b_1' \in B$ since B is a subring. Hence this term is in AB and similarly for each of the other terms. Therefore $r_1 r_2 \in AB$ and hence as before this is sufficient for AB to be a subring.

Now let $r \in R$ and consider rr_1. This is then

$$rr_1 = ra_1 b_1 + \cdots + ra_n b_n.$$

Now $ra_i \in A$ for each i since A is an ideal. Hence each summand is in AB and then $rr_1 \in AB$. Therefore AB is an ideal, since R is commutative. See the exercises for a proof that $AB \subset A \cap B$.

$\square$

Lemma 11.3.5. *Let R be a commutative ring with an identity $1 \neq 0$ and let A and B be ideals in R. If P is a prime ideal in R then $AB \subset P$ implies that $A \subset P$ or $B \subset P$.*

Proof. Suppose that $AB \subset P$ with P a prime ideal and suppose that B is not contained in P. We show that $A \subset P$. Since $AB \subset P$ each product $a_i b_j \in P$. Choose a $b \in B$ with $b \notin P$ and let a be an arbitrary element of A. Then $ab \in P$. Since P is a prime ideal this implies either $a \in P$ or $b \in P$. But by assumption $b \notin P$ so $a \in P$. Since a was arbitrary we have $A \subset P$. □

11.3.2 Maximal Ideals and Fields

Now, motivated by Lemma 11.3.3 we define a maximal ideal.

Definition 11.3.3. *Let R be a ring and I an ideal in R. Then I is a* **maximal ideal** *if $I \neq R$ and if J is an ideal in R with $I \subset J$ then $I = J$ or $J = R$.*

If R is a commutative ring with an identity this property on an ideal I is precisely what is necessary and sufficient so that R/I is a field.

Theorem 11.3.3. *Let R be a commutative ring with an identity $1 \neq 0$ and let I be an ideal in R. Then I is a maximal ideal if and only if the factor ring R/I is a field.*

Proof. Suppose that R is a commutative ring with an identity $1 \neq 0$ and let I be an ideal in R. Suppose first that I is a maximal ideal and we show that the factor ring R/I is a field.

Since R is a commutative ring with an identity the factor ring R/I is also a commutative ring with an identity. We must show then that each nonzero element of R/I has a multiplicative inverse. Suppose then that $\bar{r} = r + I \in R/I$ is a nonzero element of R/I. It follows that $r \notin I$. Consider the set $\langle r, I \rangle = \{rx + i : x \in R, i \in I\}$. This is also an ideal (see the exercises) called the ideal generated by r and I, denoted $\langle r, I \rangle$. Clearly $I \subset \langle r, I \rangle$ and since $r \notin I$ and $r = r \cdot 1 + 0 \in \langle r, I \rangle$ it follows that $\langle r, I \rangle \neq I$. Since I is a maximal ideal it follows that $\langle r, I \rangle = R$, the whole ring. Hence the identity element $1 \in \langle r, I \rangle$ and so there exist elements $x \in R$ and $i \in I$ such that $1 = rx + i$. But then $1 \in (r + I)(x + I)$ and so $1 + I = (r + I)(x + I)$. Since $1 + I$ is the multiplicative identity of R/I, then since R is commutative, $x + I$

is the multiplicative inverse of $r + I$ in R/I. Since $r + I$ was an arbitrary nonzero element of R/I it follows that R/I is a field.

Now suppose that R/I is a field for an ideal I. We show that I must be maximal. Suppose then that I_1 is an ideal with $I \subset I_1$ and $I \neq I_1$. We must show that I_1 is all of R. Since $I \neq I_1$ there exists and $r \in I_1$ with $r \notin I$. Therefore the element $r + I$ is nonzero in the factor ring R/I and since R/I is a field it must have a multiplicative inverse $x + I$. Hence $(r+I)(x+I) = rx+I = 1+I$ and therefore there is an $i \in I$ with $1 = rx+i$. Since $r \in I_1$ and I_1 is an ideal we get that $rx \in I_1$. Further, since $I \subset I_1$ it follows that $rx + i \in I_1$ and so $1 \in I_1$. If r_1 is an arbitrary element of R then $r_1 \cdot 1 = r_1 \in I_1$. Hence $R \subset I_1$ and so $R = I_1$. Therefore I is a maximal ideal. $\square$

Recall that a field is already an integral domain. Combining this with the ideas of prime and maximal ideals we obtain:

Theorem 11.3.4. *Let R be a commutative ring with an identity $1 \neq 0$. Then each maximal ideal is a prime ideal.*

Proof. Suppose that R is a commutative ring with an identity and I is a maximal ideal in R. Then from Theorem 11.3.3 we have that the factor ring R/I is a field. But a field is an integral domain so R/I is an integral domain. Therefore from Theorem 11.3.2 we have that I must be a prime ideal. $\square$

The converse is not true in general. That is, there are prime ideals that are not maximal. Consider for example $R = \mathbb{Z}$, the integers, and $I = \{0\}$. Then I is an ideal and $R/I = \mathbb{Z}/\{0\} \cong \mathbb{Z}$ is an integral domain. Hence $\{0\}$ is a prime ideal. However, $\mathbb{Z}$ is not a field so $\{0\}$ is not maximal. Note, however, that in the integers $\mathbb{Z}$ a proper ideal is maximal if and only if it is a prime ideal. We note that it can be shown that in any ring with identity maximal ideals exist.

11.4 Principal Ideal Domains and Unique Factorization

We saw that in the integers $\mathbb{Z}$ every ideal is a principal ideal. In the next chapter we will give other examples where this occurs but here we prove that this actually implies unique factorization into primes. We take the condition that every ideal is principal as a definition of a special type of integral domain.

Definition 11.4.1. *A* **principal ideal domain** *(which we will abbreviate as PID) is an integral domain in which every ideal is a principal ideal.*

In this language we have that the integers $\mathbb{Z}$ are a principal ideal domain. In the next chapter we will prove that the set of polynomials $K[x]$ with coefficients from a field K is also a principal ideal domain. We will also give examples of rings that have nonprincipal ideals. Here we show that being a PID implies unique factorization into irreducibles, which we define below.

First we must define the relevant elements in an integral domain. We must distinguish between prime elements and irreducible elements.

Definition 11.4.2. *Let R be an integral domain.*

(1) Suppose that $a, b \in R$. Then a is a **factor** *or* **divisor** *of b or a* **divides** *b if there exists a $c \in R$ with $b = ac$. We denote this, as in the integers, by $a|b$. If a is a factor of b then b is called a* **multiple** *of a.*

(2) An element $a \in R$ is a **unit** *in R if a has a multiplicative inverse within R, that is, there exists an element $a^{-1} \in R$ with $aa^{-1} = 1$ or a divides 1.*

(3) A **prime element** *of R is an element $p \neq 0$ such that p is not a unit and if $p|ab$ then $p|a$ or $p|b$.*

(4) An **irreducible** *in R is an element $c \neq 0$ such that c is not a unit and if $c = ab$ then a or b must be a unit.*

(5) a and b in R are **associates** *if there exists a unit $e \in R$ with $a = eb$.*

Now we redefine a general unique factorization domain.

Definition 11.4.3. *An integral domain D is a* **unique factorization domain** *or* **UFD** *if for each $d \in D$ then either $d = 0$, d is a unit or d has a factorization into irreducibles which is unique up to ordering and unit factors. This means that if*

$$d = p_1 \cdots p_m = q_1 \cdots q_k$$

where the p_i and q_j are irreducibles, then $m = k$ and each p_i is an associate of some q_j.

In this terminology the integers $\mathbb{Z}$ are a UFD. They are also a PID. To end this chapter we will prove that every PID is actually a UFD. In the next chapter we will present other examples of UFD's and show that not every integral domain is a unique factorization domain and also that not every UFD is a PID.

To prove this we need to show that there are several relationships in integral domains that are equivalent to unique factorization.

Definition 11.4.4. *Let R be an integral domain.*

    (1) R has property (A) if and only if for each nonunit $a \neq 0$ there are irreducible elements $q_1, \ldots, q_r \in R$ satisfying $a = q_1 \cdots q_r$.

    (2) R has property (A') if and only if for each nonunit $a \neq 0$ there are prime elements $p_1, \ldots, p_r \in R$ satisfying $a = p_1 \cdots p_r$.

    (3) R has property (B) if and only if whenever $q_1, \ldots, q_r$ and $q_1', \ldots, q_s'$ are irreducible elements of R with

$$q_1 \cdots q_r = q_1' \cdots q_s'$$

then $r = s$ and there is a permutation $\pi \in S_r$ such that for each $i \in \{1, \ldots, r\}$ the elements q_i and $q_{\pi(i)}'$ are associates (uniqueness up to ordering and unit factors).

    (4) R has property (C) if and only if each irreducible element of R is a prime element.

Notice that properties (A) and (B) together are equivalent to what we defined as **unique factorization**. Hence an integral domain satisfying (A) and (B) is a UFD. We show next that there are other equivalent formulations. We need first the following result that says in any integral domain a prime element is always an irreducible element.

Lemma 11.4.1. *Let R be an integral domain. Then any prime element $p \in R$ is an irreducible in R.*

Proof. Suppose that p is a prime and $p = ab$. We must show that either a or b is a unit. By definition of a prime we know that $p|a$ or $p|b$. Suppose that $p|a$. Thus there exists $x \in R$ such that $a = xp$. But then $a = a \cdot 1 = xp = x(ab) = a(xb)$. We know that $a \neq 0$ since p is a prime and hence by cancellation in R we get $1 = xb$. Thus b is a unit and therefore p is an irreducible. $\qquad\qquad\square$

The converse of this is not true. That is, there exists rings in which irreducible elements are not prime. Examples of these are given in the exercises at the end of the next chapter, Chapter 12. Later in this section we will prove, however, that in a PID the two concepts are equivalent.

Theorem 11.4.1. *In an integral domain R the following are equivalent:*
   (1) R is a UFD.
   (2) R satisfies properties (A) and (B).
   (3) R satisfies properties (A) and (C).
   (4) R satisfies propery (A').

Proof. As remarked before the statement of the theorem, by definition (A) and (B) are equivalent to unique factorization. We show here that (2),(3) and (4) are equivalent.

First we show that (2) implies (3).

Suppose that R satisfies properties (A) and (B). We must show that it also satisfies (C), that is, we must show that if $q \in R$ is irreducible then q is prime. Suppose that $q \in R$ is irreducible and $q|ab$ with $a, b \in R$. Then we have $ab = cq$ for some $c \in R$. If a is a unit from $ab = cq$ we get that $b = a^{-1}cq$ and $q|b$. Similary if b is a unit we get $q|a$. Therefore we may assume that neither a nor b is a unit.

If $c = 0$ then since R is an integral domain either $a = 0$ or $b = 0$ and $q|a$ or $q|b$. We may assume then that $c \neq 0$.

If c is a unit then $q = (c^{-1}a)b$ and since q is irreducible either $c^{-1}a$ or b are units. If $c^{-1}a$ is a unit then a is also a unit so if c is a unit either a or b is a unit, contrary to our assumption.

Therefore we may assume that $c \neq 0$ and c is not a unit. Since if a or b is 0 this would imply that q divides either a or b respectively, we can assume that none of a, b, c is 0 or a unit. From property (A) we have

$$a = q_1 \cdots q_r,$$

$$b = q_1' \cdots q_s',$$

$$c = q_1'' \cdots q_t'',$$

where $q_1, \ldots q_r, q_1', \ldots, q_s', q_1'', \ldots q_t''$ are all irreducibles. But since $ab = cq$ we have

$$q_1 \cdots q_r q_1' \cdots q_s' = q_1'' \cdots q_t'' \cdot q.$$

From property (B) q is an associate of some q_i or q_j'. Hence $q|q_i$ or $q|q_j'$. It follows that $q|a$ or $q|b$ and therefore q is a prime element.

That (3) implies (4) is direct.

We show that (4) implies (2).

Suppose that R satisfies property (A'). We must show that it satisfies both (A) and (B). (A) follows from (A') since prime elements are irreducible from Lemma 11.4.1.

To show that (B) also holds we now need to show that if (A') holds any irreducible is prime.

Suppose that q is irreducible. Then from (A') we have

$$q = p_1 \cdots p_r$$

with each p_i prime. Writing $q = p_1(p_2 \cdots p_r)$ it follows without loss of generality that $p_2 \cdots p_r$ is a unit since p_1 is a prime and so it is a nonunit. But if there exists $x \in R$ such that $(p_2 \cdots p_r)x = 1$ then $p_2(p_3 \cdots p_r x) = 1$ and it follows that p_2 is a unit. Similarly since R is commutative $p_3, \ldots, p_r$ are units for $i = 2, \ldots, r$. But the p_i were supposed to be primes and a prime by definition is not a unit. Thus the prime factorization of q which must exist by (A') reduces to $q = p_1$ and q is prime. Therefore (A) holds.

We now show that (B) holds. Let

$$q_1 \cdots q_r = q_1' \cdots q_s'$$

where q_i, q_j' are all irreducibles and hence primes. Then

$$q_1' | q_1 \cdots q_r$$

and so $q_1' | q_i$ for some i by extending the definition of prime (see the exercises). Without loss of generality suppose $q_1' | q_1$. Then $q_1 = a q_1'$. Since q_1 is irreducible it follows that a is a unit and q_1 and q_1' are associates. It follows then by cancelling q_1' that

$$a q_2 \cdots q_r = q_2' \cdots q_s'$$

since R is an integral domain. Property (B) holds by an argument similar to the proof of Theorem 4.4.3 (see the exercises). $\square$

We say that an ascending chain of ideals in R

$$I_1 \subset I_2 \subset \cdots I_n \subset \cdots$$

becomes **stationary** if there exists an m such that $I_r = I_m$ for all $r \geq m$.

Theorem 11.4.2. *Let R be an integral domain. If each ascending chain of principal ideals in R becomes stationary, then R satisfies property (A).*

Proof. Suppose that $a \neq 0$ is not a unit in R. Suppose that a is not a product of irreducible elements. Clearly then a cannot itself be irreducible. Hence $a = a_1 b_1$ with $a_1, b_1 \in R$ and a_1, b_1 are not units. If both a_1 and b_1 can be expressed as a product of irreducible elements then so can a. Without loss of generality then suppose that a_1 is not a product of irreducible elements.

Since $a_1 | a$ we have the inclusion of ideals $aR \subseteq a_1 R$. If $a_1 R = aR$ then $a_1 \in aR$ and $a_1 = ar = a_1 b_1 r$ which implies by cancellation that b_1 is a unit contrary to our assumption. Therefore $aR \neq a_1 R$ and the inclusion is proper. Since a_1 is not a product of irreducible elements then a_1 itself cannot be irreducible. Thus $a_1 = a_2 b_2$ with $a_1, b_2 \in R$ and a_2, b_2 are nonunits. Then just as above it can be shown that $a_1 R \subset a_2 R$ properly maybe after relabelling a_2 and b_2. By iteration then we obtain a strictly increasing chain of ideals

$$aR \subset a_1 R \subset \cdots a_n R \cdots$$

From our hypothesis on R this must become stationary, contradicting the argument above that the inclusion is proper. Therefore a must be a product of irreducibles.

$\square$

We show next that in a PID, prime and irreducible are equivalent concepts.

Lemma 11.4.2. *Let R be a principal ideal domain, i.e., PID. Then an element $p \in R$ is a prime if and only if p is an irreducible in R.*

Proof. From Lemma 11.4.1 a prime element must be irreducible.

To prove the converse let a be an irreducible element of R and suppose that $a | bc$. We must show that $a | b$ or $a | c$. Consider the set of elements in R,

$$I = \{ax + by : x, y \in R\}.$$

It is not hard to prove that I is an ideal in R (see the exercises). But since R is a PID there exists $c \in R$ such that $cR = \langle c \rangle = I$. But clearly $a \in I$ so $a = cr$ for some $r \in R$. But a was an irreducible so either c is a unit or r is a unit. If c is a unit then $I = R$ (see the exercises) and so $1 = ax + by$ for some $x, y \in R$. But then $c = acx + bcy$ and since a divides both terms on the right hand side it follows that $a | c$.

The other possibility is that r is a unit. But then $\langle a \rangle = \langle c \rangle = I$. Clearly $b \in I$ so that there exists $s \in R$ such that $b = as$. Therefore in this case $a | b$.

$\square$

Theorem 11.4.3. *Each principal ideal domain R is a unique factorization domain or every PID is a UFD.*

Proof. Suppose that R is a principal ideal domain. R satisfies property (C) by Lemma 11.4.1, so to show that it is a unique factorization domain we must show that it also satisfies property (A). From the previous theorem, Theorem 11.4.2, it suffices to show that each ascending chain of principal ideals becomes stationary. Consider such an ascending chain of principal ideals

$$a_1 R \subset a_2 R \subset \cdots \subset a_n R \subset \cdots$$

Now let

$$I = \bigcup_{i=1}^{\infty} a_i R.$$

Now I is an ideal in R (see the exercises) and hence a principal ideal since R is a PID. Therefore $I = aR$ for some $a \in R$. Since I is a union there exists an m such that $a \in a_m R$. Therefore $I = aR \subset a_m R$ and hence $I = a_m R$ and $a_i R \subset a_m R$ for all $i \geq m$. Therefore the chain becomes stationary and from Theorem 11.4.2. Therefore R satisfies property (A) which is what we needed to show.

$\square$

Since we showed that the integers $\mathbb{Z}$ are a PID we can recover the fundamental theorem of arithmetic from Theorem 11.4.3.

11.5   Exercises

EXERCISES FOR SECTION 11.1

11.1.1. Assume that R is a commutative ring. Prove that $\langle a_1, \ldots, a_n \rangle$ is an ideal in R. Note to prove that this is an ideal you have to show that it's a subring and closed under multiplication from all of R.

11.1.2. In the proof of Theorem 11.1.2 explain why if $a \neq 0$ then the principal ideal generated by a, $\langle a \rangle$, must also be nonzero.

11.1.3. Suppose that R is a ring with an identity 1 but not necessarily commutative. Let I be a left ideal in R and suppose that I contains an invertible element. Show that $I = R$. Then state and prove the same result for right ideals.

11.1.4. If I and K are ideals (left, right or two-sided) in a ring R prove that $I \cap K$ is an ideal. Generalize this to any number of ideals.

11.1.5. Find integers m, n such that $\langle m \rangle \cup \langle n \rangle$ is not an ideal in $\mathbb{Z}$.

11.1.6. Show that the rationals $\mathbb{Q}$ is not an ideal in the real numbers $\mathbb{R}$ although it is a subfield.

11.1.7. Find a subring of the direct sum $\mathbb{Z} + \mathbb{Z}$ that is not an ideal in $\mathbb{Z} + \mathbb{Z}$.

11.1.8. Recall from Problem 3.1.15 that the center of a ring R is

$$Z(R) = \{z \in R : zr = rz \text{ for all } r \in R\}.$$

Also from Problem 3.3.35, the center of $M_2(\mathbb{Z})$ is

$$\{\begin{pmatrix} a & 0 \\ 0 & a \end{pmatrix} : a \in \mathbb{Z}\}.$$

Verify this and show that the center is not an ideal in $M_2(Z)$.

EXERCISES FOR SECTION 11.2

11.2.1. Let R be a ring, not necessarily commutative and I an ideal. In the text the operation $+$ and $\cdot$ were defined on the quotient set R/I and shown to be well-defined and that $+$ was commutative. Verify that the other ring axioms hold for R/I.

11.2.2. Show that if R is commutative than so is R/I. Show also that if R has an identity 1 then so does R/I.

11.2.3. In Example 11.2.3 show that the set

$$I = \{A \in M_2(\mathbb{Z}) : \text{ all entries of A are even}\}$$

is an ideal in $M_2(\mathbb{Z})$.

11.2.4. Show that $|M_2(\mathbb{Z})/I| = 16$ where I is as defined in Problem 11.2.3.

11.2.5. Show that $9\mathbb{Z}$ is an ideal in $3\mathbb{Z}$ and then write addition and multiplication tables for the quotient ring $3\mathbb{Z}/9\mathbb{Z}$.

11.2.6. In the proof of Theorem 11.2.2 explain why $\ker(f) \neq \emptyset$.

11.2.7. Show that

$$R = \{A \in M_2(\mathbb{Z}) : A = \begin{pmatrix} x & 0 \\ y & x \end{pmatrix}\}$$

is a subring of $M_2(\mathbb{Z})$. Futher prove that $f : R \to \mathbb{Z}$ given by $f(A) = x$ is a surjective ring homomorphism.

Finally verify that $\ker(f)$ is an ideal in R by direct calculation, that is, directly from the definition of an ideal.

11.2.8. Suppose that R is a commutative ring of characteristic 2 (see Definition 3.6.1). Prove that the map $f : R \to R$ given by $f(x) = x^2$ for all $x \in R$ is a ring endomorphism.

EXERCISES FOR SECTION 11.3

11.3.1. Let R be a commutative ring with identity $1 \neq 0$ and I an ideal in R. Prove that

$$\langle r, I \rangle = \{rx + i : x \in R, i \in I\}$$

is also an ideal in R. This is called the ideal **generated by** r and I.

11.3.2. Find a positive integer x such that
  (a) $\langle x \rangle = \langle 3 \rangle \langle 4 \rangle$.
  (b) $\langle x \rangle = \langle 6 \rangle \langle 8 \rangle$.
  (c) $\langle x \rangle = \langle m \rangle \langle n \rangle$.

11.3.3. Suppose that A, B are ideals in a ring R. Prove that $AB \subset A \cap B$

11.3.4. If A, B are ideals in a commutative ring R with identity and $R = A + B$ then $A \cap B = AB$.

(HINT: If $x \in A \cap B$ start by writing $1 = a + b$ where $a \in A$ and $b \in B$.)

11.3.5. Suppose that R is a commutative ring with identity $1 \in R$ and I is an ideal in R. Prove that R/I is a field if and only if for all $r \in R$ where $r \notin I$ there exists $s \in R$ such that $rs - 1 \in I$.

11.3.6. Prove that I is a maximal ideal in the ring R if and only if R/I is simple, that is, has no proper ideals.

EXERCISES FOR SECTION 11.4

11.4.1. Show by induction that if $p \in R$ with R an integral domain is prime in R and $p|a_1 \cdots a_n$ then $p|a_i$ for at least one $i = 1, \ldots, n$.

11.4.2. Complete the proof of Theorem 11.4.1 by showing that Property (B) holds from (A') using the argument that was stated in the text and following along the same lines as the proof of Theorem 4.4.3.

11.4.3. If R is an integral domain and $a, b \in R$ show that

$$\{ax + by : x, y \in R\}$$

is an ideal in R.

11.4.4. If I is a principal ideal in an integral domain R then if $r \in I$ where r is a unit, $I = R$.

11.4.5. Suppose that R is a ring and

$$I_1 \subset I_2 \subset \cdots \subset I_n \subset \cdots$$

is any ascending chain of ideals in R. Prove that

$$I = \bigcup_{i=1}^{\infty} I_i$$

is an ideal in R.

11.4.6. Prove that in any integral domain R the product of an irreducible and a unit is an irreducible.

11.4.7. Suppose that R is an integral domain. Define the relation $a \sim b$ with $a, b \in R$ if and only if a, b are associates in R. Show that this is an equivalence relation.

11.4.8. Suppose that R is an integral domain and let $x, y \in R$ with $y \neq 0$ and x not a unit. Prove that the principal ideal $\langle xy \rangle \subset \langle y \rangle$ and $\langle xy \rangle \neq \langle y \rangle$.

Chapter 12

Polynomials and Polynomial Rings

12.1 Polynomials and Polynomial Rings

In elementary algebra a polynomial is an expression $p(x) = a_n x^n + \cdots + a_o$ where n is a nonnegative integer and the a_i are real numbers. This is then generally considered as a function on the real numbers with x as a variable. Historically much of classical algebra dealt with finding the zeros of polynomials, that is, numbers x_0 such that $p(x_0) = 0$, where $p(x)$ is a polynomial. Later in this chapter we will reexamine solving polynomial equations and describe two very important algebraic results whose proofs will be given in Chapter 15. The first of these is the fundamental theorem of algebra and the second is the proof of the insolvability by radicals of polynomial equations of degree 5 or higher. We will explain both of these in detail in due course.

In this chapter we will begin by looking at polynomials as objects in their own right and show that if we start with a commutative ring R with an identity, and consider polynomials with coefficients from R, we get a new ring, called the ring of polynomials over R. We will then show that if R is a field, then the set of polynomials with coefficients in R, which we will denote by $R[x]$, forms a unique factorization domain and a principal ideal domain. At the end of this chapter we prove that if R is a UFD then the polynomial ring $R[x]$ is also a UFD and use this to show that not every UFD is a PID.

Our starting point is to take a formal look at polynomials.

Let R be a commutative ring with an identity. Consider the set $\tilde{R}$ of functions f from the nonnegative integers $\overline{\mathbb{N}} = \mathbb{N} \cup \{0\}$ into R with only a finite number of values being nonzero. That is,

$$\tilde{R} = \{f : \overline{\mathbb{N}} \to R : f(n) \neq 0 \text{ for only finitely many } n\}$$

The set $\tilde{R}$ can be thought of as the set of all infinite sequences

$$(a_0, a_1, \ldots, a_n, \ldots)$$

of elements from R, $a_i \in R$, such that all but finitely many of the a_i are equal to zero. Of course, two such sequences are equal if and only if corresponding entries are equal. For an $f \in \tilde{R}$, $f = (a_0, a_1, \ldots, a_m, \ldots)$ the a_i can be thought of as the values of f on each element of $\overline{\mathbb{N}}$. Since all but finitely many of the $a_i = 0$, there exists an m such that $a_i = 0$ for all $i > m$. Thus $f(n) \neq 0$ for only finitely many values, i.e., $f(n)$ can be nonzero only for $n \leq m$. Also if $a_m \neq 0$, then the $f = (a_0, a_1, \ldots, a_m, 0, 0, \ldots)$ is said to have **degree** $m \in \overline{\mathbb{N}}$ written, $\deg(f) = m$. Also a_m is called the **leading coefficient** or **leading term** of f.

On $\tilde{R}$ we define the following addition and multiplication for any $m \in \overline{\mathbb{N}}$:

$$(f + g)(m) = f(m) + g(m),$$

$$(f \cdot g)(m) = \sum_{i+j=m} f(i)g(j),$$

where in the second line above the sum is taken over all i and j such that $i + j = m$. Thus it can be written

$$\sum_{i=0}^{m} f(i)g(m - i).$$

Now we let $x = (0, 1, 0, \ldots)$ and identify $(r, 0, 0, \ldots)$ with $r \in R$ and then define

$$x^0 = (1, 0, \ldots) = 1 \text{ and } x^{i+1} = x \cdot x^i.$$

Moreover, from the definition of multiplication above, it is not hard to see that

$$x^i = (0, \ldots, 0, 1, 0, \ldots)$$

where the the only nonzero entry is a 1 in the $i + 1$st spot (see the exercises for a proof). Put another way, x^i is the function $f : \overline{\mathbb{N}} \to \mathbb{R}$ such that $x^i(j) = 0$ for $j \neq i$ but $x^i(i) = 1$. Note also that since we identify $r \in R$

with the $(r, 0, 0, \ldots)$ (recall that $r = 0$ has degree $-\infty$), then if $f = (a_0, a_1, \ldots, a_n, 0, 0, \ldots)$ it follows that $rf = (ra_0, ra_1, \ldots, ra_n, \ldots)$. This is so because for all $n \in \overline{\mathbb{N}}$ we have

$$rf(n) = r(0)f(n) + r(1)f(n-1) + \cdots + r(n)f(0) = ra_n$$

since $r(i) = 0$ for all $i > 0$.

It follows that if $f = (r_0, r_1, r_2, \ldots)$ then f can be written as

$$f = \sum_{i=0}^{\infty} r_i x^i = \sum_{i=0}^{m} r_i x^i$$

if f has degree m for some $m \geq 0$ since $r_i = 0$ for all $i > m$. Further this representation is unique. The reason that it is true is that this shows all the values of the function f, that is, $f(i) = r_i$. So if any function g had the same representation then $f = g$.

We call x an **indeterminate** over R and write each element of $\tilde{R}$ as $f(x) = \sum_{i=0}^{m} r_i x^i$ with $f(x) = 0$ or $r_m \neq 0$. We also now write $R[x]$ for $\tilde{R}$. Each element of $R[x]$ is called a **polynomial** over R. The elements $r_0, \ldots, r_m$ are called the **coefficients** of $f(x)$ with r_m the **leading coefficient**. If $r_m \neq 0$ the natural number m is called the **degree** of $f(x)$ which we denote by $\deg(f(x))$. We say that $f(x) = 0$ has degree $-\infty$. The uniqueness of the representation of a polynomial implies that two nonzero polynomials are equal if and only if they have the same degree and exactly the same coefficients. A polynomial of degree 1 is called a **linear polynomial** while one of degree 2 is a **quadratic polynomial**. The set of polynomials of degree 0 together with $0 = (0, 0, \ldots, 0, \ldots)$ form a ring isomorphic to R and hence can be identified with R. Thus the ring R embeds in the set of polynomials $R[x]$. We note that the polynomial $f = (r_0, r_1, \ldots,)$ of degree m is meant to be the familiar polynomial $r_0 + r_1 x + \cdots + r_m x^m$. Addition and multiplication of polynomials are very clear in terms of this representation.

Lemma 12.1.1. *Let $f(x) = r_0 + r_1 x + \cdots + r_m x^m$ and $g(x) = s_0 + s_1 x + \cdots + s_m x^m$ be two polynomials in $R[x]$. If the degrees are unequal we add zero coefficients to make the greatest exponent the same (or to get the same exponent m. Recall: adding zero coefficients does not change the degree). We also assume that any coefficients higher than m are zero. Then:*

(a) $f(x) \pm g(x) = c_0 + c_1 x + \cdots + c_m x^m$ where $c_i = r_i \pm s_i$ for all i. Note that for $i > m$ all coefficients are zero. This shows that the set of polynomials is closed under addition and subtraction.

(b) $f(x) \cdot g(x) = c_0 + c_1 x + \cdots + c_m x^{2m}$ *where* $c_i = r_0 s_i + r_1 s_{i-1} + .. + r_i s_0$ *for all* i. *This shows that if* $i > 2m$ *then* $c_i = 0$ *showing that the set of polynomials is closed under multiplication.*

Before proving this we present an example.

EXAMPLE 12.1.1. Let $f(x) = 3 + 4x - 6x^2$ and $g(x) = 2 + x + x^3$ be two polynomials in $\mathbb{Z}[x]$. Then

$$f(x) + g(x) = (3+2) + (4+1)x + (-6+0)x^2 + (0+1)x^3 = 5 + 5x - 6x^2 + x^3$$

$$f(x)g(x) = (3)(2) + ((3)(1) + (4)(2))x + ((3)(0) + (4)(-1) + (-6)(2))x^2 +$$
$$((3)(1) + (4)(0) + (-6)(1) + (3)(1))x^3 + ((4)(1) + (-6)(0))x^4 + ((-6)(1))x^5$$
$$= 6 + 11x - 8x^2 - 3x^3 + 4x^4 - 6x^5.$$

Note that to get any terms of degree larger we would need nonzero coefficients for higher degrees in either of the polynomials.

Proof. Let $f(x) = r_0 + r_1 x + \cdots + r_m x^m$ and $g(x) = s_0 + s_1 x + \cdots + s_m x^m$ be two polynomial in $R[x]$. If the degrees are unequal we add zero coefficients to make the greatest exponent the same (or to get the same exponent m). We also assume that any coefficients higher than m are zero as in the statement.

Then from the definition of polynomial addition we have precisely that the coefficient of x^i in $f(x) \pm g(x)$ for $0 \le i \le m$ is precisely $r_i \pm s_i$. For $i > m$ the coeffcients are all zeros. This also shows that the set of polynomials is closed under addition and subtraction.

Again from the definition of multiplication we have for $0 \le i \le 2m$ the coefficient of x^i is given by

$$c_i = r_0 s_i + r_1 s_{i-1} + \cdots + r_i s_0$$

as desired. If $i > 2n$ then the coefficients are 0 because in each term at least one of the coefficients must be zero. Further, this shows closure under multiplication.   □

The following results are straightforward concerning degree.

Lemma 12.1.2. *Let* $f(x)$ *and* $g(x)$ *be nonzero polynomials in* $R[x]$. *Then:*
   (a) $f(x) + 0 = 0 + f(x) = f(x)$ *and* $f(x) \cdot 0 = 0 \cdot f(x) = 0$.
   (b) deg $(f(x)g(x)) \le degf(x) + degg(x)$. *If further* R *is an integral domain then this is an equality.*
   (c) deg $(f(x) \pm g(x)) \le Max(deg\ (f(x),\ deg\ g(x)))$.

Proof. We note that (a) is easily proved by examining the definitions (see the exercises). For the remainder of the proof we assume as in the previous lemma that neither of the polynomials is the zero polynomial. Assume that $\deg(f(x)) = m$ and $\deg(g(x)) = n$ so that $f(x)$ and $g(x)$ can be represented as

$$f(x) = a_0 + a_x + \cdots + a_m x^m \text{ and } g(x) = b_0 + b_1 x + \cdots + b_n x^n$$

where $a_i \in R$ for $i = 0, \ldots, m$ and $a_m \neq 0$ and $b_i \in R$ for $i = 0, \ldots, n$ and $b_n \neq 0$. Then using the definition of multiplication of polynomials

$$f(x)g(x) = a_m b_n x^{n+m} + \text{ terms in lower powers of } x.$$

Hence $\deg(f(x)g(x)) \leq n + m = \deg(f(x)) + \deg(g(x))$. Since $a_m \neq 0$ and $b_n \neq 0$ then if R were an integral domain $a_n b_m \neq 0$ and hence in this case $\deg(f(x)g(x)) = n + m = \deg(f(x)) + \deg(g(x))$.

Now suppose first that $m > n$. Again using the definition of addition of polynomials we have

$$f(x) \pm g(x) = a_m x^m + \text{ terms in lower powers of } x.$$

It follows that $\deg(f(x) \pm g(x)) = m = \text{Max}(\deg f(x), \deg g(x))$. An identical argument works if $n > m$.

Now suppose that $n = m$. Then again using the definition of addition of polynomials we have

$$f(x) \pm g(x) = (a_m \pm b_m) x^m + \text{ terms in lower powers of } x.$$

If $a_m \pm b_m \neq 0$ then $\deg (f(x) \pm g(x)) = m = \text{Max}(\deg f(x), \deg g(x))$. On the other hand if $a_m \pm b_m = 0$ then

$$\deg(f(x) \pm g(x)) < m = \text{Max}(\deg (f(x), \deg g(x)).$$

Therefore in every case $\deg(f(x) \pm g(x)) \leq \text{Max}(\deg (f(x), \deg g(x))$. □

Theorem 12.1.1. *Let R be a commutative ring with identity. Then the set of polynomials $R[x]$ forms a ring called the **ring of polynomials** over R. The ring R identified with 0 and the polynomials of degree 0 naturally embeds into $R[x]$. Hence we can consider R as a subring of $R[x]$. $R[x]$ is commutative if and only if R is commutative.*

Proof. Set $f(x) = \sum_{i=0}^{n} r_i x^i$ and $g(x) = \sum_{j=0}^{m} s_j x^j$. The ring properties follow directly by computation and are left to the exercises. The identification of $r \in R$ with the polynomial $r(x) = r$ provides the embedding of R into $R[x]$. From the definition of multiplication in $R[x]$ if R is commutative then $R[x]$ is commutative. Conversely if $R[x]$ is commutative then from the embedding of R into $R[x]$ it follows that R must also be commutative. Note that if R has a multiplicative identity $1 \neq 0$ then this is also the multiplicative identity of $R[x]$.

$\square$

From the theorem it follows that if R is a commutative ring with an identity then $R[x]$ is also a commutative ring with an identity. If R has zero divisors then since $R \subset R[x]$ it follows that $R[x]$ also has zero divisors. On the other hand if R is an integral domain the polynomial ring $R[x]$ is also an integral domain.

Lemma 12.1.3. *Let R be a commutative ring with an identity. Then $R[x]$ is an integral domain if and only if R is an integral domain.*

Proof. If R is not an integral domain then it has zero divisors (see the discussion above) so $R[x]$ cannot be an integral domain. If R is an integral domain we show $R[x]$ is an integral domain. Suppose that $G(x) \neq 0$ and $H(x) \neq 0$. Then $\deg(G(x)) \geq 0$ and $\deg(H(x)) \geq 0$. Since R is an integral domain the degrees add under multiplication by Lemma 12.1.2 so that $\deg(G(x)H(x)) \geq 0$. Therefore $G(x)H(x) \neq 0$ and there are no zero divisors in $R[x]$. Since $R[x]$ is a commutative ring with an identity and no zero divisors it follows that $R[x]$ is an integral domain. $\square$

If R is a field then $f(x)$ is an **irreducible polynomial** if there is no factorization $f(x) = g(x)h(x)$ where $h(x)$ and $g(x)$ are polynomials of lower degree than $f(x)$. Otherwise $f(x)$ is called **reducible**. If R is an integral domain then the irreducible polynomials are precisely the irreducible elements in the integral domain $R[x]$.

In elementary mathematics polynomials are considered as functions. We recover that idea via the concept of evaluation.

Definition 12.1.1. *Let $f(x) = r_0 + r_1 x + \cdots + r_m x^n$ be a polynomial over a commuative ring R with an identity and let $c \in R$. Then the element*

$$f(c) = r_0 + r_1 c + \cdots + r_n c^n \in R$$

*is called the **evaluation** of $f(x)$ at c.*

Definition 12.1.2. *If $f(x) \in R[x]$ and $f(c) = 0$ for $c \in R$, then c is called a **zero** of $f(x)$ in R.*

Lemma 12.1.4. *Let R be a commutative ring with identity. Then for $a \in R$ the mapping $\phi_a : R[x] \to R$ defined by $\phi_a(f(x)) = f(a)$, i.e., if $f(x) = \sum_{i=0}^{m} r_i x^i$, then $\phi_a(f(x)) = \sum_{i=0}^{m} r_i a^i$, is a ring homomorphism.*

See the exercises for the proof.

12.2 Polynomial Rings over a Field

We now consider the polynomial ring $F[x]$ where F is a field. As in Section 12.1, $F[x]$ forms a ring and since F is a field it actually forms an integral domain by Lemma 12.1.3. We will then show that there is unique factorization into irreducible polynomials in $F[x]$.

Theorem 12.2.1. *If F is a field, then $F[x]$ forms an integral domain. F can be naturally embedded into $F[x]$ by identifying each element of F with the corresponding constant polynomial. The only units in $F[x]$ are the nonzero elements of F.*

Proof. Let F be a field and hence an integral domain. It follows then from the last section that $F[x]$ is an integral domain.

If $G(x)$ is a unit in $F[x]$, then there exists an $H(x) \in F[x]$ with $G(x)H(x) = 1$. From the degrees we have $\deg G(x) + \deg H(x) = 0$ and since $\deg G(x) \geq 0$ then $\deg H(x) \geq 0$. This is possible only if $\deg G(x) = \deg H(x) = 0$. Therefore $G(x) \in F$ and $G(x)$ is a nonzero element of F. Since F is a field and $F \subset F[x]$ then any nonzero element of F is a unit in $F[x]$. $\qquad\square$

Now that we have $F[x]$ as an integral domain we proceed to show that in $F[x]$ there is unique factorization into irreducibles. If $f(x) \in F[x]$ has degree ≥ 1 and has no nonunit factors (it cannot be factored into polynomials of lower degree) then $f(x)$ is an **irreducible** in $F[x]$ or an **irreducible polynomial**. Later we will see that if R is a field then $R[x]$ is a PID so from the material in Chapter 11, primes and irreducible elements coincide. Hence in this case an **irreducible polynomial** over F is also a **prime polynomial**. Clearly, if $\deg g(x) = 1$ then $g(x)$ is irreducible.

The fact that $F[x]$ is a unique factorization domain will follow from the division algorithm for polynomials, which is entirely analogous to the division algorithm for integers.

Theorem 12.2.2. *(Division Algorithm in $F[x]$)* If $0 \neq f(x), 0 \neq g(x) \in F[x]$ then there exist unique polynomials $q(x), r(x) \in F[x]$ such that $f(x) = q(x)g(x) + r(x)$ where $r(x) = 0$ or deg $r(x) <$ deg $g(x)$. (The polynomials $q(x)$ and $r(x)$ are called respectively the quotient and remainder.)

This theorem is essentially long division of polynomials. A formal proof is based on induction on the degree of $g(x)$. We omit this but give some examples from $\mathbf{Q}[x]$. A proof is given in the exercises.

EXAMPLE 12.2.1.

(a) Let $f(x) = 3x^4 - 6x^2 + 8x - 6, g(x) = 2x^2 + 4$. Then

$$\frac{3x^4 - 6x^2 + 8x - 6}{2x^2 + 4} = \frac{3}{2}x^2 - 6 \text{ with remainder } 8x + 18.$$

Thus here $q(x) = \frac{3}{2}x^2 - 6, r(x) = 8x + 18$.

(b) Let $f(x) = 2x^5 + 2x^4 + 6x^3 + 10x^2 + 4x, g(x) = x^2 + x$. Then

$$\frac{2x^5 + 2x^4 + 6x^3 + 10x^2 + 4x}{x^2 + x} = 2x^3 + 6x + 4.$$

Thus here $q(x) = 2x^3 + 6x + 4$ and $r(x) = 0$.

12.2.1 Unique Factorization of Polynomials

Here we show that if F is a field then the polynomial ring $F[x]$ is a unique factorization domain. In this section we will give a proof that exactly mirrors the proof of unique factorization for integers. In Section 12.2.3 we will give a second proof using the fact that $F[x]$ is a principal ideal domain.

Theorem 12.2.3. *Let F be a field. Then the polynomial ring $F[x]$ is a unique factorization domain.*

Proof. The proof is an exact copy of the proof of unique factorization in the integers $\mathbb{Z}$ using degree instead of the ordering on $\mathbb{Z}$. We review the steps in proving unique factorization in $\mathbb{Z}$ (see Chapter 4) and then show how it carries over to $F[x]$. □

Step 1: Using induction we proved that every positive integer has a decomposition into primes.

Step 2: We defined the GCD of two integers and showed using the LWO that the GCD of two integers m and n could be expressed as a linear combination. That is, if $d = \gcd(m, n)$ then there exists $x, y \in \mathbb{Z}$ such that $mx + ny = d$.

Step 3: Using the fact that if $(m, n) = 1$ then $mx + ny = 1$ for some integers x, y we proved Euclid's lemma. Recall that this says that if $p|ab$ where p is a prime then $p|a$ or $p|b$. We then extended this by induction to show that if a prime p is such that $p|a_1 a_2 \cdots a_k$, then $p|a_i$ for at least one i, $1 \le i \le k$.

Step 4: Using Euclid's lemma we showed that if $n = p_1^{e_1} \cdots p_k^{e_k} = q_1^{f_1} \cdots q_t^{f_t}$ are two prime decompositions of n with distinct primes then $k = t$ and each p_i is equal to some q_j and the exponents are then equal.

These four steps combined to prove the fundamental theorem of arithmetic. We now retrace these steps in $F[x]$.

Lemma 12.2.1. *Let F be a field and suppose $deg(f(x)) > 0$. Then each such $f(x) \in F[x]$ has a decomposition into irreducible polynomials.*

Proof. The proof is by induction on the degree. Recall that a polynomial $f(x)$ is reducible if $f(x) = g(x)h(x)$ with $1 \le \deg(g(x)) < \deg((f(x))$ and $1 \le \deg((h(x)) < \deg((f(x))$. Any degree zero polynomial is an element of F and is hence a unit, since F is a field, so such polynomials do not have irreducible decompositions. Since in multiplying polynomials the degrees add, it follows immediately that a degree 1 polynomial is irreducible. Let us suppose that every polynomial of degree less than n and of degree > 0 can be decomposed into a product of irreducible polynomials, where the product may be only one factor. Suppose that $f(x)$ has degree n. If $f(x)$ is irreducible we are done. If not then $f(x) = g(x)h(x)$ with $g(x)$ and $h(x)$ nonconstant polynomials of lower degree than $f(x)$. By the inductive assumption both $g(x)$ and $h(x)$ have decompositions into irreducible polynomials and therefore so does $f(x)$. □

We next define the GCD of two polynomials. We must, however, take it to be a **monic polynomial**, that is, a polynomial whose leading coefficient is 1. In the definition we say a GCD but from the lemma immediately after it, it follows that the GCD of 2 polynomials is unique.

Definition 12.2.1. *Let F be a field and $f(x), g(x) \in F[x]$. Then a monic polynomial $d(x)$ is a greatest common divisor or GCD of $f(x)$ and $g(x)$ if*

$d(x)|f(x)$, $d(x)|g(x)$ and $d_1(x)$ is any other common divisor of $f(x)$ and $g(x)$ then $d_1(x)|d(x)$. As for integers we write $\gcd(f(x), g(x)) = (f(x), g(x))$ and we say that $f(x)$ and $g(x)$ are **relatively prime** or **coprime** if $(f(x), g(x)) = 1$.

Lemma 12.2.2. *Let F be a field. Then the GCD, $d(x)$, of two nonzero polynomials $f(x), g(x) \in F[x]$ is unique. Further, there exist polynomials $h(x), k(x) \in F[x]$ with $h(x)f(x) + k(x)g(x) = d(x)$.*

Proof. This proof again mirrors the proof of the same result in the integers. Suppose that $f(x), g(x) \in F[x]$ are nonzero. Let

$$S = \{f(x)h(x) + g(x)k(x) : h(x), k(x) \in F[x]\}.$$

The set S is nonempty (why?). Let $D(x) \in S$ be of minimal degree ≥ 0. Then

$$D(x) = f(x)h(x) + g(x)k(x)$$

for some polynomials $h(x), k(x)$. If $a \in F$ is the leading coefficient of $D(x)$, since $a \neq 0$, a^{-1} exists. Then $a^{-1}D(x) = d(x)$ is a monic polynomial and

$$d(x) = f(x)(a^{-1}h(x)) + g(x)(a^{-1}k(x)).$$

Hence $d(x) \in S$ and since $\deg(d(x)) = \deg(D(x))$ it follows that $d(x)$ is a monic polynomial of minimal degree in S. We claim that $d(x)$ is a GCD for $f(x), g(x)$. We first show that $d(x)$ is a common divisor.

By the division algorithm

$$f(x) = q(x)d(x) + r(x)$$

where $r(x) = 0$ or $\deg(r(x)) < \deg(d(x))$. If $r(x) \neq 0$ then $\deg(r(x)) < \deg(d(x))$ and

$$r(x) = f(x) - q(x)d(x).$$

However, $d(x) \in S$ so $d(x) = f(x)m(x) + g(x)n(x)$ for some polynomials $m(x), n(x) \in F[x]$. But then

$$r(x) = (1 - m(x)q(x))f(x) - q(x)n(x)g(x) \in S.$$

This contradicts the minimality of the degree of $d(x)$ in S. Therefore $r(x) = 0$ and $d(x)|f(x)$. In an identical manner, we get that $d(x)|g(x)$.

Let $d_1(x)$ be another common divisor of $f(x)$ and $g(x)$. Then as with integers $d_1(x)$ must divide any linear combination of $f(x)$ and $g(x)$ and hence

$d_1(x)|d(x)$. Therefore $d(x)$ is a GCD for $f(x)$ and $g(x)$. We finish the proof by showing that $d(x)$ is unique.

Let $d_1(x)$ be another GCD for $f(x)$ and $g(x)$. Then $d_1(x)|d(x)$ but also $d(x)|d_1(x)$. This shows that $d_1(x)$ and $d(x)$ have the same degree (see the exercises). But then $d(x), d_1(x)$ are two monic polynomials of the same degree that divide each other. Therefore they must be equal (see the exercises). $\square$

The following three lemmas complete the proof of unique factorization in $F[x]$. The proofs of these lemmas are identical to the corresponding proofs in the integers (see Chapter 4) and we leave these proofs to the exercises.

Lemma 12.2.3. *Let $p(x)$ be an irreducible polynomial in $F[x]$. If $p(x)|f(x)g(x)$ then either $p(x)|f(x)$ or $p(x)|g(x)$.*

Lemma 12.2.4. *Let $p(x)$ be an irreducible polynomial in $F[x]$. If*

$$p(x)|f_1(x)\cdots f_n(x),$$

then $p(x)|f_i(x)$ for at least one $i, 1 \leq i \leq n$.

Lemma 12.2.5. *Let F be a field. Then the decomposition of any polynomial $f(x) \in F[x]$ into irreducible polynomials is unique up to order and unit factors.*

12.2.2 Euclidean Domains

In analyzing the proof of unique factorization in both $\mathbb{Z}$ and $F[x]$, it is clear that it depends primarily on the division algorithm. In $\mathbb{Z}$ the division algorithm depended on the fact that the positive integers could be ordered and in $F[x]$ on the fact that the degrees of nonzero polynomials are nonnegative integers and hence could be ordered. This basic idea can be generalized in the following way.

Definition 12.2.2. *An integral domain D is a **Euclidean domain** if there exists a function N from $D^\star = D \setminus \{0\}$ to the nonnegative integers such that*
> *(1) $N(r_1) \leq N(r_1r_2)$ for any $r_1, r_2 \in D^\star$.*
> *(2) For all $r_1, r_2 \in D$ with $r_1 \neq 0$ there exists $q, r \in D$ such that*

$$r_2 = qr_1 + r$$

where either $r = 0$ or $N(r) < N(r_1)$.

*The function N is called a **Euclidean norm** on D.*

Therefore Euclidean domains are precisely those integral domains which allow division algorithms. In the integers $\mathbb{Z}$ define $N(z) = |z|$. Then N is a Euclidean norm on $\mathbb{Z}$ and hence $\mathbb{Z}$ is a Euclidean domain. In our treatment of the division algorithm for $\mathbb{Z}$ it was assumed that the divisor $a > 0$ (see Theorem 4.2.2). It can be shown that we can replace this with $a \neq 0$ and then use absolute values as follows: if $a, b \in \mathbb{Z}$ with $a \neq 0$ then there exist unique integers q and r such that $b = qa + r$ with $r = 0$ or $0 < r < |a|$. (See the exercises for a proof of this.) Also (1) of Definition 12.2.2. holds in $\mathbb{Z}$ since if $a, b \in \mathbb{Z}$ then $|ab| = |a||b|$ if $a \neq 0$.

On $F[x]$ with F a field define $N(p(x)) = \deg(p(x))$ if $p(x) \neq 0$. Lemma 12.1.2 shows that if $f(x), g(x)$ are nonzero polynomials in $F[x]$ then

$$\deg(f(x)g(x)) = \deg(f(x)) + \deg(g(x)).$$

This implies (1) where N is the degree. To get (2) we use the division algorithm, i.e., Theorem 12.2.2. It should be noted that in the division algorithm it was assumed that both polynomials were nonzero. But if $f(x) = 0$ and $g(x) \neq 0$, then $q(x) = r(x) = 0$. Then N is also a Euclidean norm on $F[x]$ so that $F[x]$ is also a Euclidean domain. In any Euclidean domain we can mimic the proofs of unique factorization in both $\mathbb{Z}$ and $F[x]$ to obtain the following:

Theorem 12.2.4. *Every Euclidean domain is a principal ideal domain and hence a unique factorization domain.*

Before proving this theorem we must develop some results on the *number theory* of general Euclidean domains. First some properties of the norm:

Lemma 12.2.6. *If R is a Euclidean domain then*
    (a) $N(1)$ is minimal among $\{N(r) : r \in R^{\star}\}$,
    (b) $N(u) = N(1)$ if and only if u is a unit,
    (c) $N(a) = N(b)$ for $a, b \in R^{\star}$ if a, b are associates,
    (d) $N(a) < N(ab)$ for $a, b \in R^{\star}$ unless b is a unit.

Proof. (a) From property (1) of Euclidean norms we have

$$N(1) \leq N(1 \cdot r) = N(r) \text{ for any } r \in R^{\star}.$$

(b) Suppose u is a unit. Then there exists u^{-1} with $u \cdot u^{-1} = 1$. Then

$$N(u) \leq N(u \cdot u^{-1)}) = N(1)$$

From the minimality of $N(1)$ it follows that $N(u) = N(1)$.

Conversely suppose $N(u) = N(1)$. Apply the division algorithm to get

$$1 = qu + r.$$

If $r \neq 0$ then $N(r) < N(u) = N(1)$ contradicting the minimality of $N(1)$. Therefore $r = 0$ and $1 = qu$. Then u has a multiplicative inverse and hence is a unit.

(c) Suppose $a, b \in R^*$ are associates. Then $a = ub$ with u a unit. Then

$$N(b) \leq N(ub) = N(a).$$

On the other hand $b = u^{-1}a$ so

$$N(a) \leq N(u^{-1}a) = N(b).$$

Since $N(a) \leq N(b)$ and $N(b) \leq N(a)$ it follows that $N(a) = N(b)$.

(d) Suppose $a, b \in R^*$ and $N(a) = N(ab)$. Apply the division algorithm

$$a = q(ab) + r$$

where $r = 0$ or $N(r) < N(ab)$. If $r \neq 0$ then

$$r = a - qab = a(1 - qb) \implies N(ab) = N(a) \leq N(a(1 - qb)) = N(r),$$

contradicting that $N(r) < N(ab)$. Hence $r = 0$ and $a = q(ab) = (qb)a$. Then

$$a = (qb)a = 1 \cdot a \implies qb = 1$$

by cancellation in the integral domain R. Hence b is a unit. Since $N(a) \leq N(ab)$ it follows that if b is not a unit we must have $N(a) < N(ab)$. $\square$

We can now prove Theorem 12.2.4

Proof. Let D be a Euclidean domain. We show that each ideal $I \neq D$ in D is principal. Let I be an ideal in D. If $I = \{0\}$ then $I = \langle 0 \rangle$ and I is principal. Therefore we may assume that there are nonzero elements in I. Let $a \neq 0$ be an element of I of minimal norm, that is, $N(a)$ is minimal among $N(x)$ for $x \in I$ and $x \neq 0$. We claim that $I = \langle a \rangle$. Let $b \in I$. We must show that b is a multiple of a. Now by the division algorithm

$$b = qa + r$$

where either $r = 0$ or $N(r) < N(a)$. As in $\mathbb{Z}$ and $F[x]$ we have a contradiction if $r \neq 0$. So if $r \neq 0$ then $N(r) < N(a)$. However, $r = b - qa \in I$ since I is an ideal contradicting the minimality of $N(a)$. Therefore $r = 0$ and $b = qa$ and hence $I = \langle a \rangle$. $\square$

As a final example of a Euclidean domain we consider the **Gaussian integers**

$$\mathbb{Z}[i] = \{a + bi : a, b \in \mathbb{Z}\}.$$

It was first observed by Gauss that this set permits unique factorization. To show this we need a Euclidean norm on $\mathbb{Z}[i]$.

Definition 12.2.3. *If $z = a + bi \in \mathbb{Z}[i]$ then its **norm** $N(z)$ is defined by*

$$N(a + bi) = a^2 + b^2.$$

The basic properties of this norm follow directly from the definition (see the exercises).

Lemma 12.2.7. *If $\alpha, \beta \in \mathbb{Z}[i]$ then:*
    (1) $N(\alpha)$ is an integer for all $\alpha \in \mathbb{Z}[i]$,
    (2) $N(\alpha) \geq 0$ for all $\alpha \in \mathbb{Z}[i]$,
    (3) $N(\alpha) = 0$ if and only if $\alpha = 0$,
    (4) $N(\alpha) \geq 1$ for all $\alpha \neq 0$,
    (5) $N(\alpha\beta) = N(\alpha)N(\beta)$, that is, the norm is multiplicative.

From the multiplicativity of the norm we have the following concerning primes and units in $\mathbb{Z}[i]$.

Lemma 12.2.8. *(1) $u \in \mathbb{Z}[i]$ is a unit if and only if $N(u) = 1$.*
    (2) If $\pi \in \mathbb{Z}[i]$ and $N(\pi) = p$ where p is an ordinary prime in $\mathbb{Z}$ then π is an irreducible in $\mathbb{Z}[i]$.

Proof. Certainly u is a unit if and only if $N(u) = N(1)$ by Lemma 12.2.6. But in $\mathbb{Z}[i]$ we have $N(1) = 1$ so the first part follows.

Suppose next that $\pi \in \mathbb{Z}[i]$ with $N(\pi) = p$ for some prime $p \in \mathbb{Z}$. Suppose that $\pi = \pi_1 \pi_2$. From the multiplicativity of the norm we have

$$N(\pi) = p = N(\pi_1)N(\pi_2).$$

Since each norm is a positive ordinary integer and p is a prime it follows that either $N(\pi_1) = 1$ or $N(\pi_2) = 1$. Hence either π_1 or π_2 is a unit in $\mathbb{Z}[i]$. Therefore π is an irreducible in $\mathbb{Z}[i]$. $\square$

Armed with this norm we can show that $\mathbb{Z}[i]$ is a Euclidean domain.

Theorem 12.2.5. *The Gaussian integers $\mathbb{Z}[i]$ form a Euclidean domain.*

Proof. That $\mathbb{Z}[i]$ forms a commutative ring with an identity can be verified directly and easily (see the exercises). If $\alpha\beta = 0$ then $N(\alpha)N(\beta) = 0$ and since there are no zero divisors in $\mathbb{Z}$ we must have $N(\alpha) = 0$ or $N(\beta) = 0$. But then either $\alpha = 0$ or $\beta = 0$ and hence $\mathbb{Z}[i]$ is an integral domain. To complete the proof we show that the norm N is a Euclidean norm.

From the multiplicativity of the norm we have if $\alpha, \beta \neq 0$

$$N(\alpha\beta) = N(\alpha)N(\beta) \geq N(\alpha) \text{ since } N(\beta) \geq 1.$$

Therefore property (1) of Euclidean norms is satisfied. We must now show that the division algorithm holds.

Let $\alpha = a + bi$ and $\beta = c + di$ be Gaussian integers with $\beta \neq 0$. Recall that for a nonzero complex number $z = x + iy$ its inverse is

$$\frac{1}{z} = \frac{\overline{z}}{|z|^2} = \frac{x - iy}{x^2 + y^2}.$$

Therefore as a complex number

$$\frac{\alpha}{\beta} = \alpha\frac{\overline{\beta}}{|\beta|^2} = (a + bi)\frac{c - di}{c^2 + d^2}$$

$$= \frac{ac + bd}{c^2 + d^2} + \frac{ac - bd}{c^2 + d^2}i = u + iv.$$

Now since a, b, c, d are integers u, v must be rationals. The set

$$\mathbb{Q}(i) = \{u + iv : u, v \in \mathbb{Q}\}$$

are called the **Gaussian rationals**.

If $u, v \in \mathbb{Z}$ then $u + iv \in \mathbb{Z}[i]$, $\alpha = q\beta$ with $q = u + iv$, and we are done. Otherwise choose ordinary integers m, n satisfying $|u - m| \leq \frac{1}{2}$ and $|v - n| \leq \frac{1}{2}$ and let $q = m + in$. Then $q \in \mathbb{Z}[i]$. Let $r = \alpha - q\beta$ with $r \neq 0$. We must show that $N(r) < N(\beta)$.

Working with complex absolute value we get

$$|r| = |\alpha - q\beta| = |\beta||\frac{\alpha}{\beta} - q|.$$

Now

$$\left|\frac{\alpha}{\beta} - q\right| = |(u-m)+i(v-n)| = \sqrt{(u-m)^2 + (v-n)^2} \leq \sqrt{(\frac{1}{2})^2 + (\frac{1}{2})^2} < 1.$$

Therefore

$$|r| < |\beta| \implies |r|^2 < |\beta|^2 \implies N(r) < N(\beta),$$

completing the proof.

$\square$

Since $\mathbb{Z}[i]$ forms a Euclidean domain it follows from our previous results that $\mathbb{Z}[i]$ must be a principal ideal domain and hence a unique factorization domain.

Corollary 12.2.1. *The Gaussian integers are a UFD.*

12.2.3 $F[x]$ as a Principal Ideal Domain

To obtain the fact that $F[x]$ is a UFD for F a field we copied the proof for the integers and adapted it to the situation of polynomials using degrees. Here using the division algorithm we give another proof by showing that $F[x]$ is a principal ideal domain. Then from Theorem 11.4.3 it follows that $F[x]$ is a UFD.

Theorem 12.2.6. *Let F be a field. Then the polynomial ring $F[x]$ is a principal ideal domain and hence a unique factorization domain.*

Proof. Since for a field F the polynomial ring $F[x]$ is an integral domain to show that it is a PID we must show that every ideal is principal. Let $S \subset F[x]$ be an ideal. If S contains a unit, that is, a nonzero element of F from Theorem 12.2.1 then since S is an ideal it follows that $S = F[x] = \langle 1 \rangle$ (see the exercises) and S is principal.

Let S be a proper ideal, that is, $S \neq \{0\}$ and $S \neq F[x]$. It follows that S contains no units or elements of F other than 0. Since by Lemma 11.3.4 $F[x]S \subset S$ it follows that if $S \neq \{0\}$ that S must contain polynomials of positive degree. Let $p(x) \in S$ of minimal positive degree. Since we can divide through by the leading coefficient of $p(x)$ we may assume that $p(x)$ is monic. We claim that $S = \langle p(x) \rangle$, that is, $S = p(x)F[x]$. The proof of this again mirrors the proof in the integers.

Let $f(x) \in S$. Then by the division algorithm we have

$$f(x) = q(x)p(x) + r(x)$$

with $r(x) = 0$ or $\deg(r(x)) < \deg(p(x))$. If $r(x) \neq 0$ then

$$r(x) = f(x) - q(x)p(x).$$

Since S is an ideal and $p(x) \in S$ it follows that $q(x)p(x) \in S$. Further $f(x) \in S$ and S is a subring so $f(x) - q(x)p(x) = r(x) \in S$. If $r(x)$ has positive degree this contradicts the minimality of the degree of $p(x)$. If $r(x)$ has zero degree then $r(x) \in F$ contradicting that S is a proper ideal. Therefore $r(x) = 0$ and $f(x) = q(x)p(x)$. This implies that every element of S is a multiple of $p(x)$ and hence $S = \langle p(x) \rangle$. □

12.2.4 Polynomial Rings over Integral Domains

Here we consider $R[x]$ where R is an integral domain.

Definition 12.2.4. *Let R be an integral domain. Then $a_1, a_2, \ldots, a_n \in R$ are* **coprime** *over R if the set of all common divisors of $a_1, \ldots, a_n$ consists only of units.*

Notice for example that this concept depends on the ring R. For example 6 and 9 are not coprime over the integers $\mathbb{Z}$ since $3|6$ and $3|9$ and 3 is not a unit. However, 6 and 9 are coprime over the rationals $\mathbb{Q}$. Here 3 is a unit.

Definition 12.2.5. *Let $f(x) = \sum_{i=1}^{n} r_i x^i \in R[x]$ where R is an integral domain. Then $f(x)$ is a* **primitive polynomial** *or just* **primitive** *over R if $r_0, r_1, \ldots, r_n$ are coprime in R.*

Theorem 12.2.7. *Let R be an integral domain. Then:*
    (a) The units of $R[x]$ are the units of R.
    (b) If p is a prime element of R then p is a prime element of $R[x]$.

Proof. If $r \in R$ is a unit then since R embeds into $R[x]$ it follows that r is also a unit in $R[x]$. Conversely suppose that $h(x) \in R[x]$ is a unit. Then there is a $g(x)$ such that $h(x)g(x) = 1$. Hence $\deg(h(x)) + \deg(g(x)) = \deg 1 = 0$, by Lemma 12.1.2. Since degrees are nonnnegative integers it follows that $\deg(h(x)) = \deg(g(x)) = 0$ and hence $h(x) \in R$.

Now suppose that p is a prime element of R. Then $p \neq 0$ and pR is a prime ideal in R. We must show that $pR[x]$ is a prime ideal in $R[x]$. Consider the map

$$\tau : R[x] \to (R/pR)[x] \text{ given by}$$

$$\tau(\sum_{i=0}^{n} r_i x^i) = \sum_{i=0}^{n} (r_i + pR)x^i.$$

Then τ is an epimorphism with kernel $pR[x]$ (see the exercises). Since pR is a prime ideal we know that R/pR is an integral domain by Theorem 11.3.2. It follows that $(R/pR)[x]$ is also an integral domain from Lemma 12.1.3. We show that $pR[x]$ is a prime ideal.

To see this suppose $f(x) = \sum_{i \geq 0} r_i x^i$ and $g(x) = \sum_{j \geq 0} s_j x^j$ are two polynomials in $R[x]$ and suppose that $f(x)g(x) \in pR[x]$. Then $(fg)(x) = \sum_{k \geq 0} w_k x^k$ where $w_k = \sum_{i+j=k} r_i s_j$. But if $fg \in pR[x]$ then $w_k \in pR$ for all $k \geq 0$. So then consider these elements in $(R/pR)[x]$, $\sum_{i \geq 0}(r_i + pR)x^i$ and $\sum_{j \geq 0}(s_j + pR)x^j$ and the product

$$(\sum_{i \geq 0}(r_i + pR)x^i)(\sum_{j \geq 0}(s_j + pR)x^j) = \sum_{k \geq 0}(w_k + pR)x^k.$$

But $w_k \in pR$ for all $k \geq 0$ and this implies that the product is the zero polynomial in $(R/pR)[x]$. But since $R/pR)[x]$ is an integral domain either $\sum_{i \geq 0}(r_i + pR)x^i$ or $\sum_{j \geq 0}(s_i + pR)x^j$ is the zero polynomial in $R/pR)[x]$ so that either $f(x) \in pR[x]$ or $g(x) \in pR[x]$.

Hence $pR[x]$ must be a prime ideal in $R[x]$ and therefore p is also a prime element of $R[x]$.

$\square$

Recall that each integral domain R can be embedded into a unique field of fractions K. We can use results on $K[x]$ to deduce some results in $R[x]$.

Lemma 12.2.9. *If K is a field then each nonzero $f(x) \in K[x]$ is a primitive.*

Proof. Since K is a field each nonzero element of K is a unit. Therefore the only common divisors of the coefficients of a nonzero $f(x)$ are units and hence $f(x) \in K[x]$ is primitive. $\square$

Theorem 12.2.8. *Let R be an integral domain. Then each irreducible $f(x) \in R[x]$ of degree > 0 is primitive.*

Proof. Let $f(x)$ be an irreducible polynomial in $R[x]$ and let $r \in R$ be a common divisor of the coefficients of $f(x)$. Then $f(x) = rg(x)$ where $g(x) \in R[x]$. Then $\deg(f(x)) = \deg(g(x)) > 0$ so $g(x) \notin R$. Since the units of $R[x]$ are the units of R by Theorem 12.2.7 it follows that $g(x)$ is not a unit in $R[x]$. Since $f(x)$ is irreducible it follows that r must be a unit in $R[x]$ and hence r is a unit in R. Therefore $f(x)$ is primitive. $\square$

12.3 Zeros of Polynomials

We now give some consequences relative to zeros of polynomials in $F[x]$. We first show that if $f(x)$ has a root, and $\deg f(x) > 1$, then $f(x)$ factors, that is, $f(x)$ is not irreducible.

Theorem 12.3.1. *(Factor Theorem) Let F be a field and suppose that $f(x) \in F[x]$. Suppose that $c \in F$ with $f(c) = 0$. Then*

$$f(x) = (x - c)h(x),$$

where $\deg(h(x)) < \deg(f(x))$.

Proof. Divide $f(x)$ by $x - c$. Then by the division algorithm we have

$$f(x) = (x - c)h(x) + r(x)$$

where $r(x) = 0$ or $\deg(r(x)) < \deg(x - c) = 1$. Hence if $r(x) \neq 0$ then $r(x)$ is a polynomial of degree zero and hence constant polynomial, that is, $r(x) = r$ for $r \in F$. Hence we have

$$f(x) = (x - c)h(x) + r.$$

This implies by Lemma 12.1.4 that

$$0 = f(c) = 0 \cdot h(c) + r = r$$

and therefore $r = 0$ and $f(x) = (x - c)h(x)$. Since $\deg(x - c) = 1$ we must have that $\deg(h(x)) < \deg(f(x))$.
$\square$

If $f(x) = (x - c)^k h(x)$ for some $k \geq 1$ with $h(c) \neq 0$ then c is called a **zero of order** k or **of multiplicity** k.

Corollary 12.3.1. *Let $f(x) \in F[x]$ with degree 2 or 3. Then f is irreducible if and only if $f(x)$ doesn't have a zero in F.*

Proof. Suppose that $f(x)$ is irreducible of degree 2 or 3. If $f(x)$ has a zero c then from Theorem 12.3.1 we have $f(x) = (x - c)h(x)$ with $h(x)$ of degree 1 or 2. Therefore $f(x)$ is reducible a contradiction and hence $f(x)$ cannot have a zero.

Conversely if $f(x)$ has a zero and is of degree greater than 1 then $f(x)$ is reducible by Theorem 12.3.1.   □

From this we obtain the following important theorem restricting the number of roots a polynomial over a field can have.

Theorem 12.3.2. *A polynomial of degree n in $F[x]$ can have at most n distinct roots.*

Proof. Suppose $P(x)$ has degree n and suppose $c_1, \ldots, c_n$ are n distinct roots. We show that there cannot be any other roots. From repeated application of Theorem 12.3.1,
$$P(x) = k(x - c_1) \cdots (x - c_n)$$
where $k \in F$ since the degree of $P(x)$ is n and the degree of $(x - c_1) \cdots (x - c_n)$ is also n since degrees add. Suppose c is any other root. Then $P(c) = 0 = k(c - c_1) \cdots (c - c_n)$. Since a field F has no zero divisors, one of these terms must be zero: $c - c_i = 0$ for some i, and hence $c = c_i$.   □

Besides having a maximum of n roots (with n the degree) as a consequence of unique factorization of polynomials over a field the roots of a polynomial are unique. Suppose $P(x)$ has degree n and roots $c_1, .., c_k$ with $k \leq n$. Then from the unique factorization in $F[x]$ we have
$$P(x) = (x - c_1)^{m_1} \cdots (x - c_k)^{m_k} Q_1(x) \cdots Q_t(x)$$
where $Q_i(x), i = 1, \ldots, t$ are irreducible and of degree greater than 1. The exponents m_i are called the **multiplicities** of the roots c_i. Let c be a root. Then as above,
$$(c - c_1)^{m_1} \cdots (c - c_k)^{m_k} Q_1(c) \cdots Q_t(c) = 0.$$

Now $Q_i(c) \neq 0$ for $i = 1, .., t$ since $Q_i(x)$ are irreducible of degree > 1. Therefore, $(c - c_i) = 0$ for some i, and hence $c = c_i$.

12.3.1 Real and Complex Polynomials

We now consider the underlying field to be $\mathbb{R}$ or $\mathbb{C}$ and consider real and complex polynomials, that is, polynomials in $\mathbb{R}[x]$ and $\mathbb{C}[x]$ respectively. We first need the following which depends on analysis.

Theorem 12.3.3. *A real polynomial of odd degree has a real root.*

Proof. Suppose $P(x) \in \mathbb{R}[x]$ with deg $P(x) = n = 2k + 1$ and suppose the leading coefficient $a_n > 0$ (the proof is almost identical if $a_n < 0$). (See the exercises.) Then

$$P(x) = a_n x^n + \text{ (lower terms) and } n \text{ is odd.}$$

Then:
(1) $\lim_{x \to \infty} P(x) = \lim_{x \to \infty} a_n x^n = \infty$ since $a_n > 0$.
(2) $\lim_{x \to -\infty} P(x) = \lim_{x \to -\infty} a_n x^n = -\infty$ since $a_n > 0$ and n is odd.

From (1) $P(x)$ gets arbitrarily large positively so there exists an x_1 with $P(x_1) > 0$. Similarly, from (2) there exists an x_2 with $P(x_2) < 0$. A real polynomial is a continuous real-valued function for all $x \in \mathbb{R}$. Since $P(x_1)P(x_2) < 0$, it follows from the intermediate value theorem that there exists an x_3, between x_1 and x_2, such that $P(x_3) = 0$. □

As an immediate consequence we have the following corollary:

Corollary 12.3.2. *If $P(x) \in \mathbb{R}[x]$ is irreducible and nonlinear, then its degree is even. (We will see later that it must be 2.)*

We now consider complex polynomials.

Lemma 12.3.1. *Every degree 2 complex polynomial has a root in $\mathbb{C}$.*

Proof. This is just the quadratic formula. If $P(x) = ax^2 + bx + c$, with $a, b, c \in \mathbb{C}$ and $a \neq 0$ then the roots formally are

$$x_1 = \frac{-b + \sqrt{b^2 - 4ac}}{2a}, \quad x_2 = \frac{-b - \sqrt{b^2 - 4ac}}{2a}.$$

We saw in Chapter 5 that from DeMoivre's theorem every complex number has a square root. Hence x_1, x_2 exist in $\mathbb{C}$. They of course may be the same, if $b^2 - 4ac = 0$. □

To continue we need the concept of the **conjugate of a polynomial** and some straightforward consequences of this idea.

Definition 12.3.1. *If $P(x) = a_0 + \cdots + a_n x^n$ is a complex polynomial then its* **conjugate** *is the polynomial*

$$\overline{P}(x) = \overline{a_0} + \cdots + \overline{a_n} x^n.$$

That is, the conjugate is the polynomial whose coefficients are the conjugates of those of $P(x)$.

Lemma 12.3.2. *For any $P(x), Q(x) \in \mathbb{C}[x]$:*
(1) $\overline{P(z)} = \overline{P}(\overline{z})$ if $z \in \mathbb{C}$.
(2) $P(x)$ is a real polynomial if and only if $P(x) = \overline{P}(x)$.
(3) If $P(x)Q(x) = H(x)$ then $\overline{H}(x) = (\overline{P}(x))(\overline{Q}(x))$.

Proof. (1) Suppose $z \in \mathbb{C}$ and $P(z) = a_0 + \cdots + a_n z^n$. Then using the properties of complex conjugates (see Lemma 5.4.2)

$$\overline{P(z)} = \overline{a_0 + \cdots + a_n z^n} = \overline{a_0} + \overline{a_1 z} + \cdots + \overline{a_n z^n} =$$

$$= \overline{a_0} + (\overline{a_1})(\overline{z}) + \cdots + (\overline{a_n})(\overline{z^n}) = \overline{P}(\overline{z}).$$

(2) Suppose $P(x)$ is real then $a_i = \overline{a_i}$ for all its coefficients and hence $P(x) = \overline{P}(x)$. Conversely suppose $P(x) = \overline{P}(x)$. Then $a_i = \overline{a_i}$ for all its coefficients and hence $a_i \in \mathbb{R}$ for each a_i and so $P(x)$ is a real polynomial.

(3) The proof is a computation and left to the exercises. $\qquad\square$

Lemma 12.3.3. *Suppose $G(x) \in \mathbb{C}[x]$. Then $H(x) = G(x)\overline{G}(x) \in \mathbb{R}[x]$.*

Proof. $\overline{H}(x) = \overline{G(x)\overline{G}(x)} = \overline{G}(x)\overline{\overline{G}}(x) = \overline{G}(x)G(x) = G(x)\overline{G}(x) = H(x)$. Therefore, $H(x)$ is a real polynomial. $\qquad\square$

Lemma 12.3.4. *If $f(x) \in \mathbb{R}[x]$ and $f(z_0) = 0$ then $f(\overline{z_0}) = 0$; the complex roots of real polynomials come in conjugate pairs.*

Proof. $f(z_0) = 0$ implies that $\overline{f(z_0)} = 0$. But then $\overline{f}(\overline{z_0}) = 0$. Since $f(x)$ is real, $f(x) = \overline{f}(x)$, so $f(\overline{z_0}) = 0$. $\qquad\square$

Notice that if z_0 is a root of $f(x) \in \mathbb{R}[x]$ then from the factor theorem (Theorem 12.3.1) both $(x - z_0)$ and $(x - \overline{z_0})$ divide $f(x)$.

Lemma 12.3.5. $(x - z)(x - \overline{z}) \in \mathbb{R}[x]$ *for any $z \in \mathbb{C}$.*

We leave the proof for the exercises. Notice that Theorem 12.3.3 and Lemma 12.3.1 now imply that any real polynomial of degree ≤ 3 completely factors into linear factors over $\mathbb{C}$.

12.3.2   The Fundamental Theorem of Algebra

The **fundamental theorem of algebra** is one of the most important algebraic results. This says that any nonconstant complex polynomial must have a complex zero. In general we say that a field F is **algebraically closed** if any nonconstant polynomial in $F[x]$ must have a zero within F. In this language the fundamental theorem of algebra says that the field of complex numbers is algebraically closed.

Theorem 12.3.4. *(Fundamental Theorem of Algebra) Each nonconstant polynomial $f(x) \in \mathbb{C}[x]$, where $\mathbb{C}$ is the field of complex numbers, has a zero in $\mathbb{C}$. Therefore $\mathbb{C}$ is an algebraically closed field.*

For any field F we have that $x - a$ divides $p(x)$ if $p(a) = 0$. Applying this to complex polynomials we obtain the fact that any complex polynomial must completely factor into linear factors.

Corollary 12.3.3. *A complex polynomial $p(x) \in \mathbb{C}[x]$ factors into a product of linear polynomials.*

Proof. Let $f(x) \in \mathbb{C}[x]$ and use induction on the degree. The corollary is clearly true if $\deg f(x) = 1$, since then $f(x)$ is itself linear. Suppose the result is true for all polynomials of degree $< n$ and let $\deg(f(x)) = n$. From the fundamental theorem of algebra, there exists a root $x_0 \in \mathbb{C}$, and therefore $(x - x_0)$ divides $f(x)$. Hence from Theorem 12.3.1 $f(x) = (x - x_0)g(x)$ with $\deg g(x) < n$. From the inductive hypothesis, $g(x)$ factors into linear factors, so therefore $f(x)$ does also. $\qquad\qquad\square$

There are many distinct and completely different proofs of the fundamental theorem of algebra. In [FR] twelve proofs were given covering a wide area of mathematics. In Chapter 14 we will use Galois theory to present a proof. In this section we briefly mention some of the history surrounding this theorem.

The first mention of the fundamental theorem of algebra, in the form that every polynomial equation of degree n has exactly n roots, was given by Peter Roth of Nurnberg in 1608. However, its conjecture is generally credited to Girard who also stated the result in 1629. It was then more clearly stated by Descartes in 1637 who also distinguished between real and imaginary roots. The first published proof of the fundamental theorem of algebra was then given by D'Alembert in 1746. However, there were gaps

in D'Alembert's proof and the first fully accepted proof was that given by
Gauss in 1797 in his Ph.D. thesis. This was published in 1799. Interestingly
enough, in reviewing Gauss's original proof, modern scholars tend to agree
that there are as many holes in this proof as in D'Alembert's proof. Gauss,
however, published three other proofs with no such holes. He published
second and third proofs in 1816 while his final proof, which was essentially
another version of the first, was presented in 1849.

   The fundamental theorem of algebra says that every nonconstant complex
polynomial has a complex zero. Hence every nonconstant real polynomial
must have a complex zero. However, not every nonconstant real polynomial
has a real zero although as we have seen a real polynomial of odd degree must
have a real zero. We close this section by looking at irreducible polynomials
over $\mathbb{R}$. The following is an immediate corollary of the complete factorization
of complex polynomials.

Corollary 12.3.4. *Suppose $f(x) \in \mathbb{C}[x]$ with $\deg f(x) = n$. Suppose the roots
of $f(x)$ are $x_1, x_2, \ldots, x_n$ (some may be repeated so that each x_i is counted
the number of times of its multiplicity). Then*

$$f(x) = \alpha(x - x_1) \cdots (x - x_n) \text{ with } \alpha \in \mathbb{C}.$$

Lemma 12.3.6. *A nonconstant real polynomial factors into degree 1 and
degree 2 factors over $\mathbb{R}$. Equivalently, the only irreducible real polynomials
are linear polynomials and quadratic polynomials without real roots.*

Proof. Suppose $P(x) \in \mathbb{R}[x]$, then $P(x) \in \mathbb{C}[x]$. Suppose $z_1, \ldots, z_n$ are its
complex roots, so that

$$P(x) = \alpha(x - z_1) \cdots (x - z_n)$$

where here $\alpha \in \mathbb{R}$, since α is the leading coefficient of $P(x)$. Here again each
z_i may be repeated as above. If z_i is real, then $(x - z_i)$ is a real linear factor.
If $z_i \notin \mathbb{R}$ then its complex conjugate $\overline{z_i}$ is also a root by Lemma 12.3.4. Thus
the factor $(x - \overline{z_i})$ must occur in the above factorization. But then by Lemma
12.3.5 $(x - z_i)(x - \overline{z_i})$ is a real factor of degree 2. Grouping together all factors
of this form we see that $P(x)$ is a product of real polynomials of degree 1 and
2. Thus if $P(x)$ is any real polynomial of degree > 2 it can be factored into
degree 1 and degree 2 polynomials. Thus it must be reducible. This shows
that the only irreducible real polynomials are of degree ≤ 2. Clearly linear
polynomials are irreducible. By Corollary 12.3.1 a polynomial of degree 2 is
irreducible over $\mathbb{R}$ if and only if it does not have a zero in $\mathbb{R}$.   □

Corollary 12.3.5. *An irreducible real polynomial must be of degree 1 or 2.*

12.3.3 The Rational Roots Theorem

Any odd degree real polynomial has a real root. There can be irreducible rational polynomials of every possible degree, The question arises as to when a rational polynomial has a rational root. If a rational polynomial has a rational root then multiplying through by the LCM of the denominators then a corresponding integral polynomial has a rational root. A criterion for an integral polynomial to have a rational root is given by what is called the **rational roots theorem**. The proofs are direct and left to the exercises. We will see the general version of this result in Theorem 12.4.5.

Lemma 12.3.7. *Let $p(x) = q_0 + q_1 x + \cdots + q_n x^n \in \mathbb{Q}[x]$. Let d be the LCM of the denominators of $q_0, \ldots, q_n$. Then $p_1(x) = dp(x) \in \mathbb{Z}[x]$. Suppose that $\frac{r}{s} \in \mathbb{Q}$ with $p(\frac{r}{s}) = 0$ if and only if $p_1(\frac{r}{s}) = 0$.*

Theorem 12.3.5. *(Rational Roots Theorem) Let $f(x) = a_0 + a_1 x + \cdots + a_n x^n \in \mathbb{Z}[x]$ with $\deg(f(x)) = n$. Then if $f(\frac{r}{s}) = 0$ with $r, s \in \mathbb{Z}$ and $\gcd(r, s) = 1$ then $r | a_0$ and $s | a_n$.*

12.3.4 Solvability by Radicals

The fundamental theorem of algebra and the study of the zeros of real and complex polynomials is actually part of a general development in the theory of equations. The ability to solve quadratic equations and in essence the quadratic formula, which should be familiar to the reader, was known to the Babylonians some 3600 years ago. The quadratic formula is true in any field not of characteristic 2.

Theorem 12.3.6. *Let F be a field with characteristic not equal to 2. Then the zeros of the quadratic polynomial $p(x) = ax^2 + bx + c$ with $a, b, c \in F$ and $a \neq 0$ are given by*

$$x = \frac{-b \pm \sqrt{b^2 - 4ac}}{2a}.$$

If $b^2 - 4ac$ does not have a square root within F then $p(x)$ is irreducible in $F[x]$.

Proof. The proof depends upon a method called completing the square. Suppose we have the equation

$$ax^2 + bx + c = 0$$

with $a, b, c \in F$ and $a \neq 0$ so that $a^{-1} = \frac{1}{a}$ exists in F. Then we can rearrange and write

$$x^2 + \frac{b}{a}x = -\frac{c}{a}.$$

Here $\frac{b}{a} = a^{-1}b$ which exists since $a \neq 0$ and hence a has an inverse in F.

Recall that over any field from elementary algebra we have $(x + t)^2 = x^2 + 2xt + t^2$. Therefore in a perfect square expression of a binomial the squared term on the end is the square of $\frac{1}{2}$ of the coefficient of the middle term. Applying this to our equation above we add $\frac{b^2}{4a^2}$ to both sides to obtain

$$x^2 + \frac{b}{a}x + \frac{b^2}{4a^2} = \frac{b^2}{4a^2} - \frac{c}{2a} \implies$$

$$(x + \frac{b}{2a})^2 = \frac{b^2 - 4ac}{4a^2}.$$

Taking square roots if possible we get

$$x + \frac{b}{2a} = \frac{\pm\sqrt{b^2 - 4ac}}{2a} \implies x = \frac{-b \pm \sqrt{b^2 - 4ac}}{2a}.$$

If $b^2 - 4ac$ has a square root in F then this formula provides the zeros of the polynomial. If $b^2 - 4ac$ does not have a square root in F then $p(x)$ can have no zeros in F and since it is of degree 2 it cannot factor. Therefore in this case it is irreducible.   □

Since the quadratic formula uses only field operations and square roots we say that the quadratic formula solves any quadratic polynomial **by radicals**.

Before continuing we present an example over a finite field.

EXAMPLE 12.3.1. Solve $2x^2 + x + 3 = 0$ over the finite field $\mathbb{Z}_7$.

Here $a = 2, b = 1, c = 3$ so applying the quadratic formula we obtain

$$x = \frac{-1 \pm \sqrt{1 - 4(2)(3)}}{4} = \frac{-1 \pm \sqrt{-23}}{4}.$$

In $\mathbb{Z}_7$ we have $-23 = -2 = 5$. If we check the squares mod 7 we find

$$\begin{array}{c|cccccc} x & 1 & 2 & 3 & 4 & 5 & 6 \\ x^2 & 1 & 4 & 2 & 2 & 4 & 1 \end{array}$$

Therefore 5 has no square root in $\mathbb{Z}_7$ and the above equation is nonsolvable over $\mathbb{Z}_7$. So by Corollary 12.3.1 the polynomial $2x^2 + x + 3$ is irreducible over $\mathbb{Z}_7$.

We saw in Chapter 5 that every complex number has a square root. Hence the quadratic formula then says that any degree-2 two polynomial over $\mathbb{C}$ has a root in $\mathbb{C}$. Further any degree-2 polynomial equation can be solved by radicals over $\mathbb{C}$. In the sixteenth century the Italian mathematician Niccolo Tartaglia discovered a similar formula in terms of radicals to solve cubic equations. This **cubic formula** is now known erroneously as **Cardano's formula** in honor of Cardano, who first published it in 1545. An earlier special version of this formula was discovered by Scipione del Ferro. Cardano's student Ferrari extended the formula to solutions by radicals for fourth degree polynomials. The combination of these formulas says that complex polynomials of degree 4 or less must have complex roots and further could be solved by radicals. In the seventeenth century it became clear from the elementary properties of continuous functions that all odd degree real polynomials must have real roots. All these results lent credence to the Fundamental Theorem which was mentioned by Roth in 1608 and conjectured by Girard in 1629.

From Cardano's work until the very early nineteenth century, attempts were made to find similar formulas for degree-5 polynomials. In 1805 Ruffini proved that fifth degree polynomial equations are insolvable by radicals in general. Therefore there exists no comparable formula for degree 5. Abel in 1825 - 1826 and Galois in 1831 extended Ruffini's result and proved the insolubility by radicals for all degrees 5 or greater. In doing this, Galois developed a general theory of field extensions and its relationship to group theory. This has come to be known as **Galois theory** which we will discuss in Chapter 14. At that point we will give a proof of the insolvability by radicals of fifth degree and higher polynomial equations.

12.3.5   Algebraic and Transcendental Numbers

In Chapter 5 we saw that the real number system can be partitioned into the rational numbers and the irrational numbers. Further there were uncountably many irrational numbers but only countably many rational numbers. Here we consider another partition of the real numbers into algebraic numbers and transcendental numbers

Definition 12.3.2. *An* **algebraic number** α *is an element of* $\mathbb{R}$ *which is a zero of a rational polynomial that is, a polynomial with rational coefficients. Hence an algebraic number is an* $\alpha \in \mathbb{R}$ *such that* $f(\alpha) = 0$ *for some* $f(x) \in \mathbb{Q}[x]$. *If* $\alpha \in \mathbb{R}$ *is not algebraic it is* **transcendental**.

Lemma 12.3.8. *Any* $q \in \mathbb{Q}$ *is algebraic; hence being rational implies being algebraic. However, there are algebraic numbers that are irrational.*

Proof. Let $q \in \mathbb{Q}$. Then q is a zero of the polynomial $p(x) = x - q \in \mathbb{Q}[x]$. Hence each rational number is algebraic. Consider $\sqrt{2}$. We know that this is irrational. However, it is a zero of the rational polynomial $x^2 - 2$ and hence is algebraic.   □

We will let $\mathcal{A}$ denote the totality of algebraic numbers within the real numbers $\mathbb{R}$, and $\mathcal{T}$ the set of transcendentals so that $\mathbb{R} = \mathcal{A} \cup \mathcal{T}$. We will prove the following result in Chapter 14 when we consider field extensions.

Theorem 12.3.7. *The set* $\mathcal{A}$ *of algebraic numbers forms a subfield of* $\mathbb{R}$.

Since each rational is algebraic it is clear that there are algebraic numbers. On the other hand we haven't examined the question of whether transcendental numbers really exist. To show that any particular real number is transcendental is in general extremely difficult. However, it is relatively easy to show that there are uncountably infinitely many transcendentals. To do this we need some results from the theory of sets. We just state them here. See [HA] for a discussion of them.

Lemma 12.3.9. *(1) A finite Cartesian product of countable sets is countable. (2) A union of countably many countable sets is itself countable.*

Theorem 12.3.8. *The set* $\mathcal{A}$ *of algebraic numbers is countably infinite. Therefore* $\mathcal{T}$ *the set of transcendental numbers is uncountably infinite.*

Proof. Let
$$\mathcal{P}_n = \{f(x) \in \mathbb{Q}[x] : \deg(f(x)) \le n\}.$$
Since if $f(x) \in \mathcal{P}_n$, $f(x) = q_o + q_1 x + \cdots + q_n x^n$ with $q_i \in \mathbb{Q}$ we can identify a polynomial of degree $\le n$ with an $(n+1)$-tuple $(q_0, q_1, \ldots, q_n)$ of rational

numbers. We can then establish a bijection between the set $\mathcal{P}_n$ and the set of $(n+1)$-tuples of rational numbers (see the exercises for details). Therefore the set $\mathcal{P}_n$ has the same size as the $(n+1)$-fold Cartesian product of $\mathbb{Q}$:

$$\mathbb{Q}^{n+1} = \mathbb{Q} \times \mathbb{Q} \times \cdots \times \mathbb{Q}$$

where there are $n+1$ factors. Since a finite Cartesian product of countable sets by Lemma 12.3.9 (i) is still countable it follows that $\mathcal{P}_n$ is a countable set.

Now let

$$\mathcal{B}_n = \bigcup_{p(x) \in \mathcal{P}_n} \{ \text{ roots of } p(x)\},$$

that is, $\mathcal{B}_n$ is the union of all roots in $\mathbb{C}$ of all rational polynomials of degree $\leq n$. Since each such $p(x)$ has a maximum of n roots by Theorem 12.3.2 and since $\mathcal{P}_n$ is countable it follows that $\mathcal{B}_n$ is a countable union of finite sets and hence is still countable.

Now

$$\mathcal{A} = \bigcup_{n=1}^{\infty} \mathcal{B}_n$$

so that $\mathcal{A}$ is a countable union of countable sets and is therefore countable by Lemma 12.3.9 (ii).

Since $\mathbb{R}$ is uncountably infinite the second assertion follows directly from the countability of $\mathcal{A}$. If $\mathcal{T}$ were countable then $\mathbb{R} = \mathcal{A} \cup \mathcal{T}$ would also be countable which is a contradiction. □

From Theorem 12.3.8 we know that there exist infinitely many transcendental numbers. Liouville in 1851 gave the first proof of the existence of transcendentals by exhibiting a few. He gave as one the following example. We give the proof to show how difficult it is to verify that a number is transcendental.

Theorem 12.3.9. *The real number*

$$c = \sum_{j=1}^{\infty} \frac{1}{10^{j!}}$$

is transcendental.

Proof. First of all since $\frac{1}{10^{j!}} \leq \frac{1}{10^j}$, and $\sum_{j=1}^{\infty} \frac{1}{10^j}$ is a convergent geometric series, it follows from the comparison test that the infinite series defining c converges and defines a real number. Further, since $\sum_{j=1}^{\infty} \frac{1}{10^j} = \frac{1}{9}$, it follows that $c \leq \frac{1}{9} < 1$.

Suppose that c is algebraic in order to deduce a contradiction so that $g(c) = 0$ for some rational nonzero polynomial $g(x)$. Multiplying through by the least common multiple of all the denominators in $g(x)$ we may suppose that $f(c) = 0$ for some integral polynomial

$$f(x) = \sum_{j=0}^{n} m_j x^j$$

with $m_j \in \mathbb{Z}$. Then c satisfies

$$\sum_{j=0}^{n} m_j c^j = 0$$

for some integers $m_0, \ldots, m_n$.

If $0 < x < 1$ then by the triangle inequality

$$|f'(x)| = |\sum_{j=1}^{n} j m_j x^{j-1}| \leq \sum_{j=1}^{n} |j m_j| = B$$

where B is a real constant depending only on the coefficients and the degree of $f(x)$.

Now let

$$c_k = \sum_{j=1}^{k} \frac{1}{10^{j!}}$$

be the kth partial sum for c. Then since the series for c is term by term $\leq$ a geometric series with common ratio $r = \frac{1}{10}$ we have

$$|c - c_k| = \sum_{j=k+1}^{\infty} \frac{1}{10^{j!}} < 2 \cdot \frac{1}{10^{(k+1)!}}.$$

Since the series for c is a positive series $c_k < c$ and clearly $f(x)$ is continuous and differentiable on $[c_k, c]$ we can apply the mean value theorem to $f(x)$ on $[c_k, c]$ to obtain

$$|f(c) - f(c_k)| = |c - c_k||f'(\zeta)|$$

for some ζ with $c_k < \zeta < c < 1$. Now since $0 < \zeta < 1$ we have from the above that

$$|c - c_k||f'(\zeta)| < 2B\frac{1}{10^{(k+1)!}}.$$

On the other hand, since $f(x)$ can have at most n roots by Theorem 12.3.2, it follows that for all k large enough we would have $f(c_k) \neq 0$. Since $f(c) = 0$ we have

$$|f(c) - f(c_k)| = |f(c_k)| = |\sum_{j=1}^{n} m_j c_k^j| > \frac{1}{10^{nk!}}$$

since for each j, $m_j c_k^j$ is a rational number with denominator $10^{jk!}$. However, if k is chosen sufficiently large and n is fixed we have

$$\frac{1}{10^{nk!}} > \frac{2B}{10^{(k+1)!}}.$$

To see this inequality multiply both sides of the above inequality by $10^{nk!}$. Thus the above inequality is equivalent to

$$1 > \frac{2B(10^{nk!})}{10^{(k+1)!}}.$$

But

$$\frac{2B(10^{nk!})}{10^{(k+1)!}} = 2B(10^{(n-(k+1))k!}).$$

Here the exponent of 10 becomes negative as soon as $k > (n - 1)$. Since B is fixed we can certainly make this term less than 1 for sufficiently large k. Thus we get that

$$|f(c) - f(c_k)| > \frac{2B}{10^{(k+1)!}}.$$

This contradicts the inequality from the mean value theorem. Therefore c is transcendental. $\square$

In 1873 Hermite proved that e is transcendental while Lindemann in 1882 showed that π is transcendental. Schneider in 1934 showed that a^b is transcendental if $a \neq 0$, a and b are algebraic and b is irrational.

12.4 Unique Factorization in $\mathbb{Z}[x]$

We have seen that if F is a field then the ring of polynomials $F[x]$ has unique factorization into irreducible polynomials and is hence a UFD. In this section we prove that if R is a UFD then the polynomial ring $R[x]$ is also a UFD. In particular the polynomial ring over the integers $\mathbb{Z}[x]$ is also a UFD.

To prove this we first need the following due to Gauss. However, before proving Gauss's lemma we need the following preliminary fact.

Lemma 12.4.1. *Let R be a UFD. Then an element $p \in R$ is a prime if and only if p is an irreducible in R.*

Proof. As in the proof of Lemma 11.4.1 in any integral domain R a prime must be an irreducible. This shows one direction.

Conversely suppose that $p \in R$ and $p|ab$ for $a, b \in R$ and p is an irreducible in R. Since R is a UFD the elements a, b, ab all have unique factorizations in terms of irreducibles in R. Thus if p occurs in the factorization of ab, which it must since $p|ab$ then by uniqueness, an associate of p, say q occurs in the factorization of a or of b. Since p is an associate of q this implies that $p|a$ or $p|b$ (see the exercises) proving the lemma. □

Theorem 12.4.1. *(Gauss's Lemma) Let R be a UFD and $f(x), g(x)$ primitive polynomials in $R[x]$. Then their product $f(x)g(x)$ is also primitive.*

Proof. Let R be a UFD and $f(x), g(x)$ primitive polynomials in $R[x]$. Suppose that $f(x)g(x)$ is not primitive. Since R is a UFD there must exist an irreducible $p \in R$ such that p divides each coefficient of $f(x)g(x)$. But by the above Lemma 12.4.1 p is also a prime in R. Then $p|f(x)g(x)$. Since prime elements of R are also prime elements of $R[x]$ it follows that p is also a prime element of $R[x]$ and hence $p|f(x)$ or $p|g(x)$. Therefore either $f(x)$ or $g(x)$ is not primitive, giving a contradiction.

□

To continue we need a result about GCD's in UFD's.

Lemma 12.4.2. *Let R be a UFD. Then the greatest common divisor and least common multiple of any finite set of nonzero elements of R exist.*

Proof. We prove this for two nonzero elements. For greater than two it follows easily from an induction on the number of elements (see exercises).

Let a, b be nonzero elements of R. Recall Theorem 4.4.4. If either a or b is a unit then take $\gcd(a, b) = \operatorname{lcm}(a, b)$ to be the other nonunit. If they are both units take $\gcd(a, b) = \operatorname{lcm}(a, b) = 1$. Now suppose that neither a nor b is a unit. Since R is a UFD we have unique decompositions of a, b in terms of irreducibles. That is

$$a = p_1^{e_1} \cdots p_k^{e_k},$$

$$b = p_1^{f_1} \cdots p_k^{f_k},$$

where $p_1, \ldots, p_k$ are irreducibles and as in the integers we may include zero exponents for noncommon irreducibles. Then

$$\gcd(a, b) = p_1^{\min(e_1, f_1)} \cdots p_k^{\min(e_k, f_k)}$$

and

$$\operatorname{lcm}(a, b) = p_1^{\max(e_1, f_1) \cdots p_k^{\max(e_k, f_k)}}.$$

Thus both the GCD and LCM exist. $\hspace{1cm} \square$

Theorem 12.4.2. *Let R be a UFD and K its field of fractions.*

(a) If $g(x) \in K[x]$ is nonzero then there is a nonzero $a \in K$ such that $ag(x) \in R[x]$ is primitive.

(b) Let $f(x), g(x) \in R[x]$ with $g(x)$ primitive and $f(x) = ag(x)$ for some $a \in K$. Then $a \in R$.

(c) If $f(x) \in R[x]$ is nonzero then there is a $b \in R$ and a primitive $g(x) \in R[x]$ such that $f(x) = bg(x)$.

Proof. (a) Suppose that $g(x) = \sum_{i=0}^{n} a_i x^i$ with $a_i = \frac{r_i}{s_i}$, $r_i, s_i \in R$. Set $s = s_0 s_1 \cdots s_n$. Then $sg(x)$ is a nonzero element of $R[x]$. Let d be a greatest common divisor of the coefficients of $sg(x)$ which exists by Lemma 12.4.2 and $d \neq 0$. If we set $a = \frac{s}{d}$ then $ag(x)$ is primitive and $ag(x) \in R[x]$.

(b) For $a \in K$ there are coprime $r, s \in R$ satisfying $a = \frac{r}{s}$. Suppose that $a \notin R$. Since the units in $R[x]$ are the units in R if s were a unit then $\frac{r}{s} \in R$. Hence we may assume that s is not a unit. By Lemma 12.4.1 there is a prime element $p \in R$ dividing s. Since $g(x)$ is primitive p does not divide all the coefficients of $g(x)$. However, we also have $f(x) = ag(x) = \frac{r}{s}g(x)$. Hence $sf(x) = rg(x)$ where $p|s$ and p doesn't divide r since it was assumed that r, s were coprime. Therefore p divides all the coefficients of $g(x)$ and hence $a \in R$.

(c) From part (a) there is a nonzero $a \in K$ such that $af(x)$ is primitive in $R[x]$. Then $f(x) = a^{-1}(af(x))$. From part (b) we must have $a^{-1} \in R$. Set $g(x) = af(x)$ and $b = a^{-1}$.

$\square$

Theorem 12.4.3. *Let R be a UFD and K its field of fractions. Let $f(x) \in R[x]$ be a polynomial of degree ≥ 1.*

(a) If $f(x)$ is primitive and $f(x)|g(x)$ in $K[x]$ then $f(x)$ divides $g(x)$ also in $R[x]$.

(b) If $f(x)$ is irreducible in $R[x]$ then it is also irreducible in $K[x]$.

(c) If $f(x)$ is primitive and a prime element of $K[x]$ then $f(x)$ is also a prime element of $R[x]$.

Proof. (a) Suppose that $g(x) = f(x)h(x)$ with $h(x) \in K[x]$. From Theorem 12.4.2 part (a) there is a nonzero $a \in K$ such that $h_1(x) = ah(x)$ is primitive in $R[x]$. Hence $g(x) = \frac{1}{a}(f(x)h_1(x))$. From Gauss's Lemma $f(x)h_1(x)$ is primitive in $R[x]$ and therefore from Theorem 12.4.2 part (b) we have $\frac{1}{a} \in R$. It follows that $f(x)|g(x)$ in $R[x]$.

(b) Suppose that $g(x) \in K[x]$ is a factor of $f(x)$. From Theorem 12.4.2 part (a) there is a nonzero $a \in K$ with $g_1(x) = ag(x)$ primitive in $R[x]$. Since a is a unit in K it follows that

$$g(x)|f(x) \text{ in } K[x] \text{ implies } g_1(x)|f(x) \text{ in } K[x]$$

and hence since $g_1(x)$ is primitive by part (a)

$$g_1(x)|f(x) \text{ in } R[x].$$

However, by assumption $f(x)$ is irreducible in $R[x]$. This implies that either $g_1(x)$ is a unit in R or $g_1(x)$ is an associate of $f(x)$.

If $g_1(x)$ is a unit then $g_1 \in K$ and $g_1 = ga$ and hence $g \in K$, that is, $g = g(x)$ is a unit.

If $g_1(x)$ is an associate of $f(x)$ then $f(x) = bg(x)$ where $b \in K$ since $g_1(x) = ag(x)$ with $a \in K$. Combining these it follows that $f(x)$ has only trivial factors in $K[x]$ and since by assumption $f(x)$ is nonconstant it follows that $f(x)$ is irreducible in $K[x]$.

(c) Suppose that $f(x)|g(x)h(x)$ with $g(x), h(x) \in R[x]$. Since $f(x)$ is a prime element in $K[x]$ we have that $f(x)|g(x)$ or $f(x)|h(x)$ in $K[x]$. From part (a) we have $f(x)|g(x)$ or $f(x)|h(x)$ in $R[x]$ implying that $f(x)$ is a prime element in $R[x]$.

$\square$

We can now state and prove our main result.

Theorem 12.4.4. *(Gauss) Let R be a UFD. Then the polynomial ring* $R[x]$ *is also a UFD*

Proof. By induction on degree we show that each nonunit and nonzero $f(x) \in R[x]$ is a product of irreducible elements. Since R is an integral domain so is $R[x]$, and so the fact that $R[x]$ is a UFD then follows from Theorem 11.4.1.

If $\deg(f(x)) = 0$ then $f(x) = f$ is a nonunit in R. Since R is a UFD $f(x)$ is a product of irreducible elements in R. But in a UFD any irreducible element is prime by Lemma 12.4.1 so f is a product of primes in R. However, from Theorem 12.2.7(b), each prime factor is then also prime in $R[x]$. In a UFD irreducible is equivalent to prime so therefore $f(x)$ is a product of prime elements.

Now suppose $n > 0$ and that the claim is true for all polynomials $f(x)$ of degree $< n$. Let $f(x)$ be a polynomial of degree $n > 0$. From Theorem 12.4.2 (c) there is an $a \in R$ and a primitive $h(x) \in R[x]$ satisfying $f(x) = ah(x)$. Since R is a UFD the element a is a product of prime elements in R or a is a unit in R. Since the units in $R[x]$ are the units in R and a prime element in R is also a prime element in $R[x]$ it follows that a is a product of prime elements in $R[x]$ or a is a unit in $R[x]$. Let K be the field of fractions of R. Then $K[x]$ is a UFD. Hence $h(x)$ is a product of prime elements of $K[x]$. Let $p(x) \in K[x]$ be a prime divisor of $h(x)$. Since K is a field of fractions of R each coefficient in $p(x)$ is of the form $\frac{r}{s}$ for some $r, s \in R$. Clearing all denominators we arrive at $dp(x)|dh(x)$ where now $dp(x) = q(x), d \in R$. But since $p(x)$ was prime in $K[x]$ and d is a unit in K then $q(x)$ is also a prime in $K[x]$. Moreover by Theorem 12.4.2 (a) since $q(x) \in R[x]$ is nonzero there exists a nonzero $\alpha \in K$ such that $\alpha q(x)$ is primitive. Since α is a unit in K and $q(x)$ is a prime in $K[x]$ then $\alpha q(x) \in R[x]$ is also a prime in $K[x]$. Let us just call $\alpha q(x)$, $p(x)$ and note that $p(x) \in R[x]$, $p(x)$ is primitive and a prime in $K[x]$ and $p(x)|h(x)$ in $K[x]$. From Theorem 12.4.3 (c) it follows that $p(x)$ is a prime element of $R[x]$ and further from Theorem 12.4.3 (a) that $p(x)$ is a divisor of $h(x)$ in $R[x]$. Therefore

$$f(x) = ah(x) = ap(x)g(x) \in R[x] \text{ where}$$

(1) a is a product of prime elements of $R[x]$ or a is a unit in $R[x]$,
(2) $\deg(p(x)) > 0$, since $p(x)$ is a prime element in $K[x]$,
(3) $p(x)$ is a prime element in $R[x]$,

(4) $\deg(g(x)) < \deg(f(x))$ since $\deg(p(x)) > 0$.

By our inductive hypothesis we have then that $g(x)$ is a product of prime elements in $R[x]$ or $g(x)$ is a unit in $R[x]$. Therefore the claim holds for $f(x)$ and therefore holds for all $f(x)$ by induction.

$\square$

If $R[x]$ is a polynomial ring over R we can form a polynomial ring in a new indeterminate y over this ring to form $(R[x])[y]$. It is straighforward that $(R[x])[y]$ is isomorphic to $(R[y])[x]$. We denote both of these rings by $R[x, y]$ and consider this as the ring of polynomials in two commuting variables x, y with coefficients in R.

If R is a UFD then $R[x]$ is also a UFD and hence $R[x, y]$ is also a UFD. Inductively then the ring of polynomials in n commuting variables $R[x_1, x_2, \ldots, x_n]$ is also a UFD.

Corollary 12.4.1. *If R is a UFD then the polynomial ring in n commuting variables $R[x_1, \ldots, x_n]$ is also a UFD.*

We now give a condition for a polynomial in $R[x]$ to have a zero in $K[x]$ where K is the field of fractions of R.

Theorem 12.4.5. *Let R be a UFD and K its field of fractions. Let $f(x) = x^n + r_{n-1}x^{n-1} + \cdots + r_0 \in R[x]$. Suppose that $\beta \in K$ is a zero of $f(x)$. Then β is in R and is a divisor of r_0.*

Proof. Let $\beta = \frac{r}{s}$ where $s \neq 0$ and $r, s \in R$ and r, s are coprime. Now

$$f(\frac{r}{s}) = 0 = \frac{r^n}{s^n} + r_{n-1}\frac{r^{n-1}}{s^{n-1}} + \cdots + r_0.$$

Hence it follows that s must divide r^n. Since r and s are coprime s must be a unit and then without loss of generality we may assume that $s = 1$. Then $\beta \in R$ and

$$r(r^{n-1} + r_{n-1}r^{n-2} + \cdots + r_1) = -r_0$$

and so $r|r_0$.

$\square$

We remarked earlier that although any PID must be a UFD it is not true that a UFD must be a PID. We now give two examples. Let F be a field so that from the above $F[x, y]$ is a UFD. We show that it is not a PID.

Let I be the subset of $F[x, y]$ consisting of all polynomials $p(x, y)$ with zero constant term. It is straightforward to show that I forms an ideal. Suppose $I = \langle p(x, y) \rangle$ for some $p(x, y) \in F[x, y]$. Now $x \in I$ since its constant term is zero. If $p(x, y)$ had a nonzero term with y in it then it is impossible to multiply $p(x, y)$ by a $q(x, y) \in F[x, y]$ so that $p(x, y)q(x, y) = x$ (why?). Hence $p(x, y)$ can have no nonzero terms with y in them. However, $y \in I$ so the same argument shows that $p(x, y)$ can have no nonzero terms with x in it. Hence $p(x, y)$ must be a constant but then $x \notin I$. Therefore I cannot be principal and $F[x, y]$ is a UFD but not a PID.

For another example note that since $\mathbb{Z}$ is a UFD, Gauss's theorem implies that $\mathbb{Z}[x]$ is also a UFD. However $\mathbb{Z}[x]$ is not a principal ideal domain. For example the set of integral polynomials with even constant term is an ideal but not principal. We leave the verification to the exercises.

On the other hand we saw that if K is a field $K[x]$ is a PID. The question arises as to when $R[x]$ actually is a principal ideal domain. It turns out to be precisely when R is a field.

Theorem 12.4.6. *Let R be a commutative ring with an identity. Then the following are equivalent:*

　　(a) R is a field.
　　(b) $R[x]$ is a Euclidean domain.
　　(c) $R[x]$ is a principal ideal domain.

Proof. From Section 12.2 we know that (a) implies (b) which in turn implies (c). Therefore we must show that (c) implies (a). Assume then that $R[x]$ is a principal ideal domain. Define the map

$$\tau : R[x] \to R$$

by

$$\tau(f(x)) = f(0).$$

It is easy to see that τ is a ring homomorphism with $R[x]/\ker(\tau) \cong R$. Therefore $\ker(\tau) \neq R[x]$. Since $R[x]$ is a principal ideal domain it is an integral domain. It follows that $\ker(\tau)$ must be a prime ideal since the quotient ring is an integral domain. However, since $R[x]$ is a principal ideal domain prime ideals are maximal ideals and hence $\ker(\tau)$ is a maximal ideal. Therefore $R \cong R[x]/\ker(\tau)$ is a field. $\qquad\square$

We must show that in a PID prime ideals are maximal. This follows from the following two results.

Lemma 12.4.3. *Let R be a PID. Then an ideal $I = \langle p \rangle$ is maximal in R if and only if p is irreducible. (Recall that in a PID irreducible is equivalent to prime — Lemma 11.4.1.)*

Proof. Let $I = \langle p \rangle$ be a maximal ideal in R a PID. Suppose that $p = ab$ for some $a, b \in R$. Then $\langle p \rangle \subseteq \langle a \rangle$. If $\langle p \rangle = \langle a \rangle$ then a and p are associates and hence b is a unit.

On the other hand if $\langle p \rangle \neq \langle a \rangle$ then $\langle a \rangle = \langle 1 \rangle = R$ since $\langle p \rangle$ is maximal. But then a is a unit. This shows that p is irreducible in R.

Conversely suppose that p is irreducible. Then if $\langle p \rangle \subseteq \langle a \rangle$ we must have $p = ab$. If a is a unit then $\langle a \rangle = \langle 1 \rangle = R$. If a is not a unit since p is an irreducible b must be a unit. Thus there exists $t \in R$ such that $bt = 1$. Then $pt = abt = a$ so that $\langle a \rangle \subseteq \langle p \rangle$. Thus $\langle p \rangle \subseteq \langle a \rangle$ implies that $\langle a \rangle = R$ or $\langle a \rangle = \langle p \rangle$ and $\langle p \rangle \neq R$ since p is not a unit. Hence $\langle p \rangle$ is a maximal ideal. $\square$

Lemma 12.4.4. *Let R be a PID. Let $I \neq R$ be a prime ideal in R. Then I is a maximal ideal in R.*

Proof. Since R is a PID we have $I = \langle p \rangle$ for some $p \in R$. If p were a unit the $\langle p \rangle = R$, contrary to assumption. Therefore p is a nonunit in R. Suppose that $p | ab$ for $a, b \in R$. Then $ab \in I$ and since I is a prime ideal then either $a \in I$ or $b \in I$. This implies that $p | a$ or $p | b$ and therefore p is a prime in R. Since prime and irreducible elements are equivalent in R and $I = \langle p \rangle$, Lemma 12.4.3 implies that I is a maximal ideal in R. $\square$

We now consider the relationship between irreducibles in $R[x]$ for a general integral domain and irreducibles in $K[x]$ where K is its field of fractions. This is handled by the next result called, Eisenstein's criterion.

Theorem 12.4.7. *(Eisenstein's Criterion) Let R be an integral domain and K its field of fractions. Let*

$$f(x) = \sum_{i=0}^{n} a_i x^i \in R[x]$$

be of degree $n > 0$. Let p be a prime element of R satisfying
(1) $p | a_i$ for $i = 0, \ldots, n - 1$,

(2) p does not divide a_n,

(3) p^2 does not divide a_0.

Then:

(a) If $f(x)$ is primitive then $f(x)$ is irreducible in $R[x]$.

(b) Suppose that R is a UFD. Then $f(x)$ is also irreducible in $K[x]$.

Proof. (a) Suppose that $f(x) = g(x)h(x)$ with $g(x), h(x) \in R[x]$. Suppose that

$$g(x) = \sum_{i=0}^{k} b_i x^i, b_k \neq 0 \text{ and } h(x) = \sum_{j=0}^{l} c_j x^j, c_l \neq 0.$$

Then $a_0 = b_0 c_0$. Now $p|a_0$ but p^2 does not divide a_0. This implies that either p doesn't divide b_0 or p doesn't divide c_0. Without loss of generality assume that $p|b_0$ and p doesn't divide c_0.

Since $a_n = b_k c_l$ and p does not divide a_n it follows that p does not divide b_k. Let b_j be the first coefficient of $g(x)$ which is not divisible by p. Consider

$$a_j = b_j c_0 + \cdots + b_0 c_j$$

where everything after the first term is divisible by p. Since p does not divide both b_j and c_0 it follows that p does not divide $b_j c_0$ and therefore p does not divide a_j which implies that $j = n$. Then from $j \leq k \leq n$ it follows that $k = n$. Therefore $\deg(g(x)) = \deg(f(x))$ and hence $\deg(h(x)) = 0$. Thus $h \in R$. Then from $f(x) = hg(x)$ with f primitive it follows that h is a unit and therefore $f(x)$ is irreducible in $R[x]$.

(b) Suppose that $f(x) = g(x)h(x)$ with $g(x), h(x) \in R[x]$. The fact that $f(x)$ was primitive was only used in the final part of part (a) so by the same arguments as in part (a) we may assume without loss of generality that $h \in R \subset K$. Since $f(x) \neq 0, h \neq 0$ so h is a unit in K. Therefore $f(x)$ is irreducible in $K[x]$. $\qquad\square$

We give some examples.

EXAMPLE 12.4.1. Let $R = \mathbb{Z}$ and p a prime number. Suppose that n, m are integers such that $n \geq 1$ and p does not divide m. Then $x^n \pm pm$ is irreducible in $\mathbb{Z}[x]$ and $\mathbb{Q}[x]$. To see this just consider Eisenstein's criterion applied to the polynomials $x^n \pm pm$. This polynomial is clearly primitive. Since $\mathbb{Z}$ is a UFD $x^n \pm pm$ is not only irreducible over $\mathbb{Z}$ but also irreducible over $\mathbb{Q}$. This certainly implies that $(pm)^{\frac{1}{n}}$ is irrational for if it were rational $x - (pm)^{\frac{1}{n}}$ would be a nontrivial factor over $\mathbb{Q}[x]$ by Theorem 12.3.1.

EXAMPLE 12.4.2. Let $R = \mathbb{Z}$ and p a prime number. Consider the polynomial

$$\Phi_p(x) = \frac{x^p - 1}{x - 1} = x^{p-1} + x^{p-2} + \cdots + 1.$$

This is called the **cyclotomic polynomial** over $\mathbb{Q}$.

We show that $\Phi_p(x)$ is irreducible in $\mathbb{Q}[x]$. We note first that to show $\Phi_p(x)$ is irredcuible in $\mathbb{Q}[x]$ we must only show that it is irreducible in $\mathbb{Z}[x]$ by Theorem 12.4.3 (b). Let

$$f(x) = \Phi_p(x + 1) = \frac{(x + 1)^p - 1}{(x + 1) - 1}.$$

Then using the binomial theorem to expand $(x+1)^p$ (note that the binomial theorem holds in any commutative ring with identity so it holds in $\mathbb{Z}[x]$) we get

$$\Phi_p(x + 1) = \frac{x^p + \binom{p}{1}x^{p-1} + \cdots + \binom{p}{p-1}x}{x} = x^{p-1} + px^{p-2} + \cdots + p,$$

Now it is not hard to prove that $p | \binom{p}{i}$ for $1 \le i \le p - 1$ (see the exercises) and since $\binom{p}{1} = \binom{p}{p-1} = p$ then the constant term is not divisible by p^2. Thus Eisenstein's criterion does apply to the polynomial $f(x)$ and implies that $f(x)$ is irreducible over $\mathbb{Q}$. Then if it was the case that

$$\Phi_p(x) = h(x)g(x)$$

is a nontrival factorization of $\Phi_p(x)$ in $\mathbb{Z}[x]$ then

$$\Phi_p(x + 1) = f(x) = h(x + 1)g(x + 1)$$

would give a nontrivial factorization of $f(x)$ in $\mathbb{Z}[x]$ contradicting that $f(x)$ is irreducible in $\mathbb{Z}[x]$. Therefore $\Phi_p(x)$ must be irreducible over $\mathbb{Q}$.

Theorem 12.4.8. *Let R be a UFD and K its field of fractions. Let $f(x) = \sum_{i=0}^n a_i x^i \in R[x]$ be a polynomial of degree ≥ 1. Let P be a prime ideal in R with $a_n \notin P$. Let $\overline{R} = R/P$ and let $\alpha : R[x] \to \overline{R}[x]$ be defined by*

$$\alpha\left(\sum_{i=0}^m r_i x^i\right) = \sum_{i=0}^m (r_i + P)x^i.$$

α is an epimorphism. Then if $\alpha(f(x))$ *is irreducible in* $\overline{R}[x]$ *then* $f(x)$ *is irreducible in* $K[x]$.

Proof. The proof that α is an epimorphism is straightforward and left to the exercises. By Theorem 12.4.2 (c) there is an $a \in R$ and a primitive $g(x) \in R[x]$ satisfying $f(x) = ag(x)$. Since $a_n \notin P$ and P is an ideal comparing coefficients of leading terms of $\alpha(f(x))$ and $\alpha(ag(x))$ we have that $\alpha(a) \neq 0$ and further the highest coefficient of $g(x)$ is also not an element of P. Now suppose that $\alpha(f(x))$ is irreducible in $R[x]$. If $\alpha(g(x))$ is reducible then $\alpha(f(x))$ is also reducible. Thus $\alpha(g(x))$ is irreducible. Similarly, if $g(x)$ were reducible in $R[x]$, then $\alpha(g(x))$ would be reducible in $\overline{R}[x]$. Thus $g(x)$ is irreducible in $R[x]$. However, from Theorem 12.4.3(b) $g(x)$ is irreducible in $K[x]$ so $f(x) = ag(x)$ is also irreducible in $K[x]$. $\qquad\square$

It is important to note that $\alpha(f(x))$ being reducible does not imply that $f(x)$ is reducible. For example $f(x) = x^2 + 1$ is irreducible in $\mathbb{Z}[x]$. However, choosing $P = 2\mathbb{Z}$ in the above Theorem 12.4.8 and noting that $\mathbb{Z}/2\mathbb{Z} \cong \mathbb{Z}_2$ in $\mathbb{Z}_2[x]$ we have

$$x^2 + 1 = (x + 1)^2$$

and hence $f(x)$ is reducible in $\mathbb{Z}_2[x]$.

EXAMPLE 12.4.3. Let $f(x) = x^5 - x^2 + 1 \in \mathbb{Z}[x]$. Choose $P = 2\mathbb{Z}$ so that

$$\alpha(f(x)) = x^5 + x^2 + 1 \in \mathbb{Z}_2[x].$$

Suppose that in $\mathbb{Z}_2[x]$ we have $\alpha(f(x)) = g(x)h(x)$. Without loss of generality we may assume that $g(x)$ is of degree 1 or 2.

If $\deg(g(x)) = 1$ then $\alpha(f(x))$ has a zero c in $\mathbb{Z}_2[x]$. The two possibilities for c are $c = 0$ or $c = 1$. Then

$$\text{If } c = 0 \text{ then } 0 + 0 + 1 = 1 \neq 0$$

$$\text{If } c = 1 \text{ then } 1 + 1 + 1 = 1 \neq 0.$$

Hence the degree of $g(x)$ cannot be 1.

Suppose $\deg(g(x)) = 2$. The polynomials of degree 2 over $\mathbb{Z}_2[x]$ have the form

$$x^2 + x + 1, x^2 + x, x^2 + 1, x^2.$$

The last three, $x^2 + x, x^2 + 1, x^2$ all have zeros in $\mathbb{Z}_2[x]$ so they can't divide $\alpha(f(x))$. Therefore $g(x)$ must be $x^2 + x + 1$. Applying the division algorithm we obtain

$$\alpha(f(x)) = (x^3 + x^2)(x^2 + x + 1) + 1$$

and therefore $x^2 + x + 1$ does not divide $\alpha(f(x))$. It follows that $\alpha(f(x))$ is irreducible and from the previous theorem $f(x)$ must be irreducible in $\mathbb{Q}[x]$.

12.5   Exercises

EXERCISES FOR SECTION 12.1

12.1.1. For R a commutative ring with identity and $x = x^1 = (0, 1, 0, \ldots)$, $x^0 = (1, 0, 0, \ldots)$

(a) First show using the definition of multiplication of these objects that $x^2 = x \cdot x = (0, 0, 1, 0, \ldots)$.

(b) Use induction to then show that $x^n = (0, 0 \ldots, 0, 1, 0, \ldots)$ where the first 1 appears in the $(n+1)$st position. That is, $x^m(m) = 1$ but $x^n(i) = 0$ if $i \neq n$.

12.1.2. For R a commutative ring with identity and $0 \in R[x]$ is the zero polynomial $(0 = (0, 0, 0, \ldots))$ then for all $f \in R[x]$ prove $0 + f = f + 0 = f$ and $0 \cdot f = f \cdot 0 = 0$.

12.1.3. For R a commutative ring with identity show that the multiplication of polynomials is associative.

(HINT: Take any three polynomials f, g, h and calculate the term for x^i in $f(gh)$ and in $(fg)h$. Recall that multiplication in R is associative.)

12.1.4. Prove Lemma 12.1.3.

(HINT: note that it follows from Lemma 12.1.1 (a).)

EXERCISES FOR SECTION 12.2

12.2.1. If F is a field show that a prime or irreducible polynomial $f(x) \in F[x]$ is the same thing as an irreducible element in the integral domain $F[x]$. (See Definition 11.4.2 for the definition of an arbitrary irreducible element in an integral domain.)

12.2.2. If F is a field then the proof of the division algorithm in $F[x]$ can be given along the same lines as the proof of Theorem 4.2.2.

Consider the set $S = \{f(x) - g(x)h(x) : h(x) \in F[x]\}$. This set S is nonempty (why?). Then there exists an $r(x) \in S$ with minimal degree (why must this exist?). Then there exists a $q(x) \in F[x]$ such that

$$f(x) = g(x)q(x) + r(x).$$

We must show that $\deg(r(x)) < \deg(g(x))$ if $r(x) \neq 0$. Let $g(x) = \sum_{i=0}^{m} b_i x^i$ with $b_m \neq 0$ and suppose that $r(x) = \sum_{i=1}^{n} c_i x^i$ with $c_n \neq 0$. If $n \geq m$ then consider $f(x) - q(x)g(x) - (\frac{c_n}{b_m})x^{n-m}g(x)$. This is also in S (why?) and the degree is strictly less than n. Explain why this implies $\deg(r(x)) < \deg(g(x))$. We assumed here that $r(x) \neq 0$ but the same argument works if $r(x) = 0$.

For uniqueness suppose that $f(x) = g(x)q_1(x) + r_1(x)$ where $\deg(r_1(x)) < \deg(g(x))$ or $r_1(x) = 0$. Then subtracting we obtain

$$g(x)(q(x) - q_1(x)) = r_1(x) - r(x).$$

But $\deg(r_1(x) - r(x)) < \deg(g(x))$. Explain why this implies that $q(x) = q_1(x)$. This implies that $r(x) = r_1(x)$.

12.2.3. Show that if $f(x), g(x)$ are nonzero elements of $F[x]$ where F is a field then $f(x)|g(x)$ and $g(x)|f(x)$ implies that $\deg(f(x)) = \deg(g(x))$.

12.2.4. Suppose that if $f(x), g(x)$ are nonzero elements of $F[x]$ where F is a field and that they are monic, that is, their leading coefficents are 1. Prove that $f(x)|g(x)$ and $g(x)|f(x)$ imply $f(x) = g(x)$.

(HINT: From the previous problem they have the same degree. Then show it for polynomials of degree 0 and then for polynomials of degree $n > 0$.)

12.2.5. Prove Lemma 12.2.3.

12.2.6. Prove Lemma 12.2.4.

12.2.7. Prove Lemma 12.2.5.

12.2.8. Adapt the Euclidean algorithm for the integers $\mathbb{Z}$ (see Chapter 4) to the case of $F[x]$ where F is a field, that is, to computing the GCD of two polynomials. Note here that you must use the division algorithm for polynomials and consider what happens to the degree of the remainder. Note that if the last nonzero remainder is not monic the GCD will be that polynomial divided by the inverse of the leading coefficient, that is, if $r_n(x)$ is the last nonzero remainder and if $r_n(x)$ has leading term $a \neq 0$, then replace $r_n(x)$ by $a^{-1}r_n(x)$ and show that this is the GCD.

12.2.9. Use the method developed in Problem 12.2.8 to find the GCD of each pair of polynomials over the rationals $\mathbb{Q}$ and express it as a linear combination of the two polynomials.

(a) $2x^3 - 4x^2 + x - 1$ and $x^3 - x^2 - x - 2$

(b) $x^4 + x^3 + x^2 + x + 1$ and $x^3 - 1$

12.2.10. Let $f(x) \in F[x]$ with F a field. Let $c \in F$. Prove that $f(c) = 0$ if and only if $(x - c)|f(x)$. This is called the **factor theorem**.

(HINT: Use the division algorithm in $F[x]$ together with Lemma 12.1.4.)

12.2.11. Consider $p(x) = x^4 + 3x^3 + 2x + 4 \in \mathbb{Z}_5[x]$.

(a) Show that this polynomial is reducible by finding a nonunit factor.

(HINT: Note that 1 is a zero of this polynomial over $\mathbb{Z}_5$ and use the factor theorem from Problem 12.2.10.)

(b) Factor this polynomial into irreducible factors over $\mathbb{Z}_5$ which Theorem 12.2.3 says exists. The final factorization here has all linear factors.

(HINT: Keep using the factor theorem as in part (a).)

12.2.12. Repeat Problem 12.2.11 for $x^3 + 2x^2 + 2x + 1 \in \mathbb{Z}_7[x]$.

12.2.13. Is $2x^3 + x^2 + 2x + 2$ an irreducible polynomial in $\mathbb{Z}_5[x]$? Why or why not?

12.2.14. Prove Lemma 12.2.7.

12.2.15. Modify the proof of the division algorithm in Theorem 4.2.2 to obtain the following:

Given integers a, b with $b \neq 0$ then there exist unique integers q, r such that $a = qb + r$ where either $r = 0$ or $0 < r < |b|$.

(HINT: Replace the set S in the proof of Theorem 4.2.2 by

$$S = \{m|b| : m \in \mathbb{Z} \text{ and } m|b| \leq a\}.$$

Note: $m|b| \leq a$ is equivalent to $a - m|b| \geq 0$. First show that $S \neq \emptyset$ — consider $|a||b|$. Then S must contain a largest element (why?). Say this largest element is $t|b|$. By definition of S this means $r = a - t|b| \geq 0$ (why?). Now we have

$$a = t|b| + r$$

where $r \geq 0$. But then $(t + 1)|b| = t|b| + |b| > t|b|$ because $b \neq 0$. Thus by maximality of $t|b|$, we must have $(t+1)|b| > a$ (why?). Hence $t|b|+|b| > t|b|+r$ (why?). Finally this gives $r < |b|$. Finally let

$$q = t \text{ if } b > 0$$

but
$$q = -t \text{ if } b < 0$$
so that $qb = t|b|$ if $b > 0$ and $qb = -tb = t(-b) = t|b|$ if $b < 0$. This gives the existence part (show it).

To get uniqueness suppose that $a = bq + r = bq' + r'$ where $0 \le r < |b|$ and $0 \le r' < |b|$. Then $r - r' = b(q' - q)$ but $-|b| < r - r' < |b|$. Arguing just as in the proof of Theorem 4.2.2 the only multiple of b which lies strictly between $-|b|$ and $|b|$ is 0. Thus $r - r' = 0$. and $r = r'$. This implies that $q = q'$ (why?) and we have uniqueness also.)

12.2.16. Show that the **Gaussian integers** $\mathbb{Z}[i] = \{a + bi : a, b \in \mathbb{C}\}$ forms a commutative ring with identity. Note since this set is part of the field $\mathbb{C}$ you need only show that it's a subring.

12.2.17. Prove that if F is a field and $S \subset F[x]$ is an ideal such that S contains a unit, that is, a nonzero element of F, then $S = F[x]$.

12.2.18. Let R be an integral domain and p a prime element of R. Prove that the map
$$\sigma : R[x] \to (R/pR)[x]$$
given by
$$\sigma\left(\sum_{i=0}^{n} r_i x^i\right) = \sum_{i=0}^{n} (r_i + pR)x^i$$
is an epimorphism with kernel $(pR)[x]$.

12.2.19. Show that $1 - i \in \mathbb{Z}[i]$ is irreducible in $\mathbb{Z}[i]$.
(HINT: Use Lemma 12.2.7 (2).)

12.2.20. Show that 3 is irreducible in $\mathbb{Z}[i]$.
(HINT: Suppose that $3 = xy$ with $x, y \in \mathbb{Z}[i]$ and neither x nor y a unit. Using Lemma 12.2.7 (5), show that this leads to a contradiction.)

12.2.21. Show that 2 and 5 are reducible in $\mathbb{Z}[i]$.
(HINT: Consider $(1 + i)(1 - i)$ and $(1 + 2i)(1 - 2i)$ respectively.)

EXERCISES FOR SECTIONS 12.3 AND 12.4

12.3.1. Theorem 12.3.2 asserts that a polynomial of degree n in $F[x]$ where F is a field has at most n roots. Show that it is not necessarily true if F is an arbitrary ring.

(HINT: Consider $p(x) = x^2 + 3x + 2$ over $\mathbb{Z}_6[x]$. Show that this has more than two roots in $\mathbb{Z}_6$.)

12.3.2. Let R and S be commutative rings with identity and let $\phi : R \to S$ be a ring homomorphism. Define $\Phi : R[x] \to S[x]$ by $\Phi(\sum_{i \geq 0} a_i x^i) = \sum_{i \geq 0} \phi(a_i) x^i$. Prove that Φ is a ring homomorphism.

12.3.3. Suppose that the rings R and S are isomorphic. Prove that the polynomial rings $R[x]$ and $S[x]$ are also isomorphic.

12.3.4. Complete the proof of Theorem 12.3.3 in the case where the leading coefficient $a_n < 0$.

12.3.5. Prove that if $P(x), Q(x) \in \mathbb{C}[x]$ then $\overline{PQ}(x) = \overline{P}(x)\overline{Q}(x)$.

12.3.6. Prove Lemma 12.3.5 (look at Lemma 12.3.3).

12.3.7. In the proof of Theorem 12.3.7 define a map between the sets

$$\mathcal{P}_n = \{f(x) \in \mathbb{Q}[x] : \deg(f(x)) \leq n\}$$

and $\mathbb{Q}^{n+1}$ and show it is a bijection. This shows $|\mathcal{P}_n| = |\mathbb{Q}^{n+1}|$ as defined in Chapter 2.

12.3.8. Let D be an integral domain and suppose that $a|b$ for $a, b \in D$. Prove that $c|b$ for any associate c of a.

12.3.9. Complete the proof of Lemma 12.4.2 by showing that if R is any UFD and $A = \{a_1, \ldots, a_n\}$ with $n \geq 2$ then $\gcd(A)$ and $\text{lcm}(A)$ both exists.

12.3.10. In the proof of Theorem 12.4.2 if $a \in K$ the field of fractions of the UFD R then $a = \frac{r}{s}$ where $r, s \in R$. Show that if s is a unit in R then $a \in R$.

12.3.11. Prove that if F is a field and $f(x) \in F[x]$ is a prime in $F[x]$ and $g(x) = af(x)$ where a is a unit in $F[x]$ then $g(x)$ is also a prime in $F[x]$.

12.3.12. Let $I = \{f(x) \in \mathbb{Z}[x] : f(x)$ has even constant term$\}$. Prove that I is an ideal but not a principal ideal. Since $\mathbb{Z}[x]$ is a UFD by Gauss's theorem this shows that a UFD is not necessarily a PID.

12.3.13. Let F be a field and let D be the subring of $F[x, y]$ generated by F, x^3, xy, y^3. Then:

    (a) Show that x^3, xy and y^3 are irreducibles in D.

    (b) Show that xy, x^3, y^3 and y^3 are not prime in D.

(HINT: $(x^3)(y^3) = (xy)(xy)(xy)$ so $xy|x^3 \cdot y^3$. Similar arguments show that neither x^3 nor y^3 is prime in D.)

12.3.14. An important and familiar theorem in elementary algebra is the binomial theorem. Prove that if R is any commutative ring with identity 1 and $a, b \in R$ then

$$(a+b)^n = \sum_{i=0}^{n} \binom{n}{i} a^i b^{n-i}$$

where

$$\binom{n}{i} = \frac{n!}{i!(n-i)!} = \frac{n(n-1)\cdots(n-i+1)}{i!}.$$

(HINT: Use induction on n together with the identity $\binom{n}{i} + \binom{n}{i-1} = \binom{n+1}{i}$ that can be verified by direct computation.)

12.3.15. Prove that if $p \in \mathbb{Z}$ is a prime then $p \mid \binom{p}{i}$ for $1 \le i \le p-1$.
Note: $\binom{p}{i}$ is an integer for $1 \le i \le p$. Why?

12.3.16. Prove that the map $\alpha : R[x] \to \overline{R}[x]$ given in Theorem 12.4.8 is a ring epimorphism.

12.3.18. Prove the following: Let $p(x) = q_0 + q_1 x + \cdots + q_n x^n \in \mathbb{Q}[x]$. Let d be the LCM of the denominators of $q_0, \ldots, q_n$. Then $p_1(x) = dp(x) \in \mathbb{Z}[x]$. Suppose that $\frac{r}{s} \in \mathbb{Q}$ with $p(\frac{r}{s}) = 0$ if and only if $p_1(\frac{r}{s}) = 0$.

12.3.19. Prove the rational roots theorem: Let $f(x) = a_0 + a_1 x + \cdots + a_n x^n \in \mathbb{Z}[x]$ with $\deg(f(x)) = n$. Prove that if $f(\frac{r}{s}) = 0$ with $r, s \in \mathbb{Z}$ and $(r, s) = 1$ then $r \mid a_0$ and $s \mid a_n$.

Chapter 13

Algebraic Linear Algebra

13.1 Linear Algebra

In this chapter we look at a special branch of abstract algebra called **linear algebra**. This area of algebra covers a vast array of material and is usually studied in a completely separate course. We will need several basic ideas from linear algebra in our look at Galois theory and its applications. In this chapter we will introduce those ideas that will be necessary for us.

Linear algebra is that branch of abstract algebra which studies **vector spaces** and concepts and applications related to vector spaces. These include the study of **systems of linear equations** and **linear geometry**. Vector spaces are generalizations of vector analysis in real 3-space $\mathbb{R}^3$. We will review this in the next section in which we will also look at some linear geometry in $\mathbb{R}^3$. Vector spaces will be formally defined in an abstract manner in Section 13.2.

In the study of vector spaces, linear algebra is much like group theory or ring theory. However, linear algebra is distinguished from the other parts of abstract algebra both because it developed separately and because it handles a wide array of other problems from disparate areas of mathematics. It wasn't until algebra was really abstracted that it was realized that linear and abstract algebra were really part of the same subject. Historically linear algebra grew out of two subjects, the study of linear systems of equations which arise both in algebra and differential equations and linear geometry, specifically the metric and affine geometry of Euclidean n-space.

Linear algebra plays a major role in many areas of mathematics. The set of continuous functions and the set of differentiable functions form vector spaces. Therefore linear algebra plays a large role in multivariable calculus and functional analysis. The solution sets of linear differential equations, both ordinary and partial, also form vector spaces and hence linear algebra forms one of the basic cornerstones of the theory of ordinary and partial differential equations.

Our basic assumption is that the readers have some familiarity with linear algebra especially matrix algebra and solutions of equations. We consider this chapter as an overview so most results are not proven. We refer to [F] or [HK] for any details that we have omitted. We will need several concepts from theoretical linear algebra in the final chapter on Galois theory. If students are sufficiently familiar with linear algebra this chapter can be omitted.

13.1.1   Vector Analysis in $\mathbb{R}^3$

Vector spaces are generalizations of vector analysis in real 3-space. In this subsection we review some of the basic ideas concerning vector analysis. Many of these ideas will be abstracted to general vector spaces. We assume that the reader has encountered two- and three- dimensional vectors within the study of multivariable calculus. First we recall the concept of a three-dimensional vector. This is the mathematical interpretation of **vector quantities** in physics. These are quantities that have both magnitude and direction.

Definition 13.1.1. *Suppose A, B are points in $\mathbb{R}^3$. Then a* **vector** $\vec{v} = [\vec{AB}]$ *is an equivalence class of directed line segments with $\vec{AB} \equiv \vec{CD}$ if $ABDC$ forms a parallelogram (we allow this parallelogram to collapse into a directed line segment). The* **magnitude** *or* **norm** *of $\vec{v}$ is the length of $\vec{AB}$. We denote this $|\vec{v}|$. A* **scalar** *is any real number. The vector with zero magnitude (that is, a point) is called the* **zero vector** *denoted $\vec{0}$. For a vector $\vec{v}$, the negative $-\vec{v}$ is the vector with the same magnitude as $\vec{v}$ but oppositely directed, that is, if $\vec{v} = [\vec{AB}]$ then $-\vec{v} = [\vec{BA}]$.*

Motivated by physics there are certain vector operations.

Definition 13.1.2. (Vector Addition) *If $\vec{u}, \vec{v}$ are vectors then* **vector addition** *is defined in the following way: Place the vector $\vec{v}$ beginning at the tip of $\vec{u}$. The vector sum $\vec{u} + \vec{v}$ is the vector beginning at the initial point of $\vec{u}$ and ending at the tip of $\vec{v}$. By the difference of vectors $\vec{u} - \vec{v}$ we mean $\vec{u} + (-\vec{v})$. This is pictured in Figure 13.1.*

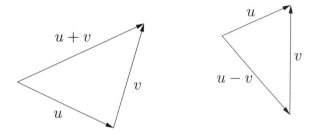

Figure 13.1 Vector Addition and Vector Subtraction

Lemma 13.1.1. *Vector addition satisfies the following properties:*
    (1) $\vec{u} + \vec{v} = \vec{v} + \vec{u}$ (commutativity).
    (2) $(\vec{u} + \vec{v}) + \vec{w} = \vec{u} + (\vec{v} + \vec{w})$ (associativity).
    (3) $\vec{v} + \vec{0} = \vec{v}$ (The zero vector is an additive identity).
    (4) $\vec{v} - \vec{v} = \vec{0}$ (Each vector has an additive inverse).
It follows that the vectors in $\mathbb{R}^3$ form an abelian group under addition.

Definition 13.1.3. *(**Scalar Multiplication**) If $\vec{v}$ is a vector and $r \in \mathbb{R}$ is a scalar then the **scalar multiple** $r\vec{v}$ is the vector with magnitude $|r||\vec{v}|$ and direction the same as $\vec{v}$ if $r > 0$ and oppositely directed if $r < 0$. If $r = 0$ then $r\vec{u} = \vec{0}$. Scalar multiplication is geometrically a stretching or contracting of the vector $\vec{v}$.*

Lemma 13.1.2. *Scalar multiplication satisfies the following properties:*
    (1) $r(\vec{u} + \vec{v}) = r\vec{u} + r\vec{v}$ for $\vec{u}, \vec{v} \in \mathbb{R}^3, r \in \mathbb{R}$.
    (2) $(r + s)\vec{u} = r\vec{u} + s\vec{u}$ for $\vec{u} \in \mathbb{R}^3, r, s \in \mathbb{R}$.
    (3) $(rs)\vec{u} = r(s\vec{u})$ for $\vec{u} \in \mathbb{R}^3, r, s \in \mathbb{R}$.
    (4) $1\vec{u} = \vec{u}$ for $\vec{u} \in \mathbb{R}^3$.

    These properties will be abstracted to define a vector space (see Section 13.2).

    A **unit vector** $\vec{u}$ is a vector with magnitude 1. If $\vec{v}$ is any nonzero vector then the vector $\frac{\vec{v}}{|\vec{v}|}$ is a unit vector in the same direction as $\vec{v}$.

In $\mathbb{R}^3$ there are certain special vectors from which all the other vectors can be built.

Definition 13.1.4. *(The Standard Basis) Let* $\mathbf{i}, \mathbf{j}, \mathbf{k}$ *denote unit vectors along the positive x,y,z axes respectively. These are called the standard unit vectors or* **standard basis** *for* $\mathbb{R}^3$.

Lemma 13.1.3. *Given a three-dimensional coordinate system with* x, y, z-*axes we have that the vector with initial point* $A = (x_1, y_1, z_1)$ *and terminal point* $B = (x_2, y_2, z_2)$ *can be written as*

$$\vec{AB} = (x_2 - x_1)\mathbf{i} + (y_2 - y_1)\mathbf{j} + (z_2 - z_1)\mathbf{k}.$$

Theorem 13.1.1. *Any* $\vec{u} \in \mathbb{R}^3$ *can be written uniquely as*

$$\vec{u} = u_1\mathbf{i} + u_2\mathbf{j} + u_3\mathbf{k}.$$

u_1, u_2, u_3 *are called the* **components** *of* $\vec{u}$. *We will also write* $\vec{u} = \langle u_1, u_2, u_3 \rangle$.

In language that we will introduce later, we say that $\mathbf{i}, \mathbf{j}, \mathbf{k}$ form a basis for $\mathbb{R}^3$. A key idea is that the vector operations can be done **componentwise**.

Theorem 13.1.2. *Vector addition and scalar multiplication are done componentwise. That is, if* $\vec{u} = u_1\mathbf{i} + u_2\mathbf{j} + u_3\mathbf{k}, \vec{v} = v_1\mathbf{i} + v_2\mathbf{j} + v_3\mathbf{k}$ *then* $\vec{u} + \vec{v} = (u_1 + v_1)\mathbf{i} + (u_2 + v_2)\mathbf{j} + (u_3 + v_3)\mathbf{k}$ *and* $r\vec{u} = ru_1\mathbf{i} + ru_2\mathbf{j} + ru_3\mathbf{k}$. *Further,* $|\vec{u}| = \sqrt{u_1^2 + u_2^2 + u_3^2}$.

EXAMPLE 13.1.1. If $\vec{u} = 3\mathbf{i} + 4\mathbf{j} + 5\mathbf{k}, \vec{v} = \mathbf{i} - \mathbf{j} + 2\mathbf{k}$ then

$$\vec{u} + \vec{v} = 4\mathbf{i} + 3\mathbf{j} + 7\mathbf{k},$$

$$8\vec{u} = 24\mathbf{i} + 32\mathbf{j} + 40\mathbf{k},$$

$$|\vec{u}| = \sqrt{3^2 + 4^2 + 5^2} = \sqrt{50}.$$

We now describe two different types of vector multiplication. The first, the **scalar product**, takes two vectors and produces a scalar. If $\vec{u}$ and $\vec{v}$ are vectors we can place them at the same initial point. The angle formed is called the **angle between the vectors**. There are actually two angles formed but we take the angle between the vectors as the smaller of the two. Thus if θ is the angle between $\vec{u}$ and $\vec{v}$ then $0 \leq \theta \leq \pi$.

Definition 13.1.5. *(The* **dot product** *or* **scalar product***) If $\vec{u}, \vec{v}$ are vectors then their* **scalar product** *or* **dot product** *denoted $\vec{u} \cdot \vec{v}$ or $\langle \vec{u}, \vec{v} \rangle$ is the scalar $|\vec{u}||\vec{v}| \cos\theta$ where θ is the angle between $\vec{u}$ and $\vec{v}$. That is,*

$$\vec{u} \cdot \vec{v} = |\vec{u}||\vec{v}| \cos\theta.$$

As with the other vector operations, the scalar product is motivated in part by a very important concept from physics, namely, **work**. If a particle is moved by a force vector **F** through a displacement given by the vector **d** then the work done is the scalar given by $W = \mathbf{F} \cdot \mathbf{d}$.

Lemma 13.1.4. *The dot product of vectors satisfies the following properties:*
    (1) $\vec{u} \cdot \vec{u} \geq 0$ and $\vec{u} \cdot \vec{u} = \vec{0}$ iff $\vec{u} = \vec{0}$.
    (2) $\vec{u} \cdot \vec{v} = \vec{v} \cdot \vec{u}$.
    (3) $\vec{u} \cdot (\vec{v} + \vec{w}) = \vec{u} \cdot \vec{v} + \vec{u} \cdot \vec{w}$.
    (4) $(r\vec{u}) \cdot \vec{v}) = r(\vec{u} \cdot \vec{v})$.
    (5) $\vec{u}$ is orthogonal (perpendicular) to $\vec{v}$ if and only if $\vec{u} \cdot \vec{v} = 0$.

As with the other vector operations, scalar product can be done componentwise.

Lemma 13.1.5. *(***The Dot Product Componentwise***) If $\vec{u} = \langle u_1, u_2, u_3 \rangle$, $\vec{v} = \langle v_1, v_2, v_3 \rangle$ then*
$$\vec{u} \cdot \vec{v} = u_1 v_1 + u_2 v_2 + u_3 v_3.$$

EXAMPLE 13.1.2. If $\vec{u} = 3\mathbf{i} + 4\mathbf{j} + 5\mathbf{k}, \vec{v} = \mathbf{i} - \mathbf{j} + 2\mathbf{k}$ then

$$\vec{u} \cdot \vec{v} = (3)(1) + (4)(-1) + 5(2) = 9.$$

This can be used to find the angle between two nonzero vectors. We have from the definition
$$\cos\theta = \frac{\vec{u} \cdot \vec{v}}{|u||v|}.$$

Therefore we can find the angle between $\vec{u}$ and $\vec{v}$.

EXAMPLE 13.1.3. If $\vec{u} = 3\mathbf{i} + 4\mathbf{j} + 5\mathbf{k}, \vec{v} = \mathbf{i} - \mathbf{j} + 2\mathbf{k}$ then

$$|\vec{u}| = \sqrt{50}, |\vec{v}| = \sqrt{6}$$

and so

$$\theta = \arccos \frac{9}{\sqrt{50}\sqrt{6}} \approx 59^0.$$

The second type of multiplication is called the **vector product** or **cross product**. This operation starts with two vectors and produces a third vector. First a bit of notation. In 3-space there are two possible orientations. If $\vec{u}$ and $\vec{v}$ are two vectors with the same initial point then a vector $\vec{w}$ perpendicular to the plane formed by $\vec{u}$ and $\vec{v}$ is said to be given by the **right-hand rule** if the four fingers of the right hand extend from $\vec{u}$ to $\vec{v}$ and the thumb extends in the direction of $\vec{w}$.

Definition 13.1.6. *(The Vector Product) If $\vec{u}, \vec{v}$ are vectors then their* **vector product** *or* **cross product** *is the vector $\vec{u} \times \vec{v} = (|\vec{u}||\vec{v}| \sin \theta)\vec{n}$ where $\vec{n}$ is a unit vector perpendicular to the plane formed by $\vec{u}, \vec{v}$ directed so that the orientation of $\vec{n}$ is determined by $\vec{u}$ and $\vec{v}$ according to the right-hand rule. Thus the vector product is a vector whose magnitude is $|\vec{u}||\vec{v}| \sin \theta$ and whose direction is perpendicular to the plane defined by $\vec{u}, \vec{v}$ and directed toward $\vec{n}$ by the right-hand rule.*

This also comes from a concept in physics. If a force vector $\mathbf{F}$ acts on a point P by a displacement vector $\mathbf{d}$ then the **torque** or angular force on P is given by the vector $\tau = \mathbf{F} \times \mathbf{d}$.

Lemma 13.1.6. *The vector product satisfies the following properties:*
> *(1) $\vec{u} \times \vec{v} = -\vec{v} \times \vec{u}$.*
> *(2) $\vec{u} \times (\vec{v} + \vec{w}) = \vec{u} \times \vec{v} + \vec{u} \times \vec{w}$.*
> *(3) $(r\vec{u}) \times \vec{v} = r(\vec{u} \times \vec{v})$.*
> *(4) $\vec{u} \times \vec{v} = \vec{0}$ if $\vec{u}, \vec{v}$ are parallel.*

We assume that the reader knows how to compute the determinant of a 3×3 matrix. We discuss determinants in Section 13.1.4. Using the properties in the above lemma we find that the cross product can be determined by a determinant.

Theorem 13.1.3. *If $\vec{u} = u_1\mathbf{i} + u_2\mathbf{j} + u_3\mathbf{k}$, $\vec{v} = v_1\mathbf{i} + v_2\mathbf{j} + v_3\mathbf{k}$ then*

$$\vec{u} \times \vec{v} = det\left(\begin{pmatrix} \mathbf{i} & \mathbf{j} & \mathbf{k} \\ u_1 & u_2 & u_3 \\ v_1 & v_2 & v_3 \end{pmatrix} \right)$$

EXAMPLE 13.1.4. If $\vec{u} = 3\mathbf{i} + 4\mathbf{j} + 5\mathbf{k}, \vec{v} = \mathbf{i} - \mathbf{j} + 2\mathbf{k}$ then

$$\vec{u} \times \vec{v} = det\begin{pmatrix} \mathbf{i} & \mathbf{j} & \mathbf{k} \\ 3 & 4 & 5 \\ 1 & -1 & 2 \end{pmatrix} = 13\mathbf{i} - \mathbf{j} - 7\mathbf{k}.$$

13.1.2 Matrices and Matrix Algebra

Closely tied to vectors are **matrices**. We will consider matrices over an arbitrary commutative ring R or field F. The elements of R or F are called **scalars**.

Definition 13.1.7. *Let R be a commutative ring. Then a* **matrix** *over R is a rectangular array of scalars.*

$$M = \begin{pmatrix} x_{11} & x_{12} & \dots & x_{1n} \\ x_{21} & x_{22} & \dots & x_{2n} \\ \dots & & & \\ x_{m1} & x_{m2} & \dots & x_{mn} \end{pmatrix}.$$

M has size $m \times n$ if there are m rows and n columns. Each component is located by **two indices** *so it is a* **two-dimensional array**. *By tradition the row index goes first so x_{ij} is the component in the ith row and jth column. An n-vector is a string of n scalars $v = (x_1, \dots, x_n)$. Each x_i is a* **component** *or* **coordinate** *of the vector. It is a* **one-dimensional array** *since it takes one* **index** *to locate any component. If $m = n$ then M is a* **square matrix**.

EXAMPLE 13.1.5.
$$M = \begin{pmatrix} 2 & 4 & 0 \\ 1 & 2 & 1 \end{pmatrix}$$

has size 2×3. The element 4 would be designated M_{12}.

We define a set of operations on matrices which make them act like vectors. These operations will make the set of matrices of a fixed size a vector space. Throughout this section we consider matrices over a field F. We let $M_{mn}(F)$ denote the set of $m \times n$ matrices over the field F.

We first define **matrix addition**. If M, N are two matrices of the same size they they are added componentwise

$$(M + N)_{ij} = M_{ij} + N_{ij}.$$

Relative to matrix addition $M_{mn}(F)$ forms an abelian group. The zero element is the **zero matrix** that has all zero entries and the additive inverse of any matrix is the matrix whose entries are the additive inverses of the origional entries.

A matrix can also be multiplied by a scalar componentwise

$$(cM)_{ij} = cM_{ij}.$$

Relative to scalar multiplication the set of matrices of a fixed size will satisfy exactly the same properties as vectors.

Matrices also allow a special type of multiplication called **matrix multiplication**. This is motivated in part by systems of linear equations and we will show this in the next section. Here we just describe the multiplication, which at first glance may seem strange. If M is $m \times n$ and N is $n \times k$ then M,N are said to be **compatible** and their matrix product MN can be formed. Notice that the number of columns of the first must equal the number of rows of the second so in general M, N might be compatible but N, M are not.

The product has size $m \times k$ and

$$(MN)_{ij} = \sum_{t=1}^{m} M_{it} N_{tj},$$

that is, the (ij)-th element of the product is the dot product of the i-th row of the first with the j-th column of the second.

EXAMPLE 13.1.6. Let

$$M = \begin{pmatrix} 2 & 4 & 0 \\ 1 & 2 & 1 \end{pmatrix}, N = \begin{pmatrix} 4 & 0 & 1 & 1 \\ 1 & 0 & -1 & 0 \\ 0 & 1 & 1 & 1 \end{pmatrix}.$$

Then M is 2×3, N is 3×4 and they are compatible. Then MN has size 2×4 and

$$MN = \begin{pmatrix} 2 & 4 & 0 \\ 1 & 2 & 1 \end{pmatrix} \begin{pmatrix} 4 & 0 & 1 & 1 \\ 1 & 0 & -1 & 0 \\ 0 & 1 & 1 & 1 \end{pmatrix} = \begin{pmatrix} 12 & 0 & -2 & 2 \\ 6 & 1 & 0 & 2 \end{pmatrix}.$$

Notice that the $(1,1)$ position has the value $12 = (2,4,0) \cdot (4,1,0) = (2)(4) + (4)(1) + (0)(0)$.

Notice also that NM cannot be formed.

For square matrices this multiplication is associative

Lemma 13.1.7. *If M, N, P are square matrices of the same size then matrix multiplication is associative $\implies (MN)P = M(NP)$ (not commutative though).*

Notice that if we consider square matrices of a fixed size, say $n \times n$, then matrix multiplication can always be carried out to obtain another $n \times n$ matrix. We denote the set of $n \times n$ matrices over a commutative ring R with identity by $M_n(R)$ and from the comments above $M_n(R)$ is closed under multiplication. Further, it is closed under addition and additive inverses. Analyzing the relationship between matrix addition and matrix multiplication we get the distributive property of multiplication over addition. We then get the following result.

Theorem 13.1.4. *Let R be a ring with an identity. Then the set of $n \times n$ matrices with components from R forms a ring denoted $M_n(R)$. Further, this ring is noncommutative but has an identity given by the $n \times n$ identity matrix.*

To show that $M_n(R)$ is a noncommutative ring we must go through the ring properties. For the final statement we describe the multiplicative identity.

Definition 13.1.8. *The $n \times n$ matrix given by*

$$I_n = \begin{pmatrix} 1 & 0 & \cdots & 0 \\ 0 & 1 & \cdots & 0 \\ \cdots & & & \\ 0 & 0 & \cdots & 1 \end{pmatrix}$$

*is called the $n \times n$ (n by n) **identity matrix**. That is, it has 1 on the main diagonal and 0 elsewhere.*

We use the notation $\delta_{ij} = 1$ if $i = j$ and $\delta_{ij} = 0$ if $i \neq j$. Then $I_n = (\delta_{ij})$.

The identity matrix acts as a multiplicative identity for the ring $M_n(R)$ of $n \times n$ matrices. That is,

$$IM = MI = M$$

for any $n \times n$ matrix M. We don't prove this but give an example.

EXAMPLE 13.1.7. Let

$$A = \begin{pmatrix} 3 & 1 & 1 \\ 0 & 1 & 0 \\ 0 & 3 & 5 \end{pmatrix}, I = \begin{pmatrix} 1 & 0 & 0 \\ 0 & 1 & 0 \\ 0 & 0 & 1 \end{pmatrix}, \text{ then}$$

$$AI = \begin{pmatrix} 3 & 1 & 1 \\ 0 & 1 & 0 \\ 0 & 3 & 5 \end{pmatrix} \begin{pmatrix} 1 & 0 & 0 \\ 0 & 1 & 0 \\ 0 & 0 & 1 \end{pmatrix} = \begin{pmatrix} 3 & 1 & 1 \\ 0 & 1 & 0 \\ 0 & 3 & 5 \end{pmatrix} = A.$$

Since the ring $M_n(R)$ has an identity, we can ask which elements are units, that is, have multiplicative inverses. In linear algebra this is usually approached by considering which matrices are **invertible**.

Definition 13.1.9. *An $n \times n$ matrix A is* **invertible** *if there exists an inverse A^{-1} for multiplication, that is,*

$$AA^{-1} = A^{-1}A = I$$

for some $n \times n$ matrix A^{-1}. In algebraic language an invertible matrix is a unit in the ring of $n \times n$ matrices.

    An invertible matrix is called a **nonsingular matrix**. *If it is noninvertible it is called* **singular**.

Important to determining which matrices are invertible is the concept of the **rank of a matrix**. This is defined in various ways but here we consider a definition in terms of what are called **elementary row operations**.

Definition 13.1.10. *An* **elementary row operation** *on a matrix M is any one of the following*
    (1) Interchange two rows — indicated if the two rows interchanged are row i and row j by R_{ij}.
    (2) Multiply a row by a constant — indicated by cR_i.
    (3) Replace a row by a sum of it and a multiple of another row. The notation $cR_i + R_j$ means multiply row i times c add it to row j and then replace row j by the result.

Using elementary row operations any matrix can be changed into a matrix of one of two very special forms called the row echelon form or the row reduced echelon form.

Definition 13.1.11. *A matrix A is in* **row echelon form** *abbreviated (ref) if it satisfies the following three conditions:*
    (1) The first nonzero entry in any nonzero row is a 1.
    (2) Each leading entry 1 in a row is to the right of the first 1 in a preceding row.

(3) All rows of 0's, if there are any, are at the bottom of the matrix.

A matrix A is in **reduced row echelon form** *abbreviated (rref) if it satisfies the conditions (1),(2),(3) above and in addition the following one:*

(4) A column containing a leading entry 1 has zeros everywhere else.

We note that the row echelon form of a matrix A is not unique but the reduced row echelon form of a matrix is unique. So any sequence of elementary row operations will lead to the same reduced row echelon form.

Lemma 13.1.8. *Any matrix A over a field F can be placed into reduced row echelon form by a sequence of elementary row operations. Further, this reduced row echelon form is unique, i.e, any sequence of elementary row operations will terminate in the same reduced row echelon form.*

It can be proved that the number of nonzero rows in a reduced row echelon form of a matrix is unique.

Definition 13.1.12. *The* **rank** *of a matrix A over a field F is the number of nonzero rows in the reduced row echelon form.*

EXAMPLE 13.1.8. Let

$$A = \begin{pmatrix} 3 & 1 & 4 & 2 \\ 2 & 7 & 3 & 8 \\ 5 & 8 & 7 & 10 \end{pmatrix}, \text{ then}$$

$$A = \begin{pmatrix} 3 & 1 & 4 & 2 \\ 2 & 7 & 3 & 8 \\ 5 & 8 & 7 & 10 \end{pmatrix} \begin{array}{c} -\frac{2}{3}R_1 + R_2 \\ \longrightarrow \\ -\frac{5}{3}R_1 + R_3 \end{array} \begin{pmatrix} 3 & 1 & 4 & 2 \\ 0 & \frac{19}{3} & \frac{1}{3} & \frac{20}{3} \\ 0 & \frac{19}{3} & \frac{1}{3} & \frac{20}{3} \end{pmatrix}$$

$$\begin{array}{c} -R_2 + R_3 \\ \longrightarrow \\ \frac{3}{19}R_2 \end{array} \begin{pmatrix} 3 & 1 & 4 & 2 \\ 0 & 1 & \frac{1}{19} & \frac{20}{19} \\ 0 & 0 & 0 & 0 \end{pmatrix} \begin{array}{c} -R_2 + R_1 \\ \longrightarrow \\ \frac{1}{3}R_1 \end{array} \begin{pmatrix} 1 & 0 & \frac{25}{19} & \frac{6}{19} \\ 0 & 1 & \frac{1}{19} & \frac{20}{19} \\ 0 & 0 & 0 & 0 \end{pmatrix}.$$

This is now in reduced row echelon form and therefore rank $A = 2$.

As we have mentioned, invertibility is closely tied to rank. In fact an $n \times n$ matrix over a field F is invertible if and only if it has rank n.

Theorem 13.1.5. *The following are equivalent for an $n \times n$ matrix A over a field F:*

(1) A is invertible.

(2) A is row equivalent to the identity matrix, that is, A can be row reduced to the identity matrix.

(3) A has rank n.

If A is an $n \times n$ matrix over a field F there is a very straightforward method to both determine if it's invertible and to find the inverse if it is. We outline it briefly and give an example and refer the reader to [Li] for a complete discussion

Step 1: Augment A by the identity matrix, that is, form $(A|I)$

Step 2: Row reduce $(A|I)$ to its reduced row echelon form

Step 3: If A has been turned into I then I has been turned into A^{-1}

Note that if rank $A < n$ then a zero row appears and then A is singular. Hence the method not only produces the inverse but also shows if the matrix is noninvertible.

EXAMPLE 13.1.9. We find the inverse of $A = \begin{pmatrix} 1 & 0 & 2 \\ 2 & -1 & 3 \\ 4 & 1 & 8 \end{pmatrix}$ Now:

$$\left(\begin{array}{ccc|ccc} 1 & 0 & 2 & 1 & 0 & 0 \\ 2 & -1 & 3 & 0 & 1 & 0 \\ 4 & 1 & 8 & 0 & 0 & 1 \end{array}\right) \xrightarrow{\text{rref}} \left(\begin{array}{ccc|ccc} 1 & 0 & 0 & -11 & 2 & 2 \\ 0 & 1 & 0 & -4 & 0 & 1 \\ 0 & 0 & 1 & 6 & -1 & -1 \end{array}\right).$$

Therefore the inverse is

$$A^{-1} = \begin{pmatrix} -11 & 2 & 2 \\ -4 & 0 & 1 \\ 6 & -1 & -1 \end{pmatrix}.$$

The details of the elementary row operations are left to the exercises.

13.1.3 Systems of Linear Equations

Part of the motivation for linear algebra comes from vector analysis and from the solution techniques for systems of linear equations. Here we briefly describe these.

Definition 13.1.13. *A* **linear equation** *in n-variables is an equation*

$$a_1 x_1 + a_2 x_2 + \cdots + a_n x_n = b$$

where $a_1, .., a_n, b$ are scalars in a field F and $x_1, .., x_n$ are variables. An $m \times n$ **system of linear equations** *is a set of m linear equations each in* *n variables:*

$$a_{11}x_1 + \cdots + a_{1n}x_n = b_1$$
$$a_{21}x_1 + \cdots + a_{2n}x_n = b_2$$
$$\cdots$$
$$a_{m1}x_1 + \cdots + a_{mn}x_n = b_m$$

A **solution** *of the above system is an* **n-vector**$(x_1, \ldots, x_n)$ *which satisfies all m equations. If there is a solution the system is* **consistent**, *if not it is* **inconsistent**. *If $m = n$ so that there are the same number of equations as variables it is called a* **square system**

The numbers of solutions to a given system of equations depends on the ground field, that is, the field over which the system is being solved. For an infinite field F we have the following.

Theorem 13.1.6. *An $m \times n$ system over an infinite field F either has one unique solution, no solution, or infinitely many solutions.*

To solve and handle systems of linear equations the system is usually formulated in terms of matrices.

An $m \times n$ system

$$a_{11}x_1 + \cdots + a_{1n}x_n = b_1$$
$$a_{21}x_1 + \cdots + a_{2n}x_n = b_2$$
$$\vdots$$
$$a_{m1}x_1 + \cdots + a_{mn}x_n = b_m$$

is equivalent to a **matrix equation**

$$AX = B$$

where

$$A = \begin{pmatrix} a_{11} & a_{12} & \cdots & a_{1n} \\ a_{21} & a_{22} & \cdots & a_{2n} \\ \vdots & & & \\ a_{m1} & a_{m2} & \cdots & a_{mn} \end{pmatrix}, X = \begin{pmatrix} x_1 \\ x_2 \\ \vdots \\ x_n \end{pmatrix}, B = \begin{pmatrix} b_1 \\ b_2 \\ \vdots \\ b_m \end{pmatrix}.$$

The matrix A is called the **matrix of coefficients** and B is the column of constants. If all $b_i = 0$ then the system is called a **homogeneous system**.

Given this matrix formulation there are several methods of solution. These are usually the meat and potatoes of a computational linear algebra course. This is not our purpose in this book. However we do mention and outline the two most common methods and give examples.

The most straightforward solution technique for a linear system $AX = B$ is called **Gaussian elimination**. This has the following form in outline.

Step 1: Augment A by B, that is, form $(A|B)$.
Step 2: Row reduce $(A|B)$ to row echelon form.
Step 3: Solve the resulting system by **back substitution**.

EXAMPLE 13.1.10. Solve

$$3x + y + z = 8$$
$$x + y = 3$$
$$3y + 5z = 21$$

Then

$$\begin{pmatrix} 3 & 1 & 1 & | & 8 \\ 1 & 1 & 0 & | & 3 \\ 0 & 3 & 5 & | & 21 \end{pmatrix} \xrightarrow{\text{ref}} \begin{pmatrix} 1 & \frac{1}{3} & \frac{1}{3} & | & \frac{8}{3} \\ 0 & 1 & -\frac{1}{2} & | & \frac{1}{2} \\ 0 & 0 & 1 & | & 3 \end{pmatrix}.$$

We now have the system

$$x + \tfrac{1}{3}y + \tfrac{1}{3}z = \tfrac{8}{3}$$
$$y - \tfrac{1}{2}z = \tfrac{1}{2}$$
$$z = 3$$

$$\implies z = 3 \text{ so } y - \frac{3}{2} = \frac{1}{2} \implies y = 2 \text{ and so } x + \frac{2}{3} + 1 = \frac{8}{3} \implies x = 1.$$

Therefore the solution is $x = 1, y = 2, z = 3$.

An extension of basic Gaussian elimination is called **Gauss-Jordan elimination**. Gauss-Jordan elimination is the method where we row reduce the augmented matrix $(A|B)$ to reduced row echelon form. Here back substitution is unnecessary.

EXAMPLE 13.1.11. Solve

$$3x + y + z = 8$$
$$x + y = 3$$
$$3y + 5z = 21$$

Here

$$A = \begin{pmatrix} 3 & 1 & 1 \\ 1 & 1 & 0 \\ 0 & 3 & 5 \end{pmatrix}, B = \begin{pmatrix} 8 \\ 3 \\ 21 \end{pmatrix}$$

From the previous example we have

$$\begin{pmatrix} 3 & 1 & 1 & | & 8 \\ 1 & 1 & 0 & | & 3 \\ 0 & 3 & 5 & | & 21 \end{pmatrix} \xrightarrow{\text{ref}} \begin{pmatrix} 1 & \frac{1}{3} & \frac{1}{3} & | & \frac{8}{3} \\ 0 & 1 & -\frac{1}{2} & | & \frac{1}{2} \\ 0 & 0 & 1 & | & 3 \end{pmatrix}$$

$$\xrightarrow{\text{rref}} \begin{pmatrix} 1 & 0 & 0 & | & 1 \\ 0 & 1 & 0 & | & 2 \\ 0 & 0 & 1 & | & 3 \end{pmatrix}$$

and hence the solution is $x = 1, y = 2, z = 3$.

If the ground field F is infinite and the system is consistent then if $n > m$ (recall that there are m equations and n unknowns) there will be an infinite number of solutions. These solutions will be functions of the remaining variables. If the ground field is finite and the system is consistent in this case there will be more than one solution but only a finite number of solutions.

EXAMPLE 13.1.12. Solve

$$3x_1 + x_2 + 4x_3 + 2x_4 = 1$$
$$2x_1 + 7x_2 + 3x_3 + 8x_4 = 8$$
$$5x_1 + 8x_2 + 7x_3 + 10x_4 = 9$$

Then

$$\implies \begin{pmatrix} 3 & 1 & 4 & 2 & | & 1 \\ 2 & 7 & 3 & 8 & | & 8 \\ 5 & 8 & 7 & 10 & | & 9 \end{pmatrix} \xrightarrow{\text{rref}} \begin{pmatrix} 1 & 0 & \frac{25}{19} & \frac{6}{19} & | & -\frac{1}{19} \\ 0 & 1 & \frac{1}{19} & \frac{20}{19} & | & \frac{22}{19} \\ 0 & 0 & 0 & 0 & | & 0 \end{pmatrix}.$$

We then have

$$x_1 + \frac{25}{19}x_3 + \frac{6}{19}x_4 = -\frac{1}{19}$$
$$x_2 + \frac{1}{19}x_3 + \frac{20}{19}x_4 = \frac{22}{19}$$
$$\implies x_1 = \frac{-1 - 25x_3 - 6x_4}{19}, \quad x_2 = \frac{22 - x_3 - 20x_4}{19}$$

Then $x_3 = x_3, x_4 = x_4$ can be any real number. We use the terminology **free variables** for x_3, x_4, and each of the others can be expressed in terms of these. The set of solutions can then be expressed as

$$x_1 = \frac{-1 - 25x_3 - 6x_4}{19},$$

$$x_2 = \frac{22 - x_3 - 20x_4}{19},$$

$$x_3 = x_3(\text{ free}),$$

$$x_4 = x_4(\text{ free}).$$

We say that the above is a **two-parameter** or **two-dimensional** family of solutions since it depends on two free parameters.

13.1.4   Determinants

An extremely important concept in dealing with matrix algebra is that of the **determinant** of a square matrix. We define this inductively.

Definition 13.1.14. *If A is an $n \times n$ matrix then the ijth* **minor** *of A denoted M_{ij} is the $(n-1) \times (n-1)$ matrix formed from A by deleting the ith row and jth column.*

EXAMPLE 13.1.13. Let

$$A = \begin{pmatrix} 3 & 1 & 4 \\ 6 & 0 & 2 \\ 4 & 2 & 1 \end{pmatrix}.$$

Then the $(1, 2)$ minor of A is

$$M_{1,2} = \begin{pmatrix} 6 & 2 \\ 4 & 1 \end{pmatrix}.$$

Definition 13.1.15. *If A is an $n \times n$ matrix over a commutative ring R with identity then its* **determinant** *denoted det(A) or $|A|$ is defined inductively by*

(1) If A is 2×2, $A = \begin{pmatrix} a & b \\ c & d \end{pmatrix}$ then

$$|A| = ad - bc.$$

The determinant is thus an element of R.

(2) If A is $n \times n$ with $n > 2$ and the determinant has been defined for all square matrices of size $< n$, as an element of R, then

$$|A| = \sum_{j=1}^{n}(-1)^{i+j}a_{ij}|M_{ij}|$$

where i is any row.

This inductive definition is called the expansion by minors across the ith row.

The following crucial result says that the value is independent of the row we choose and further we can do the expansion down any column also.

Theorem 13.1.7. *Let A be an $n \times n$ matrix. Then the value $|A|$ is independent of the row or column on which we expand. Further*

$$|A| = \sum_{i=1}^{n}(-1)^{i+j}a_{ij}|M_{ij}|$$

where j is any column.

EXAMPLE 13.1.14. Suppose $A = \begin{pmatrix} 3 & 1 & 4 \\ 6 & 0 & 2 \\ 4 & 2 & 1 \end{pmatrix}$. Then

$$|A| = \begin{vmatrix} 3 & 1 & 4 \\ 6 & 0 & 2 \\ 4 & 2 & 1 \end{vmatrix} = 3\begin{vmatrix} 0 & 2 \\ 2 & 1 \end{vmatrix} - 1\begin{vmatrix} 6 & 2 \\ 4 & 1 \end{vmatrix} + 4\begin{vmatrix} 6 & 0 \\ 4 & 2 \end{vmatrix}$$

$$= 3(-4) - 1(-2) + 4(12) = 38$$

expanding by the top row.

Determinants are closely tied to invertibility. In particular

Theorem 13.1.8. *(1) $det(I_n) = 1$ where I_n is the $n \times n$ identity matrix*

(2) If A is upper triangular with $d_1, \ldots, d_n$ the diagonal elements then $det(D) = d_1 d_2 \cdots d_n$.

(3) A is invertible if and only if $|A| \neq 0$ if and only if the rank of A is n.

(4) If A is invertible then $|A|$ is a unit in R and $|A^{-1}| = |A|^{-1}$ in R.

We next mention that determinants multiply.

Theorem 13.1.9. $|AB| = |A||B|$ *for any* $n \times n$ *matrices* A *and* B.

The set of invertible matrices over a commutative ring R with identity will form a group and the above result says that the map $A \to |A|$ is a homomorphism from $M_n(R)$ to the units in R.

Theorem 13.1.10. *The set of invertible* $n \times n$ *matrices over a commutative ring* R *with identity forms a multiplicative group called the* **general linear group** *over* R *denoted* $GL_n(R)$. *The set of matrices with determinant 1 forms a subgroup called the* **special linear group** *over* R *denoted* $SL_n(R)$.

Definition 13.1.16. *Two matrices* A, B *are* **similar** *if there is an invertible matrix* T *such that* $T^{-1}AT = B$.

Lemma 13.1.9. *Similar matrices have the same determinant. It follows that the special linear group is actually a normal subgroup of the general linear group.*

Theorem 13.1.11. *The map det:* $GL_n(R) \to U(R)$ *from the general linear group to the units of the underlying ring* R *is a homomorphism with kernel equal to* $SL_n(R)$. *That is, the determinant function is a group homomorphism from the general linear group over* R *to the group of units in* R *and the kernel is the special linear group.*

Not only is the determinant of a matrix tied to its invertibility but the inverse can be found in terms of the determinant. We need the ideas here of the transpose and adjoint of a matrix.

Definition 13.1.17. *If* $A = (a_{ij})$ *is any matrix then its* **transpose** *is* $A^t = (a_{ji})$, *that is, the rows and columns are interchanged.*

EXAMPLE 13.1.15. Let $A = \begin{pmatrix} 3 & 1 & 4 \\ 6 & 0 & 2 \\ 4 & 2 & 1 \end{pmatrix}$. Then

$$A^t = \begin{pmatrix} 3 & 6 & 4 \\ 1 & 0 & 2 \\ 4 & 2 & 1 \end{pmatrix}.$$

The following are easy consequences of the definition.

Lemma 13.1.10. *The following are true about transposes:*
   (i) $(A^t)^t = A$
   (ii) $(A + B)^t = A^t + B^t.$
   (iii) $(kA)^t = kA^t.$
   (iv) $(AB)^t = B^t A^t.$
   (v) If A is square then $\det(A^t) = \det(A).$

Definition 13.1.18. *Let A be an $n \times n$ matrix. Then the ij-**cofactor** of A is $C_{ij} = (-1)^{i+j}|M_{ij}|$ where M_{ij} is the ij-minor amd $|M_{ij}|$ is its determinant. The **matrix of cofactors** is $C = C_{ij}$. The **adjoint matrix** denoted $adj(A)$ is the transpose of the matrix of cofactors.*

EXAMPLE 13.1.16. Let $A = \begin{pmatrix} 3 & 1 & 4 \\ 6 & 0 & 2 \\ 4 & 2 & 1 \end{pmatrix}$. Then

$$C_{11} = (-1)^2 \begin{vmatrix} 0 & 2 \\ 2 & 1 \end{vmatrix}, C_{12} = (-1)^3 \begin{vmatrix} 6 & 2 \\ 4 & 1 \end{vmatrix}, C_{13} = (-1)^4 \begin{vmatrix} 6 & 0 \\ 4 & 2 \end{vmatrix}$$

$$C_{21} = (-1)^3 \begin{vmatrix} 1 & 4 \\ 2 & 1 \end{vmatrix}, C_{22} = (-1)^4 \begin{vmatrix} 3 & 4 \\ 4 & 1 \end{vmatrix}, C_{23} = (-1)^5 \begin{vmatrix} 3 & 1 \\ 4 & 2 \end{vmatrix}$$

$$C_{31} = (-1)^4 \begin{vmatrix} 1 & 4 \\ 0 & 2 \end{vmatrix}, C_{32} = (-1)^5 \begin{vmatrix} 3 & 4 \\ 6 & 2 \end{vmatrix}, C_{33} = (-1)^6 \begin{vmatrix} 3 & 1 \\ 6 & 0 \end{vmatrix}$$

$$\implies C_{11} = -4, C_{12} = 2, C_{13} = -4, C_{21} = 7, C_{22} = -13, C_{23} = -2,$$

$$C_{31} = 2, C_{32} = 18, C_{33} = -6$$

$$\implies C = \begin{pmatrix} -4 & 2 & -4 \\ 7 & -13 & -2 \\ 2 & 18 & -6 \end{pmatrix}$$

is the matrix of cofactors and

$$adj(A) = \begin{pmatrix} -4 & 7 & 2 \\ 2 & -13 & 18 \\ -4 & -2 & -6 \end{pmatrix}.$$

Combining the definition of the adjoint with matrix multiplication we get that the inverse of a matrix A is precisely the adjoint divided by the determinant.

Theorem 13.1.12. *If A is a square matrix over a field F and $|A| \neq 0$ then*

$$A^{-1} = \frac{1}{|A|} adj(A)$$

Thus to find A^{-1} find $adj(A)$ and then divide each element by the determinant. In practice, however, row reduction is a much more efficient way to find the inverse.

EXAMPLE 13.1.17. Let $A = \begin{pmatrix} 3 & 1 & 4 \\ 6 & 0 & 2 \\ 4 & 2 & 1 \end{pmatrix}$. Then

$$adj(A) = \begin{pmatrix} -4 & 7 & 2 \\ 2 & -13 & 18 \\ -4 & -2 & -6 \end{pmatrix} \text{ and } |A| = 38$$

$$\implies A^{-1} = \frac{1}{38} \begin{pmatrix} -4 & 7 & 2 \\ 2 & -13 & 18 \\ -4 & -2 & -6 \end{pmatrix} = \begin{pmatrix} \frac{-4}{38} & \frac{7}{38} & \frac{2}{38} \\ \frac{2}{38} & \frac{-13}{38} & \frac{18}{38} \\ \frac{-4}{38} & \frac{-2}{38} & \frac{-6}{38} \end{pmatrix}.$$

Applying the adjoint to the solution of square systems of equations we arrive at **Cramer's rule**.

Theorem 13.1.13. *(Cramer's Rule) If $AX = B$ is a square matrix equation then if $|A| \neq 0$ there is a unique solution $(x_1, \ldots, x_n)$ given by*

$$x_i = \frac{|\ldots B_i \ldots|}{|A|}$$

where $(\ldots B_i \ldots)$ is the matrix A with the i-th column replaced by the column of constants B.

We present a 3×3 example.

EXAMPLE 13.1.18. Solve

$$3x + y + 4z = 12$$
$$6x + 2z = 10$$
$$4x + 2y + z = 8$$

Then

$$A = \begin{pmatrix} 3 & 1 & 4 \\ 6 & 0 & 2 \\ 4 & 2 & 1 \end{pmatrix}, \ |A| = 38, \ B = \begin{pmatrix} 12 \\ 10 \\ 9 \end{pmatrix}$$

$$\implies x = \frac{\begin{vmatrix} 12 & 1 & 4 \\ 10 & 0 & 2 \\ 8 & 2 & 1 \end{vmatrix}}{|A|} = \frac{38}{38} = 1,$$

$$y = \frac{\begin{vmatrix} 3 & 12 & 4 \\ 6 & 10 & 2 \\ 4 & 8 & 1 \end{vmatrix}}{|A|} = \frac{38}{38} = 1,$$

$$z = \frac{\begin{vmatrix} 3 & 1 & 12 \\ 6 & 0 & 10 \\ 4 & 2 & 8 \end{vmatrix}}{|A|} = \frac{76}{38} = 2.$$

13.2 Vector Spaces over a Field

We now turn to general vector spaces over a field F. First we look at extending the vector operations in $\mathbb{R}^3$ to general n-dimensional real space.

13.2.1 Euclidean n-Space

The concept of vectors in two and three dimensions can be extended to an arbitrary n dimensions. Here we take certain properties of vectors in $\mathbb{R}^3$ and use them as definitions in higher dimensions.

Definition 13.2.1. *Consider the set* $\mathbb{R}^n = \{(x_1, \dots, x_n) : x_i \in \mathbb{R}\}$. *The n-tuple* $\vec{v} = (x_1, \dots, x_n)$ *is a* **vector** *in* $\mathbb{R}^n$ *and* x_i *is a* **component** *or* **coordinate** *of the vector* $\vec{v}$. *If* $\vec{u} = (x_1, \dots, x_n), \vec{v} = (y_1, \dots, y_n)$ *then*

$$\vec{u} + \vec{v} = (x_1 + y_1, \dots, x_n + y_n), \quad k\vec{u} = (kx_1, \dots, kx_n) \text{ for } k \in \mathbb{R}.$$

The vector $\vec{0} = (0, \dots, 0)$ *is the* **zero vector**.

Lemma 13.2.1. *Vector addition and scalar multiplication in* $\mathbb{R}^n$ *satisfy the same properties as in* $\mathbb{R}^3$. $\mathbb{R}^n$ *is called* **real n-space**.

On $\mathbb{R}^n$ define the **inner product** as follows: If $\vec{u} = (x_1, \ldots, x_n), \vec{v} = (y_1, \ldots, y_n)$ then

$$\langle \vec{u}, \vec{v} \rangle = x_1 y_1 + \cdots + x_n y_n = \sum_{i=1}^{n} x_i y_i$$

The **length** or **norm** of a vector $\vec{u} = (x_1, \ldots, x_n)$ is

$$|\vec{u}| = \sqrt{\sum_{i=1}^{n} x_i^2} = \sqrt{\langle \vec{u}, \vec{u} \rangle}$$

Lemma 13.2.2. *The inner product and norm on $\mathbb{R}^n$ satisfy all the same properties as the dot product and magnitude in $\mathbb{R}^3$.*

The space $\mathbb{R}^n$ together with vector addition, scalar multiplication and inner product is called **Euclidean n-space** denoted $\mathcal{E}^n$.

13.2.2   Vector Spaces

We let F be an arbitrary field and we abstract the basic properties of vectors in $\mathbb{R}^n$ to define a **vector space**.

Definition 13.2.2. *A **vector space** V over the field F is a set V together with an addition $V \times V \to V$, which for $u, v \in V$ we will write as $u + v \in V$, and a scalar multiplication $F \times V \to V$, which for $u \in V, f \in F$ we write as fu, satisfying*

(1) V is an abelian group with respect to vector addition.
(2) $f(u + v) = fu + fv$ for all $f \in F, u, v \in V$.
(3) $(f_1 + f_2)u = f_1 u + f_2 u$ for all $f_1, f_2 \in F, u \in V$.
(4) $(f_1 f_2)u = f_1(f_2 u)$ for all $f_1, f_2 \in F, u \in V$.
(5) $1u = u$ if $u \in V$.

*F is called the **ground field**. Elements of V are called **vectors**. Elements of F are **scalars**. If $F = \mathbb{R}$ or $F = \mathbb{C}$, that is, the ground field is the reals or the complexes, then V is called a **linear space**.*

*If F is not a field but only a commutative ring with identity then V is called a **module** over F.*

As with all our previous algebraic structures a definition is only as good as its examples. Examples of vector spaces abound in mathematics.

EXAMPLES OF VECTOR SPACES

To show that each of these examples is actually a vector space involves verifying the axioms in the definition. This is straightforward and we leave these verifications to the exercises.

EXAMPLE 13.2.1. If F is a field let

$$F^n = \{(f_1, \ldots, f_n) : f_i \in F\}.$$

Then F^n forms a vector space over F under componentwise operations. The proof of this is exactly the same as in showing the properties in $\mathbb{R}^n$.

The next example will play a huge role in Galois theory.

EXAMPLE 13.2.2. If F' is a subfield of F then F is a vector space over F' where addition is $+$ in F and scalar multiplication is multiplication in F.

EXAMPLE 13.2.3. $\mathbb{R}^n, \mathbb{C}^n, \mathbb{Q}^n$ for any $n \geq 1$ are vector spaces over $\mathbb{R}, \mathbb{C}$ and $\mathbb{Q}$ respectively where addition and scalar multiplication are defined componentwise as in Definition 13.2.1.

EXAMPLE 13.2.4. $C^0[a,b] = \{f : [a,b] \to \mathbb{R} : f$ is continuous on $[a,b]\}$, the set of continuous functions on $[a,b]$, is a vector space over $\mathbb{R}$ where for $f, g \in C^0[a,b]$ we have $(f + g)(x) = f(x) + g(x)$ and $(cf)(x) = c(f(x))$ for $c \in \mathbb{R}$.

EXAMPLE 13.2.5. $C^k[a,b] = \{f : [a,b] \to \mathbb{R} : f^k$ is continuous on $[a,b]\}$, that is, the set of k-times continuously differentiable functions on $[a,b]$ is a vector space over $\mathbb{R}$ with the same operations as in Example 13.2.4.

EXAMPLE 13.2.6. $F[x]$, the set of polynomials over F, is a vector space over F with addition being addition of polynomials and scalar multiplication just being the multiplication of elements of $F[x]$ by elements of F.

EXAMPLE 13.2.7. $F_n[x]$ the set of polynomials over F of degree $\leq n$ is a vector space over F with respect to the same operations as defined in Example 13.2.6.

EXAMPLE 13.2.8. $M_{mn}(F)$ the set of $m \times n$ matrices over the field F is a vector space over F. Addition is just the addition of matrices and scalar multiplication is just the multiplication of an $m \times n$ matrix by a scalar.

EXAMPLE 13.2.9. Let F be a field and $A \in M_{mn}(F)$. The set of solutions to a homogeneous system of equations $AX = 0$ is a subset of F^n (see Definition 13.1.13 and Example 13.2.1.) Thus $\{X \in F_n : AX = 0\}$ where X is an $n \times 1$ column vector, A is an $m \times n$ matirx and 0 is an $m \times 1$ column vector. This subset is then a vector space over F under the same operations as defined in Example 13.2.1.

13.2.3 Subspaces

As with all our other algebraic structures a (vector) subspace is a subset which is also a vector space.

Definition 13.2.3. $W \subset V$ *is a* **subspace** *if* $W \neq \emptyset$ *and* W *is a vector space under the same operations as* V.

The proofs of the following lemmas are analogous to the proofs for groups, rings and fields.

Lemma 13.2.3. $W \subset V$ *is a subspace if* $W \neq \emptyset$ *and* W *is closed under addition and scalar multiplication*

Lemma 13.2.4. *The intersection of an arbitrary collection of subspaces is a subspace*

EXAMPLE 13.2.10. If $n < m$ then F^n can be considered as a subspace of F^m.

EXAMPLE 13.2.11. The space $C^k[a,b]$ of k-times continuously differentiable functions on $[a,b]$ is a subspace of $C^0[a,b]$ the set of continuous functions on $[a,b]$.

EXAMPLE 13.2.12. $F_n[x]$ the set of polynomials over F of degree $\leq n$ is a subspace of $F[x]$ the set of polynomials over F.

Given a vector space V and elements $u_1, \ldots, u_n \in V$ we now define a very important subspace called the **linear span** of these elements.

Definition 13.2.4. *If* $u_1, \ldots, u_n \in V$ *then a* **linear combination** *of them is a vector* $c_1 u_1 + \cdots + c_n u_n$ *where the* c_i *are scalars.*

If $u_1, \ldots, u_n \in V$ *then their* **linear span** *denoted* $[u_1, .., u_n]$ *is the set of all linear combinations of* $u_1, \ldots, u_n$.

If U *is any nonempty subset of* V *then the linear span* $[U]$ *of* U *in* V *is the set of elements in* V *that can be expressed as finite linear combinations of elements from* U, *that is, as* $c_1 u_1 + \cdots + c_n u_n$ *with* $u_1, \ldots, u_n \in U$.

Theorem 13.2.1. $[u_1, \ldots, u_n]$ *forms a subspace called the* **subspace spanned** *by* $u_1, \ldots, u_n$. *It is the smallest subspace containing* $u_1, \ldots, u_n$.

The proof is straightforward and we leave it to the exercises.

Recall that for groups and rings direct products were an important method to both construct new groups and rings from existing ones and to analyze the structure of a given group or ring in terms of its subgroups or subrings. We have an analogous concept in vector spaces.

Definition 13.2.5. *If* U *and* W *are subspaces of* V *then their* **sum** *is*

$$U + W = \{u + w : u \in U, w \in W\}.$$

Lemma 13.2.5. $U + W$ *forms a subspace containing* U *and* W.

Definition 13.2.6. V *is the* **direct sum** *of* U, W *denoted* $V = U \oplus W$ *if* $V = U + W$ *and* $U \cap W = \{0\}$.

Lemma 13.2.6. *If* $V = U \oplus W$ *and* $v = u_1 + w_1 = u_2 + w_2$ *where* $u_1, u_2 \in U$ *and* $w_1, w_2 \in W$ *then* $u_1 = u_2, w_1 = w_2$. *That is, any element of* V *is uniquely expressible as an element of* U *plus an element of* W.

13.2.4 Bases and Dimension

In $\mathbb{R}^3$ we saw that every vector is a unique linear combination of the standard basis vectors $\mathbf{i}, \mathbf{j}, \mathbf{k}$. This can be extended to Euclidean n-space by using standard basis vectors $\vec{e}_1 = (1, 0, \ldots, 0), \vec{e}_2 = (0, 1, 0, \ldots, 0), \ldots, \vec{e}_n = (0, 0, \ldots, 1)$. We generalize this idea to abstract vector spaces. We first need the idea of independence.

Definition 13.2.7. *Let* V *be a vector space. A nonempty set of vectors* $v_1, \ldots, v_m$ *is* **dependent** *or* **linearly dependent** *if there exist scalars* $c_1, \ldots, c_m$ **not all zero** *such that* $c_1 v_1 + \cdots + c_m v_m = 0$. *If* $v_1, \ldots, v_m$ *are not dependent then they are* **independent**.

An infinite set of vectors is called independent if any finite subset is independent and dependent if there exists a finite subset that is dependent.

Lemma 13.2.7. *Let* V *be a vector space and* $v_1, \ldots, v_m \in V$. *Then* $v_1, \ldots, v_m$ *are independent if and only if* $c_1 v_1 + \cdots + c_m v_m = 0$ *implies that all the scalars are zero.*

Lemma 13.2.8. *Let V be a vector space and $v_1, .., v_m \in V$. The set $v_1, \ldots, v_m$ is independent if and only if no subset is a linear combination of the remaining ones.*

There is a very simple way to test for independence in the vector space F^n where F is a field.

Lemma 13.2.9. *Let $u_1 = (f_{11}, \ldots, f_{1n}), \ldots, u_m = (f_{m1}, \ldots, f_{mn})$ be m vectors in F^n. Then they are independent if and only if the matrix*

$$A = \begin{pmatrix} f_{11} & \cdots & f_{1n} \\ \vdots & \vdots & \\ f_{m1} & \cdots & f_{mn} \end{pmatrix}$$

has rank m.

Using independence and linear span we can define a **basis**. In general a vector space will have many bases.

Definition 13.2.8. *A **basis** for a vector space V is an independent set which spans V.*

EXAMPLE 13.2.13. (1) $\mathbf{i}, \mathbf{j}, \mathbf{k}$ form a basis for $\mathbb{R}^3$.
(2) $e_1 = (1, 0, \ldots, 0), e_2 = (0, 1, 0, \ldots, 0), \ldots, e_n = (0, 0, \ldots, 1)$ forms a basis for F^n called the **standard basis**.
(3) Consider the complex numbers $\mathbb{C}$ as a vector space over the real numbers $\mathbb{R}$. Then the elements $1, i$ constitute a basis.
(4) Let $F_n[x]$ be the set of polynomials of degree $\leq n$ over F. Then the elements $1, x, x^2, \ldots, x^n$ constitute a basis.

Theorem 13.2.2. *Let V be a vector space and $v_1, .., v_n \in V$. Then $\{v_1, \ldots, v_n\}$ forms a basis for V if and only if every $v \in V$ can be expressed **uniquely** as a linear combination of $v_1, \ldots, v_n$.*

Real n-space has a basis consisting of n-elements and in ordinary language we say that n-space is n-dimensional. As with the other concepts from Euclidean n-space we abstract this. First we have that although a basis is not unique, and in fact if F is an infinite field there are infinitely many bases for a vector space over F, the number of elements in a basis is unique.

The next result gives the relationship between independence, span and the size of a basis.

Theorem 13.2.3. *Let $v_1, \ldots, v_n$ be a basis for V. Then any set with more than n vectors is dependent and any independent set has fewer than or equal to n vectors. Any independent set with n vectors also constitutes a basis.*

Proof. Let $v_1, \ldots, v_n$ be a basis for V. We first show that every subset of V with more than n vectors is dependent. Suppose that $W = \{u_1, \ldots, u_m\}$ with $m > n$. Since $v_1, \ldots, v_n$ is a basis for V there exist scalars $c_{ij} \in F$ such that

$$u_j = \sum_{i=1}^{n} c_{ij} v_i.$$

For any scalars $a_1, \ldots, a_m \in F$ we consider

$$\sum_{j=1}^{m} a_j u_j = \sum_{j=1}^{m} a_j \left(\sum_{i=1}^{n} c_{ij} v_i \right)$$

$$= \sum_{j=1}^{m} \left(\sum_{i=1}^{n} a_j c_{ij} \right) v_i = \sum_{i=1}^{n} \left(\sum_{j=1}^{m} c_{ij} a_j \right) v_i.$$

Since $m > n$ there exist scalars $a_1, a_2, \ldots, a_m$ not all zero such that

$$\sum_{j=1}^{m} c_{ij} a_j = 0, 1 \le i \le n.$$

See Section 13.1.3 where it was mentioned that if a consistent linear system over an infinite field had more unknowns than equations then the system had an infinite number of solutions. Here the system is homogeneous, so it is consistent. Also $m > n$ implies there are more unknowns, the a_j, than equations. If the field F is infinite then there are infinitely many solutions. If not, then just take any nontrivial solution in F which must also exist.

This shows that the set W is dependent. We have actually shown the following also. If a set $W \subset V$ is linearly independent then $|W| \le n$. If the set W were linearly independent and had m elements with $m > n$ then the above argument would contradict the independence of W. Hence $m \le n$.
□

To continue we need the following.

Lemma 13.2.10. *Let V be a vector space and let S be a linearly independent subset of V. Suppose that $w \in V$ is not in the subspace spanned by S. Then the set obtained by adjoining w to S, $S \cup \{w\}$ is also linearly independent.*

Proof. (of lemma) Suppose $S = \{v_1, \ldots, v_m\}$ where the v_i are distinct and independent. Consider

$$c_1 v_1 + \cdots + c_m v_m + aw = 0$$

with not all $c_i = 0$. We claim that $a \neq 0$. Suppose $a = 0$. Then

$$c_1 v_1 + \cdots + c_m v_m = 0$$

with not all $c_i = 0$ contradicting the fact that S is independent. Therefore $a \neq 0$ and from

$$c_1 v_1 + \cdots + c_m v_m + aw = 0$$

we get that

$$w = -\frac{c_1}{a} v_1 + \cdots + -\frac{c_m}{a_n} v_m$$

contradicting that w is not in the span of S. Therefore $S \cup \{w\}$ is also independent.   □

We now complete the proof of Theorem 13.2.3.

Proof. (of Theorem 13.2.3) Again suppose that $\{v_1, \ldots, v_n\}$ is a basis for V and suppose that $\{u_1, \ldots, u_n\}$ is any other linearly independent set in V. To complete the proof we must show that $\{u_1, \ldots, u_n\}$ is also a basis. Since we assumed these vectors to be independent it suffices to show that they span V. Let $w \in V$. Suppose that w is not in the span of $\{u_1, \ldots, u_n\}$. Then Lemma 13.2.10 implies that $\{u_1, \ldots, u_n, w\}$ is independent. But this contradicts the first fact that we proved that any independent set in V has $\leq n$ elements. Therefore w must be in the span of $\{u_1, \ldots, u_n\}$ and hence this set forms a basis.   □

From this result we immediately obtain the following.

Theorem 13.2.4. *Let V be a vector space with a basis consisting of a finite number of elements. Then the cardinality of any basis is unique and equals the maximal number of independent vectors in V.*

Proof. Let V be a vector space with a basis consisting of a finite number of elements. Suppose that $\{v_1, \ldots, v_n\} = S$ is a basis. Let $T = \{u_1, \ldots, u_k\}$ be any other basis. Since S is a basis and T is a linearly independent set then from Theorem 13.2.3 we have $k = |T| \leq n = |S|$. By reversing the roles of S and T we get that $n \leq k$ so therefore $n = k$.   □

Definition 13.2.9. *The **dimension** of V denoted $dim(V)$ or $dim_F(V)$ is the unique number of elements in any basis. If $dim(V) = n < \infty$ then V is **finite-dimensional**. Otherwise V is **infinite-dimensional**.*

We note that in Theorem 13.2.4 we assumed that the vector space V had a basis. The reader may wonder if it is true that any vector space has a basis. The answer is yes if one allows the axiom of choice or Zorn's lemma. Since we do not want to get into the necessary set theory to show this we will just assume that any vector space has a basis.

Suppose we are given an independent set $S = \{u_1, \ldots, u_m\}$ in a finite-dimensional vector space V of dimension $n > m$. Let $w \in V$. If w is in the span of S for all $w \in V$ then S is a basis contradicting that $n > m$. Hence there must be a vector $w \in V$ such that $S_1 = S \cup \{w\}$ is still independent. We can continue in this manner only until $m = n$ (why?). Hence we have proved the following.

Theorem 13.2.5. *Let V be a finite-dimensional vector space and*

$$S = \{u_1, \ldots, u_m\}$$

a linearly independent set in V. Then S can be extended to a basis for V.

The following lemma provides a raft of examples of dimension.

Lemma 13.2.11. *(a) $dim(F^n) = n$.*
(b) $dim_{\mathbb{Q}}(\mathbb{R}) = \infty$.
(c) $dim_{\mathbb{R}}(\mathbb{C}) = 2$.
(d) $dim(C^0[a, b]) = \infty$.
(e) $dim(M_{mn}(F)) = mn$.
(f) $dim(F[x]) = \infty$.
(g) $dim(F_k[x]) = k$.

Proof. We prove (a) and (b) and leave the rest to the exercises. For (a) the set

$$e_1 = (1, 0, \ldots, 0), e_2 = (0, 1, 0, \ldots 0), \ldots, e_n = (0, 0, \ldots, 1)$$

is easily seen to form a basis for F^n over any field F.

For (b), to see that the reals $\mathbb{R}$ are infinite-dimensional over the rationals $\mathbb{Q}$ we have to consider that there exists transcendental elements such as π or c where $c = \sum_{j=1}^{\infty} \frac{1}{10^{j!}}$ (see Theorem 12.3.8). Recall that an element $\alpha \in \mathbb{R}$ is transcendental if there does not exist a polynomial $p(x) \in \mathbb{Q}[x]$ with $p(\alpha) = 0$. The element π is transcendental.

Suppose that $\dim_{\mathbb{Q}}(\mathbb{R}) = n < \infty$. Then any $n+1$ distinct elements would be dependent. Consider then $1, \pi, \pi^2, \ldots, \pi^n$. First no two of these are equal. Suppose that $\pi^k = \pi^m$ for two distinct k, m with $k < m$. Then $\pi^{m-k} = 1$ which would imply that π is algebraic, a contradiction. Hence $1, \pi, \pi^2, \ldots, \pi^n$ is a set with $n+1$ distinct elements so they must be dependent over $\mathbb{Q}$. Hence there exist rational numbers (scalars) $q_0, q_1, \ldots, q_n$ not all zero with

$$q_0 \cdot 1 + q_1 \cdot \pi + \cdots + q_n \cdot \pi^n = 0.$$

Consider the rational polynomial $p(x) = q_0 + q_1 x + \cdots + q_n x^n$. From the above we then have $p(\pi) = 0$, a contradiction. Therefore $\mathbb{R}$ cannot be finite-dimensional over $\mathbb{Q}$.

$\square$

Two different bases for a finite-dimensional vector space are related via a matrix called the **transition matrix**.

Definition 13.2.10. *Let $v_1, \ldots, v_n$ be a basis for V and $u_1, \ldots, u_n$ a set of n vectors in V. Then*

$$u_1 = c_{11}v_1 + \cdots + c_{1n}v_n$$
$$u_2 = c_{21}v_1 + \cdots + c_{2n}v_n$$
$$\cdots$$
$$u_n = c_{n1}v_1 + \cdots + c_{nn}x_n$$

The matrix

$$C = \begin{pmatrix} c_{11} & \cdots & c_{1n} \\ c_{21} & \cdots & c_{2n} \\ \cdots & & \\ c_{n1} & \cdots & c_{nn} \end{pmatrix}$$

is the **transition matrix** *from $v_1, \ldots, v_n$ to $u_1, \ldots, u_n$.*

Theorem 13.2.6. *Let $v_1, \ldots, v_n$ be a basis for V and $u_1, \ldots, u_n$ a set of n vectors in V. Then $u_1, .., u_n$ is also a basis if and only if the transition matrix C is invertible. That is, if the rank of C is n. The transition matrix in the other direction is C^{-1}.*

We now look at the abstraction of the idea of coordinates.

Definition 13.2.11. *Suppose $v_1, \ldots, v_n$ is a basis for V over F. If $v \in V$ then $v = c_1 v_1 + \cdots + c_n v_n$. The components $(c_1, \ldots, c_n)$ are the **coordinates** of v relative to the basis $v_1, \ldots, v_n$.*

Theorem 13.2.7. *If $dim(V) = n$ then $V \cong F^n$.*

Proof. Let $v_1, \ldots, v_n$ be a basis for V. Then any $v \in V$ can be written uniquely as

$$v = c_1 v_1 + \cdots + c_n v_n$$

with $c_1, c_2, \ldots, c_n \in F$, Then the map $T : v \rightarrow (c_1, c_2, \ldots, c_n)$ gives the isomorphism. □

If there are two different bases for a finite-dimensional vector space V over F then an element $v \in V$ will have different sets of coordinates relative to the different bases. These will be related via the transition matrix.

Theorem 13.2.8. *Let $u_1, \ldots, u_n$ and $v_1, \ldots, v_n$ be two bases for a finite-dimensional vector space over F and let C be the transition matrix from the first basis to the second. Suppose $v \in V$ and v has coordinates $(c_1, \ldots, c_n)$ relative to $u_1, \ldots, u_n$. Then the coordinates of v relative to $v_1, \ldots, v_n$ are given by $C \cdot (c_1, \ldots, c_n)^t$ where the second vector is considered as a column vector.*

13.2.5 Testing for Bases in F^n

A set of n vectors in F^n forms a basis if they are independent. This is equivalent to the fact that the square matrix having these vectors as rows has rank n which is further equivalent to its determinant being nonzero.

EXAMPLE 13.2.14. Do $\vec{u}_1 = (1, 3, 3), \vec{u}_2 = (1, 0, 1), \vec{u}_3 = (0, 0, 1)$ form a basis for $\mathbb{R}^3$?

First form

$$A = \begin{pmatrix} 1 & 3 & 3 \\ 2 & 0 & 1 \\ 0 & 0 & 1 \end{pmatrix}.$$

Then

$$|A| = 1 \begin{vmatrix} 1 & 3 \\ 2 & 0 \end{vmatrix} = -6 \neq 0.$$

Hence they are independent and do form a basis for $\mathbb{R}^3$.

Independence and dependence depend upon the ground field. If we considered the same vectors over the finite field $\mathbb{Z}_3$ then the determinant would be zero and they would be dependent and not constitute a basis for $(\mathbb{Z}_3)^3$.

In $\mathbb{R}^2$ any two non-collinear vectors form a basis. In $\mathbb{R}^3$ any three non-coplanar vectors form a basis. In general in $\mathbb{R}^n$ any n non-cohyperplanar vectors form a basis where a **hyperplane** is a subspace of dimension $n-1$.

13.3   Dimension and Subspaces

As is expected the dimension of a subspace can be no larger than the dimension of the whole vector space.

Theorem 13.3.1. *If $dim(V) = n < \infty$ and $W \subset V$ then $dim(W) \leq dim(V)$. Further, if V is finite-dimensional then $dim(V) = dim(W)$ if and only if $V = W$. The second part need not be true in infinite dimensions.*

Next there is a very nice relation between the dimension of the sum of two subspaces and the dimensions of the factors.

Theorem 13.3.2. *If U, W are finite-dimensional subspaces of V then*

$$dim(U + W) = dim(U) + dim(W) - dim(U \cap W).$$

13.4   Algebras

The set $M_n(F)$ consisting of $n \times n$ matrices over the field F is both a vector space and a ring since not only do we have have the operation of scalar multiplication but there is also a matrix multiplication. We call such a structure an **algebra**.

Definition 13.4.1. *An **algebra** V over a field F is a vector space with another binary operation $V \times V \to V$ called multiplication $\cdot$, that we denote by uv, which is also a ring with respect to $+$ and $\cdot$ and such that $f(uv) = (fu)v = u(fv)$ for all $f \in F, u, v \in V$.*

The following are straightforward examples of algebras.
   (1) Any field over a subfield.
   (2) The set of $n \times n$ matrices over a field.

13.5 Inner Product Spaces

The geometry of Euclidean n-space is determined by the extension to $\mathbb{R}^n$ of the inner product. This allows us to measure both distance, using the extension of the distance formula, and angle. In a very general setting this will lead us to **inner product spaces** which we discuss in this section. We will then give a very nice application of these ideas to approximations of continuous functions by polynomials.

Definition 13.5.1. *An **inner product** on a vector space V over the ground field $\mathbb{R}$ is a function*

$$\langle,\rangle : V \times V \to \mathbb{R}$$

satisfying:

(1) $\langle v, v \rangle \geq 0$ for all $v \in V$,
(2) $\langle v, v \rangle = 0$ if and only if $v = 0$,
(3) $\langle v, w \rangle = \langle w, v \rangle$ for all $v, w \in V$,
(4) $\langle v, u + w \rangle = \langle v, u \rangle + \langle v, w \rangle$ for all $u, v, w \in V$,
(5) $\langle tu, v \rangle = \langle u, tv \rangle = t\langle u, v \rangle$ for all $u, v \in V$, $t \in \mathbb{R}$.

*Note: This defines a **real** inner product. If the ground field is $\mathbb{C}$ then a complex inner product has (3) $\langle v, w \rangle = \overline{\langle w, v \rangle}$ the complex conjugate.*

It should also be observed that if the ground field is $\mathbb{C}$ then the equality in (5) must be stated as

$$\langle u, tv \rangle = \bar{t}\langle u, v \rangle.$$

In the case of the ground field being $\mathbb{C}$ the complex conjugate is necessary for the consistency of the conditons. Without the complex conjugate we would get for $v \neq 0$, $\langle v, v \rangle > 0$ and $\langle iv, iv \rangle = (-1)\langle v, v \rangle > 0$ which is a contradiction.

For the remainder of this chapter we will be only **considering real inner product spaces**, that is, the ground field is $\mathbb{R}$.

Definition 13.5.2. *An **inner product space** or IPS is a vector space with an inner product.*

As with every definition we must give some nontrivial examples.

EXAMPLE 13.5.1. Euclidean n-Space $\mathbb{R}^n$ with

$$\langle u, v \rangle = x_1 y_1 + \cdots + x_n y_n = \sum_{i=1}^{n} x_i y_i$$

is easily shown to be a real inner product space. (See the exercises.)

Recall that the set $C^0[a, b]$ consists of all the continuous real functions on the interval $[a, b]$. As we have mentioned, this forms a vector space over the real numbers. Using integration we can make this into an inner product space.

EXAMPLE 13.5.2. The space $C^0[a, b]$ with

$$\langle f(x), g(x) \rangle = \int_a^b f(x)g(x)dx$$

forms an inner product space.

We give a proof of this last example which will be a prominent part of our application to approximations of functions.

Proof. We must verify the five properties in the definition. Let $f(x), g(x), h(x)$ be continuous functions on the interval $[a, b]$. Then

(1) $\langle f(x), f(x) \rangle = \int_a^b (f(x)^2 dx \geq 0$ since integrating a nonnegative function will always be nonnegative.

(2) If $f(x) \equiv 0$ then clearly $\langle f(x), f(x) \rangle = 0$. Conversely, suppose that $\langle f(x), f(x) \rangle = 0$. Then

$$\int_a^b f(x)^2 dx = 0.$$

This implies that $f(x) = 0$ except possibly on a set of measure zero. But $f(x)$ is continuous so $f(x)$ must be zero everywhere.

(3) Since $f(x)g(x) = g(x)f(x)$ we have

$$\langle f(x), g(x) \rangle = \int_a^b f(x)g(x)dx = \int_a^b g(x)f(x)dx = \langle g(x), f(x) \rangle.$$

(4) Here this follows from the basic additive properties of integrals.

$$\langle f(x), g(x) + h(x) \rangle = \int_a^b f(x)(g(x) + h(x))dx = \int_a^b (f(x)g(x) + f(x)h(x))dx$$

$$= \int_a^b f(x)g(x)dx + \int_a^b f(x)h(x)dx = \langle f(x), g(x) \rangle + \langle f(x), h(x) \rangle$$

(5) If $t \in \mathbb{R}$ then

$$\langle tf(x), g(x) \rangle = \int_a^b (tf(x))g(x)dx = t \int_a^b f(x)g(x)dx = t\langle f(x), g(x) \rangle.$$

Clearly $\langle f(x), tg(x) \rangle = t\langle f(x), g(x) \rangle$ also.

$\square$

On any inner product space we can define both a **norm** or **magnitude** of a vector and then from this a **distance** or **metric**.

Definition 13.5.3. *A* **norm** *on a vector space is a function* $| \ | : V \to \mathbb{R}$ *satisfying*
     (1) $|v| \geq 0$ *for all* $v \in V$,
     (2) $|v| = 0$ *if and only if* $v = 0$,
     (3) $|tv| = |t||v|$ *for* $v \in V$ *and* $t \in \mathbb{R}$,
     (4) $|u + v| \leq |u| + |v|$ *for* $u, v \in V$.

A **normed linear space** *or NLS is a vector space with a norm.*

Lemma 13.5.1. *Any inner product space* V *is a normed linear space with* $|v| = \sqrt{\langle v, v \rangle}$ *for* $v \in V$.

The proofs of $(1) - (3)$ are straightforward and left to the exercises. We will prove (4), called the **triangle inequality** after we prove the Cauchy-Schwarz inequality.

Using the norm of a vector we can define distance.

Definition 13.5.4. *In any normed linear space* V *we define* **distance** *by* $d(u, v) = |u - v|$ *for any* $u, v \in V$.

Lemma 13.5.2. *The distance* $d : V \times V \to \mathbb{R}$ *satisfies*
     (a) $d(u, v) \geq 0$,
     (b) $d(u, v) = 0$ *if and only if* $u = v$,
     (c) $d(u, v) = d(v, u)$,
     (d) $d(u, v) \leq d(u, w) + d(w, v)$.

Proof. The proofs are straightforward exercises using the definition of a norm assuming the triangle inequality for the norm.

$\square$

In Chapter 4 we saw that a **metric space** is a set M with a distance function on it satisfying exactly the properties of the previous lemma. Hence what we have proved is the following.

Corollary 13.5.1. *In any normed linear space V if we define **distance** by $d(u,v) = |u - v|$ then V is a metric space. It follows that any inner product space is a metric space under this definition of distance.*

In order to define angles in an inner product space we need the following crucial result called the **Cauchy-Schwarz Inequality**.

Theorem 13.5.1. *(**The Cauchy-Schwarz Inequality**) If V is a real inner product space then*

$$|\langle u, v \rangle| \leq |u||v|$$

for all $u, v \in V$

Proof. If $v = \vec{0}$ then $\langle u, v \rangle = 0$ because $\langle u, 0 \rangle = \langle u, v - v \rangle = \langle u, v \rangle - \langle u, v \rangle = 0$. Of course then $\langle 0, v \rangle = \langle v, 0 \rangle = 0$. Since $|v| = 0$ the inequality holds.

We assume now that $v \neq \vec{0}$. Let $t \in \mathbb{R}$. Then

$$\langle u - tv, u - tv \rangle \geq 0.$$

However, using the properties of inner product we then have

$$\langle u - tv, u - tv \rangle = \langle u, u \rangle - 2t\langle u, v \rangle + t^2 \langle v, v \rangle \geq 0.$$

Not let $t = \frac{\langle u, v \rangle}{|v|^2}$ and substitute into the above inequality. Recall that $\langle u, u \rangle = |u|^2$ and $\langle v, v \rangle = |v|^2$. So

$$\implies |u|^2 - 2\frac{(\langle u, v \rangle)^2}{|v|^2} + \frac{(\langle u, v \rangle)^2}{|v|^4}\langle v, v \rangle \geq 0.$$

Thus

$$|u|^2 - \frac{\langle u, v \rangle^2}{|v|^2} \geq 0 \implies \langle u, v \rangle^2 \leq |u|^2|v|^2.$$

Putting $|\langle u, v \rangle|^2 = (\langle u, v \rangle)^2$ and taking square roots, since all the terms are nonnegative we get

$$|\langle u, v \rangle| \leq |u||v|,$$

completing the proof.

$\square$

To prove that the triangle inequality holds in any normed linear space, that is, to complete the proof of Lemma 13.5.1, we use the Cauchy-Schwarz inequality. By the definition of norm and properties of inner products we get

$$0 \leq |u + v|^2 = \langle u + v, u + v \rangle = |u|^2 + 2\langle u, v \rangle + |v|^2 \leq |u|^2 + 2|\langle u, v \rangle| + |v|^2$$

which by the Cauchy-Schwarz inequality

$$\leq |u|^2 + 2|u||v| + |v|^2 = (|u| + |v|)^2.$$

Thus since everything is again nonnegative we get our result by taking square roots.

As examples we give some specific instances of the Cauchy-Schwarz Inequality. We note that historically each of these examples was proved separately yet all follow from the fact that the inequality holds in any inner product space.

EXAMPLE 13.5.3. If we consider the inner product space as Euclidean n-space $\mathcal{E}^n = \mathbb{R}^n$ then we have the following version of the Cauchy-Schwarz inequality:

$$|\sum_{i=1}^{n}(x_i y_i)| \leq \sum_{i=1}^{n}|x_i||y_i| \leq (\sum_{i=1}^{n}x_i^2)^{\frac{1}{2}}(\sum_{i=1}^{n}y_i^2)^{\frac{1}{2}}$$

for any two n-vectors $(x_1, \ldots, x_n), (y_1, \ldots, y_n)$ of real numbers.

EXAMPLE 13.5.4. If we consider the inner product space as the vector space of continuous real functions $C^0[a, b]$ then we have the following version of the Cauchy-Schwarz inequality

$$|\int_a^b f(x)g(x)dx| \leq (\int_a^b f(x)^2 dx)^{\frac{1}{2}}(\int_a^b g(x)^2 dx)^{\frac{1}{2}}$$

for any two continuous real functions $f(x), g(x)$ on $[a, b]$.

Recall that in $\mathbb{R}^3$ we have for any two vectors

$$\vec{u} \cdot \vec{v} = |\vec{u}||\vec{v}| \cos \theta.$$

We can use this result as a definition of an angle in a general inner product space.

Definition 13.5.5. *In an inner product space V the* **angle** *between two nonzero vectors $u, v \in V$ is defined by*

$$angle(u, v) = \cos^{-1}\left(\frac{\langle u, v \rangle}{|u||v|}\right).$$

Notice that as a consequence of the Cauchy-Schwarz inequality, the fraction $\frac{\langle u, v \rangle}{|u||v|}$ must be in absolute value less than or equal to 1 and therefore must equal the cosine of some angle. Therefore the definition makes sense if we restrict the angle to the range $0 \le \theta < 2\pi$.

Two vectors in an inner product space would then be **orthogonal** *or* **perpendicular** *if $\langle u, v \rangle = 0$.*

There are actually two angles defined from the cosine. In general we take the angle between the vectors as the smaller of the two.

13.5.1   Banach and Hilbert Spaces

There are certain special types of normed linear spaces that play a major role in functional analysis. We just define them here. Recall that the real numbers $\mathbb{R}$ were **complete** in the sense that every Cauchy sequence converged (see Chapter 5). Cauchy sequences and convergence can be defined in any metric space and hence in any normed linear space.

Definition 13.5.6. *A* **Banach space** *is a complete normed linear space, complete in the sense that every Cauchy sequence converges. A* **Hilbert space** *is a complete inner product space.*

Euclidean n-space is the prime example of a Hilbert space.

13.5.2   The Gram-Schmidt Process and Orthonormal Bases

Recall that the standard basis $\mathbf{i}, \mathbf{j}, \mathbf{k}$ for $\mathbb{R}^3$ are all unit vectors and perpendicular to each other. We show that we can always find such a basis for an inner product space. We then use this to determine the vector in a subspace closest to a given vector.

Definition 13.5.7. *If V is a an inner product space then $u, v \in V$ are* **orthogonal** *if $\langle u, v \rangle = 0$. A set of vectors $u_1, \ldots, u_n$ in V is an* **orthogonal**

set *if* $\langle u_i, u_j \rangle = 0$ *if* $i \neq j$. *They are an* **orthonormal set** *if they are an orthogonal set and each vector has norm one. If they form a basis they are called an* **orthonormal basis**.

If V *is an infinite-dimensional inner product space then a basis* B *is an orthonormal basis of* V *if it is a basis and every finite subset is an orthonormal set.*

Lemma 13.5.3. *An orthogonal set of nonzero vectors must be independent.*

Proof. Suppose that $\{u_1, u_2, \ldots, u_n\}$ is an orthogonal set and suppose that

$$c_1 u_1 + \cdots + c_n u_n = 0.$$

We must show then that each $c_i = 0$. In the above equation take the inner product with u_i. Then

$$\langle u_i, c_1 u_1 + \cdots + c_n u_n \rangle = \langle u_i, 0 \rangle = 0$$

$$\implies c_1 \langle u_i, u_1 \rangle + c_2 \langle u_i, u_2 \rangle + \cdots + c_i \langle u_i, u_i \rangle + \cdots + c_n \langle u_i, u_n \rangle = 0.$$

However $\langle u_i, u_j \rangle = 0$ is $i \neq j$ and hence the above reduces to

$$c_i \langle u_i, u_i \rangle = 0.$$

Since $u_i \neq 0$ then $\langle u_i, u_i \rangle \neq 0$ this would imply that $c_i = 0$ and therefore the set $\{u_1, \ldots, u_n\}$ is linearly independent. $\square$

Given an orthonormal basis it is very simple to determine coordinatess of a vector v relative to this basis.

Lemma 13.5.4. *If* $\{e_1, \ldots, e_n\}$ *form an orthonormal basis for the real inner product space* V *and if* $v \in V$ *then*

$$v = \sum_{i=1}^{n} \langle v, e_i \rangle e_i$$

The $\langle v, e_i \rangle$ *are called the* **Fourier coefficients** *of* v *relative to* $e_1, \ldots, e_n$.

Proof. Suppose that $\{e_1, \ldots, e_n\}$ form an orthonormal basis for V and $v \in V$. Since $\{e_1, \ldots, e_n\}$ form a basis for V then

$$v = c_1 e_1 + \cdots + c_n e_n$$

for scalars $c_1, \ldots, c_n$. We show that $c_i = \langle v, e_i \rangle$. Take the inner product of v with e_i. Then

$$\langle v, e_i \rangle = c_1 \langle e_1, e_i \rangle + \cdots + c_n \langle e_n, e_i \rangle.$$

As in the previous lemma since $\langle e_i, e_j \rangle = 0$ if $i \neq j$ this reduces to

$$\langle v, e_i \rangle = c_i \langle e_i, e_i \rangle.$$

However $\{e_1, \ldots, e_n\}$ form an *orthonormal basis* so $|e_i| = \sqrt{\langle e_i, e_i \rangle} = 1$ and so

$$\langle v, e_i \rangle = c_i$$

as desired. $\qquad\square$

Therefore orthonormal bases are easy to work with. We now show that given a finite-dimensional inner product space we can always obtain an orthonormal basis.

Notice that if $\{x_1, \ldots, x_n\}$ is an orthogonal basis, and for $i = 1, \ldots, n$ we have $|x_i|$ as the norm of x_i, then $\{\frac{x_1}{|x_1|}, \ldots, \frac{x_n}{|x_n|}\}$ is an orthonormal basis. Hence if we can find an orthogonal basis we can find an orthonormal basis.

The **Gram-Schmidt orthogonalization procedure** given in the next theorem provides a method to change an independent set into an orthogonal set that spans the same subspace.

Theorem 13.5.2. *(Gram-Schmidt Orthogonalization Procedure) Let V be a real inner product space. If $\{x_1, \ldots, x_n, \ldots\}$ is an independent set then $\{v_1, \ldots, v_n, \ldots\}$ is an orthogonal set where the v_i are defined inductively by*

$$v_1 = x_1 \text{ and } v_{k+1} = x_{k+1} - \sum_{i=1}^{k} \frac{\langle x_{k+1}, v_i \rangle}{\langle v_i, v_i \rangle} v_i.$$

*The orthogonal set $\{v_1, v_2, \ldots, v_n, \ldots\}$ is called the **Gram-Schmidt orthogonalization** or **GSO** of $\{x_1, x_2, \ldots, x_n, \ldots\}$.*

Proof. Suppose that $\{x_1, \ldots, x_n, \ldots\}$ is an independent set and $\{v_1, \ldots, v_n, \ldots\}$ is defined as above. Consider a fixed v_j; we show by induction that it is orthogonal to any v_k with $k < j$.

We take $v_1 = x_1$. Then $v_2 = x_2 - \frac{\langle x_2, v_1 \rangle}{|v_1|^2} v_1$. But since $|x_1|^2 = \langle x_1, x_2 \rangle$, then,

$$\langle v_2, v_1 \rangle = \langle x_2, x_1 \rangle - (\langle x_2, x_1 \rangle)(\frac{\langle x_1, x_1 \rangle}{\langle x_1, x_1 \rangle}) = 0.$$

This starts the induction.

Now we fix a $j > 2$ and suppose for all $i < j$ that $\langle v_i, v_k \rangle = 0$ for all $k < i$.
We have

$$v_j = x_j - \sum_{i=1}^{j-1} \frac{\langle x_j, v_i \rangle}{\langle v_i, v_i \rangle} v_i.$$

Now let $k < j$, so that we have

$$\langle v_j, v_k \rangle = \langle x_j, v_k \rangle - \sum_{i=1}^{j-1} \frac{\langle x_j, v_i \rangle}{\langle v_i, v_i \rangle} \langle v_i, v_k \rangle.$$

By induction $\langle v_i, v_k \rangle = 0$ if $i < j$.

Thus the only $\langle v_i, v_k \rangle$ which may be nonzero is for $k = i = j - 1$. But then

$$\langle v_j, v_k \rangle = \langle x_j, v_{j-1} \rangle - \frac{\langle x_j, v_{j-1} \rangle}{\langle v_{j-1}, v_{j-1} \rangle} \langle v_{j-1}, v_{j-1} \rangle = 0$$

$\square$

If V is a finite-dimensional IPS then if $v_1, \ldots, v_n$ is a basis then $v_1, \ldots, v_n$ is an independent set that spans all of V. Hence the Gram-Schmidt process determines an orthogonal set that also spans all of V. Since orthogonal sets are independent this means that we have found an orthogonal basis for V and then, by normalizing, an orthonormal basis.

Corollary 13.5.2. *Any finite-dimensional IPS has an orthonormal basis.*

EXAMPLE 13.5.5. Find the GSO for the subspace of $\mathbb{R}^4$ spanned by $u_1 = (1, 0, 0, 2), u_2 = (2, 1, 1, 0), u_3 = (0, 2, 3, 0)$.
First of all $v_1 = u_1 = (1, 0, 0, 2)$. Then

$$v_2 = u_2 - \frac{\langle u_2, v_1 \rangle}{\langle v_1, v_1 \rangle} v_1 = (2, 1, 1, 0) - \frac{2}{5}(1, 0, 0, 2) = (\frac{8}{5}, 1, 1, -\frac{4}{5})$$

$$v_3 = u_3 - \frac{\langle u_3, v_2 \rangle}{\langle v_2, v_2 \rangle} v_2 - \frac{\langle u_3, v_1 \rangle}{\langle v_1, v_1 \rangle} v_1$$

$$\implies v_3 = (-\frac{40}{26}, \frac{27}{26}, \frac{53}{26}, \frac{20}{26}).$$

This can be made into an orthonormal basis for this subspace by dividing each vector by its magnitude.

13.5.3   The Closest Vector Theorem

Given vectors $u_1, \ldots, u_m$ in an IPS V and $v \in V$ then using the Gram-Schmidt process we can determine the vector $v^\star$ in the subspace spanned by $u_1, \ldots, u_m$ that is closest to v. In addition this vector $v^\star$ is unique. The method is given in the next result called the **closest vector theorem**.

Theorem 13.5.3. *(Closest Vector Theorem) Let W be a subspace of an IPS V and let v be a vector in V. If $\{e_1, \ldots, e_n\}$ is an orthonormal basis for W then the unique vector $w \in W$ closest to v is given by*

$$w = \sum_i \langle v, e_i \rangle e_i.$$

Proof. Any vector in the subspace spanned by $e_1, \ldots, e_n$ can be written as a linear combination of $e_1, \ldots, e_n$. Therefore we are trying to find scalars $c_1, c_2, \ldots, c_n$ that minimize

$$\langle w - (c_1 e_1 + \cdots + c_n e_n), w - (c_1 e_1 + \cdots + c_n e_n) \rangle.$$

Expanding this out using the properties of inner product together with the fact that $\{e_1, \ldots, e_m\}$ is orthornomal we get

$$\langle w - (c_1 e_1 + \cdots + c_n e_n), w - (c_1 e_1 + \cdots + c_n e_n) \rangle = |w|^2 - 2 \sum_{i=1}^{n} c_i \langle w, e_i \rangle + \sum_{i=1}^{n} c_i^2.$$

This is then equal to

$$= |w|^2 + \sum_{i=1}^{n} (\langle w, e_i \rangle - c_i)^2 - \sum_{i=1}^{n} (\langle w, e_i \rangle)^2.$$

Since the second term is nonnegative the whole expression will be minimized when this second term is zero or equivalently when $\langle w, e_i \rangle = c_i$ for each i.

Notice that if v is actually in the subspace then this is just its expression in terms of the orthonormal basis.

$\square$

The closet vector theorem actually provides a method given a subspace W and a vector $v \in V$ to find the closest vector within the subspace. This works as follows. First determine an orthonormal basis for the subspace W and then use the formula provided in the closest vector theorem. As mentioned in the proof if $v \in W$ then the closest vector to v is v itself and this will be given by the procedure.

EXAMPLE 13.5.6. Find the vector in the space spanned by $u_1 = (1, 0, 0, 2)$, $u_2 = (2, 1, 1, 0)$, $u_3 = (0, 2, 3, 0)$ in $\mathbb{R}^4$ closest to $v = (0, 1, 0, 1)$.

From the above example we found an orthogonal basis for this space as

$$v_1 = (1, 0, 0, 2), \quad v_2 = (\frac{8}{5}, 1, 1, -\frac{4}{5}), \quad v_3 = (-\frac{40}{26}, \frac{27}{26}, \frac{53}{26}, \frac{20}{26})$$

Making this orthonormal by making each norm one we have an orthonormal basis

$$e_1 = (\frac{1}{\sqrt{5}}, 0, 0, \frac{2}{\sqrt{5}}), \quad e_2 = (\frac{8}{\sqrt{82}}, \frac{1}{\sqrt{82}}, \frac{1}{\sqrt{82}}, -\frac{4}{\sqrt{82}})$$

$$e_3 = (-\frac{40}{\sqrt{5538}}, \frac{27}{\sqrt{5538}}, \frac{53}{\sqrt{5538}}, \frac{20}{\sqrt{5538}}).$$

The closest vector to v in this space is then

$$v^\star = \langle v, e_1 \rangle e_1 + \langle v, e_2 \rangle e_2 + \langle v, e_3 \rangle e_3$$

$$= \frac{1}{2268530}(-445348, 437295, 938315, 2532204).$$

13.5.4 Least-Squares Approximation

In this section we present a very nice application of both inner product spaces and the closest vector theorem. This application completely solves the **least-squares problem** for continuous functions over an interval.

Given an interval $[a, b] \subset \mathbb{R}$ it is known that any continuous function $f(x)$ on this interval can be uniformly approximated by a polynomial (see [A]). The question arises as to what is the "best" polynomial approximation for a given degree and for a given continuous function $f(x)$. We put the word *best* in quotes to indicate that there are several different criteria for defining what we mean by the best approximation. We refer the reader to [A] for a discussion of these but here only look at one such criterion called **least squares** that we can solve using the closest vector theorem. The **least-squares problem** is then the following.

Definition 13.5.8. *(The Least-Squares Problem): Given* $f(x) \in C^0[a, b]$ *find a polynomial of degree n,* $P_n(x)$ *which approximates* $f(x)$ *in the sense that it minimizes*

$$\int_a^b |f(x) - P_n(x)|^2 dx$$

over all polynomials of deg $\leq n$.

The polynomial that does this is called the **least squares approximation of degree** $\leq n$ *to* $f(x)$.

Let $\mathbb{R}_n[x]$ denote the set of real polynomials of degree $\leq n$. As mentioned earlier the set $C^0[a, b]$ consisting of all continuous functions on $[a, b]$ is an inner product space and the set $\mathbb{R}_n[x]$ is a subspace. In $C^0[a, b]$ the distance between two functions $f(x), g(x)$ is given by

$$d(f(x), g(x)) = \left(\int_a^b (f(x) - g(x))^2 dx \right)^{\frac{1}{2}}.$$

Therefore the least-squares problem consists then in minimizing the distance from $f(x)$ to the subspace $\mathbb{R}_n[x]$. To solve this we can apply the closest vector theorem. Notice first that a basis for $\mathbb{R}_n[x]$ is $1, x, x^2, \ldots, x^n$.

Given a continous function $f(x)$ on $[a, b]$ the solution to the least squares problem can be summarized as follows:

        Step (1): Find the GSO of $1, x, x^2, \ldots, x^n$ on $[a, b]$. These give a set $P_0(x), \ldots, P_n(x)$ of orthogonal polynomials.

        Step (2): Normalize the orthogonal polynomials to obtain a set of orthonormal polynomials $\phi_0(x), \ldots, \phi_n(x)$ which satisfy

$$\int_a^b \phi_i(x) \phi_j(x) dx = \delta_{ij}.$$

Step (3): The least-squares approximation $r(x)$ of degree n is given by

$$r(x) = c_0\phi_0(x) + \ldots\ldots + c_n\phi_n(x)$$

where

$$c_i = \int_a^b f(x)\phi_i(x)dx \text{ for } i = 0, 1, 2\ldots, n.$$

EXAMPLE 13.5.7. Find the least squares approximation of degree 2 to $f(x) = e^x$ on $[0, 1]$.

Step (1): We find the GSO of $1, x, x^2$ on $[0, 1]$. So

$$u_1 = 1, u_2 = x, u_3 = x^2$$

$$\implies v_1 = 1, v_2 = u_2 - \frac{\langle u_2, v_1 \rangle}{\langle v_1, v_1 \rangle}v_1$$

Now

$$\langle u_2, v_1 \rangle = \int_0^1 x dx = \frac{1}{2}, \langle v_1, v_1 \rangle = \int_0^1 1 dx = 1$$

$$\implies v_2 = x - \frac{1}{2}$$

Then

$$v_3 = u_3 - \frac{\langle u_3, v_2 \rangle}{\langle v_2, v_2 \rangle}v_2 - \frac{\langle u_3, v_1 \rangle}{\langle v_1, v_1 \rangle}v_1$$

$$\langle u_3, v_2 \rangle = \int_0^1 (x^3 - \frac{x^2}{2})dx = \frac{1}{12}$$

$$\langle u_3, v_1 \rangle = \int_0^1 x^2 dx = \frac{1}{3}$$

$$\langle v_2, v_2 \rangle = \int_0^1 (x - \frac{1}{2})^2 dx = \frac{1}{12}$$

$$\implies v_3 = x^2 - x + \frac{1}{6}$$

Therefore v_1, v_2, v_3 form an orthogonal set

Step (2): We normalize the GSO

$$|v_1| = \sqrt{\langle v_1, v_1 \rangle} = 1 \implies \phi_1 = \frac{v_1}{|v_1|} = 1$$

$$|v_2| = \sqrt{\langle v_2, v_2 \rangle} = \frac{1}{\sqrt{12}} \implies \phi_2 = \frac{v_2}{|v_2|} = \sqrt{12}(x - \frac{1}{2})$$

$$|v_3| = \sqrt{\langle v_3, v_3 \rangle} = \frac{1}{\sqrt{180}} \implies \phi_3 = \frac{v_3}{|v_3|} = \sqrt{180}(x^2 - x + \frac{1}{6})$$

Step (3): Find the Fourier coefficients

$$\langle e^x, \phi_1(x) \rangle = \int_0^1 e^x dx = 1.718$$

$$\langle e^x, \phi_2(x) \rangle = \int_0^1 e^x(\sqrt{12}(x - \frac{1}{2}))dx = .4880$$

$$\langle e^x, \phi_3(x) \rangle = \int_0^1 e^x(\sqrt{180}(x^2 - x + \frac{1}{6}))dx = .0625$$

Step (4): Combine

$$P_2(x) = \langle e^x, \phi_1(x) \rangle \phi_1(x) + \langle e^x, \phi_2(x) \rangle \phi_2(x) + \langle e^x, \phi_3(x) \rangle \phi_3(x)$$

$$P_2(x) = 1.718 + .4880\phi_2(x) + .0625\phi_3(x)$$

We note that the GSO of $1, x, x^2, \ldots$ on an interval $[a, b]$ are called the **Legendre polynomials**.

13.6   Linear Transformations and Matrices

Recall that mappings between algebraic structures that preserve the operations are called homomorphisms. Homomorphisms between vector spaces are more commonly called **linear transformations**.

Definition 13.6.1. *If V, W are vector spaces over a field F then a map $T : V \to W$ is a **linear transformation** if*
    (1) $T(u + v) = Tu + Tv$ if $u, v \in V$,
    (2) $T(fu) = fTu$ if $v \in V, f \in F$.

As with groups and rings the kernel and the image are important in studying mappings. For linear transformations we have:

Definition 13.6.2. *If* $T : U \to V$ *is a linear transformation then its* **kernel** *or* **null space** *is*

$$ker(T) = \{v \in V : Tv = 0\}.$$

Its **image** *is*

$$im(T) = \{w \in W : Tv = w \text{ for some } v \in V\}.$$

The kernel is always a substructure of the domain and the image is a substructure of the range so again we obtain, as with groups and rings.

Lemma 13.6.1. $ker(T)$ *is a subspace of* V *and* $im(T)$ *is a subspace of* W.

Since a vector space is an abelian group any subgroup is normal. Therefore the kernel of a linear transformation is actually a normal subgroup and we can build the quotient group $V/ker(T)$. The elements of the quotient group are cosets $v + ker(T)$ where $v \in V$. If we define for any scalar f the product $f(v + ker(T)) = fv + ker(T)$ then the quotient group also becomes a vector space over the ground field F called the **quotient space**. The isomorphism theorem for vector spaces is identical to the isomorphism theorems for groups and rings.

Theorem 13.6.1. *(The Isomorphism Theorem for Vector Spaces) If*

$$T : V \to W$$

is a linear transformation then

$$V/ker(T) \cong im(T).$$

See the exercises for a proof.

We now relate the dimension of the kernel and the image to the dimensions of the various spaces.

Definition 13.6.3. *The* **nullity** *of* T *is the dimension of* $ker(T)$. *The* **rank** *of* T *is the dimension of* $im(T)$

The next theorem follows from the isomorphism theorem.

Theorem 13.6.2. *Suppose that* V *is a finite-dimensional vector space. If* $T : V \to W$ *is a linear transformation then*

$$dim(V) = rank(T) + nullity(T) = dim(ker(T)) + dim(im(T)).$$

Proof. Suppose that V is a finite-dimensional vector space and $T : V \to W$ is a linear transformation. Let $n < \infty$ be the dimension of V and let $S = \{v_1, \ldots, v_m\}$ be a basis for the $N = \ker(T)$. Since S is an independent set in V we have $m \leq n$ and hence S can be extended to a basis for V by Theorem 13.2.5. Let $S_1 = \{v_1, \ldots, v_m, v_{m+1}, \ldots, v_n\}$ be a basis for V. We will show that $Tv_{m+1}, \ldots, Tv_n$ is a basis for $\text{im}(T)$. This is then sufficient to prove the theorem since then $\dim(V) = n$, the nullity of T is m and the rank of T is the dimension of the image is $n - m$.

We must show that $\{Tv_{m+1}, \ldots, Tv_n\}$ is an independent set that spans $\text{im}(T)$. Let $w \in \text{im}(T)$. Then there exists a $v \in V$ with $Tv = w$. Since $\{v_1, \ldots, v_m, v_{m+1}, \ldots, v_n\}$ form a basis for V we have $v = c_1 v_1 + \cdots + c_m v_m + c_{m+1} v_{m+1} + \cdots + c_n v_n$. Then

$$Tv = T(c_1 v_1 + \cdots + c_m v_m + c_{m+1} v_{m+1} + \cdots + c_n v_n)$$

$$Tv = c_1 Tv_1 + \cdots + c_m Tv_m + c_{m+1} Tv_{m+1} + \cdots + c_n Tv_n.$$

But $Tv_1 = \cdots = Tv_m = 0$ since $v_1, \ldots, v_m \in \ker(T)$ and hence we have

$$Tv = c_{m+1} Tv_{m+1} + \cdots + c_n Tv_n.$$

Therefore $Tv_{m+1}, \ldots, Tv_n$ span $\text{im}(T)$. We must show that they are independent.

Suppose that we have

$$c_{m+1} Tv_{m+1} + \cdots + c_n Tv_n = 0 = T(c_{m+1} v_{m+1} + \cdots + c_n v_n) = 0$$

for scalars $c_{m+1}, \ldots, c_n$. Then the vector $u = c_{m+1} v_{m+1} + \cdots + c_n v_n \in \ker(T)$. Hence there exists scalars $b_1, \ldots, b_m$ such that $u = b_1 v_1 + \cdots + b_m v_m$. Combining these two expressions we have

$$b_1 v_1 + \cdots + b_m v_m = c_{m+1} v_{m+1} + \cdots + c_n v_n$$

$$\implies b_1 v_1 + \cdots + b_m v_m - c_{m+1} v_{m+1} - \cdots - c_n v_n = 0.$$

However $v_1, \ldots, v_n$ is an independent set as a basis for V and therefore $b_1 = \ldots = b_m = c_{m+1} = \ldots = c_n = 0$. This proves that $\{Tv_{m+1}, \ldots, Tv_n\}$ is an independent set completing the proof. $\square$

13.6.1 Matrix of a Linear Transformation

Any linear transformation is uniquely determined by its action on a basis. From this it will follow that any linear transformation between finite-dimensional vector spaces over the same field F is given by a matrix with entries from F.

Theorem 13.6.3. *Suppose $T : V \to W$ is a linear transformation and $v_1, \ldots, v_n$ is a basis for V. Then T is uniquely determined by its action on $v_1, .., v_n$. Suppose $v \in V$, then*

$$v = \sum_{i=1}^{n} c_i v_i \implies Tv = \sum_{i=1}^{n} c_i T v_i.$$

Corollary 13.6.1. *If $v_1, \ldots, v_n$ is a basis for V and $w_1, \ldots, w_n$ are any vectors in W then there exists a unique linear transformation T such that $T v_i = w_i$.*

We now show the relationship between linear transformations and matrices. First we have that multiplication by a matrix provides a linear transformation. The following holds from properties of matrix multiplication.

Lemma 13.6.2. *Let $A = (a_{ij})$ be an $m \times n$ matrix over a field F. If $u \in F^n$ then u is an $n \times 1$ matrix so Au is an $m \times 1$ matrix. Hence $Au \in F^m$. The map $u \to Au$ is a linear transformation from F^n to F^m.*

We now show that the converse is also true. That is, any linear transformation between finite-dimensional vector spaces can be expressed as matrix multiplication.

Suppose $T : V \to W$ is a linear transformation and $v_1, \ldots, v_n$ is a basis for V and $w_1, \ldots, w_m$ is a basis for W. Then

$$Tv_1 = a_{11}w_1 + a_{21}w_2 + \cdots + a_{m1}w_m$$
$$Tv_2 = a_{12}w_1 + a_{22}w_2 + \cdots + a_{m2}w_m$$
$$\cdots$$
$$Tv_n = a_{1n}w_1 + a_{2n}w_2 + \cdots + a_{mn}w_m$$

Let A be the transpose of the above matrix of scalars. In other words A is the matrix whose columns are the coordinates $Tv_i, 1 \le i \le n$ relative to the basis $\{w_1, \ldots, w_m\}$.

If $v \in V$ then $v = \sum_{i=1}^{n} c_i v_i$ so that the coordinates of v relative to $v_1, .., v_n$ are $(c_1, \ldots, c_n)$. Then the coordinates of Tv relative to $w_1, .., w_m$ are

$$Av = A \begin{pmatrix} c_1 \\ \vdots \\ c_n \end{pmatrix}.$$

To see this, note that

$$Tv = T(\sum_{i=1}^{n} c_i v_i) = \sum_{i=1}^{n} c_i (Tv_i)$$

$$= \sum_{i=1}^{n} c_i (\sum_{j=1}^{m} a_{ji} w_j) = \sum_{j=1}^{m} (\sum_{i=1}^{n} a_{ji} c_i) w_j.$$

If A is the transpose as above so that A is $m \times n$ and $c = \begin{pmatrix} c_1 \\ \vdots \\ c_n \end{pmatrix}$ is the $n \times 1$ vector of coordinates, then the matrix Ac is defined and is $m \times 1$. The jth row of Ac is

$$a_{j1} c_1 + a_{j2} c_2 + \cdots + a_{jn} c_n.$$

Therefore the action of T is the same as multiplication by the matrix A. The matrix A is called the **matrix of** T relative to the bases pair $\{v_1, \ldots, v_n\}, \{w_1, .., w_m\}$.

EXAMPLE 13.6.1. Suppose $T : \mathbb{R}^3 \to \mathbb{R}^4$ by

$$T(1,0,0) = (1,3,4,0), T(0,1,0) = (1,0,1,0), T(0,0,1) = (1,2,-1,1)$$

(1) What is the matrix of T relative to the standard bases?
(2) What is $T(1,2,3)$?
Now the standard bases are

$$e_1 = (1,0,0), e_2 = (0,1,0), e_3 = (0,0,1)$$

$$E_1 = (1,0,0,0), E_2 = (0,1,0,0), E_3 = (0,0,1,0), E_4 = (0,0,0,1)$$

Then

$$Te_1 = E_1 + 3E_2 + 4E_3$$
$$Te_2 = E_1 + E_3$$
$$Te_3 = E_1 + 2E_2 - E_3 + E_4$$

Therefore the matrix of T is

$$A = \begin{pmatrix} 1 & 1 & 1 \\ 3 & 0 & 2 \\ 4 & 1 & -1 \\ 0 & 0 & 1 \end{pmatrix}.$$

To find $T(1,2,3)$ multiply by the matrix A:

$$\begin{pmatrix} 1 & 1 & 1 \\ 3 & 0 & 2 \\ 4 & 1 & -1 \\ 0 & 0 & 1 \end{pmatrix} \begin{pmatrix} 1 \\ 2 \\ 3 \end{pmatrix} = \begin{pmatrix} 6 \\ 9 \\ 3 \\ 3 \end{pmatrix}.$$

Therefore

$$T(1,2,3) = (6,9,3,3).$$

Lemma 13.6.3. *If $T : V \to W$ then $rank(T) = dim(im(T))$ is equal to the rank of any matrix for T.*

EXAMPLE 13.6.2. Suppose T is as in the previous example. Then

$$A = \begin{pmatrix} 1 & 1 & 1 \\ 3 & 0 & 2 \\ 4 & 1 & -1 \\ 0 & 0 & 1 \end{pmatrix} \to \begin{pmatrix} 1 & 1 & 1 \\ 0 & 3 & 1 \\ 0 & 3 & 5 \\ 0 & 0 & 1 \end{pmatrix} \to \begin{pmatrix} 1 & 1 & 1 \\ 0 & 3 & 1 \\ 0 & 0 & 4 \\ 0 & 0 & 0 \end{pmatrix}.$$

Therefore $rank(A)$ is 3 and therefore the rank of T is 3 and hence the nullity of T is $\{0\}$. It follows that T is one-to-one.

Lemma 13.6.4. *If $T : V \to W$ and all dimensions are finite then T is one-to-one if and only if $ker(T) = \{0\}$ which is the case if and only the nullity of T is 0. T is onto if and only if $rank(T) = dim(W)$. T is invertible if and only if any matrix for T is square and invertible.*

Lemma 13.6.5. *Suppose $T_1 : V \to W, T_2 : W \to U$. Suppose $v_1, \ldots, v_n$ is a basis for V, $w_1, .., w_m$ is a basis for W and A_1 is the matrix for T_1 relative to this basis pair and $u_1, .., u_k$ is a basis for U and A_2 is the matrix for T_2 relative to the second basis pair then $A_2 A_1$ is the matrix of $T_2 T_1 : V \to U$.*

13.6.2   Linear Operators and Linear Functionals

We consider here linear transformations that either go from a vector space to itself or from the vector space to the ground field.

Definition 13.6.4. *A linear transformation $T : V \to V$, that is, a linear transformation of the vector space to itself, is called a* **linear operator** *on V.*

Theorem 13.6.4. *Let $T : V \to V$, A its matrix relative to $v_1, \ldots, v_n$ and A' its matrix relative to $v'_1, \ldots, v'_n$. Then $A' = C^{-1}AC$ where C is the transition matrix.*

If A, B are two $n \times n$ matrices then we say that they are **similar** if there exists an invertible matrix C with $A = C^{-1}BC$.

Corollary 13.6.2. *Two matrices A, B represent the same linear operator T if and only if they are similar.*

Definition 13.6.5. *If T is a linear operator on V then $det(T)$ is the determinant of any matrix for T. Since similar matrices have the same determinant this is independent of the matrix.*

Lemma 13.6.6. *A linear operator is invertible if and only if its determinant is nonzero.*

Finally we consider linear transformations from a vector space to the ground field.

Definition 13.6.6. *A linear transformation $T : V \to F$ where F is the ground field is called a* **linear functional** *on V.*

A very prominent example of a linear functional is integration on the space of continuous functions. That this is a linear transformation follows immediately from the linearity of integration.

   EXAMPLE 13.6.3. The definite integral is a linear functional on the space of continuous functions.

13.7 Exercises

EXERCISES FOR SECTION 13.1

13.1.1. If $\vec{AB}, \vec{BC}$ are two directed line segments in $\mathbb{R}^3$ then $\vec{AB} \equiv \vec{BC}$ if and only if $ABDC$ is a parallelogram (where a parallelogram can shrink to a directed line segment). Show that this is an equivalence relation on the set of directed line segments in $\mathbb{R}^3$.

13.1.2. Put each of the following matrices over $\mathbb{Q}$ in reduced row echelon form and then tell its rank

(a) $\begin{pmatrix} 1 & 1 \\ -2 & -2 \end{pmatrix}$ (b) $\begin{pmatrix} 1 & 2 & 3 & 4 \\ 3 & 1 & 2 & 1 \\ 0 & 0 & 1 & -1 \end{pmatrix}$ (c) $\begin{pmatrix} 1 & 1 & -1 & 2 \\ 0 & 1 & -1 & 2 \\ 0 & 1 & 1 & -1 \\ 1 & 2 & 1 & -1 \end{pmatrix}$

13.1.3. For each of the matrices in Problem 13.1.2 find its inverse if possible. If it is not possible explain why not.

13.1.4. Consider each system of linear equations over $\mathbb{Q}$. Use Gaussian elimination to find all solutions if there are any.

(a) $\begin{aligned} x - y + 2z &= 5 \\ 2x + y - z &= 2 \\ 2x - y - z &= 4 \\ x + 3y + 2z &= 1 \end{aligned}$ (b) $\begin{aligned} 2x - 4y + 3z - 5w &= -3 \\ 4x + 2y + z &= 1 \\ 6x - 2y + 4z - 5w &= 1 \end{aligned}$

(c) $\begin{aligned} 3x - y + z - w &= 1 \\ 2x + 3y - z + 5w &= 7 \\ 2x + 5y - z + 12w &= 8 \\ x - 2y + 2z + w &= 5 \end{aligned}$

13.1.5. Use Cramer's rule to solve each system over $\mathbb{Q}$

(a) $\begin{aligned} 2x - y + z &= 0 \\ x + 2y - z &= 10 \\ 3x - 3y - 5z &= 2 \end{aligned}$

(b) $\begin{aligned} 3x + 6y &= 15 \\ x + 4y &= 1 \end{aligned}$

13.1.6. Use Cramer's rule to solve each system over $\mathbb{Z}_3$

$$
\begin{array}{cc}
(a) & \begin{aligned}
x + y + 2z &= 0 \\
x + y + z &= 2 \\
2x + 2y + z &= 1
\end{aligned}
\qquad
(b) & \begin{aligned}
x + y + z + w &= 2 \\
2x + y + 3z &= 2 \\
x + z + w &= 1 \\
x + 2y + z - w &= 0
\end{aligned}
\end{array}
$$

13.1.7. For each of the following matrices find the determinant and trace and then determine the inverse if it is invertible. To determine at least one inverse use the method of matrix of cofactors.

$$
(a)\ \ A = \begin{pmatrix} 3 & 4 & -1 \\ 1 & 0 & 3 \\ 2 & 5 & -4 \end{pmatrix}
\quad
(b)\ \ B = \begin{pmatrix} 3 & 1 & 5 \\ 2 & 4 & 1 \\ -4 & 2 & -9 \end{pmatrix}
\quad
(c)\ \ C = \begin{pmatrix} 1 & 0 & 1 \\ 0 & 1 & 1 \\ 1 & 1 & 0 \end{pmatrix}
$$

13.1.8. Let $AX = B$ be an $n \times n$ system of equations over the integers. Show that it has integer solutions if $det(A) = \pm 1$.

13.1.9. Recall that two matrices A, B are similar if $T^{-1}AT = B$ for some invertible matrix T.

(a) Show that similar matrices have the same determinant.

(b) Show that $\mathrm{tr}(AB) = \mathrm{tr}(BA)$.

(c) Use part (b) to show that similar matrices have the same traces.

13.1.10. Do the necessary sequence of elementary row operations for Examples 13.1.9, 13.1.10, 13.1.11 and 13.1.12.

EXERCISES FOR SECTION 13.2–13.4

13.2.1. Using Definition 13.2.1 prove Lemma 13.2.1. Here you have to go back to Lemmas 13.1.1 and 13.1.2 and verify that all the properties there hold in $\mathbb{R}^n$.

13.2.2. Prove Lemma 13.2.2. Here you have to go back to Lemma 13.1.3 and verify that all the properties there hold in $\mathbb{R}^n$.

In exercises 13.2.3–13.2.10 prove that the structure together with the given operations is a vector space over the given ground field.

13.2.3. Example 13.2.2 — F is an extension field of F' over F'.

13.2.4. Example 13.2.3 — $\mathbb{R}^n, \mathbb{C}^n, \mathbb{Q}^n$ for any n over $\mathbb{R}, \mathbb{C}$ and $\mathbb{Q}$ respectively.

13.2.5. Example 13.2.4 — $C^0[a, b]$ the set of continuous functions on $[a, b]$ over $\mathbb{R}$.

13.2.6. Example 13.2.5 — $C^k[a, b]$ the set of k-times continuously differentiable functions on $[a, b]$ over $\mathbb{R}$.

13.2.7. Example 13.2.6 — $F[x]$ the set of polynomials over F.

13.2.8. Example 13.2.7 — $F_n[x]$ the set of polynomials over F of degree $\leq n$.

13.2.9. Example 13.2.8 — $M_{mn}(F)$ the set of $m \times n$ matrices over F.

13.2.10. Example 13.2.9 — The set of solutions to a homogeneous system of equations $AX = 0$ over F.

13.2.11. Prove that $W \subset V$ is a subspace if $W \neq \emptyset$ and W is closed under addition and scalar multiplication.

13.2.12. Prove that the intersection of an arbitrary collection of subspaces is a subspace.

13.2.13. Prove that if $n < m$ then F^n can be considered as a subspace of F^m.

13.2.14. Prove that the space $C^k[a, b]$ of k-times continuously differentiable functions on $[a, b]$ is a subspace of $C^0[a, b]$ the set of continuous functions on $[a, b]$

13.2.15. Prove that $F_n[x]$ the set of polynomials over F of degree $\leq n$ is a subspace of $F[x]$ the set of polynomials over F.

13.2.16. If A is an $m \times n$ matrix then prove the vector space in Problem 13.2.10 is a subspace of F^n.

13.2.17. (a) Prove Theorem 13.2.1 that the span forms a subspace, $[u_1, \ldots, u_n]$ called the **subspace spanned** by $u_1, \ldots, u_n$. It is the smallest subspace containing $u_1, \ldots, u_n$. That is, in the following sense any subspace $W \supset \{u_1, \ldots, u_n\}$ must be such that $[u_1, \ldots, u_n] \subset W$.

(b) Show that $[u_1, \ldots, u_n]$ is the intersection of all subspaces containing $\{u_1, \ldots, u_n\}$.

13.2.18. Prove that the following are bases:

(a) i, j, k form a basis for $\mathbb{R}^3$.

(b) $e_1 = (1, 0, \ldots, 0), e_2 = (0, 1, 0, \ldots, 0), \ldots, e_n = (0, 0, \ldots, 1)$ forms a basis for F^n called the **standard basis**.

(c) Consider the complex numbers $\mathbb{C}$ as a vector space over the real numbers $\mathbb{R}$. Then the elements $1, i$ constitute a basis.

(d) Let $F_n[x]$ be the set of polynomials of degree $\leq n$ over F. Then the elements $1, x, x^2, \ldots, x^n$ constitute a basis.

13.2.19. Prove Lemma 13.2.5.

13.2.20. Prove Lemma 13.2.6.

13.2.21. Prove that if U, W are subspaces of a vector space V and $U \cup V$ is also a subspace then $U \subset W$ or $W \subset U$.

13.2.22. Prove Lemma 13.2.7.

13.2.23. Prove Lemma 13.2.8.

13.2.24. Prove Lemma 13.2.9.

13.2.25. In Example 13.2.13 prove that each set of vectors in the given vector space in (1)–(4) is a basis for that vector space.

13.2.26. Prove Theorem 13.2.2.

13.2.27. Prove Lemma 13.2.10 parts (c)–(g).

13.2.28. Prove that the map $T : V \to F^n$ given in the proof of Theorem 13.2.6 is an isomorphism.

13.2.29. Show that the three vectors $u_1 = (1,3,3), u_2 = (2,0,1).u_3 = (0,0,1)$ are not linearly independent over $\mathbb{Z}_3$ and therefore do not form a basis for $\mathbb{Z}_3^3$.

13.2.30. Show that a field F is an algebra over any subfield F' with the obvious binary operations.

13.2.31. Show that the set $M_n(F)$ of $n \times n$ matrices over a field F is an algebra over F again with the obvious binary operations.

13.2.32. Prove Theorem 13.3.2.

(HINT: By Theorem 13.3.1, $U \cap W$ has a finite basis. Using Theorem 13.2.5 extend this to a basis for U and then also to a basis for W. Then consider the set of vectors consisting of the basis for $U \cap W$ union the basis for U minus the basis of $U \cap W$ and the basis for W minus the basis for $U \cap W$. Show that this is a basis for $U + W$.)

EXERCISES FOR SECTION 13.5

13.5.1. Explain why putting the complex conjugate in (3) of Definition 13.5.1 clears up the contradiction in the note from the text.

13.5.2. Prove that Euclidean n-space, $\mathbb{R}^n$, is an IPS with the inner product defined as in Example 13.5.1.

13.5.3. Prove that if V is an IPS with $|v| = \sqrt{\langle v, v \rangle}$ then (1),(2),(3) of Definition 13.5.3 hold.

13.5.4. Prove that any inner product space V is a normed linear space with $|v| = \sqrt{\langle v, v \rangle}$.

13.5.5. In any normed linear space V we define **distance** by $d(u, v) = |u - v|$ for any $u, v \in V$. Prove that the distance $d : V \times V \to \mathbb{R}$ satisfies
   (a) $d(u, v) \geq 0$
   (b) $d(u, v) = 0$ if and only if $u = v$
   (c) $d(u, v) = d(v, u)$
   (d) $d(u, v) \leq d(u, w) + d(w, v)$.

13.5.6. Consider the IPS with norm $|v| = \sqrt{\langle v, v \rangle}$. In the proof of Theorem 13.5.2 the vector $w = u - \frac{\langle u, v \rangle}{|v|^2} v$ was introduced if $v \neq 0$. Prove that $\langle w, v \rangle = 0$.

13.5.7. Consider the IPS, $C^0[0, 1]$. For $n \in N$ let $f_n(x) = \sqrt{2} \cos(2n\pi x)$ and $g_n(x) = \sqrt{2} \sin(2n\pi x)$. These are in $C^0[0, 1]$. Show that each of the sets $\{f_n(x) : n \in \mathbb{N}\}$ and $\{g_n(x) : m \in \mathbb{N}\}$ are orthonormal sets.
   Recall a set $W = \{v_1, v_2, \ldots, \}$ is orthonormal if each vector has norm 1 and any two different vectors are orthogonal.

13.5.8. (a) Show that orthogonal vectors must be independent.
   (b) Prove that every finite-dimensional inner product space has an orthonormal basis.
   (c) Determine an orthonormal basis for the subspace V of $\mathbb{R}^3$ spanned by $u_1 = (1, 2, 3), u_2 = (2, 3, 1), u_2 = (1, 3, 2)$.
   (d) Use the results from (c) to find the vector in V closest to $(1, 5, 1)$.

13.5.9. Find the vector in the space spanned by $u_1 = (1, 1, -1, 1), u_2 = (3, 2, -1, 0)$ in $\mathbb{R}^4$ closest to $v = (0, 7, 4, 7)$.

13.5.10. (a) Find the Gram-Schmidt orthogonalization for $1, x, x^2, x^3$ in $C^0[0, 1]$.
   (b) Use the results in (a) to find the least squares approximation of degree 3 to $f(x) = \sin x$ on $[0, 1]$.

13.5.11. Find the least squares approximation of degree 2 to $f(x) = x^{1/3}$ on $[-1, 1]$. Here the vectors space is $C^0[-1, 1]$. Also use the basis $\{1, x, x^2\}$.

13.5.12. Let V be an inner product space and let W be a subspace of V. Define
$$W^\perp = \{v \in V : \langle v, w \rangle = 0 \text{ for all } w \in W\}$$

Show that $W^\perp$ is a subspace of V and $V = W \oplus W^\perp$, that is, V is the direct sum of W and $W^\perp$

EXERCISES FOR SECTION 13.6

13.6.1. Prove the isomorphism theorem for vector spaces (Theorem 13.6.1). (HINT: Use the proof for groups modified appropriately.)

13.6.2. (a) Show that the set of vectors in $\mathbb{R}^4$ of the form $\{(x, 2x, y, x+y)\}$ forms a subspace and determine a basis for it.

(b) Which of the following are independent in $\mathbb{R}^4$? Why?

$$\{v_1 = (6, 3, 7, 0), v_2 = (4, 6, 2, 1)\}$$

$$\{u_1 = (3, 7, 2, 1), u_2 = (4, 0, 1, 6), u_3 = (7, 7, 3, 7)\}$$

(c) (i) Show that $v_1 = (6, 7, 1), v_2 = (2, 0, 3), v_3 = (4, 6, 1)$ form a basis for $\mathbb{R}^3$.

(ii) Find the coordinates of $(6, 2, 6)$ relative to the above basis.

(iii) Find the transition matrix between the above basis and the standard basis.

(d) Find the dimension of the subspace of $\mathbb{R}^4$ spanned by $v_1 = (3, 6, 3, 0), v_2 = (4, 2, 1, 1), v_3 = (2, 0, 2, -2)$ and give a basis for it. Then give a general form for a vector in this subspace.

13.6.3. Prove Lemma 13.6.2.

13.6.4. Prove that similar matrices have the same determinant. (HINT: Here you may use facts about determinants stated in the text.)

13.6.5. Let $T : \mathbb{R}^3 \to \mathbb{R}^2$ be defined by

$$T(x_1, x_2, x_3) = T(\begin{pmatrix} x_1 \\ x_2 \\ x_3 \end{pmatrix}) = \begin{pmatrix} 1 & 1 & 1 \\ 1 & 2 & 3 \end{pmatrix} \begin{pmatrix} x_1 \\ x_2 \\ x_3 \end{pmatrix}.$$

Find the matrix of T with respect to the standard bases of $\mathbb{R}^3$ and $\mathbb{R}^2$.

13.6.6. Let $T : F_2[x] \to F_1[x]$ be defined by $T(p(x)) = p'(x)$, that is, the formal derivative of $p(x)$ and consider the bases $\{1, x, x^2\}$ and $\{1, x\}$ for $F_2[x]$ and $F_1[x]$ respectively. Find the matrix of T with respect to these bases.

13.6.7. Let $T : \mathbb{R}^3 \to \mathbb{R}^3$ be the linear operator defined by

$$T(1,0,0) = (1,1,0), T(0,1,0) = (2,0,1), T(0,0,1) = (1,0,1).$$

(a) Find the matrix of T with respect to the standard bases of $\mathbb{R}^3$.
(b) Find $T(1,2,3)$ using the definition of T.
(c) Find $T(1,2,3)$ using the matrix representation of T found in (a).

13.6.8. Let $T : \mathbb{R}^4 \to \mathbb{R}^3$ by

$$T(x, y, z, w) = (x + 2y + z, 3x + y + 2w, 4x + 3y + z + 2w).$$

(a) Find the matrix of T relative to the standard bases for $\mathbb{R}^4$ and $\mathbb{R}^3$.
(b) What is the rank of T and the nullity of T?
(c) Give a basis for the kernel of T and for the image of T.

13.6.9. Let $G = A \times A$ be the direct product of A with itself with $|A| = p$ (hence A is cyclic). Find how many automorphisms of G there are.

(HINT: For $(a_1, a_2) \in A \times A$ with each $a_i \in A$ use a counting argument to determine how many possibilities there are if the map has to be an automorphism. Also see Example 10.3.1.)

Chapter 14

Fields and Field Extensions

14.1 Abstract Algebra and Galois Theory

In the final two chapters we examine one of the major applications of modern abstract algebra, the Galois theory of equations. Many of the fundamental concepts in abstract algebra, such as finite group theory and field extensions were motivated by the development of this theory.

The theory has its origins in the search for solutions of polynomial equations and in particular extensions of the quadratic formula to higher-degree polynomials. The ability to solve quadratic equations and in essence the quadratic formula were known to the Babylonians some 3600 years ago. With the discovery of imaginary and complex numbers, the quadratic formula then says that any degree 2 polynomial over $\mathbb{C}$ has a zero in $\mathbb{C}$ and further this formula is given in terms of algebraic operations and radicals in the coefficients. In the sixteenth century the Italian mathematician Niccolo Tartaglia discovered a similar formula in terms of radicals to solve cubic equations. This **cubic formula** is now known erroneously as **Cardano's formula** in honor of Cardano, who first published it in 1545. An earlier special version of this formula was discovered by Scipione del Ferro. Cardano's student Ferrari extended the formula to solutions by radicals for fourth-degree polynomials. The combination of these formulas says that complex polynomials of degree 4 or less must have complex zeros given in terms of algebraic operations and radicals involving the coefficients.

From Cardano's work until the very early nineteenth century, attempts were made to find similar formulas for degree 5 polynomials. In 1805 Ruffini proved that fifth-degree polynomial equations are insolvable by radicals in general. Therefore there exists no comparable formula for degree 5. Abel in 1825–1826 and Galois in 1831 extended Ruffini's result and proved the insolubility by radicals for all degrees five or greater. In doing this, Galois developed a general theory of field extensions and its relationship to group theory. This has come to be known as **Galois theory**. In Chapter 15 we will give a survey of Galois theory and show how it leads to a proof of the insolvability of the quintic. We will also provide a proof of the fundamental theorem of algebra (see Chapter 12) as well as indicating how to prove that certain geometric constructions, such as trisecting an angle with just a ruler and straightedge, are impossible.

Galois theory depends upon the theory of algebraic field extensions which we will examine in the present chapter.

14.2 Field Extensions

If F and F' are fields with F a subfield of F', then F' is an **extension field**, or **field extension**, or simply an **extension**, of F. As we mentioned in Chapter 13 a field extension F' is then a vector space over the subfield F, where addition is the usual addition in F' and scalar multiplication is the usual field multiplication in F'. The **degree of the extension** is the dimension of F' as a vector space over F. We denote the degree by $|F' : F|$. If the degree is finite, that is, $|F' : F| < \infty$, so that F' is a finite-dimensional vector space over F, then F' is called a **finite extension** of F.

From vector space theory we easily obtain that the degrees are multiplicative. Specifically:

Lemma 14.2.1. *If $F \subset F' \subset F''$ are fields with F'' a finite extension of F, then $|F' : F|$ and $|F'' : F'|$ are also finite, and $|F'' : F| = |F'' : F'||F' : F|$.*

Proof. The fact that $|F' : F|$ and $|F'' : F'|$ are also finite follows easily from linear algebra since the dimension of a subspace must be less than the dimension of the whole vector space. If $|F' : F| = n$ with $\alpha_1, \ldots, \alpha_n$ a basis for F' over F, and $|F'' : F'| = m$ with $\beta_1, \ldots, \beta_m$ a basis for F'' over F' then the mn products $\{\alpha_i \beta_j\}$ form a basis for F'' over F. To see this let γ be any element of F''. Then since the set $\{\beta_j\}$ constitutes a basis for F'' over F'

there exist scalars $b_j \in F'$ such that

$$\gamma = \sum_{j=1}^{m} b_j \beta_j.$$

Since the set $\{\alpha_i\}$ forms a basis for F' over F there exist $a_{ij} \in F$ such that

$$b_j = \sum_{i=1}^{n} a_{ij} \alpha_i.$$

Hence

$$\gamma = \sum_{j=1}^{m} (\sum_{i=1}^{n} a_{ij} \alpha_i) \beta_j = \sum_{j=1}^{m} \sum_{i=1}^{n} a_{ij} (\alpha_i \beta_j).$$

This shows that the mn vectors $\alpha_i \beta_j$ span F'' as a vector space over F. Now we need to show that this set $\{\alpha_i \beta_j\}$ is linearly independent over F.

Suppose that

$$\sum_{i,j} c_{ij} (\alpha_i \beta_j) = 0$$

where $c_{ij} \in F$ and we have abbreviated the double sum. Then

$$\sum_{j=1}^{m} (\sum_{i=1}^{n} c_{ij} \alpha_i) \beta_j = 0$$

and $\sum_{i=1}^{n} c_{ij} \alpha_i \in F'$ since $\alpha_i \in F'$. Since the elements $\beta_j \in F''$ are linearly independent over F' we must have

$$\sum_{i=1}^{n} c_{ij} \alpha_i = 0 \text{ for all } j.$$

But since the set $\{\alpha_j\}$ is linearly independent over F it follows that $c_{ij} = 0$ for all i and all j. This shows that the mn products $\alpha_i \beta_j$ form a basis for F'' over F.

Therefore $|F'' : F| = mn = |F'' : F'||F' : F|$. $\square$

We note that the proof shows slightly more than what was stated in the lemma. In particular we assumed that F'' is a finite extension of F and implied from that F'' was a finite extension of F' and that F' was a finite extension of F. The proof shows the converse is also true, that is, if F' is a finite extension of F and F'' is a finite extension of F' then F'' is also a finite extension of F.

In the case of the lemma we say that F' is an **intermediate field** (when F and F'' are understood) and F is the **ground field**.

Lemma 14.2.2. $\mathbb{C}$ *is a finite extension of* $\mathbb{R}$, *but* $\mathbb{R}$ *is an infinite extension of* $\mathbb{Q}$.

Proof. The elements $1, i$ form a basis for $\mathbb{C}$ over $\mathbb{R}$. Hence the dimension $\mathbb{C}$ over $\mathbb{R}$ is 2 and $\mathbb{C}$ is a finite extension of $\mathbb{R}$.

This second fact depends on the existence of **transcendental numbers** (see Chapter 5). Recall that an element $r \in \mathbf{R}$ is **algebraic** (over $\mathbf{Q}$) if it satisfies some nonzero polynomial with coefficients from $\mathbf{Q}$. That is, $P(r) = 0$, where

$$0 \neq P(x) = a_0 + a_1 x + \cdots + a_n x^n \text{ with } a_i \in \mathbf{Q}.$$

Any $q \in \mathbf{Q}$ is algebraic since if $P(x) = x - q$ then $P(q) = 0$. However, many irrationals are also algebraic. For example $\sqrt{2}$ is algebraic since $x^2 - 2 = 0$ has $\sqrt{2}$ as a root. An element $r \in \mathbf{R}$ is **transcendental** if it is not algebraic.

In general we saw that it is very difficult to show that a particular element is transcendental. However, there are uncountably many transcendental elements (see Theorem 12.3.8 in Chapter 12). In Chapter 12 we also showed that the real number $c = \sum_{j=1}^{\infty} \frac{1}{10^{j!}}$ is transcendental (see Theorem 12.8.9). Other specific examples are e and π. Here we use e. Since e is transcendental, for any natural number n the set of vectors $\{1, e, e^2, \ldots, e^n\}$ must be independent over $\mathbf{Q}$, for otherwise there would be a polynomial that e would satisfy. Thus the set $\{1, e, \ldots, e^n, \ldots\}$ is also independent. Recall from Definition 13.2.7 that an infinite set of vectors is independent if any finite subset is independent. Therefore, we have infinitely many independent vectors in $\mathbf{R}$ over $\mathbf{Q}$ which would be impossible if $\mathbf{R}$ had finite degree over $\mathbf{Q}$. $\qquad\square$

14.3   Algebraic Field Extensions

Our basic approach is to study extension fields whose elements are zeros of polynomials over a fixed ground field. To this end we need the following definition.

Definition 14.3.1. *Suppose F' is an extension field of F and $\alpha \in F'$. Then α is* **algebraic over F** *if there exists a polynomial $0 \neq p(x) \in F[x]$ with $p(\alpha) = 0$. This means that the element α is a zero of a polynomial with coefficients in F. If every element of F' is algebraic over F, then F' is an* **algebraic extension** *of F. If $\alpha \in F'$ is nonalgebraic over F then α is called* **transcendental** *over F. A nonalgebraic extension is called a* **transcendental extension***.*

Lemma 14.3.1. *Every element of F is algebraic over F.*

Proof. If $f \in F$ then $p(x) = x - f \in F[x]$ and $p(f) = 0$. $\qquad\square$

The tie-in to finite extensions is via the following theorem.

Theorem 14.3.1. *If F' is a finite extension of F, then F' is an algebraic extension.*

Proof. Suppose $\alpha \in F' \setminus F$. We must show that there exists a nonzero polynomial $0 \neq p(x) \in F[x]$ with $p(\alpha) = 0$. Since F' is a finite extension, $|F' : F| = n < \infty$. This implies that there are n elements in a basis for F' over F, and hence any set of $(n+1)$ elements in F' must be linearly dependent over F. Consider then $1, \alpha, \alpha^2, \ldots, \alpha^n$. These are $(n+1)$ elements in F'. If any two of these powers were equal we would already have a polynomial over F which α satisfies (see exercises). Hence we may assume that these powers are distinct and therefore must be linearly dependent. Then there must exist elements $f_0, f_1, \ldots, f_n \in F$ not all zero such that

$$f_0 + f_1\alpha + \cdots + f_n\alpha^n = 0.$$

Let $p(x) = f_0 + f_1 x + \cdots + f_n x^n$. Then $p(x) \in F[x]$ and $p(\alpha) = 0$ (see Definition 12.1.1). $\qquad\square$

EXAMPLE 14.2.1. $\mathbb{C}$ is algebraic over $\mathbb{R}$, but $\mathbb{R}$ is transcendental over $\mathbb{Q}$. Since $\{1, i\}$ is a basis for $\mathbb{C}$ over $\mathbb{R}$, $|\mathbb{C} : \mathbb{R}| = 2$. Thus $\mathbb{C}$ being algebraic over $\mathbb{R}$ follows from Theorem 14.3.1. More directly, if $z \in \mathbb{C}$ then $p(x) = (x - z)(x - \bar{z}) \in \mathbb{R}[x]$ and $p(z) = 0$. $\mathbb{R}$ (and thus $\mathbb{C}$) being transcendental over $\mathbb{Q}$ follows from the existence of transcendental numbers such as e and π.

We note that Theorem 14.3.1 says that a finite extension must be algebraic. The converse is not true, that is, there exists algebraic extensions that are not finite extensions. In the next section we present an example. Here we mention though that if $\mathcal{A}$ is the set of algebraic numbers over $\mathbb{Q}$ and within $\mathbb{R}$ then $\mathcal{A}$ forms a subfield of $\mathbb{R}$ and hence an algebraic extension of $\mathbb{Q}$. We will show that the algebraic numbers in total do not form a finite extension of $\mathbb{Q}$.

If α is algebraic over F, it satisfies a polynomial over F and hence an irreducible polynomial over F (see the proof of the next lemma). Since F is a field, if $f \in F$ and $p(x) \in F[x]$, then $f^{-1}p(x) \in F[x]$ also. This implies that if $p(x)$ is irreducible in $F[x]$ and $p(\alpha) = 0$ with a_n the leading coefficient of $p(x)$, then $p_1(x) = a_n^{-1}p(x)$ is a monic polynomial in $F[x]$ that α also satisfies. Thus if α is algebraic over F there is a monic irreducible polynomial that α satisfies. The next result says that this polynomial is unique.

Lemma 14.3.2. *If $\alpha \in F'$ is algebraic over F, then there exists a unique monic irreducible polynomial $p(x) \in F[x]$ of minimal degree such that $p(\alpha) = 0$. This unique monic irreducible polynomial is denoted by* $\mathrm{irr}(\alpha, F)$.

Proof. Suppose $f(\alpha) = 0$ with $0 \neq f(x) \in F[x]$. Then $f(x)$ factors into irreducible polynomials. Since there are no zero divisors in a field, one of these factors, say $p_1(x)$ must also have α as a root. If the leading coefficient of $p_1(x)$ is a_n then $p(x) = a_n^{-1}p_1(x)$ is a monic irreducible polynomial in $F[x]$ that also has α as a root. Therefore, there exist monic irreducible polynomials that have α as a root. Let $p(x)$ be one such polynomial of minimal degree. It remains to show that $p(x)$ is unique. Suppose $g(x)$ is another monic irreducible polynomial with $g(\alpha) = 0$. Since $p(x)$ has minimal degree, $\deg(p(x)) \leq \deg(g(x))$. By the division algorithm

$$g(x) = q(x)p(x) + r(x)$$

where $r(x) \equiv 0$ or $\deg(r(x)) < \deg(p(x))$. Substituting α into the above we get

$$g(\alpha) = q(\alpha)p(\alpha) + r(\alpha),$$

which implies that $r(\alpha) = 0$ since $g(\alpha) = p(\alpha) = 0$. But then if $r(x)$ is not identically 0, α is a root of $r(x)$, which contradicts the minimality of the degree of $p(x)$. Therefore, $r(x) = 0$ and $g(x) = q(x)p(x)$. The polynomial $q(x)$ must be a constant (unit factor) since $g(x)$ is irreducible, but then $q(x) = 1$ since both $g(x), p(x)$ are monic. This says that $g(x) = p(x)$, and hence $p(x)$ is unique.

$\square$

Suppose $\alpha \in F'$ is algebraic over F and $p(x) = \text{irr}(\alpha, F)$. Then there exists a smallest intermediate field E with $F \subset E \subset F'$ such that $\alpha \in E$. By smallest we mean that if E' is another intermediate field with $\alpha \in E'$ then $E \subset E'$. To see that this smallest field exists, notice that there are subfields E' in F' in which $\alpha \in E'$, namely F' itself. Let E be the intersection of all subfields of F' containing α and F. E is a subfield of F' (see the exercises) and E contains both α and F. Further, this intersection is contained in any other subfield containing α and F. This smallest subfield has a very special form.

Definition 14.3.2. *Suppose $\alpha \in F'$ is algebraic over F and $p(x) = \text{irr}(\alpha, F) = a_0 + a_1 x + \cdots + a_{n-1} x^{n-1} + x^n$. Let*

$$F(\alpha) = \{f_0 + f_1\alpha + \cdots + f_{n-1}\alpha^{n-1} : f_i \in F\}.$$

On $F(\alpha)$ define addition and subtraction componentwise and define multiplication by algebraic manipulation, replacing powers of α higher than α^n by using

$$\alpha^n = -a_0 - a_1\alpha - \cdots - a_{n-1}\alpha^{n-1}.$$

Theorem 14.3.2. *The field $F(\alpha)$ forms a finite algebraic extension of F with the degree given by*

$$|F(\alpha) : F| = deg(irr(\alpha, F)).$$

*The field $F(\alpha)$ is the smallest subfield of F' that contains the root α. The set $\{1, \alpha, \ldots, \alpha^{n-1}\}$ is a basis for the vector space $F(\alpha)$ over F. A field extension of the form $F(\alpha)$ for some α is called a **simple extension** of F.*

Proof. Suppose that $\deg(\text{irr}(\alpha, F)) = n$. Recall that $F_{n-1}[x]$ is the set of all polynomials over F of degree $\leq n - 1$ together with the zero polynomial. This set forms a vector space of dimension n over F with basis $\{1, x, x^2, \ldots, x^{n-1}\}$. As defined, relative to addition and subtraction, $F(\alpha)$ is the same as $F_{n-1}[x]$, and thus $F(\alpha)$ is a vector space of dimension n over F with basis $\{1, \alpha, \ldots, \alpha^{n-1}\}$ and hence an abelian group. If $h(x) \in F[x]$ with $\deg(h(x)) \geq n$, then $h(\alpha) = h_1(\alpha)$, where $h_1(x)$ is a polynomial of degree $\leq n - 1$, obtained by replacing powers of α greater than or equal to n by combinations of lower powers using

$$\alpha^n = -a_0 - a_1\alpha - \cdots - a_{n-1}\alpha^{n-1}.$$

If $\deg(h(x)) < n$ then $h(\alpha) = h_1(\alpha)$.

Multiplication is then done via multiplication of polynomials replacing powers of α with exponents greater than or equal to n by lower powers as above. With this multiplication it follows that $F(\alpha)$ forms a commutative ring with an identity. We must show that it forms a field. To do this we must show that every nonzero element of $F(\alpha)$ has a multiplicative inverse. Suppose $0 \neq g(x) \in F[x]$. If $\deg(g(x)) < n = \deg(\mathrm{irr}(\alpha, F))$, then $g(\alpha) \neq 0$ since $\mathrm{irr}(\alpha, F)$ is the irreducible polynomial of minimal degree that has α as a root. Now suppose $g(\alpha) \in F(\alpha)$, $g(\alpha) \neq 0$. Consider the corresponding polynomial $g(x) \in F[x]$ of degree $\leq n - 1$. Since $p(x) = \mathrm{irr}(\alpha, F)$ is irreducible, it follows that $g(x)$ and $p(x)$ must be relatively prime, that is, $\gcd(g(x), p(x)) = 1$. Therefore, there exist $h(x), k(x) \in F[x]$ such that

$$g(x)h(x) + p(x)k(x) = 1.$$

Substituting α into the above we obtain using the fact that this substitution is a ring homomorphism (see Lemma 12.1.4):

$$g(\alpha)h(\alpha) + p(\alpha)k(\alpha) = 1.$$

However, $p(\alpha) = 0$ and $h(\alpha) = h_1(\alpha) \in F(\alpha)$, as explained above, so that $g(\alpha)h_1(\alpha) = 1$. It follows then that in $F(\alpha)$, $h_1(\alpha) \in F(\alpha)$ is the multiplicative inverse of $g(\alpha)$. Since every nonzero element of $F(\alpha)$ has such an inverse in $F(\alpha)$ it follows that $F(\alpha)$ forms a field. F is contained in $F(\alpha)$ by identifying F with the constant polynomials. Therefore, $F(\alpha)$ is an extension field of F. For $F(\alpha)$, we have that $\{1, \alpha, \alpha^2, \ldots, \alpha^{n-1}\}$ forms a basis, so $F(\alpha)$ has degree n over F. Therefore, $F(\alpha)$ is a finite extension and hence an algebraic extension. If $F \subset E \subset F'$ and E contains α, then clearly E contains all powers of α since E is a subfield. E then contains $F(\alpha)$, and hence $F(\alpha)$ is the smallest subfield containing both F and α. $\qquad\square$

EXAMPLE 14.2.2. Consider $p(x) = x^3 - 2$ over $\mathbb{Q}$. This is irreducible over $\mathbb{Q}$ but has the root $\alpha = 2^{1/3} \in \mathbb{R}$. The field $\mathbb{Q}(\alpha) = \mathbb{Q}(2^{1/3})$ is then the smallest subfield of $\mathbb{R}$ that contains $\mathbb{Q}$ and $2^{1/3}$. Here

$$\mathbb{Q}(\alpha) = \{q_0 + q_1\alpha + q_2\alpha^2 : q_i \in \mathbb{Q} \text{ and } \alpha^3 = 2\}.$$

We first give examples of addition and multiplication in $\mathbb{Q}(\alpha)$. Let $g = 3 + 4\alpha + 5\alpha^2$, $h = 2 - \alpha + \alpha^2$. Then

$$g + h = 5 + 3\alpha + 6\alpha^2$$

and

$$gh = 6 - 3\alpha + 3\alpha^2 + 8\alpha - 4\alpha^2 + 4\alpha^3 + 10\alpha^2 - 5\alpha^3 + 5\alpha^4 = 6 + 5\alpha + 9\alpha^2 - \alpha^3 + 5\alpha^4.$$

But $\alpha^3 = 2$, so $\alpha^4 = 2\alpha$, and then

$$gh = 6 + 5\alpha + 9\alpha^2 - 2 + 5(2\alpha) = 4 + 15\alpha + 9\alpha^2.$$

We now show how to find the inverse of h in $\mathbb{Q}(\alpha)$. Let $h(x) = 2 - x + x^2$, $p(x) = x^3 - 2$. Use the Euclidean algorithm as in Chapter 3 to express 1 as a linear combination of $h(x), p(x)$.

$$x^3 - 2 = (x^2 - x + 2)(x + 1) + (-x - 4),$$

$$x^2 - x + 2 = (-x - 4)(-x + 5) + 22.$$

This implies that

$$22 = (x^2 - x + 2)(1 + (x + 1)(-x + 5)) - ((x^3 - 2)(-x + 5))$$

or

$$1 = \frac{1}{22}[(x^2 - x + 2)(-x^2 + 4x + 6)] - [(x^3 - 2)(-x + 5)].$$

Now substituting α and using that $\alpha^3 = 2$, we have

$$1 = \frac{1}{22}[(\alpha^2 - \alpha + 2)(-\alpha^2 + 4\alpha + 6)],$$

and hence

$$h^{-1} = \frac{1}{22}(-\alpha^2 + 4\alpha + 6).$$

14.4 *F*-automorphisms, Conjugates and Algebraic Closures

In this section and in the next chapter we will be considering automorphisms of fields. An **automorphism** of a field F is an isomorphism $\phi : F \to F$. We first show that any surjective homomorphism between fields must be an isomorphism.

Lemma 14.4.1. *Let F and F' be fields and $\phi : F \to F'$ a surjective homomorphism. Then ϕ is an isomorphism.*

Proof. The lemma is proved if we can show the kernel is trivial, that is, just the zero element of F. However, the kernel is an ideal in F and the only nontrivial ideals in a field are the whole field and the zero element. Since F' is nontrivial and ϕ is onto it cannot be that $\ker(\phi) = F$. Therefore $\ker(\phi) = \{0\}$ and hence ϕ is an isomorphism.   □

It follows from the lemma that any homomorphism from a field onto itself must be an automorphism.

Definition 14.4.1. *Let F', F'' be extension fields of F. An F-isomorphism is an isomorphism $\sigma : F' \to F''$ such that $\sigma(f) = f$ for all $f \in F$. That is, an F-isomorphism is an isomorphism of the extension fields that fixes each element of the ground field. If F', F'' are F-isomorphic, we denote this relationship by $F' \underset{F}{\cong} F''$.*

Lemma 14.4.2. *Suppose that $\alpha, \beta \in F'$, where F' an extension field of F, and suppose that both are algebraic over F. If $irr(\alpha, F) = irr(\beta, F)$, then $F(\alpha)$ is F-isomorphic to $F(\beta)$.*

Proof. Define the map $\sigma : F(\alpha) \to F(\beta)$ by $\sigma(\alpha) = \beta$ and $\sigma(f) = f$ for all $f \in F$. Allow σ to be a homomorphism, that is, preserve addition and multiplication. It follows then that σ maps $f_0 + f_1\alpha + \cdots + f_n\alpha^{n-1} \in F(\alpha)$ to $f_0 + f_1\beta + \cdots + f_n\beta^{n-1} \in F(\beta)$. From this it is straightforward that σ is an F-isomorphism (see the exercises for details).   □

Definition 14.4.2. *Let F' be an algebraic extension of a field F. We say that $\alpha, \beta \in F'$ are* **conjugate** *over F if $irr(\alpha, F) = irr(\beta, F)$. Hence α, β are zeros of the same irreducible polynomial over F.*

Lemma 14.4.2 says that for conjugate elements α, β we have that $F(\alpha)$ is isomorphic to $F(\beta)$. The converse is also true, that is, an F-isomorphism must take a zero of an irreducible polynomial over F into a conjugate.

Lemma 14.4.3. *Let $\alpha, \beta \in F'$ and suppose that α, β are algebraic over F. Let $deg(irr(\alpha, F)) = n$. Suppose that the map*

$$\phi : F(\alpha) \to F(\beta)$$

be defined by

$$\phi(f_0 + f_1\alpha + \cdots + f_{n-1}\alpha^{n-1}) = f_0 + f_1\beta + \cdots + f_{n-1}\beta^{n-1}$$

where $f_i \in F$ is an F-isomorphism from $F(\alpha)$ onto $F(\beta)$. Then $\mathrm{irr}(\alpha, F) = \mathrm{irr}(\beta, F)$, that is, α and β are conjugates over F.

Proof. Let $\mathrm{irr}(\alpha, F) = a_0 + a_1 x + \cdots + a_{n-1}x^{n-1} + x^n$. Then $a_0 + a_1\alpha + \cdots + a_{n-1}\alpha^{n-1} + \alpha^n = 0$. Thus

$$\phi(a_0 + a_1\alpha + \cdots + a_{n-1}\alpha^{n-1} + \alpha^n) = \phi(0) = 0 = a_0 + a_1\beta + \cdots + a_{n-1}\beta^{n-1} + \beta^n$$

which holds since ϕ is a ring homomorphism. Thus $\mathrm{irr}(\alpha, F)(\beta) = 0$. We claim that this implies that $\mathrm{irr}(\beta, F)|\mathrm{irr}(\alpha, F)$.

To see this use the division algorithm for polynomials (Theorem 12.2.2) to divide $\mathrm{irr}(\alpha, F)$ by $\mathrm{irr}(\beta, F)$ to get

$$\mathrm{irr}(\alpha, F) = q(x)\mathrm{irr}(\beta, F) + r(x)$$

where $r(x) = 0$ or $\deg(r(x)) < \deg(\mathrm{irr}(\beta, F))$. But since $\mathrm{irr}(\alpha, F)(\beta) = \mathrm{irr}(\beta, F)(\beta) = 0$ this implies that $r(\beta) = 0$. This if $r(x) \neq 0$ it would be a polynomial of degree less than that of $\mathrm{irr}(\beta, F)$ which also has β has a zero. This contradicts the minimality of the degree of $\mathrm{irr}(\beta, F)$ (see Lemma 14.3.2). Thus $r(x) = 0$ and $\mathrm{irr}(\beta, F)|\mathrm{irr}(\alpha, F)$. Now it follows that $\mathrm{irr}(\alpha, F) = \mathrm{irr}(\beta, F)$ just as in the proof of the uniqueness of $\mathrm{irr}(\alpha, F)$ (see the proof of Theorem 14.3.2). □

We note that Lemma 14.4.3 actually says that any F-isomorphism must take a zero α of an irreducible into a conjugate because any such isomorphism must have the form given there.

EXAMPLE 14.4.1. $\mathbb{Q}(\sqrt{3})$ over $\mathbb{Q}$ has $\mathrm{irr}(\sqrt{3}, \mathbb{Q}) = x^2 - 3$. So $\sqrt{3}$ and $-\sqrt{3}$ are conjugate over $\mathbb{Q}$. Thus according to the above the

$$\phi : \mathbb{Q}(\sqrt{3}) \to \mathbb{Q}(\sqrt{3})$$

given by

$$\phi(a + b\sqrt{3}) = a - b\sqrt{3}$$

is an automorphism of $\mathbb{Q}(\sqrt{3})$ onto itself leaving $\mathbb{Q}$ fixed.

We now show that the set of algebraic elements over F actually forms a field. If $\alpha, \beta \in F'$, an extension field of F, are two algebraic elements over F, we use $F(\alpha, \beta)$ to denote $(F(\alpha))(\beta)$. $F(\alpha, \beta)$ and $F(\beta, \alpha)$ are F-isomorphic so we treat them as the same. We now show that the set of algebraic elements over a ground field is closed under the algebraic operations and from this obtain that the algebraic elements form a subfield.

Lemma 14.4.4. *Let F' be an extension field of F. If $\alpha, \beta \in F'$ are two algebraic elements over F, then the elements $\alpha \pm \beta, \alpha\beta$, and $\alpha/\beta = \alpha\beta^{-1}$, with $\beta \neq 0$, are also algebraic over F.*

Proof. Since α, β are algebraic, the subfield $F(\alpha, \beta)$ will be of finite degree over F and therefore algebraic over F (see the note after Lemma 14.2.1). Now, $\alpha, \beta \in F(\alpha, \beta)$ and since $F(\alpha, \beta)$ is a subfield, it follows that $\alpha \pm \beta, \alpha\beta$, and α/β if $\beta \neq 0$ are also elements of $F(\alpha, \beta)$. Since $F(\alpha, \beta)$ is an algebraic extension of F, each of these elements is algebraic over F. $\square$

Theorem 14.4.1. *If F' is an extension field of F, then the set of elements of F' that are algebraic over F forms a subfield of F'. This subfield is called the* **algebraic closure of** F **in** F'.

Proof. Let $A_F(F')$ be the set of algebraic elements over F in F'. $A_F(F') \neq \emptyset$ since it contains F. From the previous lemma it is closed under addition, subtraction, multiplication, and division, and therefore it forms a subfield. $\square$

Recall that an element of $\mathbb{R}$ which is algebraic over $\mathbb{Q}$ is called an **algebraic number**. Hence the set $\mathcal{A}$ of algebraic numbers in $\mathbb{R}$ is the **algebraic closure** of $\mathbb{Q}$ within $\mathbb{R}$. From Theorem 14.4.1 this forms a subfield of $\mathbb{R}$. It is clearly algebraic. However, it is not of finite degree and hence provides an example of an algebraic extension that is not a finite extension.

Theorem 14.4.2. *The set of algebraic numbers $\mathcal{A}$ forms an algebraic extension of $\mathbb{Q}$ within $\mathbb{R}$ but is not a finite extension.*

Proof. The fact that $\mathcal{A}$ is an algebraic extension is clear. We show that it is not a finite extension. Suppose that $|\mathcal{A} : \mathbb{Q}| = m < \infty$. Then as in the proof that a finite extension is algebraic (Theorem 14.3.1) it would follow that the degree of $\text{irr}(\alpha, \mathbb{Q}) \leq m$ for every $\alpha \in \mathcal{A}$. We show that this is not true. Let n be any integer > 1 and consider $x^n - 2$. Then for any such n, $x^n - 2$ is irreducible over $\mathbb{Q}$. Then $\alpha = 2^{1/n}$ satisfies $x^n = 2$ and hence α is algebraic over $\mathbb{Q}$ and $\text{irr}(\alpha, \mathbb{Q}) = x^n - 2$ (see the exercises) and so $|\mathbb{Q}(\alpha) : \mathbb{Q}| = n$ and therefore $|\mathcal{A} : \mathbb{Q}| \geq n$ (why?). Hence $|\mathcal{A} : \mathbb{Q}| \geq n$ for any positive integer n and therefore $|\mathcal{A} : \mathbb{Q}|$ is not finite. $\square$

We close this section with a final result that says that every finite extension is formed by taking successive simple extensions.

Theorem 14.4.3. *If F' is a finite extension of F, then there exists a finite set of algebraic elements $\alpha_1, \ldots, \alpha_n$ such that $F' = F(\alpha_1, \ldots, \alpha_n)$.*

Proof. Suppose $|F' : F| = k < \infty$. By Theorem 14.3.1 F' is algebraic over F. If $k = 1$ then $|F' : F| = 1$ and $F' = F(1) = F$ and we are done. If $F \neq F'$ choose an $\alpha_1 \in F', \alpha_1 \notin F$. Then $|F(\alpha_1) : F| > 1$ so that $F \subset F(\alpha_1) \subset F'$ and $|F' : F(\alpha_1)| < k$ (why?). If the degree of this extension is 1, then $F' = F(\alpha_1)$, and we are done. If not, choose an $\alpha_2 \in F', \alpha_2 \notin F(\alpha_1)$. Then as above $F \subset F(\alpha_1) \subset F(\alpha_1, \alpha_2) \subset F'$ with $|F' : F(\alpha_1, \alpha_2)| < |F' : F(\alpha_1)|$. As before, if this degree is one we are done; if not, continue. Since k is finite then from Lemma 14.2.1 and the note after it this process must terminate in a finite number of steps. To see this note that $|F' : F| = k$ implies that any linearly independent set $\{\alpha_1, \ldots, \alpha_n\}$ must have at most k elements, that is, $n \leq k$. If we can continue this process without ever getting F' in finitely many steps this would give a linearly independent set $\{\alpha_1, \ldots, \alpha_n, \ldots\}$ with more than k elements as soon as $n > k$. Therefore we must finally get an $\alpha_n \in F'$ in finitely many steps such that $F' = F(\alpha_1, \ldots, \alpha_n)$. $\square$

14.5 Adjoining Roots to Fields

In the previous section we assumed that we began with an extension field and then considered algebraic elements in that extension. The next result, due to Kronecker, is fundamental because it says that given any irreducible polynomial $f(x) \in F[x]$ we can construct an extension field F' of F in which $f(x)$ has a root.

Theorem 14.5.1. *(Kronecker's Theorem) Let F be a field and $f(x) \in F[x]$ an irreducible polynomial over F. Then there exists a finite extension F' of F where $f(x)$ has a root. Further, if α is a root in some extension F'' with $\mathrm{irr}(\alpha, F) = f(x)$, then F' is F-isomorphic to $F(\alpha)$.*

Proof. To construct the field F' we essentially mimic the construction of $F(\alpha)$ as in the last section. Suppose $f(x) = a_0 + a_1 x + \cdots + a_n x^n$ with $a_n \neq 0$. Define α to satisfy

$$a_0 + a_1 \alpha + \cdots + a_n \alpha^n = 0,$$

and also assume that $p(\alpha) \neq 0$ for any $p(x) \in F[x]$ with $\deg(p(x)) < n$.

Now define $F' = F(\alpha)$ as in the last section. That is,

$$F(\alpha) = \{f_0 + f_1\alpha + \cdots + f_{n-1}\alpha^{n-1} : f_i \in F\}.$$

Then on $F(\alpha)$ define addition and subtraction componentwise and define multiplication by algebraic manipulation, replacing powers of α higher than α^n by using

$$\alpha^n = \frac{-a_0 - a_1\alpha - \cdots - a_{n-1}\alpha^{n-1}}{a_n}.$$

$F' = F(\alpha)$ then forms a field of finite degree over F — the proof being identical to that of Section 14.3. The difference between this construction and the construction in Theorem 14.3.2 is that here α is defined to be the root and we constructed the field around it, whereas in the previous construction α was assumed to satisfy the polynomial and $F(\alpha)$ was an already existing field that contained α. The field F' constructed above is said to be constructed by **adjoining the root α to F**. The field $F(\alpha)$ is called α **adjoined to F** or F **adjoin on** α.   □

EXAMPLE 14.5.1. Let $f(x) = x^2 + 1 \in \mathbb{R}[x]$. This is irreducible over $\mathbb{R}$. We construct the field in which this has a root. Let α be an indeterminate with $\alpha^2 + 1 = 0$ or $\alpha^2 = -1$. The extension field $\mathbb{R}(\alpha)$ then has the form

$$\mathbb{R}(\alpha) = \{x + \alpha y : x, y \in \mathbb{R}, \alpha^2 = -1\}.$$

It is clear (see Chapter 5) that this field is $\mathbb{R}$-isomorphic to the complex numbers $\mathbb{C}$, that is, $\mathbb{R}(\alpha) \underset{\mathbb{R}}{\cong} \mathbb{R}(i) \underset{\mathbb{R}}{\cong} \mathbb{C}$.

The construction of the field $F(\alpha)$ in Theorem 14.5.1 is actually part of a much more general algebraic approach. This approach uses the theory of ideals in rings introduced in Chapter 11 and provides an alternative proof for Kronecker's theorem.

Recall from Chapter 11 that if R is a commutative ring then a subring $I \subset R$ is an **ideal** if $rI \subset I$ for all $r \in R$. Ideals are used to construct quotient rings. Let $I \subset R$ be an ideal then a **coset** of I in R is a subset of the form $r + I$ for r a given element in R. The set of all cosets of the ideal $I \subset R$ is denoted by R/I. Addition and multiplication are defined as follows for $r_1, r_2 \in R$:

   (1) $(r_1 + I) + (r_2 + I) = (r_1 + r_2) + I$.

(2) $(r_1 + I) \cdot (r_2 + I) = (r_1 \cdot r_2) + I$.

We recall the following crucial result (Theorem 11.2.1).

Theorem 14.5.2. *Given a commutative ring R and an ideal $I \subset R$, then the set of cosets R/I forms a commutative ring under the operations defined above. The coset $0 + I$ is the zero element of R/I, while if R has a multiplicative identity 1 then the coset $1 + I$ is the multiplicative identity for R/I. The ring R/I is called the* **quotient ring,** *or* **factor ring of R modulo the ideal I.**

To prove Kronecker's theorem we need some further ideas from Chapter 11.

Definition 14.5.1. *Let R be a commutative ring. An ideal $I \subset R$ is a* **maximal ideal** *if $I \neq R$ and if J is an ideal in R with $I \subset J$ then $J = I$ or $J = R$.*

The important result for us is the following (see Theorem 11.3.3).

Lemma 14.5.1. *Suppose R is a commutative ring with an identity. Then R/I is a field if and only if I is a maximal ideal.*

Finally let R be a commutative ring with identity and let $r \in R$. Let $\langle r \rangle = \{r_1 r : r_1 \in R\}$ be the set of all multiples of r in R. Then $\langle r \rangle$ forms an ideal called the **principal ideal generated by** r. Further let $r \in R$ and I be an ideal in R. Then

$$\langle r, I \rangle = \{rx + i : x \in R, i \in I\}$$

is called the **ideal generated by** r and I. It is easy to show that this is an ideal containing r and I (see the exercises). It follows that if $r, s, \in R$ then the ideal generated by r, s is

$$\langle r, s \rangle = \{rx + sy : x, y \in R\}.$$

Now we show how this relates to the field extension theorem and provides an alternative constructive proof of Kronecker's theorem.

Suppose that F is a field and suppose that $f(x) \in F[x]$ is an irreducible polynomial over F. The ring of polynomials $F[x]$ is a commutative ring with an identity. Consider $\langle f(x) \rangle$, the principal ideal in $F[x]$ generated by $f(x)$. Suppose $g(x) \notin \langle f(x) \rangle$, so that $g(x)$ is not a multiple of $f(x)$. Since $f(x)$

is irreducible, it follows that $(f(x), g(x)) = 1$. Thus there exist $h(x), k(x) \in F[x]$ with

$$h(x)f(x) + k(x)g(x) = 1.$$

The element on the left is in the ideal $\langle g(x), f(x) \rangle$, so the identity, 1, is in this ideal. Therefore, the whole ring $F[x]$ is in this ideal. Since $g(x)$ was arbitrary, this implies that if $f(x)$ is irreducible then the principal ideal $\langle f(x) \rangle$ is maximal.

Now let $F' = F[x]/\langle f(x) \rangle$ the quotient ring of $F[x]$ modulo the principal ideal generated by $f(x)$. Since this principal ideal is maximal it follows that F' is a field from Lemma 14.5.1. Since $F \subset F[x]$ and F consists of polynomials of degree zero together with 0, it follows that the map $F \to F'$ given by taking each $f \in F$ to its image in F' is injective (see the exercises). Therefore F' can be considered as an extension field of F. We claim that $f(x)$ has a zero in F'.

Let $\bar{x}$ be the coset of x modulo the ideal $\langle f(x) \rangle$, that is,

$$\bar{x} = x + \langle f(x) \rangle.$$

Then it can be shown that $f(\bar{x}) = \overline{f(x)}$ (see the exercises). However, since $f(x) \in \langle f(x) \rangle$ it follows that $\overline{f(x)} = \bar{0}$ in F'. Therefore $\bar{x}$ is a root of $f(x)$ wihin F'. Here the overbars represent cosets. We have therefore constructed an extension field F' of F in which the irreducible polynomial $f(x)$ has a zero.

14.6   Splitting Fields and Algebraic Closures

We have just seen that given an irreducible polynomial over a field F we could always find a field extension in which this polynomial has a root. We now push this further to obtain field extensions where a given polynomial has all its roots.

Definition 14.6.1. *If $f(x) \in F[x]$ is a nonconstant polynomial and F' is an extension field of F, then $f(x)$ **splits** in F', (F' may be F) if $f(x)$ factors into linear factors in $F'[x]$. Equivalently, this means that all the roots of $f(x)$ are in F'. A field F' is a **splitting field** for $f(x)$ over F if F' is the smallest extension field of F in which $f(x)$ splits. (A splitting field for $f(x)$ is the smallest extension field in which $f(x)$ has all its possible roots.) F'' is a **splitting field** of $\{f_i(x) \in F[x] : i \in I\}$ over F if F'' is the smallest* *extension field of F containing the roots of all $f_i(x), i \in I$. F'' is a **splitting** field over F if it is the splitting field for some set of polynomials over F.*

Theorem 14.6.1. *If $f(x) \in F[x]$ be a nonconstant polynomial, then there exists a splitting field for $f(x)$ over F.*

Proof. The splitting field F' of F is constructed by repeated adjoining of roots. Here we do an induction on the degree of $f(x)$. If $\deg f(x) = 1$ then $f(x)$ is linear and therefore its splitting field $F' = F$. Now suppose that for all fields and all polynomials of degree less than $\deg(f(x))$ there exist splitting fields. Suppose without loss of generality that $f(x)$ is irreducible over F. Since $F[x]$ is a UFD if we can find splitting fields for each of the irreducible factors then we can find a splitting field for $f(x)$. From Theorem 14.5.1 there exists a field F' containing α_1 with $f(\alpha_1) = 0$. Then $f(x) = (x - \alpha_1)g(x) \in F'[x]$ with $\deg(g(x)) = n - 1$ by the factor theorem (Theorem 12.3.1). Since $\deg(g(x)) < \deg(f(x))$ it follows by the inductive hypothesis that there exists a field F'' that contains F' and all the zeros of $g(x)$ say $\alpha_2, \dots, \alpha_m$. The splitting field for $f(x)$ over F is then $F(\alpha_1, \dots, \alpha_n)$. $\qquad\square$

EXAMPLE 14.6.1. Consider $f(x) = x^4 - x^2 - 2 \in \mathbb{Q}[x]$. Clearly $f(x) = (x^2 - 2)(x^2 + 1)$. Hence the zeros of $f(x)$ are $\pm\sqrt{2}, \pm i$. Therefore the splitting field for $f(x)$ over $\mathbb{Q}$ is

$$\mathbb{Q}(\sqrt{2}, i) = \mathbb{Q}(\sqrt{2})(i) = \{\alpha + \beta i : \alpha, \beta \in \mathbb{Q}(\sqrt{2}), i^2 = -1\}$$

$$= \{(a + b\sqrt{2} + (c + d\sqrt{2})i : a, b, c, d \in \mathbb{Q}, (\sqrt{2})^2 = 2, i^2 = -1\}$$

Later in this chapter we will give a further characterization of splitting fields. Here we now return to some ideas introduced in Section 14.1. Recall that if F' is an extension of F, the set of elements of F' which are algebraic over F forms a subfield called the algebraic closure of F in F'. More generally, we say that a field F' is **algebraically closed** if every nonconstant polynomial in $F'[x]$ has a root in F'. Note that in this language the fundamental theorem of algebra says that the complex number field $\mathbb{C}$ is algebraically closed. The next result gives several clearly equivalent formulations of being algebraically closed.

Theorem 14.6.2. *Let F be a field. Then the following are equivalent:*

(1) F is algebraically closed.
(2) Every nonconstant polynomial $f(x) \in F[x]$ splits in $F[x]$.

(3) F has no proper algebraic extensions, that is, there is no algebraic field extension E with $F \subset E$ and $F \neq E$.

Proof. Suppose first that F is algebraically closed and let $f(x)$ be any non-constant polynomial in $F[x]$. Since F is algebraically closed $f(x)$ has a zero $\alpha \in F$. By the factor theorem $f(x) = (x - \alpha)g(x)$ with $g(x) \in F[x]$. Then if $g(x)$ is also nonconstant it also has a zero $\beta \in F$ and hence $f(x) = (x - \alpha)(x - \beta)h(x)$ with $h(x) \in F[x]$. Continuing in this manner we get a factorization of $f(x) \in F[x]$ into linear factors and units. Therefore (1) implies (2).

Now suppose (2), that is, that every nonconstant polynomial in $f[x]$ splits in $F[x]$. Let E be an algebraic extension of F so that $F \subset E$. Then if $\alpha \in E$ we must have $\mathrm{irr}(\alpha, F) = x - \alpha$ because any nonconstant polynomial splits in $F[x]$. Therefore $\alpha \in F$ and since we started with an arbitrary element of F we get that $F = E$. Hence (2) implies (3).

Now suppose (3) so that F has no proper algebraic extensions. Let $f(x)$ be a nonconstant polynomial in $f[x]$. As before without loss of generality we may assume that $f(x)$ is irreducible. Then by Kronecker's theorem there exists a finite extension F' of F such that $f(\alpha) = 0$ for some $\alpha \in F'$. But F' is a finite extension of F so it is algebraic by Theorem 14.3.1 and therefore by our assumption $F' = F$. It follows that $\alpha \in F$ and so $f(x)$ has a root in F proving that F is algebraically closed. Therefore (3) implies (1) and the theorem is proved. □

Definition 14.6.2. *An extension field F' of F is an* **algebraic closure** *of F if F' is algebraic over F and F' is algebraically closed.*

EXAMPLE 14.6.2. For this example we assume the fundamental theorem of algebra – that is, that $\mathbb{C}$ is algebraically closed. Note that $\mathbb{C}$ is not the algebraic closure of $\mathbb{Q}$ since $\mathbb{C}$ is not algebraic over $\mathbb{Q}$. However, the complex algebraic numbers $A_{\mathbb{C}}$, that is, the set of complex numbers which are algebraic over $\mathbb{Q}$, is the algebraic closure of $\mathbb{Q}$. To see this, notice that $A_{\mathbb{C}}$ is algebraic over $\mathbb{Q}$ by definition. Now we show that it is algebraically closed. Let $f(x) \in A_{\mathbb{C}}[x]$. If α is a root of $f(x)$, then $\alpha \in \mathbb{C}$, and then α is also algebraic over $\mathbb{Q}$ since each element of $A_{\mathbb{C}}$ is algebraic over $\mathbb{Q}$. Therefore, $\alpha \in A_{\mathbb{C}}$ and $A_{\mathbb{C}}$ is algebraically closed. More generally, if K is an extension field of F and K is algebraically closed, then the algebraic closure of F in K is the algebraic closure of F.

Given a polynomial $f(x) \in F[x]$ we have seen that we can construct a splitting field. The next result, whose proof depends on the axiom of choice so it will not be given here, indicates that this procedure can be extended to obtain an algebraic closure for any field.

Theorem 14.6.3. *Every field F has an algebraic closure, and any two algebraic closures of F are F-isomorphic.*

14.7 Automorphisms and Fixed Fields

In our treatment of Galois theory that will be done in the next chapter, we will consider special types of algebraic extensions called Galois extensions. For these extensions there is a beautiful interplay between group theory and field theory. Here we introduce some of the ideas that we will be looking at more closely in the next chapter.

Throughout this section we consider F to be a field and F' a finite and hence algebraic extension of F.

Lemma 14.7.1. *The set of F-automorphisms of F' forms a group under composition of functions as the operation. We will denote this by $Aut_F(F')$.*

The total set of automorphisms of an algebraic structure forms a group under composition of functions (see the exercises) and hence $\text{Aut}_F(F')$ is a subset of $\text{Aut}(F')$. Therefore to show $\text{Aut}_F(F')$ is a group we must only show that it is a subgroup of $\text{Aut}(F')$. Hence what must be done is show that $\text{Aut}_F(F')$ is nonempty and closed under composition and inverse. The details are straightforward and we leave them to the exercises.

We now prove a theorem that we will see in the next chapter as one of the cornerstones of Galois theory.

Theorem 14.7.1. *Let F' be a finite extension field of F and let $G = Aut_F(F')$. Then to each intermediate field $F \subset K \subset F'$ there corresponds a subgroup H of G and to each subgroup H of G there corresponds an intermediate field.*

Proof. Let K be an intermediate field. Consider the set

$$H = \{\phi \in G : \phi(k) = k \text{ for all } k \in K\}.$$

Since the identity element fixes everything in K it follows that H is nonempty. It is straightforward that the composition of two elements in H is still in H and that the inverse of an element in H is still in H. Therefore H is a subgroup.

Now let H be a subgroup of G and consider the set

$$K = \{x \in F' : \phi(x) = x \text{ for all } \phi \in H\}.$$

Each element of G fixes all the elements in F and therefore $F \subset K$ and hence $K \neq \emptyset$. We claim K is a field proving the theorem. Suppose $x_1, x_2 \in K$ and $\phi \in H$. Then

$$\phi(x_1 \pm x_2) = \phi(x_1) \pm \phi(x_2) = x_1 \pm x_2$$

and hence $x_1 \pm x_2 \in K$. In an identical manner $x_1 x_2^{-1} \in K$ if $x_2 \neq 0$. Therefore K is a subfield of F' and we have

$$F \subset K \subset F'$$

and hence K is an intermediate field proving the theorem.   □

The field K is called the **fixed field** for the subgroup H.

EXAMPLE 14.7.1. Consider $f(x) = x^3 - 2 \in \mathbb{Q}[x]$ and $\mathbb{Q}(2^{1/3})$. Then $f(x)$ factors in $\mathbb{Q}(2^{1/3})$ into a linear term and an irreducible quadratic. Further

$$|\mathbb{Q}(2^{1/3}) : \mathbb{Q}| = 3$$

since the irreducible polynomial that $2^{1/3}$ satisfies over $\mathbb{Q}$ is $x^3 - 2$. To see that this is irreducible over $\mathbb{Q}$ we use Eisenstein's criterion (Chapter 12). Moreover

$$\mathbb{Q}(2^{1/3}) = \{a_0 + a_1 + a_2\alpha^2 : \alpha^3 = 2, a_0, a_1, a_2 \in \mathbb{Q}\}.$$

However, $f(x)$ does not split in $\mathbb{Q}(2^{1/3})$ because this field does not contain the other two roots of $f(x)$,

$$\alpha_1 = \frac{-2^{1/3} + \sqrt{3} \cdot 2^{1/3}i}{2}, \quad \alpha_2 = \frac{-2^{1/3} - \sqrt{3} \cdot 2^{1/3}i}{2}.$$

But $f(x) = x^3 - 2$ does split in $\mathbb{Q}(2^{1/3}, i\sqrt{3})$. Thus $\mathbb{Q}(2^{1/3}, i\sqrt{3})$ is a splitting field for $f(x)$ as it contains all the roots of $f(x)$ and it is the smallest subfield of $\mathbb{R}$ that contains these. Also $|\mathbb{Q}(2^{1/3}, i\sqrt{3}) : \mathbb{Q}| = 6 = |\mathbb{Q}(2^{1/3}, i\sqrt{3}) :$

$\mathbb{Q}(2^{1/3})||Q(2^{1/3}) : \mathbb{Q}|$ where $|\mathbb{Q}(2^{1/3}, i\sqrt{3}) : \mathbb{Q}(2^{1/3})| = 2$ since $\text{irr}(i\sqrt{3}, \mathbb{Q}(2^{1/3}))$ $= x^2 + 3$. Of course

$$\mathbb{Q}(2^{1/3}, i\sqrt{3}) = Q(2^{1/3})(i\sqrt{3}) = \{a_0 + a_1\alpha : \alpha^2 = -3, a_0, a_1 \in \mathbb{Q}(2^{1/3})\}.$$

By Lemma 14.4.2 if $\phi : \mathbb{Q}(2^{1/3}, i\sqrt{3}) \to \mathbb{Q}(2^{1/3}, i\sqrt{3})$ is an automorphism with

$$\phi(\alpha_1) = \phi(\frac{-2^{1/3} + \sqrt{3} \cdot 2^{1/3}i}{2}) = \alpha_2 = \frac{-2^{1/3} - \sqrt{3} \cdot 2^{1/3}i}{2}$$

and such that $\phi(f) = f$ for all $f \in \mathbb{Q}(2^{1/3})$ then this automorphism has fixed field $\mathbb{Q}(2^{1/3})$. Of course this also implies that $\phi(2^{1/3}) = 2^{1/3}$. It can be shown that

$$\text{Aut}_{\mathbb{Q}}(\mathbb{Q}(2^{1/3}, i\sqrt{3}) \cong S_3$$

(see the exercises for the details).

For Galois extensions (see the next chapter for a definition of this) this correspondence between subgroups and intermediate fields becomes much more precise and will culminate in what is called the fundamental theorem of Galois theory. We will see this in the next chapter.

We close this section by giving a result that will also be used in the next chapter. It tells us when an isomorphism of a field can be extended to an algebraic field extension. Since its proof depends upon Zorn's lemma we omit the proof here. For a proof see [F] and [CFR].

Theorem 14.7.2. *(Isomorphism Extension Theorem) Let K be an algebraic extension of F and let ϕ be an isomorphism of F onto a field E, that is, $\phi : F \to E$. Let $\overline{E}$ be an algebraic closure of E. Then ϕ can be extended to an isomorphism $\overline{\phi}$ of K onto a subfield of $\overline{E}$, i.e., $\overline{\phi}(x) = \phi(x)$ for all $x \in F$.*

14.8   Finite Fields

In this section we examine **finite fields**, that is, fields F with finitely many elements. We have seen that if p is a prime then the modular ring $\mathbb{Z}_p$ is a finite field. Here we prove that any finite field F has order p^n for some prime p and some integer n. Further given a prime p and an integer n we show that there exists a finite field of order p^n and further this field is unique up to isomorphism.

Recall that the characteristic of a ring R denoted $\text{char}(R)$ is the least positive integer n such that $nr = 0$ for all $r \in R$. If no such n exists then

R has characteristic zero. If R has an identity 1 then the characteristic is the smallest n such that $n \cdot 1 = 0$. Further for an integral domain the characteristic must be zero or a prime p. Rings of characteristic zero (see Chapter 3) contain isomorphic copies of the integers $\mathbb{Z}$ and hence must be infinite. It follows that any finite field F must have characteristic a prime p. We use this to prove the following.

Theorem 14.8.1. *Let F be a finite field. Then there exists a prime p such that $|F| = p^n$ for some integer $n \geq 1$.*

Proof. Let F be a finite field. Then as explained above it must have characteristic $p > 0$. Then F can be considered as an extension field of the finite field $\mathbb{Z}_p$. It follows from Section 14.1 that F is a vector space over $\mathbb{Z}_p$. If it were infinite dimensional there would be infinitely many independent elements but there are only finitely many elements in F. Hence F is a finite dimensional vector space over $\mathbb{Z}_p$. Suppose that the dimension is $n > 0$. Then there is a finite basis $\alpha_1, \dots, \alpha_n$ for F over $\mathbb{Z}_p$. Therefore each $f \in F$ can be written as a linear combination $f = c_1\alpha_1 + \cdots + c_n\alpha_n$ with each $c_i \in \mathbb{Z}_p$. There are p choices for each c_i and therefore p^n choices for each f. □

Since a finite field is a finite extension of $\mathbb{Z}_p$ then from Theorem 14.2.1 we obtain.

Corollary 14.8.1. *Any finite field F of characteristic p is an algebraic extension of $\mathbb{Z}_p$.*

From Theorem 6.4.7 we have that for any field K any finite subgroup of the multiplicative group $K^\star$ is a cyclic group. If F is a finite field then certainly its multiplicative group is a finite group. Therefore as a corollary of Theorem 6.4.7 we get the following. In addition to the multiplicative group being cyclic it follows directly that the field F must be a simple extension of $\mathbb{Z}_p$, in particular $F = \mathbb{Z}_p(g)$ where g is a generator of the multiplicative group.

Corollary 14.8.2. *Let F be a finite field. Then the multiplicative group $F^\star$ is cyclic. Hence $F = \mathbb{Z}_p(g)$ where g is a generator of $F^\star$.*

Corollary 14.8.3. *Any finite field F is a simple extension of $\mathbb{Z}_p$.*

If F is a finite field with order p^n then its multiplicative subgroup $F^\star$ has order $p^n - 1$. Then from Lagrange's theorem each nonzero element to the power $p^n - 1$ is the identity and hence as field elements each element $x \in F$ satisfies $x^{p^n} - x = 0$. Therefore we have the result.

Lemma 14.8.1. *Let F be a field of order p^n. Then each $\alpha \in F$ is a zero of the polynomial $x^{p^n} - x$. In particular if $\alpha \neq 0$ then α is a zero of $x^{p^n - 1} - 1$.*

If F is a finite field of order p^n, it is a finite extension of $\mathbb{Z}_p$. Since, as we have already mentioned, the multiplicative group is cyclic we must have $F = \mathbb{Z}_p(\alpha)$ for some $\alpha \in F$. From this we obtain that for a given possible finite order there is only one finite field up to isomorphism.

Theorem 14.8.2. *Let F_1, F_2 be finite fields with $|F_1| = |F_2| = p^n$ for some prime p and n a positive integer. Then $F_1 \cong F_2$.*

Proof. We first note that it is sufficient to show that any field F with p^n elements is isomorphic to $\mathbb{Z}_p[x]/\langle f(x) \rangle$ where $f(x)$ is an irreducible polynomial of degree n in $\mathbb{Z}_p[x]$ (see the exercises). Let F be any field of p^n elements. The prime subfield of F is $\mathbb{Z}_p$ (see Corollary 11.2.2. Also explain why the characteristic of F is p.) Hence we may consider $\mathbb{Z}_p \subset F$. But by Corollary 14.8.3 above, F is a simple extension of $\mathbb{Z}_p$. So there exists $\alpha \in F$ such that $F = \mathbb{Z}_p(\alpha)$. Let $f(x) = \mathrm{irr}(\alpha, \mathbb{Z}_p)$. Since $|F| = p^n$, it is clear that $deg(f(x)) = n$ (see the exercises). If we use the same construction as in the alternate proof of Kronecker's theorem given in Section 14.5 we get a field $F' = \mathbb{Z}_p[x]/\langle f(x) \rangle$ with p^n elements (why?). We note that here we took $f(x) = \mathrm{irr}(\alpha, \mathbb{Z}_p)$ but any irreducible polynomial of degree n in $\mathbb{Z}_p[x]$ will work. So the field F' we get does not really depend upon the choice of $f(x)$. Now it is straightforward to show that the map

$$\phi : F = \mathbb{Z}_p(\alpha) \to F' = \mathbb{Z}_p[x]/\langle f(x) \rangle$$

given by $\phi(a_0 + a_1\alpha + \cdots + a_{n-1}\alpha^{n-1}) = a_0 + a_1 x + \cdots + a_{n-1}x^{n-1} + \langle f(x) \rangle$ where $a_i \in \mathbb{Z}_p$ for all i is an isomorphism (see the exercises). Thus $F \cong \mathbb{Z}_p[x]/\langle f(x) \rangle$. $\square$

Hence there is, up to isomorphism, only one finite field of order p^n. We denote this by $GF(p^n)$ standing for **Galois field of order p^n**. We now show that given a prime p and a natural number n there does exist a finite field of order p^n. First we need the following lemma.

Lemma 14.8.2. *Let $g(x) \in \mathbb{Z}_p[x]$ be the polynomial $g(x) = x^{p^n} - x$ where p is a prime. Let F be a splitting field for $g(x)$. Then the zeros of $g(x)$ are distinct and therefore there are p^n zeros of $g(x)$.*

Proof. We show that each zero α of $g(x)$ has multiplicity 1 in $g(x)$, that is, $g(x)$ does not have a factor of the form $(x - \alpha)^2$. The root 0 clearly has multiplicity 1. Suppose that $\alpha \neq 0$ is a zero of $g(x)$. Then α is also a zero of $g_1(x) = x^{p^n - 1} - 1$ and hence $x - \alpha$ is a factor of $g_1(x)$. Then by a direct computation

$$g_2(x) = \frac{g_1(x)}{x - \alpha} = x^{p^n - 2} + \alpha x^{p^n - 3} + \alpha^2 x^{p^n - 4} + \cdots + \alpha^{p^n - 2}.$$

The polynomial $g_2(x)$ has $p^n - 1$ terms and in computing $g_2(\alpha)$ each term is given by

$$\alpha^{p^n - 2} = \frac{\alpha^{p^n - 1}}{\alpha} = \frac{1}{\alpha}.$$

Therefore

$$g_2(\alpha) = (p^n - 1)\frac{1}{\alpha} = -\frac{1}{\alpha}$$

since the characteristic of the field is the prime p. Therefore $g_2(\alpha) \neq 0$ and hence $(x - \alpha)^2$ does not divide $g(x)$. Therefore all the zeros must be distinct and hence there are p^n of them.

$\square$

Theorem 14.8.3. *Let p be a prime and $n > 0$ a natural number. Then there exists a field F of order p^n.*

Proof. Given a prime p consider the polynomial $g(x) = x^{p^n} - x \in \mathbb{Z}_p[x]$. Let K be the splitting field of this polynomial over $\mathbb{Z}_p$. From Lemma 14.8.2 the roots of $g(x)$ are distinct and hence there are p^n of them.

Let F be the set of p^n distinct zeros of $g(x)$ within K. Let $a, b \in F$. Since

$$(a \pm b)^{p^n} = a^{p^n} \pm b^{p^n} \text{ and } (ab)^{p^n} = a^{p^n} b^{p^n}$$

it follows that F forms a subfield of K. Since the degree of $g(x)$ is p^n there are exactly p^n elements in F and hence F is a field with p^n elements. $\square$

14.9 Transcendental Extensions

In Chapter 12 (Section 12.3.5) we examined transcendental and algebraic elements in the real number system $\mathbb{R}$ and showed that there do exist transcendental elements. In the language of this chapter this means that the real numbers $\mathbb{R}$ are a **transcendental extension** of the rationals $\mathbb{Q}$. From Kronecker's theorem we have seen that given a field F either F is algebraically closed or there exist a nontrivial algebraic extension F' of F. Here by nontrivial we mean that the degree $|F' : F|$ is greater than 1. In this section we show that given any field F whatsoever, algebraically closed or not, there exists a **transcendental extension** F' of F. Note that since finite extensions are algebraic, a transcendental extension must have infinite degree.

Let F be a field and consider $F[x]$ the ring of polynomials with coefficients in F. From the work in Chapter 12, $F[x]$ is an integral domain. It is further a unique factorization domain but we don't need that fact here. Recall from Section 5.2.1 that given an integral domain D we can always build a field F containing D. This field F is called the **field of fractions** of D and is constructed in an analogous manner as the rationals are constructed from the integers (see Section 5.2.1). If D is already a field its field of fractions corresponds to D itself.

Now the polynomial ring $F[x]$ is an integral domain but not a field (why?). We denote its field of fractions by $F(x)$ and we call it the **field of rational functions** over F. Since $F[x]$ is not a field we have $F[x] \neq F(x)$. The field $F(x)$ is a transcendental extension of F.

Theorem 14.9.1. *Let F be any field. Then the field of rational functions over F, $F(x)$, is a transcendental extension of F.*

Proof. Recall that $F \subset F[x]$ so $F \subset F(x)$ and hence $F(x)$ is an extension field of F. We must only show then that $F(x)$ is transcendental. To do this we must show that there exists a transcendental element. We claim that the element x is transcendental. Recall that in the construction of the field of fractions or field of quotients of an integral domain D the elements are equivalence classes of pairs (fractions) (a, b) with $a, b \in D$ and $b \neq 0$. From this construction the zero element of the field of fractions is precisely the class of $(0, b)$ where $b \neq 0$. The elements of D are considered to be inside the field of quotients because the mapping $D \to E$ where E is the field of quotients of D given by $d \mapsto (d, 1)$ is an embedding. Thus an element $d \in D$

corresponds to $(d, 1)$. Hence it is the zero element of the field of fractions if and only if $d = 0$.

Now suppose that x were algebraic over F. Then there exists a polynomial $f(x) \in F[x]$ with $f(x) = 0$ in $F(x)$. But then $f(x) = 0$ in $F[x]$ which it is not. Therefore x is transcendental over F.

$\square$

EXAMPLE 14.9.1. If we assume that we know that e is transcendental over $\mathbb{Q}$ then the field $\mathbb{Q}(e)$ is isomorphic to the field of rational functions over $\mathbb{Q}$ in the indeterminate x, that is, quotients of polynomials with coefficients in x.

Note that if we consider the complex numbers $\mathbb{C}$ by the fundamental theorem of algebra they are algebraically closed. Hence there are no nontrivial algebraic extensions of $\mathbb{C}$. However, by Theorem 14.9.1 there are nontrivial transcendental extensions of $\mathbb{C}$.

Lemma 14.9.1. *Let F' be an extension field of the field F and let $\alpha \in F'$. Consider the evaluation map $\phi_\alpha : F[x] \to F$ (see Lemma 12.1.4) Then ϕ_α is injective if and only if α is transcendental over F. Further ϕ_α fixes all elements of F.*

Proof. By definition α is transcendental over F if and only if it is not a zero of any polynomial with coefficients in F, that is, for $f(x) \in F[x]$, $f(\alpha) = \phi_\alpha(f(x)) \neq 0$. But this is true if and only of $\ker(\phi_\alpha) = \{0\}$, that is, ϕ_α is injective. $\square$

We note that since the transcendental extension $F(x)$ of F is the field of fractions then this means that

$$F(x) = \{\frac{f(x)}{g(x)} : f(x), g(x) \in F[x], g(x) \neq 0\}.$$

Thus from Example 14.9.1 $\mathbb{Q}(e)$ is

$$\mathbb{Q}(e) = \{\frac{a_0 + a_1 e + \cdots + a_n e^n}{b_0 + b_1 e + \cdots + b_k e^k} : a_i, b_i \in \mathbb{Q}, \; b_0 + b_1 e + \cdots + b_k e^k \neq 0\}.$$

What this basically says is that a transcendental element is just an indeterminate over the ground field, that is, α is transcendental over F if and only if $F(\alpha) \cong F(x)$.

14.10 Exercises

EXERCISES FOR SECTIONS 14.1–14.3

14.1.1. Verify that if $z \in \mathbb{C}$ then $P(x) = (x - z)(x - \bar{z}) \in \mathbb{R}[x]$ where $\bar{z}$ is the complex conjugate.

14.1.2. Prove that the intersection of any collection of subfields of a field is itself a field.

14.1.3. Show that in the proof of Theorem 14.3.1 if any two of $1, \alpha, \alpha^2, \ldots, \alpha_n$ are equal then we already have a polynomial $p(x)$ such that $p(\alpha) = 0$.

14.1.4. Show that if F_i is a field for $1 \leq 1 \leq n$ and F_{i+1} is a finite extension of F_i then F_n is a finite extension of F_1 and

$$|F_n : F_1| = |F_n : F_{n-1}| \cdots |F_2 : F_1|.$$

(HINT: Use Lemma 14.2.1 and induction on n.)

14.1.5. Prove that if F' is an extension of F, $\alpha \in F'$ is algebraic over F and $\beta \in F(\alpha)$ then $\deg(\mathrm{irr}(\beta, F)) | \deg(\mathrm{irr}(\alpha, F))$.

(HINT: Use the fact that $F \subseteq F(\beta) \subseteq F(\alpha)$ and Theorem 14.3.2 and Lemma 14.2.1.)

14.1.6. Use Problem 14.1.5 to show that there is no element in $\mathbb{Q}(\sqrt{2})$ that is a zero of $x^3 - 2$.

(HINT: What is $\mathrm{irr}(2^{1/3}, \mathbb{Q})$? Note that $\deg(\mathrm{irr}(\sqrt{2}, \mathbb{Q})) = 2$. But by Problem 14.1.5 if there were a $\beta \in \mathbb{Q}(\sqrt{2})$ such that $\beta^3 - 2 = 0$, what would that mean about $\deg(\mathrm{irr}(\beta, \mathbb{Q}))$ and $\deg(\mathrm{irr}(2^{1/3}, \mathbb{Q}))$?)

14.1.7. Let $2^{1/3}$ be the real cube root of 2 and $2^{1/2}$ be the positive square root of 2. Then by Problem 14.1.6, $2^{1/3} \notin Q(2^{1/2})$.

(a) Show that $|Q(2^{1/2}, 2^{1/3}) : \mathbb{Q}(2^{1/2})| = 3$.

(b) Write bases for $Q(2^{1/2})$ over $\mathbb{Q}$ and for $\mathbb{Q}(2^{1/2}, 2^{1/3})$ over $\mathbb{Q}(2^{1/2})$.

(c) Use Lemma 14.2.1 to write a basis of $\mathbb{Q}(2^{1/2}, 2^{1/3})$ over $\mathbb{Q}$.

(d) Find $|\mathbb{Q}(2^{1/2}, 2^{1/3}) : \mathbb{Q}|$.

14.1.8. (a) Show using the basis you found in Problem 14.1.7 for $\mathbb{Q}(2^{1/2}, 2^{1/3})$ over $\mathbb{Q}$ that $2^{1/6} \in \mathbb{Q}(2^{1/2}, 2^{1/3})$.

(b) Show that $x^6 - 2$ is irreducible over $\mathbb{Q}$.

(c) Thus we have $\mathbb{Q} \subseteq \mathbb{Q}(2^{1/6}) \subseteq \mathbb{Q}(2^{1/2}, 2^{1/3})$. Use Lemma 14.2.1 to show that this implies that $\mathbb{Q}(2^{1/6}) = \mathbb{Q}(2^{1/2}, 2^{1/3})$.

(HINT: For (a) note that $2^{7/6} = 2 \cdot 2^{1/6}$. For (b) apply Eisenstein's criterion with $p = 2$ (Theorem 12.4.7). For (c) use the fact that if E is an extension of F then $|E : F| = 1$ implies that $E = F$.)

14.1.9. Let $\alpha = 2^{1/3}$ be the real root of $x^3 - 2 \in \mathbb{Q}[x]$.

(a) Use the basis which you found for $\mathbb{Q}(2^{1/2}, 2^{1/3})$ over $\mathbb{Q}(2^{1/2})$ to write $\mathbb{Q}(2^{1/2}, 2^{1/3})$ over $\mathbb{Q}(2^{1/2})$ as a set in the fashion of Example 14.2.2.

(b) Consider $f = 2^{1/6} + 5\alpha - \sqrt{2}\alpha^2, g = \sqrt{2} - 2\alpha + \sqrt{2}\alpha^2$. First indicate why these are elements of $\mathbb{Q}(2^{1/2}, 2^{1/3})$ over $\mathbb{Q}(2^{1/2})$ and then find $g + h, gh, h^{-1}$.

14.1.10. Show that in the proof of Theorem 14.3.2, if we call $p(x) = \mathrm{irr}(\alpha, F)$, then the multiplication of two polynomials $f(\alpha), g(\alpha) \in F(\alpha)$ defined there can also be defined as follows: take the product $f(\alpha)g(\alpha)$ to be the unique remainder given by the division algorithm when $f(\alpha)g(\alpha)$ is divided by $p(\alpha)$.

EXERCISES FOR SECTIONS 14.4 AND 14.5

14.4.1. Complete the proof of Lemma 14.4.2 by showing that the map $\sigma : F(\alpha) \to F(\beta)$ defined there is a bijection.

(HINT: You need to use the bases of $F(\alpha)$ and $F(\beta)$ over F.)

14.4.2. Show that if F' is an extension field of F and α, β are algebraic over F then

$$F(\alpha, \beta) \underset{F}{\cong} F(\beta, \alpha).$$

14.4.3. In the proof of Kronecker's theorem it is said that one can mimic the construction in the proof of Theorem 14.3.2 to show that $F(\alpha)$ as defined in the proof of Theorem 14.45.1 is a field of finite degree over F. Go through the details of this for $F(\alpha)$ as defined in Kronecker's theorem.

14.4.4. In the proof of Theorem 14.4.2 show that $\mathrm{irr}(\alpha, \mathbb{Q}) = x^n - 2$.

(HINT: Use Eisenstein's criterion from Chapter 12.)

14.4.5. (a) Let R be a commutative ring with identity, $I \subset R$ an ideal and $r \in I$. Prove that $\langle r, I \rangle$ as defined in the text forms an ideal containing r and I.

(b) Show using (a) that if $r, s \in R$ then

$$\langle r, s \rangle = \{rx + sy : x, y \in R\}$$

is also an ideal. This is called the **ideal generated by** r and s and contains r and s as long as R has an identity.

14.4.6. (a) Show that there exists an irreducible polynomial of degree 3 in $\mathbb{Z}_3[x]$.

(b) Let F' be an extension of a finite field $\mathbb{Z}_p$ with p a prime. Let $\alpha \in F'$ be algebraic over $\mathbb{Z}_p$ of degree n. How do you know that such an F' and α exist?

(c) Prove that $|\mathbb{Z}_p(\alpha)| = p^n$.

(d) Use the results of (a) and (c) to show that there exists a field with 27 elements.

14.4.7. Prove that the map $\phi : F \to F' = F[x]/\langle f(x)\rangle$ defined in the second proof of Kronecker's theorem by $\phi(f) = f + \langle f(x)\rangle$ is one-to-one. (HINT: use degrees.)

14.4.8. Again let $F, F', f(x)$ be as in Problem 14.4.7. Suppose $f(x) = a_0 + a_1 x + \cdots + a_n x^n$ where $a_i \in F$ and let $\overline{x} = x + \langle f(x)\rangle$. Prove that

$$f(\overline{x}) = \overline{f(x)} = f(x) + \langle f(x)\rangle.$$

(HINT: Consider how you do operations in the quotient ring

$$F' = F[x]/\langle f(x)\rangle.)$$

14.4.9. In Example 14.5.1 it was shown that $\mathbb{C} \underset{\mathbb{R}}{\cong} \mathbb{R}(\alpha)$ where $\alpha^2 = -1$ using the method of Kronecker's theorem (Theorem 14.5.1). Go back and re-do this using the method of the proof of Kronecker's theorem which uses Theorem 14.5.2. Here note that $f(x) = x^2 + 1 \in \mathbb{R}[x]$ is well known to have no zeros in $\mathbb{R}$. Thus it is irreducible (why?). Then consider $F' = \mathbb{R}[x]/\langle x^2 + 1\rangle$. Show that this is a field, that $\mathbb{R}$ embeds into it and that $\alpha = x + \langle x^2 + 1\rangle$ is then a root of $f(x)$ in F'.

14.4.10. Consider $f(x) = x^4 - 5x^2 + 6 \in \mathbb{Q}[x]$. This factors as $(x^2 - 3)(x^2 - 2)$. Explain why each factor is irreducible over $\mathbb{Q}$.

14.4.11. In each case show that the given number $\alpha \in \mathbb{C}$ is algebraic over $\mathbb{Q}$ by finding a suitable polynomial.

(a) $1 + \sqrt{3}$ (b) $\sqrt{5} + \sqrt{7}$ (c) $1 - i$

14.4.12. (a) Show that $f(x) = x^2 + x + 1 \in \mathbb{Z}_2[x]$ is irreducible.

(b) Use Kronecker's theorem to show that there must exist a field with four elements.

(c) Write addition and multiplication tables for the field you found in (b).

(HINT: For (b) see Problem 14.4.9.)

EXERCISES FOR SECTIONS 14.6 – 14.9

14.6.1. Let H be any algebraic structure. We suppose that H has two binary operations $*$ and $\circ$ defined on it but it can have any number. Let $\text{Aut}(H)$ be the set of all automorphism of H where an automorphism is an isomorphism $\phi : H \to H$. Prove that $\text{Aut}(H)$ is a group under composition of functions.

(HINT: To show that ϕ^{-1} preserves the operations if ϕ does let $a, b \in H$. We must show that $\phi^{-1}(a * b) = \phi^{-1}(a) * \phi^{-1}(b)$. But $\phi^{-1}(a * b) = c \in H$ if and only if $\phi(c) = a*b$. If $\phi^{-1}(a) = d, \phi^{-1}(b) = e$ then $\phi(d) = a, \phi(e) = b$ and proceed from here. Since ϕ preserves $*$ we have $\phi(d*e) = \phi(d)*\phi(e) = a*b$. So $\phi^{-1}(a * b) = d * e$ (why?). But $d * e = \phi^{-1}(a) * \phi^{-1}(b)$ why? This shows it for $*$. Now do the same for $\circ$.)

14.6.2. Now prove that $\text{Aut}_F(F')$ is a subgroup of $\text{Aut}(F')$. (From Problem 14.6.1 we know that $\text{Aut}(F')$ is a group under composition.

14.6.3. Show that the polynomial $f(x) = x^4 - 5x^2 + 6 \in \mathbb{Q}[x]$, splits in $\mathbb{Q}(\sqrt{2}, \sqrt{3})$. Using degrees show that this must be the splitting field for $f(x)$.

14.6.4. (a) Consider $\mathbb{Q}(\sqrt{2}.\sqrt{3})$. If we view this as $\mathbb{Q}(\sqrt{2})(\sqrt{3})$ then explain why $\sigma_1(a + b\sqrt{2}) = a - b\sqrt{2}$ for $a, b \in \mathbb{Q}(\sqrt{3})$ is an automorphism of $\mathbb{Q}(\sqrt{2}, \sqrt{3})$.

(b) What is the fixed field of the automorphism σ_1 from part (a)?

(c) Similarly let $\sigma_2(a + b\sqrt{3}) = a - b\sqrt{3}$ with $a, b \in \mathbb{Q}(\sqrt{2})$ is an automorphism. Let $\sigma_3 = \sigma_1\sigma_2$. What is the fixed field of σ_3?

14.6.5. Continuing from Problem 14.6.4 let $F' = \mathbb{Q}(\sqrt{2}, \sqrt{3})$ and let G be the group of all automorphisms of F'. What is its fixed field?

(HINT: Consider a basis for $\mathbb{Q}(\sqrt{2}, \sqrt{3})$ and the effect of σ_1 and σ_2 on the non-identity elements in the basis.)

14.6.6. The group G in Problem 14.6.5 is $\text{Aut}_\mathbb{Q}(F')$. In the notation of Problem 14.6.5, the elements of this group are $\{1, \sigma_1, \sigma_2, \sigma_3\}$.

(a) Write a multiplication table for this group.

(b) Identify this group.

(c) What do you note about $|G|$ and $|F' : \mathbb{Q}|$?

14.6.7. (a) Consider Example 14.7.1. Show that $x^3 - 2$ is irreducible over $\mathbb{Q}$.

(b) Show that $\mathrm{Aut}_{\mathbb{Q}}(2^{1/3}, i\sqrt{3}) \cong S_3$.

(HINT: For (b) From the note after Lemma 14.4.3 what must any automorphism in $\mathrm{Aut}_{\mathbb{Q}}(\mathbb{Q}(2^{1/3}, i\sqrt{3})$ do to any root of $x^3 - 2$.)

14.6.8. Determine whether or not there are finite fields for each of the following orders: (A calculator might be helpful here.)

(a) 83521   (b) 50625   (c) 78125

14.6.9. This exercise gives an alternate construction of finite fields. Let p be a prime and let $f(x)$ be an irreducible polynomial of degree n over $\mathbb{Z}_p$. Then

(a) Explain why $\langle f(x) \rangle$ is a maximal ideal in $\mathbb{Z}_p[x]$.
(b) Now show that $F = \mathbb{Z}_p[x]/\langle f(x) \rangle$ is a field.
(c) Finally show that $|F| = p^n$.

14.6.10. Prove that if it is true that any finite field of p^n elements is isomorphic to $\mathbb{Z}_p[x]/\langle f(x) \rangle$ where $f(x)$ is an irreducible polynomial of degree n in $\mathbb{Z}_p[x]$, then if F_1, F_2 are finite fields with $|F_1| = |F_2| = p^n$ then $F_1 \cong F_2$.

14.6.11. Show that in the proof of Theorem 14.8.2 then the degree of $\mathrm{irr}(\alpha, \mathbb{Z}_p)$ is equal to n.

14.6.12. Show that the map ϕ defined in the proof of Theorem 14.8.2 is an isomorphism.

(HINT: To show that it preserves products of elements $g(\alpha), h(\alpha) \in \mathbb{Z}_p(\alpha)$ use the result of Problem 14.1.10 in the exercises for Sections 14.1–14.3.)

14.6.13. Tell whether the following $\alpha \in \mathbb{C}$ are algebraic or transcendental over the given field.

(a) $\alpha = \sqrt{\pi}$ over $\mathbb{Q}(\pi)$.
(b) $\alpha = \pi^2$ over $\mathbb{Q}$.
(c) $\alpha = e^2$ over $\mathbb{Q}(e^3)$.

14.6.14. Let F' be an extension field of F and let $\alpha \in F'$ be transcendental over F. Prove that every element of $F(\alpha)$ that is not in F must also be transcendental over F.

(HINT: Use the description given after Lemma 14.9.1 of what any element in the transcendental extension $F(\alpha)$ looks like.)

Chapter 15

A Survey of Galois Theory

15.1 An Overview of Galois Theory

Galois theory is that branch of mathematics that deals with the interplay of the algebraic theory of fields, the theory of equations and finite group theory. Much of the foundation for Galois theory, involving algebraic extensions of fields, was introduced in the previous sections. This theory was introduced by Evariste Galois about 1830 in his study of the insolvability by radicals of quintic (degree 5) polynomials, a result proved somewhat earlier by Ruffini and independently by Abel. Galois was the first to see the close connection between field extensions and permutation groups. In doing so he initiated the study of finite groups. He was the first to use the term group, as an abstract concept although his definition was really just for a closed set of permutations. The method Galois developed not only facilitated the proof of the insolvability of the quintic and higher powers but led to other applications and to a much larger theory as well. In this chapter, however, we will only present a survey of the theory and then give some applications including a proof of the fundamental theorem of algebra.

The main idea of Galois theory is to associate to certain special types of algebraic field extensions called **Galois extensions** a group called the **Galois group**. The properties of the field extension will be reflected in the properties of the group, which are somewhat easier to examine. Thus, for example, solvability by radicals can be translated into a group property called solvability of groups. Showing that for every polynomial of degree 5 or greater, there exists a field extension whose Galois group is not solvable proves that there cannot be a general formula for solvability by radicals for such polynomials. We give an outline of a proof of this in Section 15.6.

The tie-in to the theory of equations is as follows: If $f(x) = 0$ is a polynomial equation over some field F, we can form the splitting field K. This is usually a Galois extension, and therefore has a Galois group called the **Galois group of the equation**. As before, properties of this group will reflect properties of this equation. Galois theory depends in part on the theory of finite groups that we examined in Chapters 8, 9 and 10. In Section 15.2 we introduce the properties of normality and separability of field extensions that define Galois extensions and in the subsequent section develop the Galois group and its construction. We next summarize all the results in the **fundamental theorem of Galois theory** which describes the interplay between the Galois groups and Galois extensions. With all the machinery in place, in Section 15.5 we will present a proof of the fundamental theorem of algebra. Finally, we close the chapter by giving two additional applications of Galois theory. The first is a sketch of the proof of the insolvability of the quintic, while the second is a discussion of certain geometric ruler and compass constructions and their algebraic interpretations. Since our aim is to arrive rather quickly at the main results of Galois theory many of the more difficult proofs along the way will be omitted. As a general reference for proofs left out see [F] and [CFR].

15.2   Galois Extensions

Galois theory deals with certain special types of finite algebraic extensions. In particular, we need two special properties – separability and normality. Normality is the simpler one so we discuss that first. For the remainder of the chapter all extensions are to be considered finite extensions.

Definition 15.2.1. K *is a* **normal extension** *of a ground field F if K is a splitting field over F.*

Several facts about normal extensions are crucial for us. These are given in the next theorem which we state without proof.

Theorem 15.2.1. *Suppose K is a normal extension of F and suppose*

$$F \subset E \subset K \subset \overline{F},$$

where $\overline{F}$ is an algebraic closure of F. Then:

(1) Any automorphism of $\overline{F}$ leaving F fixed maps K onto itself and is thus an automorphism of K leaving F fixed. Thus any isomorphism of K within $\overline{F}$ leaving F fixed is actually an automorphism of K.

(2) Every irreducible polynomial in $F[x]$ having a root in K splits in K.

(3) K is a normal extension of E.

The other major property is separability. This concerns multiplicity of roots.

Definition 15.2.2. *If α is a root of $f(x)$ then α has* **multiplicity** *$m \geq 1$ if $f(x) = (x - \alpha)^m g(x)$, where $g(\alpha) \neq 0$. If $m = 1$, then α is a* **simple root** *otherwise it is a* **multiple root**.

Now, suppose K is a finite extension of F and $\alpha \in K$. Then α is **separable** *over F if α is a simple root of $irr(\alpha, F)$. K is a* **separable extension** *if every $\alpha \in K$ is separable over F.*

Thus in a separable extension of F, if $\alpha \notin F$, then α is not a multiple nonsimple root of its irreducible polynomial. Although separability is an essential property for general Galois extensions it will not play a major role here since for the applications we will present the ground fields worked with are extensions of $\mathbb{Q}, \mathbb{R}$, or $\mathbb{C}$. All these fields have characteristic zero and this forces any finite extension to be separable.

We recall some facts about the characteristic of rings and fields that were introduced in Chapters 3, 5 and 11. We refer the reader back to those chapters for proofs of the results below.

Definition 15.2.3. *A field F has* **characteristic** *n if n is the least positive integer such that $n \cdot 1 = 0$ in F. We denote this by char $F = n$. If no such n exists, we say that F has* **characteristic zero** *denoted by char $F = 0$.*

EXAMPLE 15.2.1. char $\mathbb{Q}$ = char $\mathbb{R}$ = char $\mathbb{C} = 0$, and thus any extension of these has characteristic zero. On the other hand, char $\mathbb{Z}_p = p$, and thus any extension of $\mathbb{Z}_p$ also has characteristic p.

Lemma 15.2.1. *The characteristic of a field is zero or a prime.*

Lemma 15.2.2. *(see Corollary 11.2.2) If char $F = 0$, then F contains a subfield isomorphic to $\mathbb{Q}$. If char $F = p$, then F contains a subfield isomorphic to $\mathbf{Z_p}$. In particular, a field of characteristic zero must be infinite.*

The relevance of characteristic zero to separability is the following theorem.

Theorem 15.2.2. *Any finite extension of a field of characteristic zero must be a separable extension.*

In fact, any finite extension of a finite field is also separable so the only bad cases are infinite fields of characteristic p. For our purposes, what is important is that any extension of $\mathbb{Q}, \mathbb{R}$, or $\mathbb{C}$ is separable. Separable extensions are essential to the interplay between field extensions and group theory because of the following two results, for the second of which we give the proof.

Theorem 15.2.3. *Let $F \subset K \subset \overline{F}$ where $\overline{F}$ is an algebraic closure of F. If K is a finite separable extension of F, then the number of isomorphisms of K fixing F onto a subfield of $\overline{F}$ is finite and equal to the degree $|K : F|$.*

Theorem 15.2.4. *(Primitive Element Theorem) If K is a finite separable normal extension of F, then K is a simple extension. That is, $K = F(\alpha)$ for some $\alpha \in K$.*

Proof. Since K is a finite extension, $K = F(\alpha_1, \ldots, \alpha_r)$ for some elements $\alpha_1, .., \alpha_r \in K$ by Theorem 14.4.3. By induction, it is enough to show that if $K = F(\alpha, \beta)$, then $K = F(\gamma)$ for some $\gamma \in K$.

Case (1): F is a finite field.

In this case since $|K : F| < \infty$ we have $|K| < \infty$. But then by Theorem 6.4.7 the set of nonzero elements in K, $K^\star$, forms a cyclic group. Let γ be a generator of this group. Then clearly $K = F(\gamma)$.

Case (2): F is an infinite field.

Here let $n = |K : F|$. Since K is separable, Theorem 15.2.3 implies there exist exactly n distinct isomorphisms $\sigma_1, \ldots, \sigma_n$ of K fixing F onto some subfield, say E, of $\overline{F}$ the algebraic closure of F. Form the polynomial

$$p(x) = \prod_{i \neq j} (\sigma_i(\alpha) + x\sigma_i(\beta) - \sigma_j(\alpha) - x\sigma_j(\beta)).$$

We first claim that $p(x)$ is not the zero polynomial as an element of $E[x]$. To see this note that if $p(x) = 0$ (the zero polynomial) then at least one of the factors above must be zero since $\overline{F}[x]$ is an integral domain. This forces for some $i, j \in \{1, 2, \ldots, n\}$ with $i \neq j$ that $\sigma_i(\alpha) = \sigma_j(\alpha)$ and $\sigma_i(\beta) = \sigma_j(\beta)$. But since $K = F(\alpha, \beta)$ if the above is true then $\sigma_i = \sigma_j$ on K. This contradicts the fact that the $\sigma_i, i = 1, \ldots, n$ were distinct isomorphisms of K. Thus $p(x)$ is not the zero polynomial in $\overline{F}[x]$ and hence in $E[x]$. Hence, since F is infinite, there exists a $c \in F$ with $p(c) \neq 0$.

We next claim that the elements $\sigma_i(\alpha + c\beta), i = 1, \ldots, n$, are distinct. In order to see this, consider the evaluation map

$$0 \neq p(c) = \prod_{i \neq j} (\sigma_i(\alpha) + c\sigma_i(\beta) - \sigma_j(\alpha) - c\sigma_j(\beta)).$$

Hence it must be true that for all $i \neq j$, $\sigma_i(\alpha) + c\sigma_i(\beta) \neq \sigma_j(\alpha) + c\sigma_j(\beta)$. But since $c \in F$ we have $c = \sigma_i(c) = \sigma_j(c)$ and this forces $\sigma_i(\alpha + c\beta) \neq \sigma_j(\alpha + c\beta)$. Thus all $\sigma_i(\alpha + c\beta)$ are distinct.

It follows that $F(\alpha + c\beta)$ has degree at least n over F. To see this consider $\mathrm{irr}(\alpha + c\beta, F)$. Suppose that

$$f(x) = \mathrm{irr}(\alpha + c\beta, F) = x^k + a_1 x^{k-1} + \cdots + a_{k-1}x + a_k$$

where $a_i \in F$ for all i. We note that since each σ_i is an automorphism of K leaving F fixed that if $f(d) = 0$ then

$$\sigma_i(f(d)) = \sigma_i(0) = 0 = \sigma_i(d)^k + \cdots + a_k = f(\sigma_i(d)).$$

So that if d is a root of $f(x)$ then so is $\sigma_i(d)$. But $\alpha + c\beta$ is a root of $f(x)$ and so all of $\sigma_i(\alpha + c\beta)$ are roots for all $i = 1, 2, \ldots, n$. Since the $\sigma_i(\alpha + c\beta)$ are all distinct this shows that $\deg(f(x)) \geq n$. Thus $|F(\alpha, \beta) : F| = \deg(f(x)) \geq n$.

But $F(\alpha + c\beta) \subset F(\alpha, \beta)$, and $F(\alpha, \beta)$ has degree n over F. Therefore, $F(\alpha, \beta) = F(\alpha + c\beta)$. If $K = F(\alpha)$, α is called a **primitive element** of K over F. $\qquad \square$

With these properties introduced we can define Galois extensions.

Definition 15.2.4. *A **Galois extension** of F is a finite separable normal extension, that is, a finite separable splitting field over F.*

Notice that if F has characteristic zero, then a Galois extension is just a finite extension splitting field over F. Suppose that F is a field of characteristic zero, so that all finite extensions are separable, and suppose that

$f(x) \in F[x]$ is an irreducible polynomial. If K is the splitting field for $f(x)$ over F, then K is a Galois extension of F. K is called the **Galois extension of F relative to $f(x)$**.

We close this section by summarizing what we know so far about Galois extensions.

Suppose $F \subset E \subset K \subset \overline{F}$ with K Galois over F and $\overline{F}$ an algebraic closure of F. Then:

(1) K is also Galois over E.

(2) The number of automorphisms of K fixing F is equal to the degree $|K : F|$.

(3) Any isomorphism of K within $\overline{F}$ fixing F is actually an automorphism of K.

(4) K is a simple extension of F.

15.3   Automorphisms and the Galois Group

We now introduce the Galois group and discuss the interplay between the group theory and the field extensions. We suppose that K is a finite extension of F. Recall from Section 14.7 that $\text{Aut}(K)$ forms a group under composition and that the elements of $\text{Aut}(K)$ that fix F form a subgroup. If K is Galois over F, this subgroup is the Galois group of K over F. In 14.6, we called this $\text{Aut}_F(K)$. Since here we have additional hypotheses we make the following definition.

Definition 15.3.1. *Let K be a Galois extension of F. Then the group of automorphisms of K that fix F is called the* **Galois group** *of K over F, denoted by $Gal(K/F)$. If H is a subgroup of $Gal(K/F)$, we let K^H denote the elements of K fixed by H. That is, K^H = the fixed field of H (see Theorem 14.7.1).*

Since K is Galois over F, it is separable and normal. Then from Theorem 15.2.1 (1) and Theorem 15.2.3 we have:

Lemma 15.3.1. $|Gal(K/F)| = |K : F|$.

If E is an intermediate field, then K is also Galois over E. Those automorphisms in $\text{Gal}(K/F)$ that also fix E form a subgroup. Thus $\text{Gal}(K/E)$ is a subgroup of $\text{Gal}(K/F)$. Conversely, if H is a subgroup of $\text{Gal}(K/F)$, then K^H is an intermediate field and $\text{Gal}(K/K^H) = H$.

Lemma 15.3.2. *Suppose $F \subset E \subset K$ with K Galois over F. Then:*

(1) K is Galois over E and $\text{Gal}(K/E)$ is a subgroup of $\text{Gal}(K/F)$.

(2) If H is a subgroup of $\text{Gal}(K/F)$, then $E = K^H$ is an intermediate field and $\text{Gal}(K/E) = H$.

Proof. Part (1) is clear. We have already done most of the work in Theorem 14.7.1.

For part (2) we must show three things: that K^H is a field, that $F \subset K^H$, and that $\text{Gal}(K/K^H) = H$. Since $H \subset \text{Gal}(K/F)$, every element of F is fixed by each element of H. Therefore, $F \subset K^H$. The fact that K^H is an intermediate field between K and H follows from Theorem 14.7.1. Finally, if $E = K^H$, then $\text{Gal}(K/E)$ consists of those automorphisms of K that fix E. But by definition E consists of those elements of K fixed by elements of H. Therefore, $\text{Gal}(K/E) = H$. $\qquad\square$

Corollary 15.3.1. *The map $\tau : H \to K^H$ from subgroups H of $\text{Gal}(K/F)$ to intermediate fields between K and F is a bijection.*

From the previous results we have that if $F \subset E \subset K$ with K Galois over F, then K is Galois over E. The question naturally arises as to when E is also Galois over F. E is separable over F, so the question then becomes when is E a normal extension of F. This has the simple and elegant answer that E is a normal extension of F if and only if the corresponding subgroup of $\text{Gal}(K/F)$ is a normal subgroup of $\text{Gal}(K/F)$.

Lemma 15.3.3. *Suppose $F \subset E \subset K$ with K Galois over F and suppose $E = K^H$. Then E is Galois over F if and only if H is a normal subgroup of $\text{Gal}(K/F)$. In this case*

$$\text{Gal}(E/F) \cong G/H \cong \text{Gal}(K/F)/\text{Gal}(K/E).$$

For a proof of this lemma we refer to [F] and [CFR]. We have the following theorem of Artin.

Theorem 15.3.1. *(Artin) Let K be a field and G a finite group of automorphisms of K with $|G| = n$. Assume that K is a finite extension of the fixed field K^G. Then K is a Galois extension of $F = K^G$ and the Galois group of K over F is G.*

In order to prove Artin's theorem we need the following lemma whose proof we give as an exercise.

Lemma 15.3.4. *Let F be a field and $\sigma \in Aut(F)$. Then we can extend σ to an automorphism of $F[x]$ in a natural manner by taking*

$$\sigma(\sum_{i \geq 0} a_i x^i) = \sum_{i \geq 0} \sigma(a_i) x^i.$$

We note that in this lemma we are calling the map which results by the same notation as the original map, i.e., σ.

Proof. (Artin's Theorem) We define a polynomial over F to be **separable** if it is nonconstant, i.e., of degree ≥ 1, and if each of its irreducible factors has only simple roots in its splitting field. (Recall $F[x]$ is a UFD.) If α is a root of a separable polynomial $f(x)$ over F then $\mathrm{irr}(\alpha, F)|f(x)$. To see this, just recall that $\mathrm{irr}(\alpha, F)$ is the monic polynomial of minimal degree which α satisfies. If $\mathrm{irr}(\alpha, F)$ did not divide $f(x)$ then the division algorithm for polynomials applied to $\mathrm{irr}(\alpha, F)$ divided into $f(x)$ would give a contradiction to the minimality of degree.

Thus since $\mathrm{irr}(\alpha, F)|f(x)$ and $f(x)$ is separable by definition $\mathrm{irr}(\alpha, F)$ can have only simple roots. This means that α is a simple root of $\mathrm{irr}(\alpha, F)$ and hence α is separable over F.

Now suppose $F = K^G$ and let α be an arbitrary element of K. Since the group G is finite, $|G| = n$, there exist a maximal set of elements, $\{g_1, \ldots, g_r\}$ with $r \leq n$, such that $g_1(\alpha), \ldots, g_r(\alpha)$ are all distinct.

We claim that if $g \in G$, then $\{gg_1(\alpha), \ldots, gg_r(\alpha)\}$ is a permutation of the set $\{g_1(\alpha), \ldots, g_r(\alpha)\}$, that is,

$$\{gg_1(\alpha), \ldots, gg_r(\alpha)\} = \{g_1(\alpha), \ldots, g_r(\alpha)\}.$$

To see this note that since G is a group under composition, each gg_i for $i = 1, \ldots, r$ is an automorphism in G of K. If it were the case that any $gg_i(\alpha) \notin \{g_1(\alpha), \ldots, g_r(\alpha)\}$ this would contradict the maximality of the set

$$\{g_1(\alpha), \ldots, g_r(\alpha)\}.$$

Thus

$$\{gg_1(\alpha), \ldots, gg_r(\alpha)\} \subset \{g_1(\alpha), \ldots, g_r(\alpha)\}.$$

But since $g_i(\alpha), i = 1, \ldots, r$ are distinct and g is an automorphism this implies that the $gg_1(\alpha), \ldots, gg_r(\alpha)$ must also be distinct. Hence there are r elements in the set $\{gg_1(\alpha), \ldots, gg_r(\alpha)\}$ and thus the two sets are equal. It follows that the set $\{gg_1(\alpha), \ldots, gg_r(\alpha)\}$ is just a permutation of $\{g_1(\alpha), \ldots, g_r(\alpha)\}$.

We note that $\alpha \in \{g_1(\alpha), \ldots, g_r(\alpha)\}$ because we showed that for any $g \in G$

$$\{gg_1(\alpha), \ldots, gg_r(\alpha)\} = \{g_1(\alpha), \ldots, g_r(\alpha)\}.$$

If we were to take $g = g_1^{-1}$ then the very first entry would be α. Therefore, α is a root of the polynomial

$$f(x) = \prod_{i=1}^{r}(x - g_i(\alpha)) \in K[x].$$

Moreover by construction, $f(x)$ is a separable polynomial over K. Further, $f(x)$ is fixed by any $g \in G$. By Lemma 15.3.4, since any $g \in G$ is an automorphism of K, it can be extended to an automorphism of $K[x]$ as indicated in the statement of the lemma. Calling the extension of g also g we get

$$g(f(x)) = \prod_{i=1}^{r}(x - gg_i(\alpha)) = \prod_{i=1}^{r}(x - g_i(\alpha)) = f(x).$$

Since $f(x)$ is fixed by any $g \in G$ it follows that the coefficients of $f(x)$ lie in $F = K^G$. Since the $g_i(\alpha)$ are distinct, as previously mentioned, $f(x)$ is separable over F. Hence we have shown that every $\alpha \in K$ is a root of a separable polynomial of degree $\leq n$ with coefficients in F. Therefore, K is a separable extension of F.

Further, $f(x)$ splits into linear factors over K. For K to be a splitting field of $f(x)$ over F we need to show that K is the smallest extension of F in which $f(x)$ splits. Suppose then that $f(x)$ splits in an extension field K_1 with $F \subseteq K_1 \subset K$. Thus there exists a $\beta \in K \setminus K_1$. Since $f(x) = \prod_{i=1}^{r}(x - g_i(\alpha)) \in K_1[x]$ splits in K_1 all its zeros $g_1(\alpha), \ldots, g_r(\alpha) \in K_1$. But take any $g_0 \in G$. It is an automorphism of K so there exists $\alpha_0 \in K$ such that $g_0(\alpha_0) = \beta$. Since the α that was used in the construction of the set $\{g_1(\alpha), \ldots, g_r(\alpha)\}$ was arbitrary we now use α_0. But $g_0(\alpha_0) = \beta \notin K_1$. Hence $g_0(\alpha_0) \notin \{g_1(\alpha_0), \ldots, g_r(\alpha_0)\}$ since this is a subset of K_1. This contradicts the maximality of $\{g_1, \ldots, g_r\}$. Thus K is a splitting field for $f(x)$ over F. Therefore K is a normal extension of F. Since K is separable and normal over F, K is a Galois extension over F since we assumed that K is a finite extension of F.

Since K is a finite, separable, normal extension of F, it is a simple extension by Theorem 15.2.4, say $K = F(\gamma)$ for some $\gamma \in K$. But by what we have just proved γ is a root of a polynomial of degree $\leq n$, and thus $|K : F| \leq n$. But $G \subset \mathrm{Gal}(K/F)$, since G consists of automorphisms of K fixing F. Therefore, since K is Galois over F, by Lemma 15.3.1,

$$|K : F| = |\mathrm{Gal}(K/F)| \geq |G| = n.$$

It follows that $|K : F| = n$, and G must be the whole Galois group $\mathrm{Gal}(K/F)$.

$\square$

The results we have just outlined are the main results in Galois theory. In the next section we summarize them and give examples. For now, we examine the Galois group of a polynomial. We suppose here and in the remainder of the section that F has characteristic zero, so that all finite extensions are separable.

Definition 15.3.2. *Let $f(x)$ be a polynomial over F and let K be the splitting field for $f(x)$. Then K is Galois over F, and the corresponding Galois group of K over F is called the* **Galois group of the polynomial**.

We examine this Galois group in more detail. Let K be any algebraic extension of F. Suppose $\alpha \in K$ with $\mathrm{irr}(\alpha, F)$ its irreducible polynomial. Any other root $\overline{\alpha}$ of $\mathrm{irr}(\alpha, F)$ is called a **conjugate** of α. Now, suppose K is Galois over F and $\sigma \in \mathrm{Gal}(K/F)$. Since σ fixes F it also fixes $\mathrm{irr}(\alpha, F) \in F[x]$(see Lemma 15.3.4) and hence $\sigma(\alpha)$ must be another root of $\mathrm{irr}(\alpha, F)$. Therefore, any $\sigma \in \mathrm{Gal}(K/F)$ maps elements to their conjugates. Now, let $f(x) \in F[x]$ be irreducible, with roots $\alpha_1, \ldots, \alpha_n$ in K. If $\sigma \in \mathrm{Gal}(K/F)$, σ fixes $f(x)$ and so maps this set of roots onto itself. Therefore, we have two facts: any $\sigma \in \mathrm{Gal}(K/F)$ permutes the roots of irreducible polynomials, and conjugates of roots of irreducible polynomials are also roots. We also have the following:

Lemma 15.3.5. *If K is Galois over F with $|K : F| = n$, then $\mathrm{Gal}(K/F)$ is a subgroup of the symmetric group S_n.*

Since $|\mathrm{Gal}(K/F)| = n$ by Cayley's theorem it is a permutation group on itself and so a subgroup of S_n. What is important is how the permutations from $\mathrm{Gal}(K/F)$ are obtained. Since K is a finite extension,

$K = F(\alpha_1, \ldots, \alpha_r)$ for elements $\alpha_1, \ldots, \alpha_r \in K$. Those permutations that map $\alpha_1, \ldots, \alpha_r$ onto conjugates will lead to automorphisms and will give the elements of the Galois group. We illustrate this with an example.

EXAMPLE 15.3.1. Consider the splitting field of $x^4 + 1$ over $\mathbb{Q}$. In $\mathbb{C}$ this factors as $(x^2 + i)(x^2 - i)$, so this polynomial has four roots in $\mathbb{C}$ namely

$$\omega_1 = \frac{1+i}{\sqrt{2}} = \sqrt{i}, \qquad \omega_2 = \frac{1-i}{\sqrt{2}} = \overline{\omega_1}$$

$$\omega_3 = \frac{-1+i}{\sqrt{2}} = i\omega_1, \qquad \omega_4 = \frac{-1-i}{\sqrt{2}} = \overline{\omega_3}.$$

These roots can be easily found by DeMoivre's theorem (see Chapter 5). Notice that ω_1 is a primitive eighth root of unity. Recall that this means the order of ω_1 is 8 in the multiplicative group $\mathbb{C}^*$ of nonzero complex numbers. Further $\omega_1^3 = \omega_3, \omega_1^5 = \omega_4, \omega_1^7 = \omega_2, \omega_1^8 = 1$. Therefore, ω_1 is a primitive element of K over $\mathbb{Q}$. By Eisenstein's criterion (as used in Example 12.4.2 — see the exercises) $x^4 + 1$ is irreducible over $\mathbb{Q}$ so $\mathrm{irr}(\omega_1, \mathbb{Q}) = x^4 + 1$, and $K = \mathbb{Q}(\omega_1)$. It follows that $|K : \mathbb{Q}| = 4$. Let us examine the Galois group. Since ω_1 is a primitive root an automorphism σ is completely determined by its action on ω_1. There are exactly four elements in the Galois group and four possible conjugates of ω_1. Mapping ω_1 onto each will determine the four different automorphisms. Suppose $\sigma_1(\omega_1) = \omega_1$. Then σ_1 fixes $\omega_1^3, \omega_1^5, \omega_1^7$, and thus σ_1 fixes K and is therefore the identity. Suppose $\sigma_2(\omega_1) = \omega_2 = \omega_1^7$. Then $\sigma_2(\omega_2) = \sigma_2(\omega_1^7) = \omega_1^{49} = \omega_1$. Similarly, $\sigma_2(\omega_3) = \omega_4, \sigma_2(\omega_4) = \omega_3$. Therefore, the automorphism σ_2 is given by the permutation

$$\omega_1 \to \omega_2, \omega_2 \to \omega_1, \omega_3 \to \omega_4, \omega_4 \to \omega_3.$$

The same type of analysis shows that the two remaining automorphisms are given by the permutations

$$\sigma_3 : \omega_1 \to \omega_3, \omega_2 \to \omega_4, \omega_3 \to \omega_1, \omega_4 \to \omega_2.$$

$$\sigma_4 : \omega_1 \to \omega_4, \omega_2 \to \omega_3, \omega_3 \to \omega_2, \omega_4 \to \omega_1.$$

Checking, we see that each of these permutations has order 2. The group $G = \mathrm{Gal}(K/\mathbb{Q})$ then has the presentation

$$G = \langle \sigma_2, \sigma_3; \sigma_2^2 = \sigma_3^2 = (\sigma_2\sigma_3)^2 = 1 \rangle.$$

It is easy to see that this group is then abelian and equal to $\mathbb{Z}_2 \times \mathbb{Z}_2$, the direct product of two finite cyclic groups of order 2. This group is usually referred to as the Klein 4-group. Another way to see the structure of $\mathrm{Gal}(K/\mathbb{Q})$ is as follows. Notice that since $\omega_1 = \sqrt{i} \in K$, then $i \in K$. It follows that $1 + i \in K$, and hence $\frac{1+i}{\omega_1} = \sqrt{2} \in K$. Therefore, $\mathbb{Q}(i, \sqrt{2}) \subset K$. Further $|\mathbb{Q}(i, \sqrt{2}) : \mathbb{Q}| = 4$, since this is the extension of $\mathbb{Q}(\sqrt{2})$ by adjoining i which also has degree 2. It then follows that $K = \mathbb{Q}(i, \sqrt{2})$. Now, $\mathrm{irr}(i, \mathbb{Q}) = x^2 + 1$, so the conjugates of i are $\pm i$, while $\mathrm{irr}(\sqrt{2}, \mathbb{Q}) = x^2 - 2$, so the conjugates of $\sqrt{2}$ are $\pm\sqrt{2}$. Therefore, the four possible automorphisms in $\mathrm{Gal}(K/\mathbb{Q})$ are given by:

$$\sigma_1 : i \to i, \sqrt{2} \to \sqrt{2}, \text{ the identity automorphism}$$

$$\sigma_2 : i \to -i, \sqrt{2} \to \sqrt{2};$$

$$\sigma_3 : i \to i, \sqrt{2} \to -\sqrt{2};$$

$$\sigma_4 : i \to -i, \sqrt{2} \to -\sqrt{2}.$$

It is very clear here that each automorphism has order 2 and again we get the Klein 4-group as $\mathrm{Gal}(K/\mathbb{Q})$.

15.4   The Fundamental Theorem of Galois Theory

We now summarize without proof the results of the last two sections into one theorem called the fundamental theorem of Galois theory.

Theorem 15.4.1. *(Fundamental Theorem of Galois Theory) Let K be a Galois extension of F with Galois group $G = \mathrm{Gal}(K/F)$. For each intermediate field E, $F \subset E \subset K$, let $\tau(E)$ be the subgroup of G fixing E, i.e., $\tau(E) = \mathrm{Gal}(K/E)$. Then:*

(1) τ is a bijection between intermediate fields containing F and subgroups of G.

(2) If H is a subgroup of G and $E = K^H$, then $\tau(E) = H$.

(3) K is Galois over E, and $\mathrm{Gal}(K/E) = \tau(E)$.

(4) $|G| = |K : F|$.

(5) $|E : F| = [G : \tau(E)]$. That is, the degree of an intermediate field over the ground field is the index of the corresponding subgroup in the Galois group.

(6) E is Galois over F if and only if $\tau(E)$ is a normal subgroup of G. In this case

$$Gal(E/F) \cong G/\tau(E) \cong Gal(K/F)/Gal(K/E).$$

(7) The lattice of subfields of K containing F is the inverted lattice of subgroups of $Gal(K/F)$.

Parts (1) through (4) have already been discussed. Part (5) follows from the following theorem together with Theorem 15.2.3 which gives the equality of the number of isomorphisms of E fixing F within $\overline{F}$ with the degree $|E : F|$. We also need the isomorphism extension theorem (see Chapter 14, Theorem 14.7.2.)

Theorem 15.4.2. *Let K be a finite normal extension of F and let E be an extension of F where*

$$F \subset E \subset K \subset \overline{F}$$

where $\overline{F}$ is an algebraic closure of F. Then K is a finite normal extension of E and $Gal(K/E)$ is precisely the subgroup of $Gal(K/F)$ consisting of all those automorphisms that leave E fixed. Further two automorphisms $\sigma_1, \sigma_2 \in Gal(K/F)$ induce the same isomorphism of E onto a subfield of $\overline{F}$ if and only if they are in the same left coset of $Gal(K/E)$ in $Gal(K/F)$.

For proofs of this theorem and part (6) see [F] and [CFR].

We say more about part (7). By the lattice of subfields of K over F we mean the complete collection of intermediate fields and their interrelationships by inclusion. Similarly for the lattice of subgroups. One is the inverted lattice of the other, since in the field case if E is a subfield, the degree $|K : E|$ is the order of $Gal(K/E)$, while its index is the degree of $|E : F|$. From Lagrange's theorem (Chapter 6),

$$|G| = [G : \tau(E)]|\tau(E)|.$$

But then from above

$$|G| = |E : F||Gal(K/E)|.$$

Thus we see that for fixed G if $|E : F|$ gets bigger, i.e., the field E gets bigger, then the $|\text{Gal}(K/E)|$ must get smaller and vice versa. Therefore, the subfields are from above in the lattice while the subgroups are from below.

The standard notation for lattices is that objects higher contain objects lower and if K is an extension field of the field F while G is the group of all automorphisms of K which leave F fixed, we have Figure 15.1, where $\text{Gal}(K/K) = \{1\}$, the trivial group. Here 1 means the identity map.

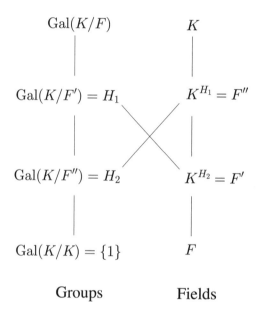

Figure 15.1. Lattice of Groups and Fields

So the inverse inclusion relation is as follows:

$$K^{H_1} \subset K^{H_2} \Leftrightarrow H_2 \subset H_1.$$

This is easy to see. Suppose $H_2 \subset H_1$ and let $x \in K^{H_1}$. This implies that $\sigma(x) = x$ for all $\sigma \in H_1$. Since $H_2 \subset H_1$ this implies $\sigma(x) = x$ for all $\sigma \in H_2$ so that $x \in K^{H_2}$ and therefore $K^{H_1} \subset K^{H_2}$.

Conversely suppose that $K^{H_1} \subset K^{H_2}$ and let $\sigma \in H_2$. This implies that $\sigma(x) = x$ for all $x \in K^{H_2}$ and hence $\sigma(x) = x$ for all $x \in K^{H_1}$. By definition K^{H_1} consists of all elements of K left fixed by precisely the elements of H_1 and therefore $\sigma \in H_1$. It follows that $H_2 \subset H_1$.

We then illustrate this with three examples.

EXAMPLE 15.4.1. Consider the splitting field K of $x^4 + 1$ over $\mathbb{Q}$. As we saw in example 15.3.1, $K = \mathbb{Q}(i, \sqrt{2})$. Then $|K : \mathbb{Q}| = 4$, and the Galois group is $\mathbb{Z}_2 \times \mathbb{Z}_2$. There are then four automorphisms in $\text{Gal}(K/\mathbb{Q})$, given by:

$$1 : i \to i, \sqrt{2} \to \sqrt{2},$$

$$\sigma : i \to i, \sqrt{2} \to -\sqrt{2},$$

$$\tau : i \to -i, \sqrt{2} \to \sqrt{2},$$

$$\sigma\tau : i \to -i, \sqrt{2} \to -\sqrt{2}.$$

Each of these has order 2, and therefore there are five total subgroups of $G = \text{Gal}(K/\mathbb{Q})$, namely,

$$\{1\}, H_1 = \{1, \tau\}, H_2 = \{1, \sigma\tau\}, H_3 = \{1, \sigma\}, G.$$

We exhibit the five intermediate fields. The fixed field of G is precisely $\mathbb{Q}$, while the fixed field of $\{1\}$ is all of K. Now, consider H_3. Since $i \in K$ and $\mathbb{Q}(i) \subset K$, $|K : \mathbb{Q}(i)| = 2$, so $\mathbb{Q}(i)$ will correspond to a subgroup of index 2 and thus order 2. Now, σ fixes i but not $\sqrt{2}$. Therefore, σ fixes $\mathbb{Q}(i)$ but not all of K, so the fixed field of H_3 is $\mathbb{Q}(i)$. Similarly, $\mathbb{Q}(\sqrt{2}) \subset K$, and then $\mathbb{Q}(\sqrt{2})$ is the fixed field of H_1. Finally, since $i, \sqrt{2} \in K$, it follows that $i\sqrt{2} \in K$, so $\mathbb{Q}(i\sqrt{2}) \subset K$. Since $\text{irr}(i\sqrt{2}, \mathbb{Q}) = x^2 + 2, |\mathbb{Q}(i\sqrt{2}) : \mathbb{Q}| = 2$. The automorphism $\sigma\tau$ fixes $i\sqrt{2}$; hence $\mathbb{Q}(i\sqrt{2})$ is the fixed field of H_2. We illustrate these relationships in Figures 15.2 and 15.3.

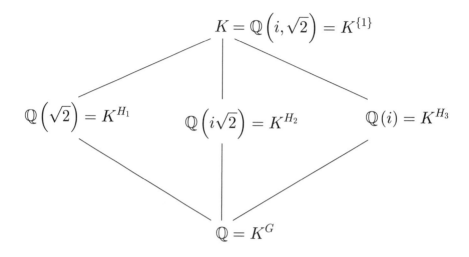

Figure 15.2. Subfield Lattice for $\mathbb{Q}(i, \sqrt{2})$

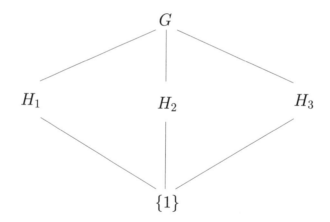

Figure 15.3. Subgroup Lattice of the Galois Group

EXAMPLE 15.4.2. Let $K = \mathbb{Q}(\sqrt{2}, \sqrt{3})$. This is the splitting field of $(x^2 - 2)(x^2 - 3)$ over $\mathbb{Q}$, so K is Galois over $\mathbb{Q}$. Since $|\mathbb{Q}(\sqrt{2}) : \mathbb{Q}| = 2$ and $\sqrt{3} \notin \mathbb{Q}(\sqrt{2})$, we have $|K : \mathbb{Q}| = 4$. The Galois group can then be described by the four automorphisms

$$1 : \sqrt{2} \to \sqrt{2}, \sqrt{3} \to \sqrt{3},$$

$$\sigma : \sqrt{2} \to \sqrt{2}, \sqrt{3} \to -\sqrt{3},$$

$$\tau : \sqrt{2} \to -\sqrt{2}, \sqrt{3} \to \sqrt{3},$$

$$\sigma\tau : \sqrt{2} \to -\sqrt{2}, \sqrt{3} \to -\sqrt{3}.$$

It is easy to see that this group is $\mathbb{Z}_2 \times \mathbb{Z}_2$ and is isomorphic to the Galois group of the previous example. Thus different field extensions of the same ground field can have isomorphic Galois groups. The corresponding subgroups and fixed fields are given below:

$$H_0 = \{1\} \Rightarrow \text{ fixed field is } K = \mathbb{Q}(\sqrt{2}, \sqrt{3}), \tag{15.1}$$

$$H_1 = \{1, \sigma\} \Rightarrow \text{ fixed field is } \mathbb{Q}(\sqrt{2}), \tag{15.2}$$

$$H_2 = \{1, \tau\} \Rightarrow \text{ fixed field is } \mathbb{Q}(\sqrt{3}), \tag{15.3}$$

$$H_3 = \{1, \sigma\tau\} \Rightarrow \text{ fixed field is } \mathbb{Q}(\sqrt{2}\sqrt{3}) = \mathbb{Q}(\sqrt{6}), \tag{15.4}$$

$$H_4 = G \Rightarrow \text{ fixed field is } \mathbb{Q}. \tag{15.5}$$

EXAMPLE 15.4.3. As a somewhat more complicated example, consider the splitting field K of $x^4 - 2$ over $\mathbb{Q}$. We have over $\mathbb{C}$ the factorization $x^4 - 2 = (x^2 - \sqrt{2})(x^2 + \sqrt{2})$, so if $\omega = 2^{1/4}$, the four roots are $\omega, i\omega, -\omega, -i\omega$. Therefore, $i\omega/\omega = i \in K$ so $\mathbb{Q}(i, \omega) \subset K$. But $x^4 - 2$ splits in $\mathbb{Q}(i, \omega)$, so $K = \mathbb{Q}(i, \omega)$. Now, $\mathbb{Q}(\omega)$ has degree 4 over $\mathbb{Q}$, and $i \notin \mathbb{Q}(\omega)$, since $\omega \in \mathbb{R}$. Therefore, $|K : \mathbb{Q}(\omega)| = 2$, since we are adjoining i. Hence $|K : \mathbb{Q}| = |K : \mathbb{Q}(\omega)||\mathbb{Q}(\omega) : \mathbb{Q}| = 8$.

The four conjugates of ω are $\omega, i\omega, -\omega, -i\omega$, while the conjugates of i are $\pm i$. Let σ and τ be the automorphisms of K given by

$$\sigma : \omega \to i\omega, i \to i,$$

$$\tau : \omega \to \omega, i \to -i.$$

Then the eight automorphisms in $\mathrm{Gal}(K/\mathbb{Q})$ are

$$1 : \omega \to \omega, i \to i, \tag{15.6}$$
$$\sigma : \omega \to i\omega, i \to i, \tag{15.7}$$
$$\sigma^2 : \omega \Rightarrow -\omega, i \Rightarrow i, \tag{15.8}$$
$$\sigma^3 : \omega \Rightarrow -i\omega, i \Rightarrow i, \tag{15.9}$$
$$\tau : \omega \to \omega, i \to -i, \tag{15.10}$$
$$\sigma\tau : \omega \to i\omega, i \to -i, \tag{15.11}$$
$$\sigma^2\tau : \omega \to -\omega, i \to -i, \tag{15.12}$$
$$\sigma^3\tau : \omega \to -i\omega, i \to -i. \tag{15.13}$$

A computation shows that $\sigma^4 = 1, \tau^2 = 1$ and $\sigma\tau = \tau\sigma^3$. The group is then D_4, the dihedral group of order 8 that represents the symmetries of a square. This group has the presentation

$$D_4 = \langle \sigma, \tau ; \sigma^4 = 1, \tau^2 = 1, \sigma\tau = \tau\sigma^3 \rangle.$$

There are ten total subgroups of D_4. We list them below with the corresponding fixed fields. Notice that $\omega^2 = \sqrt{2} \in K$, so $\mathbb{Q}(i, \sqrt{2}) \subset K$.

$$H_1 = \{1\} \Rightarrow \text{ fixed field is } K = \mathbb{Q}(i, \omega) \tag{15.14}$$
$$H_2 = \{1, \sigma, \sigma^2, \sigma^3\} \Rightarrow \text{ fixed field is } \mathbb{Q}(i), \tag{15.15}$$
$$H_3 = \{1, \sigma^2\} \Rightarrow \text{ fixed field is } \mathbb{Q}(i, \sqrt{2}), \tag{15.16}$$
$$H_4 = \{1, \tau\} \Rightarrow \text{ fixed field is } \mathbb{Q}(\omega), \tag{15.17}$$
$$H_5 = \{1, \sigma\tau\} \Rightarrow \text{ fixed field is } \mathbb{Q}(\omega + i\omega), \tag{15.18}$$
$$H_6 = \{1, \sigma^2\tau\} \Rightarrow \text{ fixed field is } \mathbb{Q}(i\omega), \tag{15.19}$$
$$H_7 = \{1, \sigma^3\tau\} \Rightarrow \text{ fixed field is } \mathbb{Q}(\omega - i\omega), \tag{15.20}$$
$$H_8 = \{1, \sigma^2, \tau, \sigma^2\tau\} \Rightarrow \text{ fixed field is } \mathbb{Q}(\sqrt{2}), \tag{15.21}$$
$$H_9 = \{1, \sigma^2, \sigma\tau, \sigma^3\tau\} \Rightarrow \text{ fixed field is } \mathbb{Q}(i\sqrt{2}), \tag{15.22}$$
$$H_{10} = D_4 \Rightarrow \text{ fixed field is } \mathbb{Q}. \tag{15.23}$$

15.5   A Proof of the Fundamental Theorem of Algebra

In this section we use the Galois theory that we have developed to provide a proof of the fundamental theorem of algebra. That is, the result that every nonconstant complex polynomial has a complex zero.

Theorem 15.5.1. *(Fundamental Theorem of Algebra) The complex number field $\mathbb{C}$ is algebraically closed. That is, any nonconstant complex polynomial has a root in $\mathbb{C}$.*

Proof. If $f(x) \in \mathbb{C}[x]$, we can form the splitting field K for $f(x)$ over $\mathbb{C}$. The fundamental theorem of algebra asserts that K must be $\mathbb{C}$ itself, and hence the fundamental theorem of algebra is equivalent to the fact that any nontrivial Galois extension of $\mathbb{C}$ must be $\mathbb{C}$.

Now K is a finite separable extension of $\mathbb{C}$ so it also a finite separable extension of $\mathbb{R}$ and therefore a simple extension. Hence $K = \mathbb{R}(\alpha)$. Let $g(x) = \text{irr}(\alpha, \mathbb{R})$ and let K_1 be the splitting field of $g(x)$ over $\mathbb{R}$. K_1 is then a Galois extension of $\mathbb{R}$. We show that either $K_1 = \mathbb{R}$ or $K_1 = \mathbb{C}$ and hence, since $\mathbb{C} \subset K \subset K_1$ we have $K = \mathbb{C}$.

Suppose K is any finite extension of $\mathbb{R}$ with $|K : \mathbb{R}| = 2^m q, (2, q) = 1$. If $m = 0$, then K is an odd-degree extension of $\mathbb{R}$. However, odd-degree real polynomials always have a real root (see Chapter 12, Theorem 12.3.3), and therefore $\text{irr}(\alpha, \mathbb{R})$ is irreducible only if its degree is one. But then $\alpha \in \mathbb{R}$ and $K = \mathbb{R}$. Therefore, if K is a nontrivial extension of $\mathbb{R}, m > 0$. This shows more generally that there are no odd-degree finite extensions of $\mathbb{R}$.

Now suppose that K is a degree-2 extension of $\mathbb{C}$. Then $K = \mathbb{C}(\alpha)$ with $\deg \text{irr}(\alpha, \mathbb{C}) = 2$. But using the quadratic formula, complex quadratic polynomials always have roots in $\mathbb{C}$. This is a contradiction. Therefore, $\mathbb{C}$ has no degree 2 extensions. Now let K be a Galois extension of $\mathbb{R}$. Suppose $|K : \mathbb{R}| = 2^m q, (2, q) = 1$. From the argument above we must have $m > 0$. Let $G = \text{Gal}(K/\mathbb{R})$ be the Galois group. Then by Theorem 15.4.1 (4) $|G| = 2^m q, m > 0, (2, q) = 1$. Thus G has a 2-Sylow subgroup of order 2^m and index q. This would correspond to an intermediate field E with $|K : E| = 2^m$ and $|E : \mathbb{R}| = q$. However, then E is an odd-degree finite extension of $\mathbb{R}$. It follows that $q = 1$ and $E = \mathbb{R}$. Therefore, $|K : \mathbb{R}| = 2^m$ and $|G| = 2^m$. Now, $|K : \mathbb{C}| = 2^{m-1}$ and suppose $G_1 = \text{Gal}(K/\mathbb{C})$. This is a 2-group. If it were not trivial, then from Theorem 10.3.6 there would exist a subgroup of order 2^{m-2} and index 2. This would correspond to an intermediate field E of degree 2 over $\mathbb{C}$. However from the argument above $\mathbb{C}$ has no degree-2 extensions. It follows then that G_1 is trivial, that is, $|G_1| = 1$, so $|K : \mathbb{C}| = 1$ and $K = \mathbb{C}$, completing the proof. $\qquad\square$

We have actually proved a more general result. In the above proof, outside of Galois theory, we used two facts: odd-degree real polynomials always have real roots, and degree-2 complex polynomials have complex roots. Using these two facts as properties we could prove the following generalization.

Theorem 15.5.2. *Let K be an ordered field in which all positive elements have square roots. Suppose further that each odd-degree polynomial in $K[x]$ has a root in K. Then $K(i)$ is algebraically closed, where $i = \sqrt{-1}$ is a root of the irreducible polynomial $x^2 + 1 \in K[x]$.*

15.6   Some Applications of Galois Theory

In this final section of the book we present some very nice applications of Galois theory. First we give a proof of the insolvability of the quintic. That is, we show that there can be no formula to solve any fifth-degree or higher polynomial equation by radicals. Next we show the impossibility of certain ruler and compass constructions such as squaring the circle and trisecting an angle.

15.6.1   The Insolvability of the Quintic

Galois theory was developed primarily as a tool for handling the proof of the insolvability by radicals of quintic polynomials. In this section we outline the proof of this. In order to apply the Galois theory we must translate solvability by radicals into a group property, that is, a property that must be satisfied by the corresponding Galois group.

In Chapter 10 we introduced solvable groups and their properties. Here we recall the definition (see 10.5.1) and basic properties.

Definition 15.6.1. *G is a **solvable group** if it has a finite series of subgroups*

$$G = G_1 \supset G_2 \supset G_3 \cdots \supset G_n \supset G_{n+1} = \{1\},$$

with G_{i+1} a normal subgroup of G_i and G_i/G_{i+1} abelian.

Such a series as in the definition is called a **normal series** for G, and the terms G_i/G_{i+1} are called the **factors** of the series. The definition can then be put concisely as: A group G is solvable if it has a normal series with abelian factors. It can be shown that the class of solvable groups is closed under subgroups, factor groups, and finite direct products. That is, if G, H, are solvable groups, then so are any subgroups of G or H; any factor groups of G or H, and the direct product $G \times H$ (see Theorem 10.5.1 and Corollary 10.5.1).

Now, we must determine what is meant by solvability by radicals in terms of field extensions.

Definition 15.6.2. *Let K be an extension field of the field F. Then K is an* **extension of** *F by* **radicals** *if there exist elements $\alpha_1, \ldots, \alpha_r \in K$ and positive integers $n_1, \ldots, n_r$ such that $K = F(\alpha_1, \ldots, \alpha_r)$, with $\alpha_1^{n_1} \in F$, and $\alpha_i^{n_i} \in F(\alpha_1, \ldots, \alpha_{i-1})$, for $i = 2, \ldots, r$. A polynomial $f(x) \in F[x]$ is* **solvable by radicals over** *F if the splitting field K of $f(x)$ over F is an extension by radicals of F.*

The key result tying these two concepts – solvability of groups and solvability by radicals – is the following.

Theorem 15.6.1. *Suppose K is a Galois extension of F with char $F = 0$. Then K is an extension of F by radicals if and only if $\mathrm{Gal}(K/F)$ is a solvable group.*

In order to prove this theorem we need the following lemma.

Lemma 15.6.1. *Let G be a finite abelian group. Then G has a normal series with cyclic factors of prime order, that is, there exists a normal series*

$$G = G_1 \supset G_2 \supset \cdots \supset G_n \supset G_{n+1} = \{1\}$$

such that $|G_i/G_{i+1}|$ is a prime for all $i = 1, \ldots, n$.

Proof. The proof is by induction on $|G|$, the order of G. If $|G| = 1$ then G is a normal series without factors so the result is vacuously true. If $|G| = p$, a prime, then $G = G_1 \subset G_2 = \{1\}$ is a normal series and the result follows.

Now suppose $|G| > 1$ and $|G|$ is not prime. Further suppose that the result holds for all abelian groups of order less than $|G|$. Let $g \in G$ with $g \neq 1$ be any nontrivial element of G and consider $H = \langle g \rangle$. Then since

H is nontrivial we have that both H and G/H are abelian groups of order less than $|G|$ so we have by our inductive hypothesis that both have normal series with prime order factors. Hence we have

$$G/H = \overline{G_1} \supset \overline{G_2} \supset \cdots \supset \overline{G_r} = H/H$$

where $\overline{G_i}/\overline{G_{i+1}}$ is cyclic of prime order. Then by the correspondence theorem (Theorem 7.3.4) and Corollary 7.3.2 there exist subgroups G_i of G containing H such that $\overline{G_i} = G_i/H$ for $i = 1, \ldots, r$ (here $G_1 = G$ and $G_r = H$). Thus we have

$$G = G_1 \supset G_2 \supset \cdots \supset G_r = H. \qquad (1)$$

The third isomorphism theorem (Theorem 7.3.3) implies that

$$G_i/G_{i+1} \cong (G_i/H)(G_{i+1}/H) = \overline{G_i}/\overline{G_{i+1}}.$$

Thus each factor $G_i/G_{i+1}, i = 1, 2, \ldots, r-1$ in the series (1) is cyclic of prime order.

The inductive hypothesis also implies that H has such a normal series as well,

$$G_r = H \supset G_{r+1} \supset G_{r+2} \supset \cdots \supset G_n \subset G_{n+1} = \{1\}. \qquad (2)$$

Further this series (2) has prime order factors. Taking series (1) followed by series (2) and removing the second occurrence of H we get the required series for G.

$\square$

We will also use the following result.

Lemma 15.6.2. *Let F be a field of characteristic zero and $\overline{F}$ be an algebraic closure of F. Let $a \in F$ and let K be the splitting field in $\overline{F}$ of $x^n - a$ over F. Then $\mathrm{Gal}(K/F)$ is a solvable group.*

Proof. To prove this lemma we consider two cases.

Case (1): F contains the nth roots of unity.

Case (2): F does not contain the nth roots of unity.

In Case (1) recall that the nth roots of unity form a cyclic group of order n (see Chapter 5) with respect to the multiplication within F. If ω is a

primitve *nth* root of unity, that is, it is a generator of this cyclic group, then the set of nonzero elements $F^\star$ in F contains the cyclic subgroup

$$\langle \omega \rangle = \{1, \omega, \omega^2, \ldots, \omega^{n-1}\}.$$

Hence these are distinct elements of $F^\star$.

Suppose that $b \in \overline{F}$ is a root of $(x^n - a) \in F[x]$. Then all the roots of $x^n - a$ are $b, \omega b, \ldots, \omega^{n-1}b$. (This fact follows from the fact that the ω's form a group under the field multiplication — see the exercises for details.) Note that the splitting field $K = F(b)$ is an extension by radicals of the ground field F since $b^n = a \in F$.

Since $K = F(b)$ it follows that an automorphism $\sigma \in \text{Gal}(K/F)$ is determined by its value on b because $F(b)$ consists of linear combinations of powers of b with coefficients in F and any such automorphism σ fixes elements in F. Moreover, this value $\sigma(b)$ must be a conjugate of b. Now if $\sigma(b) = \omega^i b$ and if for $\tau \in \text{Gal}(K/F)$ we have $\tau(b) = \omega^j b$ and

$$\sigma\tau(b) = \sigma(\omega^j b) = \omega^j \sigma(b) = \omega^j \omega^i b$$

since $\omega^j \in F$. Similarly $\tau\sigma(b) = \omega^i \omega^j b$. Thus $\sigma\tau = \tau\sigma$ and $\text{Gal}(K/F)$ is abelian and hence solvable.

In Case (2), F does not contain a primitive nth root of unity. Let us again say that $\omega \in \overline{F} \setminus F$ is such a primitive root. Again let $b \in \overline{F}$ be a root of $x^n - a \in F[x]$. Since both b and $b\omega$ are roots of $x^n - a$ they must lie in the splitting field K. Thus $\omega = \omega b \cdot b^{-1} \in K$. Hence K contains all the nth roots of unity. Let us put $F' = F(\omega)$. We have

$$F \subset F' \subset K$$

where $K = F(\omega, b)$. Note that F' is a normal extension of F since it is a splitting field of $x^n - 1 \in F[x]$. Note further that the splitting field K is an extension by radicals of the ground field F since $F' = F(\omega)$ is for $\omega^n - 1 \in F$, $K = F'(b)$ is for the same reasons as before. Thus in either case our splitting field here is always an extension by radicals of F. This will be important for the proof of Theorem 15.6.1.

Since $F' = F(\omega)$ it follows that an automorphism $\gamma \in \text{Gal}(F'/F)$ is determined by $\gamma(\omega) = \omega^i$ for some i since all roots of $x^n - 1$ are powers of ω. If $\eta \in \text{Gal}(F'/F)$ and $\eta(\omega) = \omega^j$ then

$$\gamma\eta(\omega) = \gamma(\omega^j) = (\gamma(\omega))^j = \omega^{ij}.$$

Similarly $\eta(\gamma)(\omega) = \omega^{ji} = \omega^{ij}$. It follows that $\mathrm{Gal}(F'/F)$ is abelian. Then by the fundamental theorem of Galois theory

$$\{1\} \subset \mathrm{Gal}(K/F') \subset \mathrm{Gal}(K/F)$$

is a normal series since K is Galois over F (it is the splitting field of $x^n - a$ over F). The fundamental theorem then implies that $\mathrm{Gal}(K/F')$ is normal in $\mathrm{Gal}(K/F)$. From the first part of the proof we see that $\mathrm{Gal}(K/F')$ is abelian and again using the fundamental theorem of Galois theory

$$\mathrm{Gal}(K/F)/\mathrm{Gal}(K/F') \cong \mathrm{Gal}(F'/F).$$

However we just showed that $\mathrm{Gal}(F'/F)$ is abelian and therefore the above is a solvable series for $\mathrm{Gal}(K/F)$.

$\square$

We now return to the proof of Theorem 15.6.1.

Proof. (Theorem 15.6.1) Most of the proof is a straightforward consequence of the fundamental theorem of Galois theory together with the lemmas above. First let K be a Galois extension of F by radicals. We show that the Galois group must be a solvable group. By definition there exist elements $\alpha_1, \ldots, \alpha_r \in K$ and positive integers $n_1, \ldots, n_r$ such that $K = F(\alpha_1, \ldots, \alpha_r)$, with $\alpha_1^{n_1} \in F$, and $\alpha_i^{n_i} \in F(\alpha_1, \ldots, \alpha_{i-1})$, for $i = 2, \ldots, r$. For each $i = 1, \ldots, r$ let $K_i = F(\alpha_1, \ldots, \alpha_{i-1})(\alpha_i)$. Hence we have the following chain of field extensions,

$$F \subset K_1 \subset K_2 \subset \cdots \subset K_r = K.$$

Let $n = n_1 n_2 \cdots n_r$ and suppose that ω is a primitve nth root of unity as in Lemma 15.6.2. From the theory of cyclic groups (see Chapter 6), the group $\langle \omega \rangle$ contains a primitive n_ith root of unity for all n_i that divide n. If F does not contain ω then we make our first step in the chain of radical extensions $K_1 = F(\alpha_1)$ where $\alpha_1 = \omega$. As in the proof of Lemma 15.6.2 K_1 is the splitting field of $x^n - 1 \in F$ over F and hence K_1 is Galois over F and it is also true that $K_1 = F(\alpha_1)$ where $\alpha_1^n \in F$. Hence the requirements for an extension of F by radicals are met. Further from the proof of Lemma 15.6.2 we have that $\mathrm{Gal}(K_1/F)$ is abelian. Thus we may assume that F contains the nth roots of unity for the rest of the proof in this direction.

We assume that we have the following chain of radical field extensions,

$$F \subset K_1 \subset K_2 \subset \cdots \subset K_r = K.$$

and that F contains a primitive nth root of unity.

For $i = 1, \ldots, r$, let G_i be the subgroup of $G = \mathrm{Gal}(K/F)$ corresponding to K_i under the map τ of Theorem 15.4.1. That is, $G_i = \tau(K_i) = \mathrm{Gal}(K/K_i)$. We show that this provides a solvable series for G, that is, a normal series with abelian factors. By the correspondence between subfields and subgroups we have the chain of subgroups

$$G = G_0 \supset G_1 \supset G_2 \supset \cdots \supset G_r = \{1\}.$$

Notice the chain goes in the reverse direction. Since K is a finite extension of F each K_i is a finite extension of F and $G = \mathrm{Gal}(K/F)$ is finite. Not only is K_i a finite extension of F but also $|\mathrm{Gal}(K_i/F)| < \infty$. The reason for this is that from the Isomorphism Extension Theorem (Theorem 14.7.2) each $\sigma \in \mathrm{Gal}(K_i/F)$ can be extended to an isomorphism of K leaving F fixed. But since K is Galois over F, Theorem 15.2.1 (1) implies that any such isomorphism ia an automorphism of K. Therefore

$$|\mathrm{Gal}(K_i/F)| \leq |\mathrm{Gal}(K/F)| = |G| < \infty.$$

Then by Artin's theorem (Theorem 15.3.1) it follows that K_i is a Galois extension of F for each $i = 1, 2, \ldots, r$. But then since

$$F \subset K_i \subset K_{i+1}$$

and K_{i+1} is Galois over F it must be true by Lemma 15.3.2 that K_{i+1} is Galois over K_i for each i. It then follows from the fundamental theorem of Galois theory that G_{i+1} is a normal subgroup of G_i and hence this series of subgroups is a normal series. Since we are assuming that F contains a primitive nth root of unity and $n_{i+1}|n$ from the previous remark it must contain a primitive n_{i+1}th root of unity. Hence, exactly as in the proof of Lemma 15.6.2, the Galois group of K_{i+1} over K_i, $\mathrm{Gal}(K_{i+1}/K_i)$ is abelian (see the exercises for a detailed proof). Recall that $G_i = \tau(K_i) = \mathrm{Gal}(K/K_i)$. Then using (6) from the fundamental theorem of Galois theory (Theorem 15.4.1) we have that

$$G_i/G_{i+1} = (\mathrm{Gal}(K/K_i))/(\mathrm{Gal}(K/K_{i+1}) \cong \mathrm{Gal}(K_{i+1}/K_i).$$

This implies that the factor groups G_{i+1}/G_i are abelian. Hence the factors in the normal series of subgroups are abelian and this normal series is actually a solvable series. Therefore the Galois group G must be solvable.

Now conversely assume that the K is a Galois extension of F with Galois group $\text{Gal}(K/F)$. Thus $\text{Gal}(K/F)$ is a finite group. Suppose further that $\text{Gal}(K/F)$ is solvable. We show that K is an extension of F by radicals. Note that in the proof we constructed the subgroups in a normal series by using the extension fields given in Galois theory as the fixed fields of the subgroups. We reverse the direction here. Let

$$Gal(K/F) = G = G_0 \supset G_1 \supset \cdots \supset G_n = \{1\}.$$

Each factor G_i/G_{i+1} is a finite abelian group. Hence by Lemma 15.6.1 we may take a refinement so that each factor is cyclic of prime order (see the exercises). Therefore we may assume that

$$Gal(K/F) = G = G_0 \supset G_1 \supset \cdots \supset G_n = \{1\}$$

is a normal series with cyclic factors of prime order. For each $i = 0, \ldots, n$ let $K_i = K^{G_i}$. Hence we have the chain of subfields

$$F = K_0 \subset K_1 \subset \cdots \subset K_n = K$$

where each K_i is a Galois extension of K_{i-1} by the same reasoning as in the first part of this proof with Artin's thoerem. Furthermore, just as above by Theorem 15.4.1 (6) $\text{Gal}(K_{i=1}/K_i) \cong G_i/G_{i+1}$ so it is a cyclic group of prime order. The proof is completed if we show that each K_i can be formed from K_{i-1} by adjoining a root of an element thus giving an extension by radicals.

Just as in the first half of this proof let $d_i = |K_i : K_{i-1}|$ and $d = d_1 d_2 \cdots d_m$. Here each d_i ia a prime where repetitions are allowed. If F does not contain the dth roots of unity we make the first step in our extension by radicals $K_1 = F(\alpha_1)$ where α_1 is a primitive dth root of unity. Then as in the first part of the proof we may assume that all fields contain the dth roots of unity and in particular K_i contains the d_ith roots of unity.

Next we claim that for each $i = 1, \ldots, n$ there exists $b_i \in K_i$ such that $K_i = K_{i-1}(b_i)$ where $b_i^{d_i} \in K_{i-1}$ and $d_i = |K_i : K_{i-1}| = |\text{Gal}(K_i/K_{i-1})|$. This will prove the second part of the theorem.

To minimize the use of subscripts let $b = b_i$ and let $d_i = p$ a prime be the order of the cyclic group $\text{Gal}(K_i/K_{i-1})$. Let $a \in K_i \setminus K_{i-1}$ and let ω

be a primitive pth root of unity so that $\omega \in K_{i-1}$ and $\omega^p = 1$. Let σ be a generator of $\text{Gal}(K_i/K_{i-1})$. Thus $\sigma^p = 1$ where here 1 is the identity map. Set

$$b = a + \omega\sigma^{-1}(a) = \omega^2\sigma^{-2}(a) + \cdots + \omega^{p-1}\sigma^{-(p-1)}(a).$$

Thus $b \in K_i$.

We claim that $\sigma(b) = \omega b$. Note that

$$\sigma(b) = \sigma(a) + \omega a + \omega^2\sigma^{-1}(a) + \cdots + \omega^{p-1}\sigma^{-(p-2)}(a).$$

But

$$\omega b = \omega a + \omega^2\sigma^{-1}(a) + \cdots + \omega^{p-1}\sigma^{-(p-2)}(a) + \sigma^{-(p-1)}(a).$$

This is true because $\omega^p = 1$. Also observe that $\sigma^{-(p-1)}(a) = \sigma^{-p+1}(a) = \sigma(a)$ since σ^{-p} is the identity map. It follows that we have

$$\omega b = \sigma(a) + \omega a + \omega^2\sigma^{-1}(a) + \cdots + \omega^{p-1}\sigma^{-(p-2)}(a)$$

establishing our claim that $\sigma(b) = \omega b$.

We next note that for every integer j, $\sigma^j(b^p) = b^p$ (see the exercises for a proof). Since $\text{Gal}(K_i/K_{i-1}) = \langle\sigma\rangle$, this implies that b^p is in the fixed field of $\text{Gal}(K_i/K_{i-1})$. Thus $b^p \in K_{i-1}$.

Finally since K_{i-1} contains the pth roots of unity, $b \in K_i$ then $K_{i-1}(b) \subseteq K_i$ is the splitting field of $x^p - b^p$. By Exercise 14.1.5 we get that $\deg(\text{irr}(b, K_{i-1}))$ must divide $K_i : K_{i-1}| = p$, a prime. Since we are assuming that this is a nontrivial extension this can happen only if

$$|K_{i-1}(b) : K_{i-1}| = \deg(\text{irr}(b, K_{i-1})) = p = |K_i : K_{i-1}|.$$

Therefore $K_i = K_{i-1}(b)$ where $b^p \in K_{i-1}$. This completes the proof. $\square$

Therefore, to show that it is not possible in general to solve a polynomial of degree 5 or greater by radicals, we must show that for any $n \geq 5$ there exists a polynomial of degree n whose Galois group is not solvable. The Galois group is always contained in a symmetric group. The following was proved in Chapter 10 (Theorem 10.5.3).

Theorem 15.6.2. *For any $n \geq 5$ the symmetric group S_n is not a solvable group.*

Therefore, we could show the insolvability by radicals by exhibiting for each $n \geq 5$ a polynomial whose Galois group is the whole symmetric group S_n. To do this we need the concept of a **symmetric polynomial**.

Definition 15.6.3. *Let $y_1, \ldots, y_n$ be indeterminates over a field F. A polynomial $f(y_1, \ldots, y_n) \in F[y_1, \ldots, y_n]$ is a* **symmetric polynomial** *in $y_1, \ldots, y_n$ if $f(y_1, \ldots, y_n)$ is unchanged by any permutation σ of $\{y_1, \ldots, y_n\}$, that is, $f(y_1, \ldots, y_n) = f(\sigma(y_1), \ldots, \sigma(y_n))$.*

If $F \subset F'$ are fields and $\alpha_1, \ldots, \alpha_n$ are in F', then we call a polynomial $f(\alpha_1, \ldots, \alpha_n)$ with coefficients in F **symmetric** *in $\alpha_1, \ldots, \alpha_n$ if $f(\alpha_1, \ldots, \alpha_n)$ is unchanged by any permutation σ of $\{\alpha_1, \ldots, \alpha_n\}$.*

EXAMPLE 15.6.1. Let F be a field and $f_0, f_1 \in F$. Let

$$h(y_1, y_2) = f_0(y_1 + y_2) + f_1(y_1 y_2).$$

There are two permutations on $\{y_1, y_2\}$, namely, $\sigma_1 : y_1 \to y_1, y_2 \to y_2$ and $\sigma_2 : y_1 \to y_2, y_2 \to y_1$.

Applying either one of these two to $\{y_1, y_2\}$ leaves $h(y_1, y_2)$ invariant. Therefore, $h(y_1, y_2)$ is a symmetric polynomial.

Definition 15.6.4. *Let $x, y_1, \ldots, y_n$ be indeterminates over a field F (or elements of an extension field F' over F). Form the polynomial*

$$p(x, y_1, \ldots, y_n) = (x - y_1) \cdots (x - y_n).$$

The ith **elementary symmetric polynomial** *s_i in $y_1, \ldots, y_n$ for $i = 1, \ldots, n$, is $(-1)^i a_i$, where a_i is the coefficient of x^{n-i} in $p(x, y_1, \ldots, y_n)$.*

EXAMPLE 15.6.2. Consider y_1, y_2, y_3. Then

$$p(x, y_1, y_2, y_3) = (x - y_1)(x - y_2)(x - y_3)$$

$$= x^3 - (y_1 + y_2 + y_3)x^2 + (y_1 y_2 + y_1 y_3 + y_2 y_3)x - y_1 y_2 y_3.$$

Therefore, the three elementary symmetric polynomials in y_1, y_2, y_3 over any field are

1. $s_1 = y_1 + y_2 + y_3$.

2. $s_2 = y_1 y_2 + y_1 y_3 + y_2 y_3$.

3. $s_3 = y_1 y_2 y_3$.

In general, the pattern of the last example holds for $y_1, \ldots, y_n$. That is,

$$s_1 = y_1 + y_2 + \cdots + y_n$$

$$s_2 = y_1 y_2 + y_1 y_3 + \cdots + y_{n-1} y_n$$

$$s_3 = y_1 y_2 y_3 + y_1 y_2 y_4 + \cdots + y_{n-2} y_{n-1} y_n$$

$$\vdots$$

$$s_n = y_1 \cdots y_n.$$

The importance of the elementary symmetric polynomials is that any symmetric polynomial can be built up from the elementary symmetric polynomials. This is called the **fundamental theorem of symmetric polynomials**. A proof can be found in [FR] or [CFR].

Theorem 15.6.3. *(Fundamental Theorem of Symmetric Polynomials) If P is a symmetric polynomial in the indeterminates $y_1, .., y_n$ over F, that is, $P \in F[y_1, .., y_n]$ and P is symmetric, then there exists a unique $g \in F[y_1, .., y_n]$ such that $P(y_1, \ldots, y_n) = g(s_1, .., s_n)$. That is, any symmetric polynomial in $y_1, \ldots, y_n$ is a polynomial expression in the elementary symmetric polynomials in $y_1, .., y_n$.*

Now to show the insolvability by radicals of any polynomial of degree 5 or greater let $y_1, \ldots, y_n$ be transcendental elements over $\mathbb{Q}$ and let $K = \mathbb{Q}(y_1, \ldots, y_n)$. Let $s_1, \ldots, s_n$ be the elementary symmetric polynomials in $y_1, \ldots, y_n$ and let $F = \mathbb{Q}(s_1, \ldots, s_n)$. Then it can be shown that K is a Galois extension of F and $\mathrm{Gal}(K/F) \cong S_n$. This was what was essentially proved by Galois.

Theorem 15.6.4. *Let $n \geq 5$ and let $y_1, \ldots, y_n$ be transcendental elements over $\mathbb{Q}$. Then the polynomial $f(x) = \prod_{i=1}^{n}(x - y_i)$ is not solvable by radicals over $F = \mathbb{Q}(s_1, \ldots, s_n)$, where $s_1, \ldots, s_n$ are the elementary symmetric polynomials in $y_1, \ldots, y_n$.*

For a proof of Theorem 15.6.4 see [F]. Note that in terms of the history that we mentioned previously Ruffini and independently Abel proved the insolvability of the quintic by exhibiting a polynomial whose Galois group is the nonsolvable group S_5. Galois proved the general result above.

15.6.2   Some Ruler and Compass Constructions

As a final application we indicate the impossibility of certain geometric ruler (straightedge) and compass constructions. Greek mathematicians in the classical period posed the problem of finding geometric constructions using only ruler and compass to double a cube, trisect an angle, and square a circle. The Greeks were never able to prove that such constructions were impossible but were able to construct solutions to these problems only using other techniques, including conic sections. In 1837, Pierre Wantzel, using algebraic methods was able to prove the impossibility of either trisecting an angle or doubling a cube. With the proof that π is transcendental (done by Lindemann in 1882 – see Section 12.3.5) Wantzel's method could be applied to showing that squaring the circle is also impossible. We will outline the algebraic method. As in the insolvability of the quintic, we must translate into the language of field extensions. As a first step we define a **constructible number**.

Definition 15.6.5. *Suppose we are given a line segment of unit length. An $\alpha \in \mathbb{R}$ is* **constructible** *if we can construct a line segment of length $|\alpha|$ in a finite number of steps from the unit segment using a ruler and compass.*

Recall from elementary geometry that using a ruler and compass, it is possible to draw a line parallel to a given line segment through a given point, to extend a given line segment, and to erect a perpendicular to a given line at a given point on that line. Our first result is that the set of all constructible numbers forms a subfield of $\mathbb{R}$.

Theorem 15.6.5. *The set C of all constructible numbers forms a subfield of $\mathbb{R}$. Further, $\mathbb{Q} \subset C$.*

Proof. Let C be the set of all constructible numbers. Since the given unit length segment is constructible, we have $1 \in C$. Therefore, $C \neq \emptyset$, and thus to show that it is a field we must show that it is closed under the field operations. Suppose α, β are constructible. We must show then that $\alpha \pm \beta, \alpha\beta$, and α/β for $\beta \neq 0$ are constructible. If $\alpha, \beta > 0$, construct a line segment of length $|\alpha|$. At one end of this line segment extend it by a segment of length $|\beta|$. This will construct a segment of length $\alpha + \beta$. Similarly, if $\alpha > \beta$, lay off a segment of length $|\beta|$ at the beginning of a segment of length $|\alpha|$. The remaining piece will be $\alpha - \beta$. By considering cases we can do this in the same manner if either α or β or both are negative. These constructions are pictured in Figure 15.4. Therefore, $\alpha \pm \beta$ are constructible.

Figure 15.4. Constructibility of $\alpha \pm \beta$

In Figure 15.5 we show how to construct $\alpha\beta$. Let the line segment $\overline{OA}$ have length $|\alpha|$. Consider a line L through O not coincident with $\overline{OA}$. Let $\overline{OB}$ have length $|\beta|$ as in the diagram. Let P be on ray $\overline{OB}$ so that $\overline{OP}$ has length 1. Draw $\overline{AP}$ and then find Q on ray $\overline{OA}$ such that $\overline{BQ}$ is parallel to $\overline{AP}$. From similar triangles we then have

$$\frac{|\overline{OP}|}{|\overline{OB}|} = \frac{|\overline{OA}|}{|\overline{OQ}|} \Rightarrow \frac{1}{|\beta|} = \frac{|\alpha|}{|\overline{OQ}|}.$$

Then $|\overline{OQ}| = |\alpha||\beta|$, and so $\alpha\beta$ is constructible.

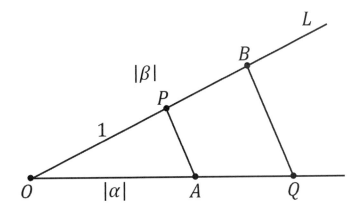

Figure 15.5. Constructibility of $\alpha\beta$

A similar construction, pictured in Figure 15.6, shows that α/β for $\beta \neq 0$ is constructible. Find $\overline{OA}, \overline{OB}, \overline{OP}$ as above. Now, connect A to B and let $\overline{PQ}$ be parallel to $\overline{AB}$. From similar triangles again we have

$$\frac{1}{|\beta|} = \frac{|\overline{OQ}|}{|\alpha|} \Rightarrow \frac{|\alpha|}{|\beta|} = |\overline{OQ}|.$$

Hence α/β is constructible.

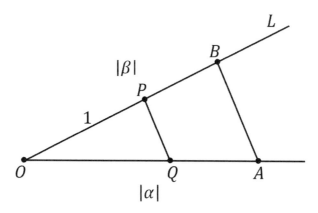

Figure 15.6. Constructibility of α/β

Therefore, C is a subfield of $\mathbb{R}$. Since char $C = 0$ and $\mathbb{Q}$ is the prime subfield of $\mathbb{R}$ (see Corollary 11.2.2) it follows that $\mathbb{Q} \subset C$.   □

Let us now consider analytically how a constructible number is found in the plane. Starting at the origin and using the unit length and the constructions above, we can locate any point in the plane with rational coordinates. That is, we can construct the point $P = (q_1, q_2)$ with $q_1, q_2 \in \mathbb{Q}$. Using only ruler and compass, any further point in the plane can be determined in one of the following three ways.

(1) The intersection point of two lines each of which passes through two known points each having rational coordinates.

(2) The intersection point of a line passing through two known points having rational coordinates and a circle whose center has rational coordinates and whose radius squared is rational.

(3) The intersection point of two circles each of whose centers has rational coordinates and each of whose radii is the square root of a rational number.

Analytically, the first case involves the solution of a pair of linear equations each with rational coefficients and thus only leads to other rational numbers. In cases 2 and 3 we must solve equations of the form $x^2+y^2+ax+by+c = 0$, with $a, b, c \in \mathbb{Q}$. These will then be quadratic equations over $\mathbb{Q}$, and thus the solutions will either be in $\mathbb{Q}$ or in a quadratic extension $\mathbb{Q}(\sqrt{\alpha})$ of $\mathbb{Q}$. Once a real quadratic extension of $\mathbb{Q}$ is found, the process can be iterated. Conversely it can be shown that if α is constructible, so is $\sqrt{\alpha}$. We thus can prove the following theorem.

Theorem 15.6.6. *If γ is constructible with $\gamma \notin \mathbb{Q}$, then there exists a finite number of elements $\alpha_1, \ldots, \alpha_r \in \mathbb{R}$ with $\alpha_r = \gamma$ such that for $i = 1, \ldots, r, \mathbb{Q}(\alpha_1, .., \alpha_i)$ is a quadratic extension of $\mathbb{Q}(\alpha_1, .., \alpha_{i-1})$. In particular, $|\mathbb{Q}(\gamma) : \mathbb{Q}| = 2^n$ for some $n \geq 1$.*

Therefore, the constructible numbers are precisely those real numbers that are contained in repeated quadratic extensions of $\mathbb{Q}$. We now use this idea to show the impossibility of the three mentioned construction problems.

EXAMPLE 15.6.3. It is impossible to **double the cube**. This means that it is impossible, given a cube of given side length, to construct using a ruler and compass, a side of a cube having double the volume of the original cube. Let the given side length be 1, so that the original volume is also 1. To double this we would have to construct a side of length $2^{1/3}$. However $|\mathbb{Q}(2^{1/3}) : \mathbb{Q}| = 3$ since $\mathrm{irr}(2^{1/3}, \mathbb{Q}) = x^3 - 2$. This is not a power of 2 so $2^{1/3}$ is not constructible.

EXAMPLE 15.6.4. It is impossible to **trisect an angle**. This means that it is impossible in general to trisect a given angle using only a ruler and compass. An angle θ is constructible if and only if a segment of length $|\cos \theta|$ is constructible. Since $\cos(\pi/3) = 1/2$, $\pi/3$ is constructible. We show that it cannot be trisected by ruler and compass. The following trigonometric identity can be proved

$$\cos(3\theta) = 4\cos^3(\theta) - 3\cos(\theta).$$

Let $\alpha = \cos(\pi/9)$. From the above identity we have $4\alpha^3 - 3\alpha - \frac{1}{2} = 0$. The polynomial $4x^3 - 3x - \frac{1}{2}$ is irreducible over $\mathbb{Q}$ (see the exercises) so $\mathrm{irr}(\alpha, \mathbb{Q}) = x^3 - \frac{3}{4}x - \frac{1}{8}$. It follows that $|\mathbb{Q}(\alpha) : \mathbb{Q}| = 3$, and hence α is not constructible. Therefore, the corresponding angle $\pi/9$ is not constructible. Therefore, $\pi/3$ is constructible, but it cannot be trisected.

EXAMPLE 15.6.5. It is impossible to **square the circle**. That is, it is impossible in general, given a circle, to construct using ruler and compass a square having area equal to that of the given circle. Suppose the given circle has radius 1. It is then constructible and would have an area of π. A corresponding square would then have to have a side of length $\sqrt{\pi}$. But π is transcendental, so $\sqrt{\pi}$ is not constructible.

The other great construction problem solved by Galois theory was the construction of regular n-gons. A regular n-gon will be constructible for $n \geq 3$ if and only if the angle $2\pi/n$ is constructible, which is the case if and only if $\cos(2\pi/n)$ is constructible. The algebraic study of the constructibility of regular n-gons was initiated by Gauss in the early part of the nineteenth century. For more details on the constructibility of regular n-gons see [CFR].

15.6.3   Algebraic Extensions of $\mathbb{R}$

The final thing we look at is algebraic extensions of the reals. Suppose F is a field extension of the real numbers $\mathbb{R}$ of finite degree. From the proof of the fundamental theorem of algebra it follows that either $F = \mathbb{R}$ or $|F : \mathbb{R}| = 2$ and $F \cong \mathbb{C}$. Thus the fundamental theorem of algebra can also be phrased in the following way.

Theorem 15.6.7. *If F is a field extension of $\mathbb{R}$ of finite degree with $|F : \mathbb{R}| > 1$, then $|F : \mathbb{R}| = 2$ and $F \cong \mathbb{C}$.*

If we extend somewhat the definition of a field we can obtain a further classification of extensions of the reals. Recall that a **division ring** is an algebraic structure with the same properties as a field but in which multiplication is not necessarily commutative. That is, it is a ring with an identity where every nonzero element has a multiplicative inverse. Thus every field is a division ring but as we will see below there are noncommutative division rings. In Chapter 5 we saw that a noncommutative division ring is called a **skew field**. If F is a field and $F \subset D$ where D is a division ring, then D is still a vector space over F. Further there is a multiplication in D so that each nonzero element has a multiplicative inverse. The identity in F must be an identity for the division ring. In this case D is called a **division algebra** over F. Here the elements of F will commute with all elements in D.

We now give a method to construct a class of skew fields. Suppose F is a field in which no sum of nontrivial squares can be zero. That is, if $x_1^2 + \cdots + x_n^2 = 0$ with $x_1, .., x_n \in F$ then $x_i = 0$ for $i = 1, .., n$. A field satisfying this is called a **totally real field**. Consider now a vector space $\mathcal{H}_F$ of dimension 4 over F with basis $1, i, j, k$. We identify 1 with the identity in F and then build a multiplication on $\mathcal{H}_F$ by defining the products of the basis elements. Let $i^2 = j^2 = k^2 = -1$ and $ijk = -1$. This will completely define, using associativity, all products of basis elements. For example, from $ijk = -1$ we have $ij = -k^{-1} = k$ since $k^2 = -1$. Then $ik = i(ij) = (ii)j = -j$. It is easy to show that $ij = -ji$, so that this product is noncommutative. A general element of $\mathcal{H}_F$ has the form $f_0 + f_1 i + f_2 j + f_3 k$. Multiplication of elements like this is done by algebraic manipulation using the defined products of basis elements. We thus get a product on $\mathcal{H}_F$. It is a straightforward computation to prove the following theorem (see Chapter 5).

Theorem 15.6.8. *For a totally real field F, $\mathcal{H}_F$ forms a division algebra of degree 4 over F. $\mathcal{H}_F$ is called the* **quaternion algebra** *over F.*

The only difficulty in the proof is to show the existence of inverses. This is done just as in the complex numbers and we leave it to the exercises. The **quaternions** $\mathcal{H}$ are the quaternion algebra over the reals. That is, $\mathcal{H} = \mathcal{H}_\mathbb{R}$. This algebra consists of all elements $r_0 + r_1 i + r_2 j + r_3 k$ with $r_i \in \mathbb{R}$. Identifying $\mathbb{R}$ with the first component we get that $\mathbb{R} \subset \mathcal{H}$. Therefore, $\mathcal{H}$ is a finite *skew* field extension of $\mathbb{R}$. The following theorem says that this is the only one.

Theorem 15.6.9. *Let D be a finite-dimensional division algebra over $\mathbb{R}$ of degree greater than 1. If D is a field, then $|D : \mathbb{R}| = 2$ and $D \cong \mathbb{C}$. If D is a skew field, then $|D : \mathbb{R}| = 4$ and $D \cong \mathcal{H}$.*

Proof. If D is a field the result is just the reformulation of the fundamental theorem of algebra so we consider the case where D is a skew field. We outline the proof below and leave the details to the exercises.

(1) If D is a skew field extension of $\mathbb{R}$, it must have degree at least 4. As in the field case every element is a root of a polynomial with coefficients in $\mathbb{R}$. If the degree is 3, there is a root in $\mathbb{R}$, and if the degree is 2, the resulting extension must be commutative and thus isomorphic to $\mathbb{C}$.

(2) If D has degree four over $\mathbb{R}$, then it has a basis $1, e_1, e_2, e_3$. We can show that the nonreal basis elements must satisfy quadratic polynomials over $\mathbb{R}$ and thus can be chosen to behave like the imaginary unit i. Thus $e_1^2 = e_2^2 = e_3^2 = -1$. By the noncommutativity of D we can obtain that $e_1 e_2 = e_3$ and thus $D \cong \mathcal{H}$.

(3) If D is a division algebra over $\mathbb{R}$ of degree $n + 1$, then as above there is a basis $1, e_1, \ldots, e_n$ with $e_i^2 = -1, i = 1, .., n$. We can then show that for

each pair i, j with $i \neq j$ we have $e_i e_j + e_j e_i \in \mathbb{R}$. From this we can then show that we can replace e_2 by e'_2 with $e_1 e'_2 + e'_2 e_1 = 0$. (If $e_1 e_2 + e_2 e_1 = 2c \in \mathbb{R}$ set $e'_2 = (e_2 + c e_1)/\sqrt{1 - c^2}$.) We can then show that this leads to a contradiction if $n > 3$. (We may choose $e_3 = e_1 e_2$. Then let $a_{ij} = e_i e_j + e_j e_i$, and so $a_{12} = 0$. Then $-2e_4 = a_{14} e_1 + a_{24} e_2 + a_{34} e_3$, violating the independence of the basis.) $\qquad\qquad\qquad\qquad\qquad\qquad\qquad\qquad\qquad\qquad\qquad\quad\square$

15.7   Exercises

EXERCISES FOR SECTION 15.2

15.2.1. Illustrate the primitive element theorem by showing that $\mathbb{Q}(\sqrt{2} + \sqrt{3}) = \mathbb{Q}(\sqrt{2}, \sqrt{3})$.

(HINT: First show that $\mathbb{Q}(\sqrt{2} + \sqrt{3}) \subset \mathbb{Q}(\sqrt{2}, \sqrt{3})$. Then show that

$$|\mathbb{Q}(\sqrt{2}, \sqrt{3}) : \mathbb{Q}| = |\mathbb{Q}(\sqrt{2} + \sqrt{3}) : \mathbb{Q}| = 4.)$$

15.2.2. Let y be an indeterminate over $\mathbb{Z}_p$ where p is a prime. Consider the transcendental extension $E = \mathbb{Z}_p(y)$ of $\mathbb{Z}_p$. Let $t = y^p$ and consider the subfield $F' = \mathbb{Z}_p(t) \subset E$.

(a) Show that E is algebraic over F'.

(HINT: Consider $f(x) = x^p - t \in F'[x]$ and note that it has a zero in E.)

(b) Show that $E = F'(y)$.

(HINT: Show that $\mathrm{irr}(y, F') = x^p - t$.)

(c) Show that y is not separable over F'.

(HINT: Consider $x^p - t = x^p - y^p$.)

15.2.3. Suppose that K is a finite extension field of the field F. Let $F \subset F' \subset K$. Show that K is separable over F if and only if K is separable over F' and F' is separable over F.

15.2.4. Give an example of an $f(x) \in \mathbb{Q}[x]$ that has no zeros in $\mathbb{Q}$ but whose zeros in $\mathbb{R}$ all have multiplicity 2. Explain why this is consistent with Theorem 15.3.2.

(HINT: Consider $f(x) = x^4 - 4x^2 + 4 \in \mathbb{Q}[x]$.)

EXERCISES FOR SECTION 15.3

15.3.1. Let F be a field and $\sigma \in \mathrm{Aut}(F)$. Extend σ to a map $\sigma : f[x] \to F[x]$ by

$$\sigma(\sum_{i=0}^{n} a_i x^i) = \sum_{i=0}^{n} \sigma(a_i) x^i.$$

Show that this map is an automorphism of $F[x]$.

15.3.2. From Example 15.3.1 show that:

(a) the polynomial $x^4 + 1$ is irreducible over $\mathbb{Q}$.

(b) the complex roots of the polynomial $x^4 + 1 \in \mathbb{Q}[x]$ are $\omega_1, \omega_2, \omega_3, \omega_4$ as given in the text.

(c) show that $\sigma_1, \sigma_2, \sigma_3, \sigma_4$ as given in the examples are the four automorphisms in $\mathrm{Gal}(K/Q)$.

(d) show that each of $\sigma_2, \sigma_3, \sigma_4$ has order 2 in in $\mathrm{Gal}(K/Q)$.

(HINT: (a) Use Eisenstein's criterion as in Example 12.4.2.)

15.3.3. Let E be the splitting field of $f(x)$ over F where $\deg(f(x)) = n$. Give bounds on $|E : F|$.

EXERCISES FOR SECTION 15.4

In problems 15.4.1 through 15.4.8 let $K = \mathbb{Q}(\sqrt{3}, \sqrt{5}, \sqrt{7})$ be a Galois extension of $\mathbb{Q}$. We use the notation of the fundamental theorem of Galois Theory (Theorem 15.5.1) so τ is the map between fixed fields and subgroups of the Galois group. In each case, compute the indicated number.

15.4.1. $|K : \mathbb{Q}|$

15.4.2. $|\mathrm{Gal}(K/\mathbb{Q})|$

15.4.3. $|\tau(\mathbb{Q})|$

15.4.4. $|\tau(\mathbb{Q}(\sqrt{5}, \sqrt{7})|$

15.4.5. $|\tau(\mathbb{Q}(\sqrt{15})|$

15.4.6. $|\tau(\mathbb{Q}(\sqrt{105})|$

15.4.7. $|\tau(\mathbb{Q}(\sqrt{15} + \sqrt{7})|$

15.4.8. $|\tau(K)|$

15.4.9. Prove (1) of the fundamental theorem (Theorem 15.4.1), that is, the mapping $\tau : E \to Gal(K/E)$ is a bijection from the set of all intermediate fields between K and F and the set of all subgroups of $\mathrm{Gal}(K/F)$.

15.4.10. In Example 15.4.3 explain why the Galois group of the polynomial $f(x) = x^4 - 2$ over $\mathbb{Q}$ is D_4, the dihedral group of order 8, the group of a symmetries of a square.

15.4.11. Describe the Galois group of $f(x) = x^4 - 1 \in \mathbb{Q}[x]$ over $\mathbb{Q}$.

15.4.12. We say that the field K which is a finite normal extension of F is **abelian over** F if $\mathrm{Gal}(K/F)$ is abelian. Prove that if K is abelain over F and E is a normal extension of F such that $F \subset E \subset K$ then K is abelian over E and E is abelian over F.

EXERCISES FOR SECTIONS 15.5 AND 15.6

15.5.1. Complete the proof of Theorem 15.6.1.

15.5.2. Using the results mentioned in the text give a proof of Theorem 15.6.2.

(HINT: First prove that $x^2 + 1$ is irreducible over K and then follow the proof of Theorem 15.5.1.)

15.5.3. Go through all other cases not considered in the text to show that if α, β are constructible numbers then so are $\alpha \pm \beta$.

15.5.4. Prove that $f(x) = 4x^3 - 3x - \frac{1}{2}$ is irreducible over $\mathbb{Q}$.

(HINT: Consider $g(x) = 2f(x) \in \mathbb{Z}[x]$ and apply Corollary 12.3.1 along with the rational roots theorem Theorem 12.3.5.)

15.5.5. Show that the quaternion algebra $\mathcal{H}$ is a division algebra.

(HINT: To prove the existence of inverses of nonzero elements follow the discussion in Chapter 5.)

15.5.6. Let F be a field with $\overline{F}$ an algebraic closure of F. Let $a \in F$ and $\omega \in F$ where ω is a primitive nth root of unity. Then if $b \in \overline{F}$ is a root of $(x^n - a) \in F[x]$ prove that all the roots of $x^n - a$ are precisely $b, \omega b, \omega^2 b, \ldots, \omega^{n-1} b$. Thus $F(b)$ is a splitting field of $x^n - a$ over F.

15.5.7. The idea of this exercise is needed in the proof of Theorem 15.6.1. Let H be a normal subgroup of G and G/H a finite abelian group. Show using the correspondence theorem (Theorem 7.3.4), Corollary 7.3.2 and the third isomorphism theorem (Theorem 7.3.3) how Lemma 15.6.1 implies that there exists subgroups of G, $G_1, \ldots, G_n$ containing H such that

$$G = G_0 \supset G_1 \supset \cdots \supset G_n = H$$

where G_{i-1}/G_i is cyclic of prime order for all $i = 1, \ldots, n$.

15.5.8. In the first part of the proof of Theorem 15.6.1 since $K_{i+1} = K_i(\alpha_{i+1})$ where $\alpha_{i+1}^{n_{i+1}} \in K_i$ put $\alpha_{i+1}^{n_{i+1}} = a$ and also since K_i contains the primitive n_{i+1}th roots of unity, prove that $\mathrm{Gal}(K_{i+1}/K_i)$ is abelian.

(HINT: Since K_{i+1} is the splitting field of $x^{n_{i+1}} - a$ over K_i use the same reasoning as in the proof of Lemma 15.6.2 Case (1).)

15.5.9. In the seond part of the proof of Theorem 15.6.1 prove that for any integer j we have $\sigma^j(b^p) = b^p$.

15.5.10. Fill in the missing details in the proof of Theorem 15.6.8.

Bibliography

REFERENCES ON GENERAL ABSTRACT ALGEBRA

[Ar] M. Artin, *Algebra*, Prentice-Hall, Englewood Cliffs, New Jersey, 1991.

[CFR] C. Carstensen, B. Fine and G. Rosenberger, *Abstract Algebra*, De-Gruyter, Berlin, 2011.

[CR] C. Curtis and I. Reiner, *Representation Theory of Finite Groups and Associative Algebras*, Wiley Interscience, New York, 1966.

[FR 1] B. Fine and G. Rosenberger, *The Fundamental Theorem of Algebra*, Springer, New York, 2000.

[F] J. Fraleigh, *A First Course in Abstract Algebra*, 7th Edition, Addison-Wesley, New York, 2003.

[H 1] I. Herstein, *Topics in Algebra*, Blaisdell, New York, 1964. 1975.

[KR] M. Kreuzer and S. Robiano, *Computational Commutative Algebra I and II*, Springer-Verlag, Berlin, 1999.

[L] S. Lang, *Algebra*, Addison-Wesley, New York, 1965.

[H 1] S. MacLane and G. Birkhoff, *Algebra*, Macmillan, New York, 1967. 1975

[Str] G. Stroth, *Algebra*, W. DeGruyter, Berlin, 1998.

REFERENCES ON LINEAR ALGEBRA

[An] H. Anton, *Elementary Linear Algebra*, 10th Edition, Wiley, New York, 2005.

[F] J. Fraleigh and J. Beauregard, *A First Course in Linear Algebra*, Addison-Wesley, New York, 1983.

[H] P. Halmos, *Finite Dimensional Vector Spaces*, Van Nostrand, Amsterdam, 1958.

[HK] K. Hoffman and R. Kunze, *Linear Algebra*, Prentice-Hall, Englewood Cliffs, New Jersey, 1968.

[Li] S. Lipschutz, *Linear Algebra*, Schaum's Outline Series.

REFERENCES ON GROUP THEORY AND RELATED TOPICS

[GB 1] G. Baumslag, *Topics in Combinatorial Group Theory*, Birkhauser, Boston, 1993.

[Ba] O. Bogopolski, *Introduction to Group Theory*, European Mathematical Society, 2008.

[CgrRR] T. Camps, V. gr. Rebel and G. Rosenberger, *Einfuhrung in die kombinatorische und die geometrische Gruppentheorie*, Heldermann Verlag, Berlin, 2008.

[AMG] A. Gaglione, *Group Theory*, Government Printing Office (GPO), 1992.

[G] D. Gorenstein, "The Enormous Theorem," *Scientific American*, 256, 1985, 104-115, 1977.

[RH] R. Hirshon, "On Cancellations in Groups" *Amer. Math. Monthly*, 76, 1969, 1037-1039, 1977.

[LS] R. Lyndon and P. Schupp, *Combinatorial Group Theory*, Springer-Verlag, New York, 1977.

[MKS] W. Magnus, A. Karrass, and D. Solitar, *Combinatorial Group Theory*, Wiley, New York, 1966.

[N] P. J. Nahim, *An Imaginary Tale: the Story of $\sqrt{-1}$* , Princteon University Press, Princeton, 1998.

[Ro] D.J.S. Robinson, *A Course in the Theory of Groups*, Springer-Verlag, New York, 1982.

[R] J. Rotman, *Group Theory*, 3rd Edition, Wm. C. Brown, Dubuque, Iowa, 1988.

REFERENCES ON NUMBER THEORY

[Ap] T. M. Apostol, *Introduction to Analytic Number Theory*, Springer-Verlag, New York, 1976.

[Ba] A. Baker, *Transcendental Number Theory*, Cambridge University Press, Cambridge, 1975.

[FR] B. Fine and G. Rosenberger, *Number Theory: An Introduction via the Distribution of Primes*, Birkhauser, Boston, 2006.

[HW] G. H. Hardy and E. M. Wright, *An Introduction to the Theory of Numbers*, Clarendon Press, Oxford, 5th Edition, 1979.

[Ko] N. Koblitz, *A Course in Number Theory and Cryptography*, Springer, New York, 1987.

[NZ] I. Niven and H. S. Zuckerman, *The Theory of Numbers*, 4th Edition, John Wiley, New York, 1980.

[O] O. Ore, *Number Theory and Its History*, McGraw-Hill, New York, 1949.

[PD] H. Pollard and H. Diamond, *The Theory of Algebraic Numbers*, Carus Mathematical Monographs, 9, Math. Assoc. of America, 1975.

REFERENCES ON ANALYSIS AND COMPLEX ANALYSIS

[Ah] L. Ahlfors, *Complex Analysis*, Springer-Verlag, New York, 1968.

[A] K. Atkinson, *An Introduction to Numerical Analysis*, Wiley, New York, 1972.

[Ba] R. G. Bartle, *The Elements of Real Analysis*, Wiley, New York, 1964.

[GL] E. A. Grove and G. Ladas, *Introduction to Complex Variables*, Houghton-Mifflin, New York, 1974.

[Ka] S. Katok, *p-Adic Analysis Compared with Real*, American Mathematical Society, 2007.

REFERENCES ON GEOMETRY

[L] R. Lyndon, *Groups and Geometry*, Cambridge University Press, Cambridge, 1985.

[PJ] W. Prenowitz and M. Jordan, *Basic Concepts of Geometry*, Blaisdell Publishing, New York, 1965.

[WW] E. Wallace and S. West, *Roads to Geometry*, Prentice-Hall, Englewood Cliffs, New Jersey, 1992.

Index